GAS SERVICES

Owen Smith

GAS
SERVICES

PLUMBING SKILLS Series

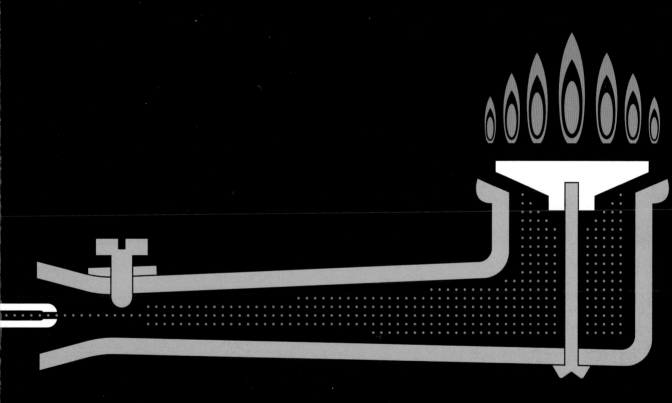

3e

Gas Services
3rd Edition
Owen Smith

Content manager: Sandy Jayadev
Content developer: Talia Lewis
Senior project editor: Nathan Katz
Cover designer: Justin Lim
Text designer: Justin Lim
Permissions/Photo researcher: Debbie Gallagher
Editor: Marta Veroni
Proofreader: James Anderson
Indexer: Julie King
Art direction: Linda Davidson
Typeset by KnowledgeWorks Global Ltd.

Any URLs contained in this publication were checked for currency during the production process. Note, however, that the publisher cannot vouch for the ongoing currency of URLs.

Third edition published in 2022

© 2022 Cengage Learning Australia Pty Limited

Copyright Notice

This Work is copyright. No part of this Work may be reproduced, stored in a retrieval system, or transmitted in any form or by any means without prior written permission of the Publisher. Except as permitted under the *Copyright Act 1968*, for example any fair dealing for the purposes of private study, research, criticism or review, subject to certain limitations. These limitations include: Restricting the copying to a maximum of one chapter or 10% of this book, whichever is greater; providing an appropriate notice and warning with the copies of the Work disseminated; taking all reasonable steps to limit access to these copies to people authorised to receive these copies; ensuring you hold the appropriate Licences issued by the Copyright Agency Limited ("CAL"), supply a remuneration notice to CAL and pay any required fees. For details of CAL licences and remuneration notices please contact CAL at Level 11, 66 Goulburn Street, Sydney NSW 2000, Tel: (02) 9394 7600, Fax: (02) 9394 7601
Email: info@copyright.com.au
Website: www.copyright.com.au

For product information and technology assistance,
  in Australia call **1300 790 853**;
  in New Zealand call **0800 449 725**

For permission to use material from this text or product, please email
aust.permissions@cengage.com

National Library of Australia Cataloguing-in-Publication Data
ISBN: 9780170439244
A catalogue record for this book is available from the National Library of Australia.

Cengage Learning Australia
Level 7, 80 Dorcas Street
South Melbourne, Victoria Australia 3205

Cengage Learning New Zealand
Unit 4B Rosedale Office Park
331 Rosedale Road, Albany, North Shore 0632, NZ

For learning solutions, visit **cengage.com.au**

Printed in China by 1010 Printing International Limited.
5 6 7 25 24

BRIEF CONTENTS

Part A	Gas fundamentals	1 GS
CHAPTER 1	Fuel gases	2 GS
CHAPTER 2	Units of measurement and gas industry terms	11 GS
CHAPTER 3	Gas distribution systems	29 GS
CHAPTER 4	Gas constituents and characteristics	43 GS
CHAPTER 5	Gas industry workplace safety	55 GS
CHAPTER 6	Combustion principles	81 GS
CHAPTER 7	LPG Basics	111 GS
Part B	**Units of competency**	**123 GS**
CHAPTER 8	Size consumer gas piping systems	124 GS
CHAPTER 9	Install gas piping systems	147 GS
CHAPTER 10	Install gas pressure control equipment	185 GS
CHAPTER 11	Purge consumer piping	225 GS
CHAPTER 12	Calculate and install natural ventilation for Type A gas appliances	241 GS
CHAPTER 13	Install and commission Type A gas appliances	271 GS
CHAPTER 14	Install Type A gas appliance flues	311 GS
CHAPTER 15	Disconnect and reconnect Type A gas appliances	337 GS
CHAPTER 16	Install LPG storage of aggregate storage capacity up to 500 litres	359 GS
CHAPTER 17	Install LPG systems in caravans, mobile homes and mobile workplaces	383 GS
CHAPTER 18	Install LPG systems in marine craft	403 GS
CHAPTER 19	Install gas sub-meters	423 GS
CHAPTER 20	Install LPG storage of aggregate storage capacity exceeding 500 litres and less than 8 kL	441 GS
CHAPTER 21	Maintain Type A gas appliances	457 GS
CHAPTER 22	Install Type B gas appliance flues	501 GS
APPENDIX	Advanced skills – flue sizing	543 GS
	Glossary	571 GS
	Index	576 GS

CONTENTS

Guide to the text — xii GS
Guide to the online resources — xv GS
Preface — xvii GS
Acknowledgements — xx GS
About the author — xx GS
Unit conversion table — xxi GS

Part A Gas fundamentals — 1 GS

CHAPTER 1 Fuel gases — 2 GS
Overview — 3 GS
The 'first' fuel gas – towns gas — 3 GS
Natural gas (NG) — 4 GS
Liquefied petroleum gas (LPG) — 5 GS
Simulated natural gas (SNG) — 6 GS
Tempered liquefied petroleum (TLP) — 6 GS
Compressed natural gas (CNG) — 6 GS
Summary — 6 GS
Get it right — 7 GS
Worksheets — 9 GS

CHAPTER 2 Units of measurement and gas industry terms — 11 GS
Overview — 12 GS
Units of volume — 12 GS
Units of heat — 12 GS
Units of mass — 13 GS
Units of pressure — 14 GS
Measuring gas pressure — 15 GS
Terms used in gas pressure measurement — 18 GS
Unit conversion — 21 GS
Summary — 22 GS
Get it right — 23 GS
Worksheets — 25 GS

CHAPTER 3 Gas distribution systems — 29 GS
Overview — 30 GS
The distribution of natural gas — 30 GS
The distribution of LPG — 35 GS
Summary — 37 GS
Get it right — 39 GS
Worksheets — 41 GS

CHAPTER 4	**Gas constituents and characteristics**	**43 GS**
	Overview	44 GS
	Constituents of fuel gases	44 GS
	Characteristics of fuel gases	44 GS
	Summary	49 GS
	Get it right	51 GS
	Worksheets	53 GS
CHAPTER 5	**Gas industry workplace safety**	**55 GS**
	Overview	56 GS
	General safety evaluations	56 GS
	Gas leaks	57 GS
	Respiratory hazards	59 GS
	Basic electrical safety	61 GS
	Electrical test equipment	66 GS
	Summary	70 GS
	Get it right	71 GS
	Worksheets	73 GS
CHAPTER 6	**Combustion principles**	**81 GS**
	Overview	82 GS
	Basic combustion principles	82 GS
	Burner design	84 GS
	Flame characteristics	92 GS
	Summary	100 GS
	Get it right	101 GS
	Worksheets	103 GS
CHAPTER 7	**LPG basics**	**111 GS**
	Overview	112 GS
	Storing LPG	112 GS
	Boiling liquid expanding vapour explosion (BLEVE)	115 GS
	Using butane	116 GS
	Summary	117 GS
	Get it right	119 GS
	Worksheets	121 GS
Part B	**Units of competency**	**123 GS**
CHAPTER 8	**Size consumer gas piping systems**	**124 GS**
	Overview	125 GS
	Identify job requirements	125 GS
	Selecting products using AS/NZS 5601	125 GS
	How to identify a certified or approved product	125 GS
	Size gas piping systems	127 GS
	Applying higher pressure drop	133 GS
	Sizing proprietary brand piping systems	134 GS
	How to size multistage gas supply systems	135 GS
	Clean up the job	135 GS
	Summary	136 GS
	Get it right	137 GS
	Worksheets	139 GS

CHAPTER 9	**Install gas piping systems**	**147 GS**
	Overview	148 GS
	Identify gas piping system requirements	148 GS
	Piping in concealed locations	150 GS
	Underground piping requirements	152 GS
	Gas load	154 GS
	Material quantities	154 GS
	Prepare for installation	156 GS
	Install and test piping system	159 GS
	Purging procedures	168 GS
	Clean up the job	170 GS
	Summary	170 GS
	Get it right	171 GS
	Worksheets	173 GS
CHAPTER 10	**Install gas pressure control equipment**	**185 GS**
	Overview	186 GS
	Pressure control fundamentals – regulators	186 GS
	Identify requirements for gas pressure control equipment	200 GS
	Determining regulator venting requirements	201 GS
	Determining over-pressure protection requirements	205 GS
	Prepare for installation	208 GS
	Install and commission control and regulating equipment	209 GS
	Clean up the job	213 GS
	Summary	214 GS
	Get it right	215 GS
	Worksheets	217 GS
CHAPTER 11	**Purge consumer piping**	**225 GS**
	Overview	226 GS
	Purge requirements for large volume gas installations	226 GS
	Purging types	226 GS
	Prepare for purging	226 GS
	Tools and equipment required for purging	227 GS
	Identify purge requirements	229 GS
	Check installation compliance	230 GS
	Perform a leakage test	230 GS
	Purging a sub-meter	230 GS
	Purging calculations and planning	230 GS
	Carry out and test purge operation	233 GS
	Clean up the job	235 GS
	Summary	236 GS
	Get it right	237 GS
	Worksheets	239 GS
CHAPTER 12	**Calculate and install natural ventilation for Type A gas appliances**	**241 GS**
	Overview	242 GS
	Type A gas appliance ventilation basics	242 GS
	Prepare for work	244 GS
	Identify natural ventilation requirements	248 GS
	Permanent ventilation	250 GS
	Mechanical ventilation	252 GS
	Specific ventilation calculations	253 GS
	Install ventilation and test appliance	256 GS
	Appliance and flue testing	257 GS
	Clean up the job	259 GS

Summary	260 GS
Get it right	261 GS
Worksheets	263 GS

CHAPTER 13 Install and commission Type A gas appliances — 271 GS

Overview	272 GS
Identify gas appliance installation requirements	272 GS
General installation requirements	272 GS
Appliance isolation and connection	273 GS
Specific installation requirements	274 GS
A note on using annealed copper loops	277 GS
Commercial catering equipment	278 GS
Continuous flow water heaters (instantaneous water heaters)	280 GS
Storage water heaters	280 GS
Space heaters	282 GS
Overhead radiant tube heaters	286 GS
Overhead radiant heaters	286 GS
Decorative flame effect fires	287 GS
Prepare for installation	288 GS
Install and commission appliance	288 GS
Protecting a combustible surface	288 GS
Commissioning	291 GS
Clean up the job	294 GS
Summary	294 GS
Get it right	295 GS
Worksheets	297 GS

CHAPTER 14 Install Type A gas appliance flues — 311 GS

Overview	312 GS
Flue types	312 GS
Prepare for flue installation	314 GS
Identify flue requirements	315 GS
Flue components	315 GS
Flue materials	315 GS
Heat loss	316 GS
Flue installation and terminal locations	319 GS
Install and test flue	321 GS
Flue pipe roof penetration flashings	323 GS
Fan-assisted flue installations	323 GS
Testing for spillage from flued appliances	324 GS
Clean up the job	328 GS
Summary	329 GS
Get it right	331 GS
Worksheets	333 GS

CHAPTER 15 Disconnect and reconnect Type A gas appliances — 337 GS

Overview	338 GS
Prepare for work	338 GS
Identify appliance requirements	340 GS
Disconnect and reconnect equipment	344 GS
Electrical safety check	344 GS
Isolate gas supply	345 GS
Disconnect the appliance	345 GS
Hose assemblies	345 GS
Test operation of equipment	346 GS
Check all associated appliance connections	347 GS
Appliance commissioning	347 GS

Test and adjust burner pressure	348 GS
Appliance operational settings	348 GS
Instruct customer on operation	350 GS
Clean up the job	350 GS
Summary	351 GS
Get it right	353 GS
Worksheets	355 GS

CHAPTER 16 Install LPG storage of aggregate storage capacity up to 500 litres — **359 GS**

Overview	360 GS
Cylinder basics	360 GS
Vaporisation calculation – the DLK formula	362 GS
Calculating cylinder storage requirements	363 GS
Cylinder locations and installation requirements	366 GS
Prepare for installation	371 GS
Install and test LPG storage system	371 GS
Clean up the job	372 GS
Summary	373 GS
Get it right	375 GS
Worksheets	377 GS

CHAPTER 17 Install LPG systems in caravans, mobile homes and mobile workplaces — **383 GS**

Overview	384 GS
Identify LPG system requirements	384 GS
System layout and gas load	385 GS
Caravan piping	389 GS
Prepare for installation	390 GS
Install LPG system, including flue and ventilation	391 GS
Clean up	396 GS
Summary	396 GS
Get it right	397 GS
Worksheets	399 GS

CHAPTER 18 Install LPG systems in marine craft — **403 GS**

Overview	404 GS
Essential terminology	404 GS
Identify LPG system requirements	404 GS
System layout and gas load	405 GS
Marine craft piping	408 GS
Prepare for installation	409 GS
Determine and install LPG gas systems	409 GS
Clean up	415 GS
Summary	415 GS
Get it right	417 GS
Worksheets	419 GS

CHAPTER 19 Install gas sub-meters — **423 GS**

Overview	424 GS
Meter applications	424 GS
Determination of system requirements	424 GS
Ventilation of sub-meter enclosures	427 GS
Prepare for sub-meter installation	427 GS
Install and test gas sub-meter	428 GS
Clean up the job	434 GS
Summary	435 GS
Get it right	437 GS
Worksheets	439 GS

CHAPTER 20	**Install LPG storage of aggregate storage capacity exceeding 500 litres and less than 8 kL**	**441 GS**
	Overview	442 GS
	Tank basics	442 GS
	Identify installation requirements	443 GS
	Vaporisation calculation – the DLK formula	443 GS
	What are vaporisers?	444 GS
	Calculating tank storage requirements	444 GS
	Prepare for installation	448 GS
	Install and test LPG tank storage system	448 GS
	Clean up the job	449 GS
	Summary	450 GS
	Get it right	451 GS
	Worksheets	453 GS
CHAPTER 21	**Maintain Type A gas appliances**	**457 GS**
	Overview	458 GS
	Pre-maintenance preparation	458 GS
	Basic control systems	464 GS
	Safety devices	464 GS
	Temperature control – thermostats	474 GS
	Ignition systems	479 GS
	Combination controls	482 GS
	Carry out maintenance	483 GS
	Clean up the job	487 GS
	Summary	488 GS
	Get it right	489 GS
	Worksheets	491 GS
CHAPTER 22	**Install Type B gas appliance flues**	**501 GS**
	Overview	502 GS
	Type B gas appliances	502 GS
	Basic combustion principles	503 GS
	Flue types	505 GS
	Prepare for flue installation	506 GS
	Identify flue requirements	508 GS
	Flue components – atmospheric burner systems	508 GS
	Flue components – forced-draught burner systems	509 GS
	Flue materials	510 GS
	Heat loss in natural draught flue systems	511 GS
	Flue installation and terminal locations	512 GS
	Atmospheric gas appliance flue sizing	513 GS
	Install the flue	522 GS
	Flue pipe roof penetration flashings	523 GS
	Clean up the job	524 GS
	Summary	525 GS
	Get it right	527 GS
	Worksheets	529 GS
APPENDIX	**Advanced skills – flue sizing**	**543 GS**
Glossary		571 GS
Index		576 GS

Guide to the text

As you read this text you will find a number of features in every chapter that will enhance your study of plumbing and help you to understand how the theory is applied in the real world.

PART AND CHAPTER-OPENING FEATURES

Part sections divide the book by function. Gain an insight into the basic principles in **Part A**. Explore the Certificate III gas-specific units in **Part B**.

PART A
GAS FUNDAMENTALS

Part A of this text will help you build a strong understanding of the fundamental gas principles required to safely and efficiently undertake gas-fitting installations. Work through this section in partnership with your teacher and peers in preparation for installation practice.

PART B
UNITS OF COMPETENCY

Part B of this text allows you to explore the requirements of the AS/NZS 5601 Gas Installation Standards in greater depth. With interpretation and guidance, you should apply these requirements to your gas installation practice in a methodical and safe manner.

Identify the key concepts you will engage with through the **Learning objectives** at the start of each chapter.

UNITS OF MEASUREMENT AND GAS INDUSTRY TERMS

Learning objectives

Areas addressed in this chapter include:
* an outline of the most common units of measurement used in the gas industry
* how to measure gas pressure using:
  – a manometer
  – a digital manometer
* key terms used in gas pressure measurement
* how to convert from imperial to SI units.

The content of this chapter provides the vital underpinning knowledge that will enable you to proceed confidently with learning and activities in subsequent sections.

FEATURES WITHIN CHAPTERS

Engage actively with the learning by completing the practical activities in the **NEW Learning task** boxes.

LEARNING TASK 1.1 FUEL GASES

For the gases listed below, state where it is commonly used and why?
- Natural gas (NG)
- Liquefied petroleum gas (LPG)
- Simulated natural gas (SNG)
- Tempered liquefied petroleum (TLP)
- Compressed natural gas (CNG)

FEATURES WITHIN CHAPTERS

From experience boxes explain the responsibilities of employees, including the skills they need to acquire and the real-life challenges they may face at work to enhance their employability skills on the job site.

FROM EXPERIENCE

A gas appliance that is not operating at the correct pressure is dangerous! In certain circumstances, gas that is burning at the wrong pressure can kill people. Safe installation and maintenance of gas appliances relies upon your thorough understanding of gas pressure.

Know your code icons highlight where Plumbing Standards are addressed to strengthen knowledge and hone research skills.

REFER TO AS/NZS 5601.1 APPENDIX B 'CONVERSION FACTORS'

NEW Caution boxes highlight important advice on safe work practices for plumbers by identifying safety issues and providing urgent safety reminders.

Under increased pressure, acetylene has been known to combust at even higher percentages. A leaking acetylene hose is not something to be ignored!

Green tip boxes highlight the applications of sustainable technology, materials or products relevant to plumbers and the plumbing industry.

GREEN TIP

Landfill sites produce high levels of methane and carbon dioxide. These harmful greenhouse gases can be captured from landfill waste and turned into a sustainable renewable energy source.

NEW How to boxes highlight a theoretical or practical task with step-by-step walkthroughs.

HOW TO

To size the vent, follow these steps.
1. Determine the vent valve nominal size (in this case, DN 20) and find this in the left-hand column of the sizing table.
2. Look horizontally across to the first number in the sizing table. The figure in the middle of the table represents the maximum distance you can run the sized vent line, as indicated in the top row. In this example, you can run a DN 20 vent a maximum distance of 10 m. Alternatively you could run a DN 25 vent a maximum of 25 m or a DN 32 vent a maximum distance of 64 m.

Learn how to complete mathematical calculations with **Example** boxes that provide worked equations.

EXAMPLE 2.1

Electric water heaters often use a 3.6 kW element to heat the water in the cylinder. This is approximately equivalent to a 12.96 MJ/h gas burner (3.6 kW × 3.6 MJ = 12.96 MJ/h).

EXAMPLE 2.2

If you were to quote for the replacement of a 5 kW electric heater with an equivalent gas appliance, you would need to consider one of at least 18 MJ/h input (5 kW × 3.6 MJ = 18 MJ/h).

END-OF-CHAPTER FEATURES

At the end of each chapter you will find several tools to help you to review, practise and extend your knowledge of the key learning objectives.

Review your understanding of the key chapter topics with the **NEW Summary**.

SUMMARY
In this chapter you have been briefly introduced to the fuel gases used in Australia. Following the widespread use of towns gas (TG) around Australia for more than a century, TG was gradually replaced by natural gas (NG) and liquefied petroleum gas (LPG), which remain the two most commonly used fuel gases found today. Alternative sources of methane gas in the form of biogas and landfill gas are increasingly likely to further supplement NG energy supply options into the future.

After you have worked through the chapter, reinforce the practical component of your training with the **NEW Get it right** section.

GET IT RIGHT

The photo below shows an incorrect practice when determining the correct fuel gas required for an appliance. The photo shows a natural gas boosted solar hot water system.

Identify the incorrect method and provide reasoning for your answer.

Worksheets give you the opportunity to test your knowledge and consolidate your understanding of the chapter competencies.

COMPLETE WORKSHEET 1

Worksheet icons indicate in the text when a student should complete an end-of-chapter worksheet.

WORKSHEET 1

| To be completed by teachers |
| Student competent ☐ |
| Student not yet competent ☐ |

Student name: _____

Enrolment year: _____

Class code: _____

Competency name/Number: _____

Task
Review the sections on *Towns gas* through to *Compressed natural gas* and answer the following questions.

1 What are the two main fuel gases used in Australia today?
   i _____
   ii _____

2 Describe the difference between biogas and landfill gas.
   _____
   _____
   _____

3 What gases are to be found in your own local work area? Name the companies that distribute the gas.
   _____
   _____
   _____

4 Give the full name of each of these abbreviations:
   CNG _____
   NG _____
   TG _____
   LPG _____
   TLP _____
   SNG _____

Guide to the online resources

FOR THE INSTRUCTOR

Cengage is pleased to provide you with a selection of resources that will help you prepare your lectures and assessments. These teaching tools are accessible via cengage.com.au/instructors for Australia or cengage.co.nz/instructors for New Zealand.

MINDTAP

Premium online teaching and learning tools are available on the *MindTap* platform, the personalised eLearning solution.

MindTap is a flexible and easy-to-use platform that helps build student confidence and gives you a clear picture of their progress. We partner with you to ease the transition to digital – we're with you every step of the way.

The *Cengage Mobile App* puts your course directly into students' hands with course materials available on their smartphone or tablet. Students can read on the go, complete practice quizzes or participate in interactive real-time activities. *MindTap* is full of innovative resources to support critical thinking and help your students move from memorisation to mastery!

The *Series MindTap for Plumbing* is a premium purchasable eLearning tool. Contact your Cengage learning consultant to find out how *MindTap* can transform your course.

SOLUTIONS MANUAL

The **Solutions manual** provides detailed answers to every question in the text.

MAPPING GRID

The **Mapping grid** is a simple grid that shows how the content of this book relates to the units of competency needed to complete the Certificate III in Plumbing.

WEBLINKS

References from the text are provided as easy-to-access **Weblinks** and can be used in presentations or for further reading for students.

INSTRUCTORS' CHOICE RESOURCE PACK

This optional, purchasable pack of premium resources provides additional teaching support, saving time and adding more depth to your classes. These resources cover additional content with an exclusive selection of engaging features aligned to the text. Contact your Cengage learning consultant to find out more.

COGNERO® TEST BANK

A bank of questions has been developed in conjunction with the text for creating quizzes, tests and exams for your students. Create multiple test versions in an instant and deliver tests from your LMS, your classroom, or wherever you want, using Cognero. **Cognero test generator** is a flexible online system that allows you to import, edit and manipulate content from the text's test bank or elsewhere, including your own favourite test questions.

POWERPOINT™ PRESENTATIONS

Cengage **PowerPoint lecture slides** are a convenient way to add more depth to your lectures, covering additional content and offering an exclusive selection of engaging features aligned to the textbook, including teaching notes with mapping, activities, and tables, photos and artwork.

ARTWORK FROM THE TEXT

Add the **Digital files** of graphs, pictures and flow charts into your learning management system, use them in student handouts, or copy them into your lecture presentations.

FOR THE STUDENT

MINDTAP

MindTap is the next-level online learning tool that helps you get better grades!

MindTap gives you the resources you need to study – all in one place and available when you need them. In the *MindTap Reader*, you can make notes, highlight text and even find a definition directly from the page.

If your instructor has chosen *MindTap* for your subject this semester, log in to *MindTap* to:
- get better grades
- save time and get organised
- connect with your instructor and peers
- study when and where you want, online and mobile
- complete assessment tasks as set by your instructor.

When your instructor creates a course using *MindTap*, they will let you know your course link so you can access the content. Please purchase *MindTap* only when directed by your instructor. Course length is set by your instructor.

PREFACE

Welcome to the gas industry! In commencing your studies and skills development as a gas fitter, you are entering into a trade that plays a major role in the day-to-day workings of a modern economy. From heating and cooking appliances in domestic homes, commercial cooking appliances in restaurants, right through to industrial processes that depend upon gas, it is the gas fitter who makes it happen and keeps it safe. As such, gas fitting is held in high professional regard throughout industry and the community at large. However, such regard must be earned. Being licensed as a gas fitter means you are entrusted with the health and safety of the entire community. You are asked to commit to lifelong learning and an uncompromising adherence to the codes and standards that underpin the safe installation and maintenance of gas appliances and systems.

How to use this Text – Students

The intent of this text is to provide an instructional guide to basic, installation-level trade gas fitting. Coupled with quality technical instruction and relevant field experience, you will be able to use this text not only in a classroom situation, but also as a resource for self-instruction where required during and after your formal training. The following points are relevant in the use of this text:

- Although most gas fitters are also plumbers, for the purposes of clarity in this text we will use the term gas fitter.
- Most gas-fitting students will already be engaged in the industry as a plumbing apprentice, some may be qualified plumbers wishing to learn new skills in the gas industry, while others may be gas-fitting specialists outside of the plumbing trade in those jurisdictions where such work categories are licensed. Regardless of student background, the content in this text is based upon the assumption that basic plumbing terminology, literacy and numeracy skills have been satisfied through prior study and industry experience, so this foundational knowledge is not covered here. However, it is important to note that where you find difficulties in any such areas you should let your teacher and/or supervisor know at the earliest opportunity so that appropriate assistance can be provided.
- This text cannot be read in isolation from the current version of the AS/NZS 5601 Gas Installations Standards. When working through each chapter you will note numerous references to the Standards and you must form the habit of cross-referencing where indicated so that you gain a full understanding of the relationship between work practices and the Standards that guide them. The health and safety of the community depends upon compliant gas installations, completed by true professionals. Unless you are prepared to embrace the consistent use of Standards in all aspects of plumbing and gas fitting you should consider a career change.

The text itself is structured to build your knowledge step-by-step, working from general/foundation subjects towards more specific and technical detail as you progress. The book is broken into two main sections.

Part A – Gas fundamentals

Part A covers the important general and foundational information that will enable a more complete understanding of installation practice. As a gas fitter, your aim is to become a skilled professional, able to solve problems associated with the installation and commissioning of gas appliances in a seemingly endless variety of situations. Only a deep and thorough understanding of basic principles and broader industry knowledge will enable you to carry out installations with confidence, safety and flexibility.

Part B – Installation practice

This section will guide you through key installation requirements and considerations supported by regular questioning and cross-references to the AS/NZS 5601 Gas Installation Standards. Based around the gas-specific Units of Competency from the Certificate III in Plumbing qualification, Part B covers a range of gas fitting subjects including pipe sizing, ventilation calculations, flue sizing, regulator venting and many others.

While the author has placed each chapter in a generally accepted sequence of progression, some of these subjects can be selected in an alternative order depending upon your own learning progress, style and input from your gas teacher. They are deliberately designed to be self-instructional and are supported with questioning, standards references and activities.

Pre-reading

At the beginning of each chapter a brief overview of its content is provided and where applicable, you may also be directed to pre-read another earlier chapter before commencing the new one. This is necessary so that you will have the foundational knowledge required to complete the new subject area. Do not skip on this step as you may find yourself having difficulty completing subsequent work.

Worksheets and Exercises

Each chapter contains a series of exercises and worksheets designed to help you gain a better understanding of both the text and Standards. Read the chapter in full first, then in reference to each question work back to the relevant section and find the answer. Try and avoid the practice of simply going to the worksheets first then flicking through the pages to just find the answer; too much supporting information is missed and your learning process too shallow. Getting the chapter done quickly is pointless if you do not build your knowledge. Becoming a good gas fitter doesn't happen automatically, it also requires your personal discipline, effort and application with the assistance of a good teacher to develop the required skills.

How to use this Text – Teachers

The key purpose of this text is to provide interpretive support of the AS/NZS 5601 Gas Installation Standards. Every effort has been made to accurately align this edition with the current version of AS/NZS 5601 as available at the time of publication, to assist teachers and students in both the classroom and in the field. However, as new versions of the Standards may be introduced at any time during the life of this edition, gas fitting teachers must be alert to changes and ensure that all students make necessary notes and amendments in the text where appropriate. The following points are worth consideration:

- Due to the irregular release of new versions and amendments to the gas Standards, actual numbered sections and cross-references from the AS/NZS 5601 are deliberately not included in this text. Only the written title of each section or clause is included. However, it is a good practice to ask students to insert their own clause number and page number notes to help build their familiarity with the Standard as they work between each resource.
- While the author has tried to use the most commonly accepted trade terminology it is inevitable that variations in terms and descriptions will occur between each State and Territory. Where applicable teachers should ask students to make any necessary changes in the text as required.
- Two primary delivery modes are employed by Registered Training Organisations (RTO) and each may require a slightly different use of this book:
  - Block Release format – the daily attendance of students at an RTO over a period of weeks enables application of a progressive and gas-specific learning process through holistic delivery projects and assessments. This text is structured in a manner that permits the sequential completion of chapters moving from fundamental principles in Part A to more

complex processes and application in Part B as student knowledge increases. Where a particular RTO requires a slightly different order of completion then this should be easily achieved so long as any required pre-reading is attended to.

– Day Release format – This delivery format presents particular challenges to teachers due to the often inconsistent and short duration of student attendance. Each gas fitting Unit of Competency includes common skills and knowledge content that if included each time would become repetitive and cumbersome in a printed format. Therefore, this foundation material has been isolated within Part A of the text leaving the core content of Part B stripped down to primary requirements. Students should be guided as to the importance of vital Part A pre-reading before attempting each of the Part B units during their day attendance. Notes at the beginning of each chapter indicate what pre-reading may be required.

- Lastly, Units of Competency are primarily written as assessment outcome documents and were never designed to represent how you should sequence the teaching process. Basic learning theory is based upon the incremental layering of more complex principles over a strong foundational knowledge framework. A textbook is an important tool that a teacher can use but cannot cover everything and as such must be employed in hand with demonstration, practical application and interpretation through your own anecdotal trade experience.

A note of thanks

I wish to thank the many lecturers who gave valuable feedback in the preparation of this 3rd edition. Their helpful suggestions, concerning many aspects of the technical content, were drawn from broad experience in the plumbing trade. Special thanks also to my own teachers in the former Eagle Farm and Yeronga TAFE campuses in QLD who (now many years ago) provided me with the strong fundamentals in my own initial trade training that still apply today. In particular, I would also like to thank Noel Roney formerly of Swinburne University of Technology (TAFE) in Victoria for his valuable input during the writing of the original edition of this text. More recently, Garry Waters with the Master Plumbers at PICAC has been a great help and source of advice and valuable critique.

Lastly, due recognition must be given to David Graham from TasTAFE in Hobart. Having worked with David for many years both 'on the tools' and within TAFE, I have come to rely on his advice and critique for this and previous editions. It is such support that enables this text to be continually improved and updated. Many thanks.

Owen Smith

ACKNOWLEDGEMENTS

The author and Cengage Learning would like to thank the following reviewers for their incisive and helpful feedback:
- David Graham – TasTAFE
- Darren Schiavello – Victoria University Polytechnic
- Ben Freer – Central Regional TAFE
- Adam Bird – Victoria University Polytechnic
- Anthony Pingnam – TAFE NSW
- Stuart Peck – MPA Skills
- Kevin Ledwidge – Peer Training.

Every effort has been made to trace and acknowledge copyright. However, should any infringement have occurred, the publishers tender their apologies and invite copyright owners to contact them.

ABOUT THE AUTHOR

Owen Smith has been a plumbing and gas-fitting teacher, and manager in the TAFE sector since 2001. He holds a Bachelor Degree in Adult and Vocational Education, was a foundation member of the National Plumbing and Services Training Advisory Group (NPSTAG) and is also the author of *Basic Plumbing Service Skills: Roof Plumbing*. Outside of work he can generally be found out in the shed restoring and modifying classic motorcycles.

UNIT CONVERSION TABLE

Pressure	Multiply by...	Equals...
kPa	0.145	psi (lbs/in^2)
psi (lbs/in^2)	6.895	kPa
kPa	4	inches WG
inches WG	0.25	kPa
kPa	10	mb
mb	0.1	kPa
inches Hg	13.6 × 0.25	kPa
Heat energy & power	**Multiply by...**	**Equals...**
MJ	947.8	BTU
BTU	0.001055	MJ
kWh	3.6	MJ
MJ	0.2778	kWh
kWh	3412	BTU
BTU	0.2931	kWh
MJ/m^3	26.76	BTU/cu.ft
BTU/cu.ft	37.37	MJ/m^3
Volume	**Multiply by...**	**Equals...**
m^3	35.32	cu.ft
cu.ft	0.128	m^3
m^3	1000	L
Imp. Gallon	4.546	L
L	0.22	Imp. Gallon
US Gallon	3.785	L
L	0.2642	US Gallon
Imp. Gallon	0.8326	US Gallon
US Gallon	1.201	Imp. Gallon
Area	**Multiply by...**	**Equals...**
mm^2	0.01	cm^2
m^2	10.764	ft^2
ft^2	0.0929	m^2
Length	**Multiply by...**	**Equals...**
m	3.281	ft
ft	0.3048	m

Abbreviations			
kPa	(kilopascals)	m^3	(cubic metres)
psi	(pounds per square inch)	cu.ft	(cubic feet)
inches WG	(inches water gauge – also "WG)	m^2	(square metres)
inches Hg	(inches mercury – also "Hg)	mm^2	(square millimetres)
mb	(millibars)	ft^2	(square feet)
MJ	(megajoules)	L	(litres)
kWh	(kilowatt hour)		
BTU	(British thermal units)		

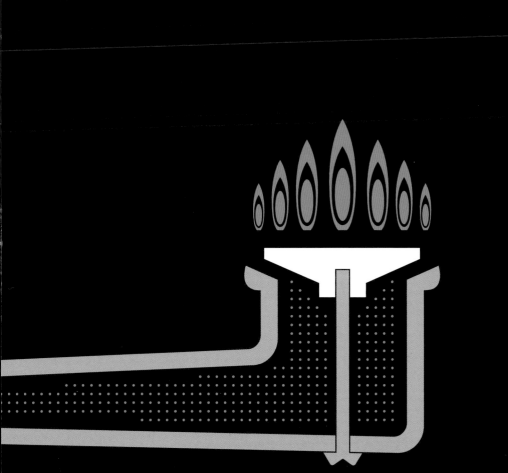

1 FUEL GASES

Learning objectives

This chapter describes the various types of gases used in the gas industry. These are:
- natural gas (NG)
- liquefied petroleum gas (LPG)
- simulated natural gas (SNG)
- tempered liquefied petroleum (TLP)
- compressed natural gas (CNG).

PART A

1

Overview

Collectively, the term applied to the various gases used in the gas industry is fuel gases. Over the years, many types of gases have been used for various reasons. Of all these gases, it is currently natural gas (NG) and liquefied petroleum gas (LPG) that you can expect to work with regularly. Other types of gases may now be obsolete or used only in particular circumstances.

It is important to note that each type of gas has different physical characteristics that affect how it is installed and used. An appliance designed for one type of gas cannot be directly used with a different gas because they burn differently. Therefore, it is important for a gas fitter to have a general understanding of all gases so that they can accommodate different installation and commissioning requirements.

In the following sections, you will find a description of each type of gas, standard abbreviations, where the gases are sourced from, their history of use and their current status.

The 'first' fuel gas – towns gas

The use of towns gas (TG) has been discontinued in Australia since the mid-1990s. Its brief inclusion here is simply to provide some historical background to the use of reticulated gases in Australia, and to provide some comparative context to the use of our modern fuel gases.

Originally known as coal gas because of its production process, TG was first used in a large-scale reticulation system in Sydney as early as 1841. Although initially its principal use was for street lighting, it gradually came to be used in many domestic, commercial and industrial applications.

The gas itself was originally manufactured through a process called 'carbonising of coal'. Coal was placed in a horizontal or vertical furnace called a retort, where it was heated to very high temperatures in the absence of air (see Figure 1.1). This allowed gas to be extracted from

FIGURE 1.1 A vertical retort TG manufacturing plant

the coal without the coal itself burning. This gas was drawn out of the retort and subjected to an extensive refining process to remove impurities.

The gas was stored close to the plant in large storage containers sometimes known as 'gasometers', a common sight on city skylines for decades. Towns gas contained a large amount of moisture and as a result was known as a 'wet gas'. This characteristic of TG necessitated that gas fitters install all piping with a graded fall so that condensate could be drawn off via tailpipes at the lowest point of the system whenever necessary. The cross-section of a gasometer in Figure 1.2 shows how the gas container floated upon a layer of

 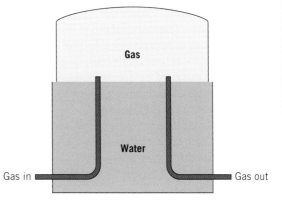

FIGURE 1.2 Photograph and cross-section of a variable volume gas container (gasometer)

water and rose up and down according to the amount of gas in the vessel.

Another key feature of TG is that it contained a high percentage of carbon monoxide, a highly poisonous and flammable substance requiring particular care from the gas fitter during installation and servicing.

The refining process for TG was quite expensive in terms of both labour and raw materials, although the introduction of more efficient vertical retorts and catalytic reforming technologies using hydrocarbons such as 'naphtha' in the 1960s helped to reduce processing costs. Despite this, the increasing use of natural gas, combined with the dramatic oil price rises of the 1970s, resulted in TG gradually being phased out. The last city in Australia to use TG was Launceston, where it was used until the manufacturing plant was shut down in 1996.

Although TG is no longer used in Australia, you will still occasionally encounter appliances and pipe installations that were originally designed for TG and were subsequently converted for use with other gases, though this is becoming increasingly rare. Regardless, this is your industry and there is a value in knowing the historical background of your profession.

Natural gas (NG)

Natural gas (NG) is the most widely used fuel gas in Australia, having gradually replaced TG since the 1960s. As the name implies, this gas occurs naturally rather than being manufactured from another substance. Natural gas is classed as a gaseous fossil fuel and, like other fossil fuels such as coal and oil, it is found in concentrations within the upper layer of Earth's crust. Fossil fuels primarily originated millions of years ago when large numbers of dead animals and plants built up and were covered over by deep sedimentary layers and then subjected to extreme pressure and heat, which gradually converted their solid remains to liquid and gaseous hydrocarbons. Fossil fuels are regarded as non-renewable resources because of this extremely long development period.

Natural gas primarily consists of a hydrocarbon known as methane, found sometimes in association with oil or coal, or in pockets of gas formed within sedimentary uplifts in the Earth's crust (see Figure 1.3). This gas can be extracted by drilling down through the layers of rock, refined to remove impurities and piped to locations where the gas is to be used.

Australia is fortunate in having significant reserves of natural gas; with large gas fields situated in Victoria's Gippsland Basin, the Cooper Basin in central Australia and the Northwest Shelf off the coast of Western Australia.

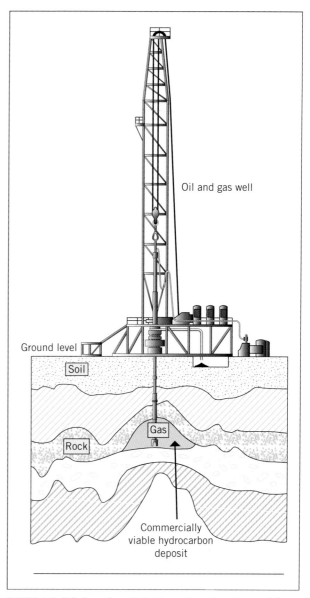

FIGURE 1.3 NG deposits within the sedimentary layers of the Earth's crust (not to scale)

GREEN TIP

In its raw state, methane is a potent greenhouse gas that contributes significantly to global warming. For this reason it is imperative that gas fitters minimise at all times the unnecessary release of methane. However, when burnt efficiently, natural gas produces approximately 45 per cent less carbon dioxide than coal and 30 per cent less than petroleum. This makes natural gas an important transitional fuel in the effort to reduce overall greenhouse gas emissions as we move towards a low carbon emitting economy.

Other sources of methane

The bulk demand for natural gas is met through standard production and refining technology; however, methane is also produced from other sources to meet niche demand or satisfy environmental demands.

Biogas

Fossil fuel reserves of natural gas are not the only source of methane. Any decomposing organic matter produces methane gas. The methane produced from the decomposition of non-fossil organic material is known as biogas. Swamps and marshlands are common sources of biogas and in the future may become more significant sources of energy.

Even a simple compost heap kept by a home gardener produces methane gas. In fact, many years ago it was not uncommon for some people in country areas to generate electricity using an internal combustion engine running on methane gas produced from the decomposition of kitchen scraps and animal waste within a biogas 'digestor'. Figure 1.4 depicts the process of biogas extraction. Today the biogas industry is growing strongly in Australia and becoming a key part of the process of energy production diversification from point-of-use consumers right through to large waste-to-energy generation plants.

Landfill gas

The waste matter buried in large landfill sites typically produces approximately 50 per cent methane and 50 per cent carbon dioxide, as organic waste products decompose. Unless these potent greenhouse gases are captured, they escape from the landfill site directly into the atmosphere. This is environmentally harmful and a waste of valuable energy. To combat this problem, many local authorities are building and operating landfill gas-to-energy sites in which methane is captured and burnt on-site, generating power that can be connected directly into the electricity grid or used to power onsite facilities (see Figure 1.5).

FIGURE 1.5 LPG is stored in cylinders that come in many shapes and sizes

GREEN TIP

Landfill sites produce high levels of methane and carbon dioxide. These harmful greenhouse gases can be captured from landfill waste and turned into a sustainable renewable energy source.

Liquefied petroleum gas (LPG)

Liquefied petroleum gas (LPG) is the name given to a mixture of hydrocarbon gases that principally include propane and butane. LPG can either be manufactured as part of the crude oil refining process or extracted directly from oil and gas deposits.

Under normal atmospheric conditions, LPG will vaporise, but when stored and transported within steel

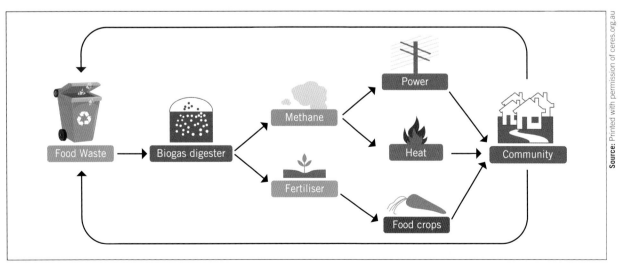

FIGURE 1.4 The process of biogas extraction and application

cylinders at moderate pressure the fuel is maintained in a concentrated liquid form. This characteristic of LPG allows for its use with portable appliances, as an automotive fuel, and wherever reticulated gas distribution systems do not exist. LPG can be stored in a range of cylinders and tanks to suit particular site requirements (see Figure 1.5).

Although produced in Australia as early as the 1950s, demand for LPG was limited to the small portable appliance market and overall it was largely regarded as a waste product of the refinery and extraction process. However, because of the oil crisis of the 1970s and the subsequent growth in the automotive use of LPG, demand and production increased dramatically, and LPG now forms a very significant part of the gas energy market.

Simulated natural gas (SNG)

Simulated natural gas (SNG) is produced in a special plant that mixes vaporised propane and air to provide a gas that has similar combustion characteristics as natural gas. Gas network operators can choose to provide SNG to entire suburbs in areas where the natural gas supply network is yet to reach. Supplying SNG enables a gas company to build a longer-term customer base in the expectation of eventual connection to natural gas. For this reason, SNG is often termed a holding gas.

A small, compact SNG plant may also be coupled to propane cylinders or a tank to provide commercial and small industrial customers with an alternative energy source in the event of normal gas supply outage. Many businesses cannot afford to be without gas for even short periods, and therefore these small SNG units can be used as an emergency back-up in the event of any supply loss.

Tempered liquefied petroleum (TLP)

During the period when natural gas was gradually being rolled out across Australia, tempered liquefied petroleum gas (TLP) was produced as a temporary simulated replacement for TG. In a similar manner to the production of SNG, TLP was made by mixing propane and air in specific proportions to create a gas that had the combustion characteristics of TG. Gas suppliers were then able to maintain a gas supply to customers despite the closing down of TG production facilities.

Compressed natural gas (CNG)

Compressed natural gas (CNG) is the same as normal, reticulated natural gas, but it is stored in some form of container under pressure. This enables CNG to be used for automotive and marine use on an increasing scale around the world. CNG use can offer considerable savings over the use of petrol and diesel fuels, and it is thought to provide longer engine life.

CNG requires a greater fuel storage space in comparison to LPG, restricting its use in standard passenger vehicles due to the relatively large tank size and limited travel range. However, this drawback does not affect its use in larger transport vehicles. Many trucking and bus fleets around Australia have been converted to CNG use.

Although a number of corporate car fleets are currently powered by CNG, there would need to be significant expansion of the current refuelling infrastructure and provision of home refill options before we are likely to see any broader use of CNG.

> **LEARNING TASK 1.1 FUEL GASES**
>
> For the gases listed below, state where it is or has commonly been used and why.
> - Natural gas (NG)
> - Liquefied petroleum gas (LPG)
> - Simulated natural gas (SNG)
> - Tempered liquefied petroleum (TLP)
> - Compressed natural gas (CNG).

COMPLETE WORKSHEET 1

SUMMARY

In this chapter you have been briefly introduced to the fuel gases used in Australia. Following the widespread use of towns gas (TG) around Australia for more than a century, TG was gradually replaced by natural gas (NG) and liquefied petroleum gas (LPG), which remain the two most commonly used fuel gases found today. Alternative sources of methane gas in the form of biogas and landfill gas are increasingly likely to further supplement NG energy supply options into the future.

GET IT RIGHT

This photo shows an incorrectly connected gas appliance. Based on your introductory knowledge of fuel gases, study the image and determine what the basic problem is.

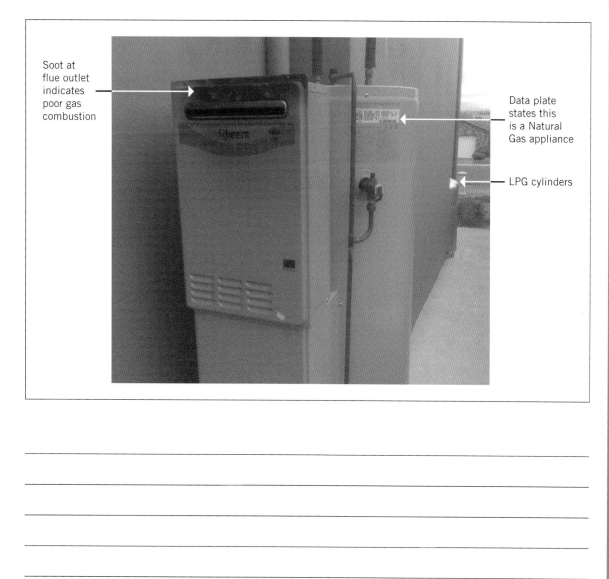

WORKSHEET 1

Student name: _____

Enrolment year: _____

Class code: _____

Competency name/Number: _____

To be completed by teachers

Student competent ☐

Student not yet competent ☐

Task

Review the sections on *Towns gas* through to *Compressed natural gas* and answer the following questions.

1. What are the two main fuel gases used in Australia today?

    i _____

    ii _____

2. Describe the difference between biogas and landfill gas.

    _____

    _____

    _____

3. What gases are to be found in your own local work area? Name the companies that distribute the gas.

    _____

    _____

    _____

4. Give the full name of each of these abbreviations:

    CNG _____

    NG _____

    TG _____

    LPG _____

    TLP _____

    SNG _____

5. Define a holding gas. Name two such gases.

    _____

    _____

    _____

6 What are the two primary constituents of SNG?

   i _____

   ii _____

7 Why was TG considered poisonous?

   _____

   _____

   _____

8 Which two gases are classed collectively as LPG?

   i _____

   ii _____

9 What hydrocarbon makes up most of natural gas?

   _____

   _____

   _____

10 Natural gas is classed as a fossil fuel. Describe what this means in your own words.

   _____

   _____

   _____

UNITS OF MEASUREMENT AND GAS INDUSTRY TERMS

Learning objectives

Areas addressed in this chapter include:
- an outline of the most common units of measurement used in the gas industry
- how to measure gas pressure using:
  - a manometer
  - a digital manometer
- key terms used in gas pressure measurement
- how to convert from imperial to SI units.

The content of this chapter provides the vital underpinning knowledge that will enable you to proceed confidently with learning and activities in subsequent sections.

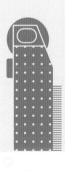

Overview

During the 1970s Australia began the gradual transfer from the old imperial measurement system to the metric-based International System of Units (abbreviated to SI). Most industries, including the gas and petrochemical sectors, had officially converted by 1974. The SI system of measurement is simple and consistent, and is used by most countries around the globe, with the notable exception of the United States of America.

A key feature of the SI system is the consistent use of the base 10 or its multiples: 10 mm to a centimetre, 100 cm to a metre and 1000 m to a kilometre. Compare this to imperial measurement in which there are 12 inches to a foot, 3 feet to a yard and 1760 yards to a mile!

Imperial measurements were inconsistent and did not follow obviously predictable patterns. For example, in Commonwealth countries there were 40 fluid ounces in a quart and 4 quarts to a gallon. In the US, however, there were 32 fluid ounces in a quart and as a result, a US gallon was equivalent to 0.883 British gallons!

In this chapter, we will review the most commonly used SI units of measurement relevant to you as a gas fitter. In addition, a section has been included on unit conversion. Although it is now increasingly rare to find older gas appliances and system components that were originally rated using imperial units, on those occasions where this does happen you will need to be able to convert these units to the modern SI equivalent.

This chapter assumes a working knowledge of basic mathematical principles and will only cover those areas that are used regularly by gas fitters. The worksheets at the end of this and other chapters will give you the opportunity to apply these basic processes.

The terms, methods and instruments used to measure gas pressure are covered later in this chapter.

Units of volume

All gas quantities must be measured so that consumption can be calculated and charged out to the consumer. You will also need to understand volume calculations in order to determine inert gas purging quantities, meter sizing or pipe sizing test times, but more on this later!

Reticulated gas volume is usually measured in litres (L) and cubic metres (m^3). A metric consumer-billing meter is graduated in both litres and cubic metres. This is based upon one simple relationship:

There are 1000 L in 1 m^3.

Imagine if you had to fill up a box that measured 1 m × 1 m × 1 m with water, using a one-litre measuring jug from your kitchen. You would need to empty that jug 1000 times to fill the box to the very top (see **Figure 2.1**).

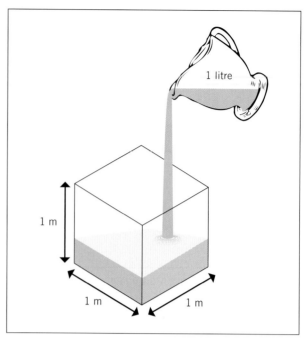

FIGURE 2.1 There are 1000 litres in one cubic metre of volume

It does not matter what substance we are talking about; all you need to remember is that every cubic metre is made up of 1000 litres. See how this works by looking at the following list of figures:

1000 L = 1 m^3
2000 L = 2 m^3
3000 L = 3 m^3
4500 L = 4.5 m^3
5750 L = 5.75 m^3

Can you see the relationship between litres and cubic metres? Divide the litres by 1000 and the answer is in cubic metres:

Litres / 1000 = m^3

You can also reverse the conversion by using multiplication:

m^3 × 1000 = litres

LPG can be measured in litres and cubic metres, but it is more common to talk of LPG in relation to its liquid weight in kilograms.

Imperial equivalents

Prior to the introduction of the SI system, volume was measured in cubic feet (ft^3) and cubic yards (yds^3). Consumer billing meters were measured more specifically in cubic feet per revolution.

Units of heat

When hydrocarbon-based fuel gases are oxidised during the combustion process, energy is released in the form of heat. A gas fitter needs to understand how heat is measured so that certain calculations relating to consumption, flow capacities and efficiency comparisons can be carried out.

The base unit used to measure the amount of energy expended during combustion is the joule (J). The joule is actually a relatively small amount of heat and, in general, the actual amount measured is in thousands of joules.

There are 1000 joules in 1 kilojoule (kJ).
There are 1000 kilojoules in 1 megajoule (MJ).

You will find that all modern Australian gas appliance data plates will record the appliance gas input in MJ per hour.

EXAMPLE 2.1

A gas cooktop may have a total gas input of 30 MJ/h, while a continuous flow hot water heater may have a gas input of 188 MJ/h.

EXAMPLE 2.2

When 1 m^3 of natural gas is burnt under test conditions, approximately 38 MJ of heat energy is released.

Although gas appliances will normally measure heat input in terms of MJ/h, you will find that electrical appliances use the SI unit of the watt (W) or its multiple, the kilowatt (kW). A watt is actually a unit of power, equal to one joule of energy per second. It is important to know about kilowatts because you will often need to compare the energy use of gas appliances with that of electrical appliances.

1 kilowatt hour = 3.6 MJ/h

EXAMPLE 2.3

Electric water heaters often use a 3.6 kW element to heat the water in the cylinder. This is approximately equivalent to a 12.96 MJ/h gas burner (3.6 kW × 3.6 MJ = 12.96 MJ/h).

EXAMPLE 2.4

If you were to quote for the replacement of a 5 kW electric heater with an equivalent gas appliance, you would need to consider one of at least 18 MJ/h input (5 kW × 3.6 MJ = 18 MJ/h).

Imperial equivalents

Prior to the introduction of SI units, gas appliance input was normally measured in British Thermal Units (BTU). Some older commercial and industrial appliances are still in operation and you will occasionally come across this unit of measurement. The BTU is actually a very small measure of heat and, as a result, often ends up as a very large number when energy is measured.

1 MJ = 947.817 BTU

EXAMPLE 2.5

A small simmer burner on a gas cooktop may be rated at 3 MJ/h. This would be the same as 2843.5 BTU.

EXAMPLE 2.6

A 427 MJ/h boiler found in a car wash would have a BTU rating of 404 717.86 BTU.

LEARNING TASK 2.1 UNITS OF HEAT

In your workplace, campus or through the internet, locate a gas appliance and establish the MJ/h rating.
Convert the MJ/h rating into the following units of heat:
1 BTU
2 kW

Units of mass

In the SI system of units, the mass of an object is measured in kilograms (kg).

1000 grams (g) = 1 kilogram (kg)
1000 kg = 1 tonne

While the kilogram is commonly used to measure the weight or mass of any given object, it is specifically discussed in this text in relation to LPG. It is customary to measure and talk of natural gas in cubic metres, but LPG is generally measured in kilograms because when contained within a cylinder, it is in liquid form.

LPG is stored in variously sized steel cylinders that are named according to the actual weight of LPG within the vessel. Therefore, a 45-kg cylinder means a specifically sized cylinder designed to hold 45 kg of LPG. Larger 190-kg cylinders are often installed for commercial uses. Once again, this is only the weight of the actual gas held in a liquid form inside the cylinder. The gross, or total, weight of both cylinder and gas is a lot more.

Imperial equivalents

Before the introduction of SI units, the principal measure of mass was the pound. Although pounds are still in use in the US, the use of the imperial pound ceased in Australia many years ago. However, you should note that it was equivalent to 453.592 g. LPG was also measured in pounds and the modern 45-kg cylinder used to be called a '100-pound' cylinder.

Fortunately, you will rarely encounter this now and only need to be aware what the old imperial unit was.

Units of pressure

The understanding and measurement of pressure is at the core of a gas fitter's knowledge. A gas appliance cannot work efficiently or safely unless it has the correct gas supply and operating pressures.

> **FROM EXPERIENCE**
>
> A gas appliance that is not operating at the correct pressure is dangerous! In certain circumstances, gas that is burning at the wrong pressure can kill people. Safe installation and maintenance of gas appliances relies upon your thorough understanding of gas pressure.

Pressure is simply defined as 'a force applied to a surface'. You can exert pressure on an object by pushing it. You can place your hand over the outlet of a running tap and actually feel the pressure of the water trying to escape. When gas is kept within a container or pipe, the molecules of the gas are in a state of constant random motion and continually bounce against the walls of the pipe. It is this action that we are able to measure as pressure.

The SI derived unit of measurement for pressure is the pascal (Pa). This unit of pressure is very small and, for general gas fitting work, it is more common to use its multiple, the kilopascal (kPa).

1000 Pa = 1 kPa

> **EXAMPLE 2.7**
>
> A gas fitter may set the working pressure of an LPG cooker burner to 2.75 kPa (or 2750 Pa).

> **EXAMPLE 2.8**
>
> The natural gas distributor may supply gas to a domestic house at 500 kPa (or 500 000 Pa).

When pressure becomes very high, it is customary to change to the next multiple, the megapascal (MPa).

1000 kPa = 1 MPa

This level of pressure is usually found when measuring the high-pressure gas supply from its production source to a town or city.

> **EXAMPLE 2.9**
>
> A pipeline operator may supply gas to a major city at 5 MPa (or 5000 kPa).

Other forms of pressure measurement

At times you will be required to commission older appliances that have been manufactured in Europe. Rather than use SI units such as kilopascals, some European countries may use a system of pressure measurement based on 'millibars' (mbar) and 'bars'. This system was based upon a standard for atmospheric pressure at sea level and is often still used in weather forecasts. One atmosphere (atm) is equal to 1.01325 bar or 101.325 kPa.

However, when converting between bar and kPa, you only need to remember the following simple relationship:

1000 mbar = 1 bar = 100 000 Pa = 100 kPa

> **EXAMPLE 2.10**
>
> An industrial gas regulator may have a pressure outlet rating of 0.3 bar. As there is 100 kPa in 1 bar, 0.3 bar would equate to 30 kPa.

Imperial equivalents

Older appliances and those with components sourced from the US will often state pressure ratings in pounds per square inch (psi). A pound is an imperial measure of mass that is approximately 0.454 kg. Therefore, psi describes a unit of pressure resulting from the application of 1 pound applied over an area of 1 square inch. One psi is equivalent to approximately 6.895 kPa.

> **EXAMPLE 2.11**
>
> During the inspection of an older LPG installation, you notice that the first stage regulator has a maximum outlet pressure of 10 psi. By multiplying this by 6.895 kPa, you will determine that the maximum outlet pressure is actually 68.95 kPa.

Another form of measurement uses the height of a column of mercury in a glass U-tube as a means of measuring pressure. This was known as 'inches of mercury' (inHg or ″Hg). As pressure was applied to one side of the U-tube, the mercury was forced up the other side and a reading of pressure could be taken. Many older industrial appliances used this form of measurement prior to the introduction of modern test equipment and the SI system.

> **LEARNING TASK 2.2 UNITS OF PRESSURE**
>
> You encounter an old pressure gauge that has a pressure reading of 30 psi. Convert the psi reading to kPa.

Measuring gas pressure

From this point on in your training and throughout the following chapters in this text, frequent references will be made in respect to gas pressure and its relationship to safe appliance operation; therefore, it is important at this point to look at how a gas fitter measures gas pressure, the actual instruments used and the particular terminology associated with gas pressure measurement.

GREEN TIP

Most modern gas appliances are high efficiency products that have an important part to play in reducing atmospheric pollutants and greenhouse gas emissions. Correct measurement of gas pressure is vital for their correct and efficient operation.

There are many different instruments available to the gas fitter and your selection will depend upon a number of factors including range of gas pressure, type of test to be applied and personal preference. Approved categories of test instruments are also listed in your AS/NZS 5601.1 Gas Installations Standard.

REFER TO AS/NZS 5601.1 APPENDIX E2 'TEST INSTRUMENTS'

Of these, we will examine the following two instruments:
- the water manometer (also known as a water gauge or U-tube)
- the digital manometer.

Manometers

A water manometer (see Figure 2.2) is one of the simplest pressure-measurement instruments used by a gas fitter. It consists of a tube formed into the shape of a U and filled to halfway with water. A silicone, plastic or rubber hose is connected on one side of the manometer; the other side left open and exposed to atmospheric pressure.

Manometers come in many shapes and sizes but are one of the most indispensable items in a gas fitter's tool kit.

The series of pictures in Figures 2.3 to 2.5 on the next page will illustrate how you can use a manometer to measure gas pressure.

If you look carefully, you will notice that the water level in the clear manometer tube is curved downwards in a concave shape. This phenomenon is called the meniscus. When measuring between each water level, take your measurement from the bottom of each meniscus (see Figure 2.5).

Most manometers are manufactured with a central scale against which the water level can be compared

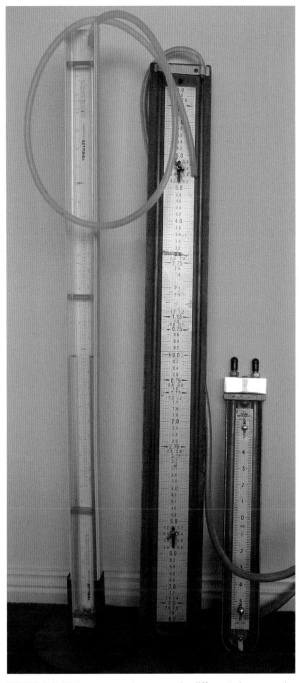

FIGURE 2.2 Water manometers come in different shapes and sizes

and the pressure determined. However, this does require that the water levels be accurately filled and checked to match the zero mark every time the device is used, or you will get an inaccurate reading. Most gas fitters find it far easier to simply use a separate ruler.

Two different types of scales are found on water gauge manometers. In Figure 2.6 you can see that both manometers are connected to a gas line measuring 1 kPa pressure but are read in different ways. Although the scale on each of these two manometers is read differently, you can see that the actual pressure is still the same.

You can see that the manometer is not yet subjected to gas pressure and both sides of the tube are exposed to atmospheric pressure (101.325 kPa). As a result, the water level is equal. The addition of simple green food colouring to the water helps the gas fitter to read the manometer.

FIGURE 2.3 A manometer connected to the test point of the appliance gas control

When connected to the gas supply, pressure is applied to the water on one side of the manometer, forcing the water level down until equilibrium is achieved between the gas pressure and water density. Ensure that the manometer is exactly vertical.

Using a steel rule, the gas fitter measures the distance between the water levels and applies the following simple rule:
100 mm = 1 kPa.

FIGURE 2.4 A manometer connected to the operating appliance with pressure applied to one side of the tube

It must be emphasised that accurate direct reading of a manometer using the scales requires that your water levels align with the zero mark before reading the pressure. However, if your water levels were not

What is the pressure in kPa?

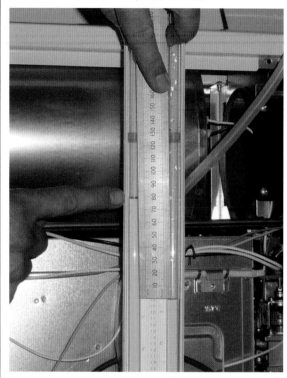

For consistency and speed, most gas fitters dispense with the use of a scale altogether and simply measure between each water level with a tape or steel rule. While it is a matter of personal preference, for the purposes of clarity this will be the method advocated throughout this text.

FIGURE 2.5 A gas fitter uses a steel rule to measure the distance between each water level on a manometer

aligned with zero before taking a reading you can still add the reading above the line and below the line together and divide in half to arrive at the actual gauge pressure.

EXAMPLE 2.12

If the gas fitter measured 200 mm between the two water levels on the manometer, the pressure would be 2 kPa.

EXAMPLE 2.13

If the measurement read was 290 mm, then the pressure would be 2.9 kPa. The pressure read off the manometer in such a manner is known as gauge pressure.

In summary, the advantages of using a manometer are that it:
- is simple
- does not require technical calibration checks
- is accurate.

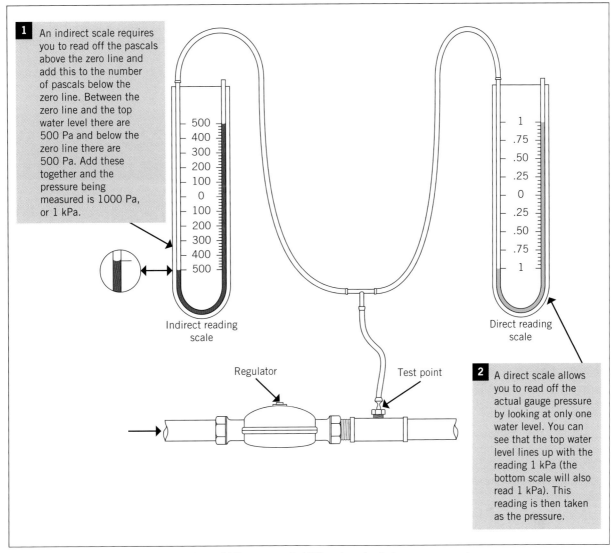

FIGURE 2.6 Two manometers connected to a 1 kPa gas supply (different scales but same pressure)

The disadvantages are that:
- the pressure measurement range is restricted by the height of the gauge
- the water level must be checked and maintained regularly.

LEARNING TASK 2.3 MANOMETERS

What are the two measurement scales found on manometers?

Inches water gauge scale

Before the introduction of SI measurements such as kPa, gas pressure was measured in inches water gauge ("WG). Early manometers were scaled in inches but used in the same way as you would today. One inch is equivalent to 25.4 mm.

Some old appliance data plates will still quote burner pressures in "WG. Table 2.2 provides a simple method of converting these units to kPa.

EXAMPLE 2.14

Before the introduction of SI units, a gas fitter might have set the working pressure of an LPG appliance to 11 "WG. This converts to the more familiar 2.75 kPa.

An old natural gas appliance data plate would often state the required gas pressure as 4 "WG. Today this would be 1 kPa.

Digital manometer

A digital manometer is a handheld electronic device used for measuring gas pressure in much the same way as a water gauge manometer. For basic pressure testing, a tube from the manometer is connected to the pressure source and the other opening on the instrument is left open to atmospheric pressure. As shown in Figure 2.7, gauge pressure can be easily determined from the digital display. The clear silicone tube is connected from the spigot of the manometer to the appliance regulator test point.

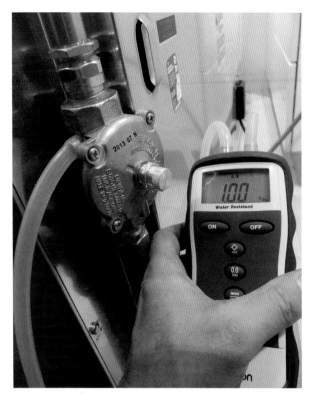

FIGURE 2.7 A digital manometer being used to measure the burner pressure of a gas cooking range

A wide range of digital manometers is available, each with different feature levels and operating ranges. Digital manometers are particularly useful with appliance servicing, and should be part of every gas fitter's instrument kit.

In summary, the advantages of using a digital manometer are that it:
- is compact
- is accurate
- is easy to read
- normally features a choice of measurement units (i.e. SI and imperial)
- can usually read a wider range of pressures than a water gauge manometer.

Disadvantages are few but include:
- occasional battery failure (keep spares on hand)
- it must be calibrated regularly to ensure that accuracy is maintained.

A note on low pressure Bourdon gauges

A Bourdon gauge is a compact measuring instrument that features a needle indicator and printed card face. Many types of Bourdon gauges are used throughout the plumbing and gas fitting industry, an example being the familiar gauges found on oxy-acetylene regulators. Prior to the introduction of digital manometers, a low pressure Bourdon gauge was the primary alternative to the standard U-tube manometer. The low-pressure Bourdon gauge in Figure 2.8 was designed specifically for gas fitting use and is housed in a compact protective box.

Although Bourdon gauges are compact and simple to read, they have a number of disadvantages. The internal mechanism is prone to damage from sudden pressure increases; they require regular calibration and they are comparatively inaccurate and require longer test times to ensure correct readings.

> Low pressure Bourdon gauges are no longer recommended for use by gas fitters, and references to them in the AS/NZS 5601 Gas Installation Standards have been removed.

FIGURE 2.8 Low pressure Bourdon gauges are no longer recommended for gas system or appliance pressure testing

Terms used in gas pressure measurement

There are a number of terms you need to be familiar with when carrying out gas pressure measurement tasks. Some are defined within the AS/NZS 5601 and others are in general use throughout the industry.

Defined terms from the Gas Installation Standards

Some terms commonly used in the gas industry are also defined terms within AS/NZS 5601 Gas Installation Standards. This means that they have a specific meaning and should be used only in the manner described. As you begin to work through and become familiar with AS/NZS 5601, you will notice that many words and phrases are written in *italics*. This is done to alert you to the fact that such a word or phrase has been defined and that its meaning can be found in Section 1 of AS/NZS 5601.1. To assist you to become more familiar with the Gas Installation Standards, some examples are included below.

This text is focused primarily on the requirements for installation of Type A gas appliances. A Type A gas appliance is a defined term within your Standard and refers to an appliance for which a certification scheme exists. This means it has been manufactured to meet strict performance standards and once certified is able to be installed by a licensed Type A gas fitter. Confirm this definition within your Standard.

REFER TO AS/NZS 5601.1 SECTION 1 'APPLIANCE'

Using instruments such as water and digital manometers, a gas fitter is able to conduct certain pressure tests on gas appliances to check performance or adjust the appliance. Classifications of pressure types and pressure tests have particular names that indicate the nature of the test itself. They must also be taken in a certain manner.

Pressure

In order for gas to flow, its pressure must exceed the pressure of the atmosphere. This combination of atmospheric pressure and gas flow pressure is known as absolute pressure. However, for general gas fitting we only measure the amount above that of atmospheric pressure. This reading is known as 'gauge pressure'.

REFER TO AS/NZS 5601.1 SECTION 1 'PRESSURE'

Operating pressure

The term operating pressure is also defined in AS/NZS 5601 and describes the pressure to which any part of the gas installation is subjected under normal operating conditions. It should not be confused with the term 'burner pressure'.

REFER TO AS/NZS 5601.1 SECTION 1 'OPERATING PRESSURE'

Normal operating conditions for a pipe installation include both dynamic and static pressures. Be aware that the static (lock-up) pressure of an installation is normally slightly higher than gas pressure under flow conditions.

EXAMPLE 2.15

A natural gas installation may be designed to provide a supply pressure of 3 kPa, but in the lock-up condition, it is likely to measure 3.1 or 3.2 kPa.

Rated working pressure

The gas supply to any section or component of an installation must not be subjected to pressure in excess of certain limits. This maximum pressure is known as the rated working pressure.

REFER TO AS/NZS 5601.1 SECTION 1 'RATED WORKING PRESSURE'

Maximum over-pressure

When a gas system is operating or is in the static condition, the pressure in any part of the system must not exceed what is known as the maximum over-pressure limit.

Identifying and preventing over-pressure conditions is a key aspect of a gas fitter's job, to ensure that system components are not subjected to pressures above those they are designed to handle.

REFER TO AS/NZS 5601.1 SECTION 1 'MAXIMUM OVER-PRESSURE'

Other useful key terms

Some terms are not specifically defined within the Gas Installation Standards, but because of their widespread use within the industry every gas fitter must understand them. Some examples are below.

Minimum inlet pressure

Each gas type and appliance has a particular burner pressure at which it is designed to operate. You must ensure each gas appliance is supplied with the minimum required gas pressure so that it can operate as designed. This is known as the minimum inlet pressure, and although not directly defined it is still a prescribed minimum within the Gas Installation Standards.

TABLE 2.1 Minimum inlet pressure

Gas type	Min. pressure at appliance inlet (kPa)
NG and SNG	1.13
LPG	2.75

REFER TO AS/NZS 5601.1 SECTION 5 'MINIMUM PRESSURE AT APPLIANCE INLET'

LEARNING TASK 2.4 MINIMUM INLET PRESSURE

In reference to the minimum pressures for NG and LPG appliances in Table 2.1 above, locate and provide a standard reference from AS/NZS 5601.1.

Importantly, these pressures must be available at each appliance, with all appliances connected to consumer piping and running at maximum gas consumption. However, while these are the prescribed minimum inlet pressures, you are able to provide higher inlet pressure where required. Burner pressures will be either the same as the inlet pressure or slightly less, depending upon internal pressure losses and the nature of the appliance itself.

Burner pressure

The burner pressure is the dynamic pressure measured while the burner is in operation. Gas is being used by the appliance burners and must do so at a certain pressure in order to combust efficiently. Burner pressure is measured at (or as close as possible to) the burner while it is operating. You must ensure that the manometer is connected between the appliance regulator and the burner itself. **Figure 2.9** shows how the burner pressure must be read as close to the burner as possible and in accordance with manufacturer's instructions. For example, the manufacturer's instructions for the storage water heater in **Figure 2.9** indicated that the burner pressure must be read from a test point beneath the combination control. Here the gas fitter checks burner pressure with a digital manometer while observing the actual flame with a mirror that you can see in the bottom of the picture.

Lock-up pressure

The static pressure of a consumer gas line is measured with the supply turned on and no appliances operating. This is also known as the lock-up pressure.

The manometer is connected to a suitable test point between the outlet of the meter and the inlet of the appliance regulator (see **Figure 2.10**).

When gas appliances are turned off, the supply regulators on both natural gas and LPG systems are designed to lock up and prevent any more gas from entering the consumer piping. This is called the lock-up pressure and must not exceed certain limits.

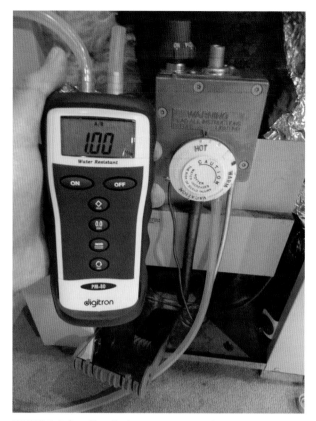

FIGURE 2.9 Reading the burner pressure of a storage water heater

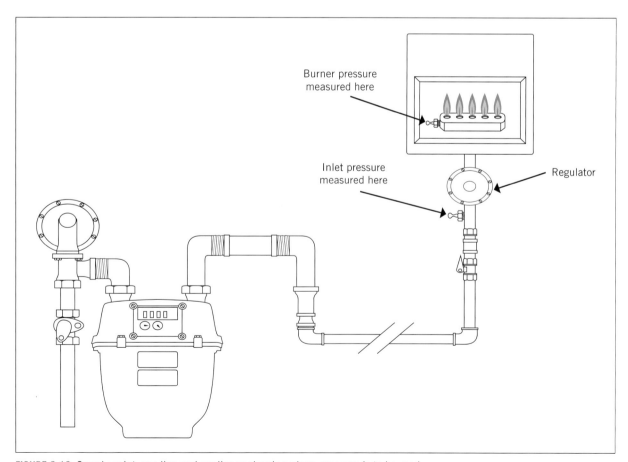

FIGURE 2.10 Supply point, gas line and appliance showing where pressure is to be read

Unit conversion

The need to be able to convert old or unusual units of measurement to modern SI units will continue for some time to come, particularly given the longevity of many gas appliances. Therefore, you will need to be able to convert these units as and when required. Table 2.2 lists some of the most common conversion factors you might need in general gas fitting practice.

TABLE 2.2 Common conversion calculations needed by a gas fitter

To convert...	Action...	Answer
"WG	Divide by 4	kPa
kPa	Multiply by 4	"WG
"Hg	Multiply by 13.6 × 0.25	kPa
Psi	Multiply by 6.895	kPa
Mb	Multiply by 0.1	kPa
kW	Multiply by 3.6	MJ
BTU	Divide by 947.817	MJ

A conversion chart that includes some additional calculations is also included in the Gas Installation Standards.

REFER TO AS/NZS 5601.1 APPENDIX B 'CONVERSION FACTORS'

In a practical sense, many of the units and terms that you have encountered in this chapter will be found on appliance data plates. Combined with manufacturer's instructions, the appliance data plate is a principal source of information for a gas fitter. All certified gas appliances must have a data plate that contains information relevant to the commissioning of the appliance.

Information found on data plates may include the:
- certification/approval number
- model and serial numbers
- burner pressure
- maximum inlet pressure (or supply pressure)
- maximum gas input in MJ/h
- gas type
- jet or injector size/s.

Figure 2.11 gives some examples of data plates found on various appliances, both old and new. All Type A gas appliances must have a data plate, and a gas fitter must always check this for key information prior to installation or servicing.

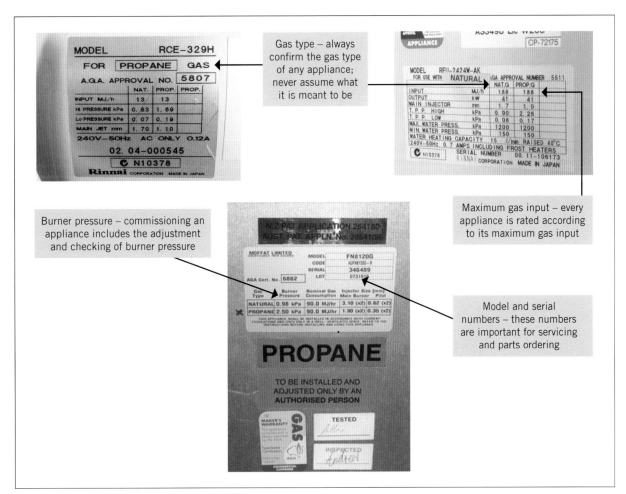

FIGURE 2.11 Examples of some common appliance data plates

LEARNING TASK 2.5: MEASURING GAS PRESSURE

In your workplace or campus, and under supervision from your class teacher or workplace supervisor, locate a gas appliance and connect a manometer to the appliance, then check that the appliance's pressures are correct in accordance with manufacturer's guidelines and determine the following:

1. the inlet operating pressure for the gas appliance and its unit of measurement
2. the burner head pressure where applicable and state its unit of measurement.
3. Where and how is this data located?

SUMMARY

In this chapter you have been introduced to some fundamental concepts that underpin everything you do in your training as a gas fitter. You must now learn to apply the correct SI units of measurement applicable to gas fitting work at all times and be able to interpret such measurement requirements from appliance data plates and installation specifications. Although almost all imperial era Type A appliances have been replaced or removed over the years, it is possible that you will still occasionally encounter such an installation and so you must be able to deduce the SI unit testing and performance requirements using relevant conversion factors.

You have also seen how gas pressure is measured with the use of both water and digital manometers. A gas fitter uses such instruments on every job. As your training progresses and armed with this basic knowledge you will soon be able to carry out a range of pressure tests on installation pipework and appliances.

 COMPLETE WORKSHEET 1

GET IT RIGHT

The photo below shows an incorrect practice when determining the pressure of an appliance using a manometer filled with red dye.

Has the manometer in the image been filled with red-coloured water to the correct level? Provide reasoning for your answer.

_____

_____

_____

_____

WORKSHEET 1

Student name: _____

Enrolment year: _____

Class code: _____

Competency name/Number: _____

To be completed by teachers
Student competent ☐
Student not yet competent ☐

Task

Review the sections on *Units of volume* through to *Unit conversion* and answer the following questions.

1. How many litres of gas are equivalent to a volume of 12.3 m^3?

   _____

   _____

2. A transmission main is pressurised to 7.5 MPa. What would be the pressure in kilopascals?

   _____

   _____

   _____

3. What are three key advantages to be found in using a water gauge manometer?

   i   _____

   ii _____

   iii _____

4   Determine the pressure from the following manometer readings:

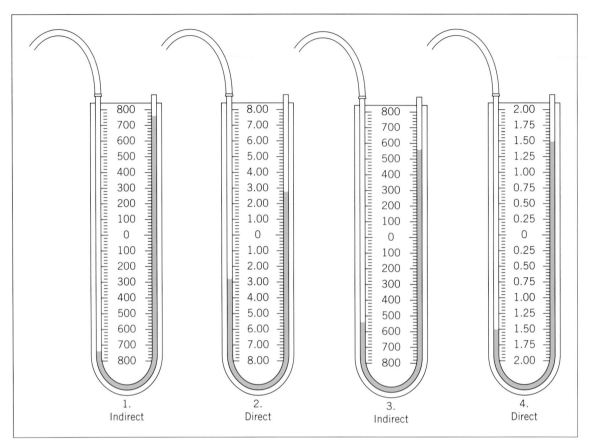

| | 1. Indirect | 2. Direct | 3. Indirect | 4. Direct |

5   Complete the following conversion list:

4 "WG   = _____ kPa

7 kW   = _____ MJ

15 psi   = _____ kPa

4 "Hg   = _____ kPa

30 000 BTU   = _____ MJ

6   What is the definition of the term *rated working pressure*? Cite a standards clause reference to support your answer.

_____

_____

_____

7   What do you understand to be the meaning of the term *burner pressure*?

    _____
    _____
    _____

8   If available in your workplace or campus, find the data plates on the following appliances and record the burner pressure and maximum gas input of each:

    An LPG storage water heater _____ kPa _____ MJ/h

    An NG upright cooker _____ kPa _____ MJ/h

    An LPG instantaneous water heater _____ kPa _____ MJ/h

    An NG cooktop _____ kPa _____ MJ/h

9   What quality control test would you need to conduct regularly if using a digital manometer?

    _____
    _____
    _____

10 What is gauge pressure?

    _____
    _____
    _____

11 A gas fitter uses a steel rule to measure the pressure on a manometer. A reading of 280 mm indicates a pressure of _____ kPa.

GAS DISTRIBUTION SYSTEMS

Learning objectives

Areas addressed in this chapter include:
- the distribution of natural gas
- the component parts of a natural gas supply system:
  - well head
  - refinery
  - trunk main
  - city gate
  - reticulation main
  - service line
  - service regulator
  - consumer billing meter
- the distribution of LPG.

Before attempting the units of competency that will be detailed within Part B of this text, you must understand the basic distribution methods used for both natural gas and LPG. This chapter provides the necessary background context detailing how the gas you are working with actually gets to the customer. Understanding this is important in dealing with the retail gas distributors in your local area, and has a direct influence on how your installations are carried out.

Overview

Unless trained and authorised to do so, the permitted scope of work for a Type A gas fitter only extends from the outlet of the consumer billing meter, or the outlet of the first regulator on an LPG cylinder. However, this work must be done with a broader knowledge of the distribution system that transports the gas to this point in the first place. You will find yourself working in close contact with gas distributors as a normal part of your daily duties.

In this chapter, you will review the infrastructure that provides natural gas and LPG to the end consumers.

REFER TO AS/NZS 5601 INTRODUCTION AND SCOPE

The distribution of natural gas

When you complete a natural gas installation and finally turn on an appliance, you have actually connected your job to an almost continuous piping system that has carried this gas hundreds, or perhaps even thousands of kilometres from its original source. Natural gas is normally a reticulated gas, which simply means it is transported via a pipe network from the refinery to the consumer's billing meter. Since the 1960s, natural gas pipelines have gradually extended across every state in Australia and are often interlinked to form an integrated gas network (see Figure 3.1).

The gas supply infrastructure can be broken down into its component parts. Figure 3.2 shows a simple generic representation of a natural gas supply system. Refer to this layout as each section is discussed below.

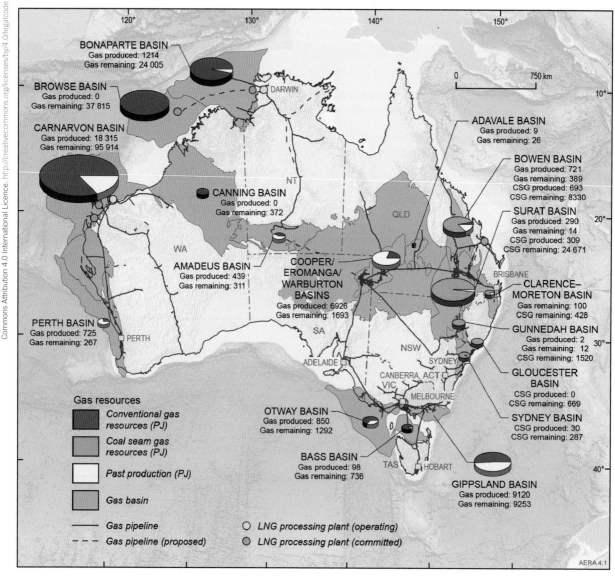

FIGURE 3.1 Major existing and proposed gas transmission pipelines around Australia

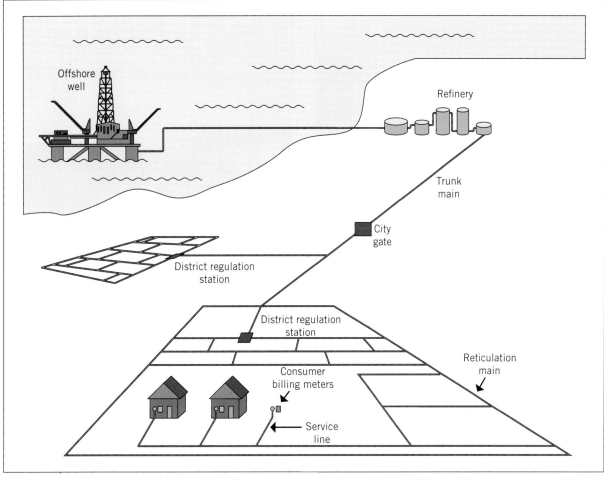

FIGURE 3.2 Bird's eye view of a natural gas supply system from well head to billing meter

This overview of a gas distribution network shows how gas is transported from its source all the way to the end consumer.

Well head

The term well head describes the infrastructure sitting atop an oil or gas well that acts to control the extraction and distribution of the hydrocarbon product after it leaves Earth's crust (see Figure 3.3). Most natural gas is extracted as a by-product of oil production. As the oil leaves its deep underground reservoir, it undergoes massive decompression and the light carbon-chain molecules within the oil come out of solution in the form of methane and other gaseous by-products. The gas and oil are then normally piped direct to the refinery.

Refinery

At the refinery (see Figure 3.4), the gas is separated from the crude oil and is subjected to processes that include the removal of moisture, solids and the heavy metal mercury. It is also checked for chemical consistency so that it produces a standard heat output from every cubic metre of gas.

FIGURE 3.3 A well head at an oil and gas field

FIGURE 3.4 A gas and oil refinery

Trunk main

After the gas is processed, it leaves the refinery via the high-pressure trunk main, also known as the transmission main. The trunk main is often owned by a separate entity, generally called 'the pipeline operator'. Actual business and regulatory arrangements may vary, but normally the pipeline operator will purchase the gas from the refinery, transport it via its pipeline and sell it on to the gas distributor.

So that large volumes of gas can be transported long distances and to meet consumer demand, trunk mains operate at very high pressures. The pressure will vary according to the distance the gas has to travel and the volume of gas required. A transmission pressure of 5000 kPa is not uncommon; for example, the pressure in the line crossing Bass Strait between Victoria and Tasmania has been, at times, in excess of 14 000 kPa!

These high pressures permit the use of pipes of smaller diameter than those designed to carry water or oil (see Figure 3.5). The high pressure also acts to provide a supply buffer, ensuring that regardless of fluctuations in demand, sufficient gas is always available at the point of consumption. This means that in most cases there is no need to provide large gas storage facilities, unlike those commonly found in all cities in the days of reticulated towns gas.

Occasionally, a large industrial customer that plans to use significant volumes of gas will be supplied directly from the trunk main itself via a pressure-reducing let-down station. The rest of the gas is piped up to the city gate, ready for distribution.

City gate

The city gate is the point at which the trunk main ends and the actual reticulation main begins (see Figure 3.6). This installation may also be known as a 'let-down station', a term relating to the pressure reduction process that occurs here. This and a number of other functions happen at the city gate:
- Ownership changes – the ownership of the gas normally changes from the pipeline operator to the distributor and/or retailer.

FIGURE 3.5 Maintenance work being carried out on a gas trunk main in western NSW

- Pressure is reduced – the transmission pressure must be reduced to the required reticulation pressure. When gas is brought down from extremely high pressures, it becomes colder and may freeze altogether. Therefore, the gas is passed through large gas-fired heat exchangers during the let-down process.
- Gas is measured – the gas is measured and its chemical properties checked.
- Odorant is added – in most instances, a mercaptan-based chemical odorant is added to the gas at this point so that the gas is detectable by smell in the event of a leak. This process may sometimes happen at the refinery.

Reticulation main

The reticulation main or supply main is the system of pipes that carries the gas to the consumer. Often, a main backbone line running at a high pressure (possibly 1000 kPa) will exit from the city gate with sections branched off to service domestic, commercial and industrial customers. These branch lines will be pressurised according to the expected demand in a given area. Gas reticulation pressures are often broken down into the following categories:

FIGURE 3.6 A city gate station

- low pressure – up to and including 7 kPa
- medium pressure – over 7 kPa and up to and including 200 kPa (210 kPa in some areas)
- high pressure – over 200 kPa (210 kPa) and up to and including 1050 kPa.

Figure 3.7 shows a steel reticulation main being installed through a major urban centre. This 'backbone' main is designed to carry 1000 kPa and is to supply 500 kPa branches to adjacent suburban streets.

In the early days of gas distribution, reticulation mains were made from large cast iron or steel pipes and ran at very low pressures. Although most networks have been converted to new piping and higher pressures, some areas still have low-pressure regions. Today, modern gas infrastructure is usually installed with a high-pressure supply and carried in pipes made from steel, nylon or polyethylene. Your teacher or supervisor will discuss the delivery pressures relevant to your local area.

Where different pressures are required, the pressure is controlled by high-pressure regulators situated in district regulator stations (DRS). These are often installed in underground pits and can usually be identified by signage and characteristic vents. Figure 3.8 shows a DRS pit. Regulators in this particular pit are adjusted to reduce a backbone supply line pressure of 1000 kPa to the suburban distribution pressure of 500 kPa.

Service line

The branches that run from the gas main in the street to the consumer billing meter are called service lines.

FIGURE 3.7 A steel reticulation main being installed through a major urban centre

FIGURE 3.8 A district regulator station pit

When a customer decides to connect to natural gas, the main is exposed and a tapping saddle installed. The service line is connected to this and terminates at the meter riser. As most modern installations use plastic pipes, service lines must be changed to a copper or steel riser before coming out of the ground.

Figure 3.9 shows a horizontal boring process for a polyethylene (PE) service line. The machine bores a hole at the required depth to an access pit some distance away. The PE pipe is then connected to the last drill shaft and, as the drill sections are withdrawn and removed, the PE pipeline is dragged back through the hole. In this way, long sections of pipe can be installed without digging open trenches.

FIGURE 3.9 Trenchless horizontal boring to install a polyethylene service line

Service regulator

In order to provide sufficient gas to all network consumers, modern reticulation mains are run at much higher pressures than can be used within the consumer piping system. To control and limit this pressure from the reticulation main, a service regulator is installed at the end of the service line (see Figure 3.10). The service regulator outlet pressure is adjusted and set by the distributor to suit the supply requirements of the consumer.

FIGURE 3.10 A service regulator fitted to a residential meter installation

EXAMPLE 3.1

A distributor may provide a 200 kPa supply to all domestic consumers. This pressure may be reduced at the service regulator to 2.7 kPa for use in the consumer piping.

FROM EXPERIENCE

Gas fitters are not permitted to adjust a service regulator unless authorised to do so by the gas distributor.

The service regulator reduces the high inlet pressure from the reticulation main down to the required low-pressure gas supply for the consumer.

Consumer billing meter

The consumer billing meter is the last key component of the gas reticulation system. This device can normally be found at the boundary of the property, or it may be attached to – or sited adjacent to – the building itself (see Figure 3.11).

FIGURE 3.11 Examples of domestic and commercial/industrial consumer billing meter assemblies

After the distributor/retailer has purchased gas from the pipeline operator, it sells it on to the consumer. The meter measures the volume of gas consumed on the premises. Just as for an electrical power supply, a regular reading is taken of gas consumption and the customer billed by the company.

LEARNING TASK 3.1 GAS DISTRIBUTION SYSTEMS

Natural gas reticulation systems consist of different components that make up the entire gas network. Using the words listed below, order the list of the components from original source to the consumer.

Reticulation main	Consumer billing meter
Well head	Service regulator
Trunk main	Service line
Refinery	City gate

The distribution of LPG

Liquefied petroleum gas (LPG) is primarily stored and distributed in steel tanks and cylinders, which are normally situated on the premises of each consumer. You may have seen such cylinders adjacent to the walls of houses and businesses.

When stored in a steel container under pressure, LPG exists as a very concentrated form of energy in the liquid state. Only when an appliance burner is running does it change from a liquid to its gaseous form, allowing vapour to be drawn from the cylinder. This feature of LPG means that it is very useful in providing a fuel gas source to those areas not serviced by a reticulated gas main.

Source

LPG occurs naturally in combination with other underground hydrocarbon deposits and can be extracted during the refining of oil and gas. The butane and propane gases that are the principal components of LPG are separated from other substances, dried, checked for consistency and stored in large tanks ready for distribution.

Delivery

Depending upon where it is sourced, bulk LPG is delivered to storage terminals around the country by ship, train or truck. Many tonnes of LPG are stored in these terminals so that a ready supply of gas is available in the regional distribution area. Gas consumers receive their gas supply in one of two ways:
- exchange cylinders
- in-situ fill cylinders or tanks.

Exchange cylinders

Once a customer has emptied a cylinder, they contact the supplier who will then arrange for a replacement cylinder to be brought out and exchanged for the empty one. This is a common supply method for low-volume consumers. Exchange cylinders come in many sizes and are normally labelled according to the amount of gas they hold in kilograms. Your gas supplier will normally provide exchange cylinders in 13.5 kg and 45 kg sizes.

Figure 3.12 shows a domestic LPG 2 × 45-kg exchange cylinder installation. In some states and territories, protective hoods must be installed over cylinders. In the region in which this picture was taken, a hood is not required, but cylinders must be restrained and regulator vents must face downwards.

In-situ cylinders or tanks

In-situ cylinders and tanks are those that are kept on site and regularly topped up with a special fill hose from an LPG tanker. The customer normally does not need to contact the supplier for a refill.

Figure 3.13 shows in-situ fill cylinders serving two separate commercial kitchens. A gas tanker will

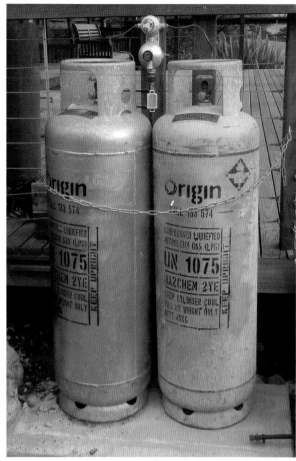

FIGURE 3.12 A domestic LPG 2 × 45-kg exchange cylinder installation

FIGURE 3.13 In-situ fill cylinders serving two separate commercial kitchens

regularly fill each cylinder up on site by connecting a fill hose attachment to the cylinder inlet valve.

In-situ fill cylinder supply is normally provided to customers with higher volume consumption, or in areas where a large number of installations warrants a truck-fill approach. Once again, the storage cylinder is named according to its weight in LPG. In-situ fill cylinders can be 45 kg, 90 kg, 190 kg and 210 kg.

Where high gas consumption is anticipated, it is more economical to install a tank for storage rather than multiple cylinders (see **Figure 3.14**). Tanks are usually

FIGURE 3.14 An LPG tank installation

specified by the supplier, often based upon customer and gas fitter estimates of gas use. Because tanks hold a lot more than cylinders, they are normally identified by the amount of LPG they hold in tonnes. For example, a service station may be supplied with a 9-tonne tank. This refers to the weight of propane contained within the tank and is sized according to the service station's consumption patterns.

Connection

Gas vapour is drawn off from the cylinders or tanks, with the pressure controlled by LPG regulators. Exchange cylinder customers pay a cost per cylinder each time they need a refill and therefore a metering system is not required. Customers with in-situ fill cylinders and tanks have the amount of gas supplied measured by a meter mounted on the tanker itself.

LEARNING TASK 3.2 — GAS DISTRIBUTION SYSTEMS

1. For your local area determine what fuel gas is predominantly used domestically and why.
2. Establish where the fuel gas is originally sourced from.

For natural gas areas, a supply pipeline diagram may be available through local gas utilities websites.

COMPLETE WORKSHEET 1

SUMMARY

In this section you have been introduced to the distribution systems relevant to both natural gas and LPG fuel gases. Key points included:

Natural gas
- extraction from a hydrocarbon deposit at the well head
- refinery processing
- transport of the gas from the refinery via the high pressure trunk main
- pressure reduction at the city gate
- reticulation to industrial, commercial and domestic customers
- measurement of gas used at the consumer billing meter

LPG
- sourced as a by-product of hydrocarbon extraction from beneath the ground
- stored in major terminals around the country
- distribution via cylinder exchange or filling of in-situ cylinders/tanks by a gas tanker truck

Understanding the basic distribution requirements of both gases has a direct impact upon your installations, particularly in relation to pressure supply, meter and cylinder siting.

GET IT RIGHT

The photo below shows an incorrect practice when installing or maintaining a gas meter connection. Identify the incorrect method and provide reasoning for your answer.

_____
_____
_____
_____
_____

WORKSHEET 1

Student name: _____

Enrolment year: _____

Class code: _____

Competency name/Number: _____

To be completed by teachers

Student competent ☐

Student not yet competent ☐

Task

Review the sections on *The distribution of natural gas* and *The distribution of LPG* and answer the following questions.

1. i List the companies that distribute natural gas in your area.

   _____
   _____
   _____

   ii List the companies that distribute LPG in your area.

   _____
   _____
   _____

2. What distribution pressure is normally supplied up to the service regulator of a domestic consumer in your area?

   _____
   _____

3. What does the term 'reticulation main' refer to?

   _____
   _____
   _____
   _____

4. In your own region, what distribution pressures match the following terms?

   Low pressure _____ kPa

   Medium pressure _____ kPa

   High pressure _____ kPa

5   Is a 90-kg cylinder classed as an exchange cylinder or in-situ fill cylinder?
_____
_____

6   Why are large storage facilities or buffer tanks no longer required for a modern natural gas supply system?
_____
_____
_____

7   From where is the natural gas used in your region originally sourced?
_____
_____

8   Referring to this chapter and the glossary, define the following terms:

Service line
_____
_____

Trunk main
_____
_____

In-situ cylinder
_____
_____

9   Describe four processes that occur at the city gate.
_____
_____
_____

10 When are you permitted to adjust the installation service regulator?
_____

GAS CONSTITUENTS AND CHARACTERISTICS

Learning objectives

Areas addressed in this chapter include:
- the constituents of fuel gases
- the characteristics of fuel gases:
  - relative density
  - flammability range
  - air requirements
  - flame speed
  - ignition temperature
  - toxicity
  - odour
  - heat value.

A knowledge of what makes up fuel gases and their particular characteristics is required for all gas fitting work and so this chapter and worksheets should be completed before working through the units of competency contained within Part B of this text.

Overview

All fuel gases have different physical characteristics. As a gas fitter, you must be aware of these characteristics in order to safely and effectively carry out your installation and servicing tasks in the industry. Installing an appliance on the wrong type of gas can be extremely dangerous for both you and the appliance owner.

Being aware of the physical characteristics of each gas gives you a deeper understanding of why gas supply systems, appliance designs and combustion settings differ quite markedly from each other.

We will examine gas constituents and characteristics by comparing three principal gases: natural gas, liquefied petroleum gas and towns gas. Although towns gas is no longer used in Australia, references are still provided where applicable so that you can gain a better understanding of modern fuel gases through such a comparison, and gain a deeper appreciation of why each of these gases is fundamentally different to the others.

Note: you may encounter variations in some of the values and figures presented on the following page. The properties of each fuel gas may vary according to its source, where it is used and the way it was measured. Some of these figures have also been rounded down for clarity. Always check with your instructor or gas supplier to source figures relevant to your particular region.

Constituents of fuel gases

In Table 4.1 you will see each of our three fuel gases – natural gas (NG), liquefied petroleum gas (LPG) and towns gas (TG) – and a breakdown of their different constituents.

TABLE 4.1 Fuel gases and a breakdown of their different constituents

Constituent	NG (%)	Commercial LPG (%)	TG (%)
Methane	90.90	–	3.8
Propane	1.020	97.6	0.3
Butane	0.180	2.4	0.5
Carbon monoxide	–	–	16.4
Carbon dioxide	2.570	–	12.0
Nitrogen	0.320	–	12.2
Oxygen	0.001	–	0.6
Hydrogen	–	–	50.0
Other trace hydrocarbons	5.009	–	6.2

Let's summarise some of the key features of this table. Looking at the NG column, you can see that it is primarily made up of the hydrocarbon methane, with the remaining constituents being smaller percentages of other gases. These figures will vary depending upon where the gas was sourced.

In comparison you can see that commercial LPG is made up from only propane and butane. The percentage will alter depending upon where it is used; for instance, in some areas LPG supplied for automotive use will have a higher percentage of butane.

Towns gas (TG) was different again. You will recall that TG was manufactured from coal or oil and so, depending upon the nature of the feedstock and the manufacturing process, these figures could differ quite considerably. However, you can see that TG had two notable constituent gases not shared by LPG or NG – hydrogen and carbon monoxide. Both these gases are highly flammable and, it is important to note, carbon monoxide is very toxic.

LEARNING TASK 4.1 CONSTITUENTS OF FUEL GASES

Referring to Table 4.1, what is the principle constituent of:
- NG?
- LPG?

Characteristics of fuel gases

In the previous section you have seen how each of these gases differ in the way they are constituted (made up). Although you may not need to remember the specific make-up of fuel gases on a daily basis, the differences between the gases have an effect on how each gas behaves, and this is where the knowledge of the characteristics of fuel gases becomes vital to you as a gas fitter.

FROM EXPERIENCE

Advanced gas fitters licensed to work on Type B industrial gas appliances will use their intimate knowledge of gas characteristics in calculations required for safe installation and commissioning.

Each of the following sections describes a characteristic or phenomenon of fuel gases that has a direct bearing on your work as a gas fitter.

Relative density

You may have seen that oil floats on the top of water. This is because each substance has a different density. Because oil is less dense than water, it floats on top and does not mix with the water itself. When scientists compare the density of substances against that of water, they often use the term specific gravity. At a given temperature, they give water a numerical value of 1 and then compare other substances against it. In our example, the fuel oil that you may find floating in a slick on the ocean has a specific gravity of 0.89. As this figure is less than 1, it means that the oil is less dense than water, enabling it to float. In comparison, a lead bar has a specific gravity of 11.34, so it will sink in water.

Differences in density can be found among all substances, including gases. When comparing the density of gases, it is now more common to use the term relative density (RD). In this instance, rather than comparing the density of gases against that of water, scientists prefer to compare them to the density of air at sea level atmospheric pressure (101.325 kPa) and 20°C. Once again, they give air a value of 1 and compare all other gases against this.

In Figure 4.1, note that both TG and NG have a relative density of 0.6, which, being less than 1, makes these gases lighter than air. This means that if NG were to be released from a gas line it would not only float away but also disperse (spread out) in the air.

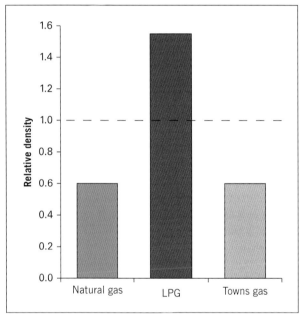

FIGURE 4.1 The density of the three gases relative to that of air (which has a value of 1)

The graph also shows that LPG has a relative density of 1.55, indicating that this gas is heavier than air. Although all gas leaks are dangerous, heavier-than-air gases are particularly hazardous. Escaping gas can build up at low points in the appliance or floor and will remain there until physically dispersed. LPG takes a long time to disperse naturally. In this situation, the gas may reach a point where it could ignite and explode. This is why leaking LPG accumulating, for example, in the bilge of a boat or the cellar of a building is considered extremely dangerous.

Flammability range

Three key elements are required for any combustion to take place:
- oxygen (air)
- heat (ignition)
- fuel.

The fire triangle shown in Figure 4.2 graphically demonstrates this relationship. If any one of these elements is removed from the triangle, combustion will not happen.

FIGURE 4.2 The fire triangle

This diagram provides a simple understanding of how substances burn, but as a gas fitter you need to know more about combustion. You need to be aware that it is not just the presence of these three elements in the combustion process, but the proportions or degree to which each element is present, in particular, that creates combustion.

Each fuel gas will burn only within a certain range of flammability. If too much or too little gas is present in an air–gas mixture, combustion will not take place. Before going any further, you need to understand two important terms relating to the range of flammability.

- Lower explosive limit (LEL): The LEL of an air–gas mixture is minimum percentage of gas in air at which ignition will take place and is normally measured as a percentage of gas in air. For example, the LEL of NG is 5 per cent gas and 95 per cent air. Therefore, at this point the mixture can be lit. Under this limit, combustion cannot be supported. *Not enough gas, no ignition.*
- Upper explosive limit (UEL). The maximum percentage of gas in air that can support combustion is known as the UEL. The UEL of NG in air is 15 per cent gas and 85 per cent air. Once the gas percentage exceeds this amount, combustion cannot be supported. *Too much gas, no ignition.*

These two terms are combined to describe the flammability range of a gas. In Figure 4.3, the flammability or explosive range of each gas is seen in red. Outside of this range, combustion cannot be supported. For example, NG with an LEL of 5 per cent and a UEL of 15 per cent has a flammability range of 5–15 per cent.

REFER TO AS/ANZ 5601.1 'DEFINITIONS – EXPLOSIVE LIMIT'

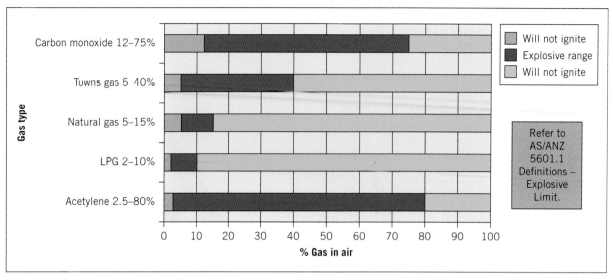

FIGURE 4.3 The flammability range of gases

Natural gas and LPG have a similar range, but this range is quite narrow in comparison to TG, which had a range of 5–40 per cent.

For comparison, Figure 4.3 includes two other gases with very wide ranges of flammability. Acetylene, a familiar gas used for welding and brazing, is flammable at atmospheric pressure through a range of 2.5–80 per cent.

You will also see the flammability range for carbon monoxide (CO). This gas is a highly toxic by-product of poor or incomplete combustion. What is often less well-known is that CO is also highly flammable, having a combustion range of 12–75 per cent. A build-up of CO is therefore not only extremely poisonous but also very explosive.

 Under increased pressure, acetylene has been known to combust at even higher percentages. A leaking acetylene hose is not something to be ignored!

When gas fitters adjust the gas pressure and aeration of a burner, they are actually ensuring that the air–gas mixture is able to combust within this range of flammability.

Air requirements

To understand flammability ranges you must also be aware of the amount of air required for combustion. This is measured in cubic metres (m³) and expressed as a simple ratio. Therefore, if LPG requires 24 m³ of air to enable the complete combustion of 1 m³ of gas, this relationship can be written as 24:1. The air requirements of each fuel gas are listed below:

- LPG   24:1
- NG   9.5:1
- TG   4.5:1

Figure 4.4 illustrates the amount of air required to combust fuel gases. Each cube in the diagram represents one cubic metre of gas or air. You can see that more air is required to combust LPG than for NG.

 REFER TO AS/NZS 5601.1 SECTION 1 'EXPLOSIVE LIMIT'

During the commissioning of a Type A gas appliance, the gas fitter will often need to adjust the burner aeration, ensuring that enough air is able to mix with the gas for complete combustion. Some appliances, such as space heaters and water heaters, may be installed inside a building. Once again, the gas fitter must calculate the amount of air required by all appliances in the room and provide additional ventilation if required.

Flame speed

The flame speed of a gas (also called the 'burning velocity') describes the rate at which a flame propagates through an air–gas mixture, usually measured in millimetres per second. Just as each gas has different constituents, each has a different flame speed. Various laboratory test methods are used to determine flame speed. One procedure measures the speed of flame propagation through a glass tube filled with an air–gas mixture over a given distance. Below you will see the approximate flame speed of each gas:

- LPG   460 mm/s
- NG   400 mm/s
- TG   1050 mm/s

It is not very obvious to the casual observer, but when a burner is lit, the flame actually burns *back* towards the burner itself (see Figure 4.5). In order for the flame to remain anchored on the burner head, the flame speed must equal the speed of the air–gas mixture issuing from the burner. A stable flame has a burning

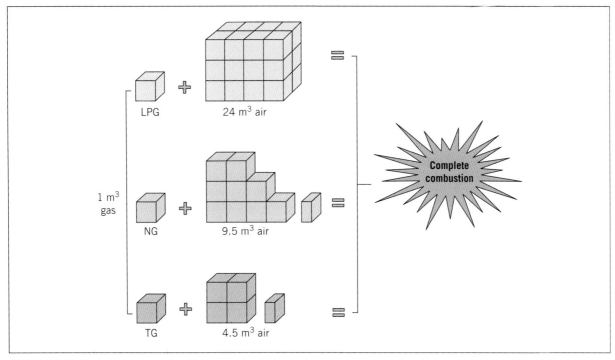

FIGURE 4.4 The amount of air required to combust fuel gases

velocity equal to the mixture speed coming out of the burner.

The arrows in Figure 4.5 indicate the relationship between the speed of the flame burning back towards the burner head and the speed of the air–gas mixture coming out of the burner. You can see that both LPG and NG burn considerably slower than TG. Unless modified, burners that have been designed to accommodate the flame speed of TG are generally unsuitable for use with slow burning gases such as NG or LPG, as they do not burn fast enough to stabilise on the burner head.

You will learn more about the relationship between flames and burners in Chapter 6, 'Combustion principles'.

Ignition temperature

Ignition temperature is usually defined as the lowest temperature at which the combustion of an air–gas mixture is self-sustaining. Once lit at this temperature, combustion will continue without any additional heat source. A number of variables relating to the combustion environment and fuel composition will affect the actual ignition temperature, but the ignition temperatures shown below are generally acceptable:

LPG   450°C
NG   630°C
TG   400°C

The ignition temperature of NG will also vary according to the source of the gas. However, the exact ignition temperatures become a significant factor only for a Type B gas fitter licensed in the installation and commissioning of industrial appliances. For general installation of Type A gas appliances, you primarily need to understand the importance of correctly adjusted and performing ignition systems. Unless the ignition system of the appliance is functioning correctly, hazardous situations may arise through the release of unburnt gases or the production of incompletely burnt gas by-products.

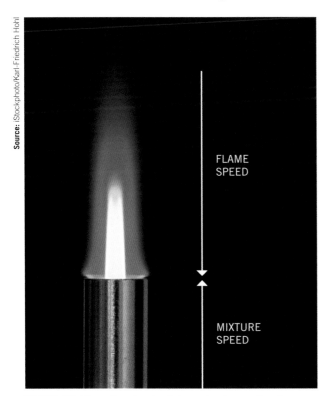

FIGURE 4.5 Flame burning back towards a burner head

Toxicity

Both NG and LPG are not toxic. However, TG was highly toxic as it contained a high percentage of the poisonous gas carbon monoxide. If you refer back to Table 4.1, you will see that neither NG nor LPG contain carbon monoxide and, therefore, breathing in the raw gas is not considered poisonous.

However, all fuel gases will displace oxygen in air. This means that a gas leak in a room will cause a reduction in the available oxygen, leading to possible suffocation. For this reason, fuel gases are regarded as asphyxiates.

Odour

Natural gas and LPG may not be toxic but they smell as if they should be! If you have ever smelled fuel gas while having a BBQ cylinder filled or perhaps at home while trying to light a cooker burner, you will know that raw gas has a very strong smell. However, prior to the refining process NG, LPG and TG do not have any characteristic smell. They are all naturally odourless.

It would be very dangerous to use fuel gases if we could not detect them by smell. Therefore, gas suppliers add a chemical odorant to all fuel gases. This additive is based upon a group of 'thiol' chemicals collectively known as 'mercaptans'. Most mercaptan chemicals have a very persistent, strong and repulsive odour, and for this reason are ideal for gas leak detection. Generally, certain types of mercaptans are used for each form of fuel gas.

- *Ethyl mercaptan* is added to LPG and is often described as smelling characteristically of leeks and onions.
- *Methyl mercaptan*, otherwise known as methanethiol, has a rotten cabbage odour and is used in NG.
- *Butyl mercaptan* derivatives are to be found in skunk secretions! In the past, this type of mercaptan was used by some gas suppliers to provide the smell to TG.

Mercaptans are flammable and are normally consumed during the combustion process. In their concentrated state mercaptans are highly poisonous, but when added to fuel gases they are very dilute and are not considered to present any hazard when breathed in under normal circumstances.

Gas suppliers will add mercaptan immediately after refining, and in some areas it is added to the gas at the city gate. LPG suppliers will often add mercaptan at their main bulk storage depots.

The amount of mercaptan added is calculated to ensure that the gas can be detected by smell at approximately 20 per cent of the LEL. This means that the gas can be smelled well before it has reached its LEL.

It is impossible to determine the amount of gas present just from the smell alone, as it may smell the same regardless of the air–gas mixture. If you can smell raw gas, you should always treat it as a dangerous situation!

Heat value

The **heat value** of a gas is an important measurement that you must remember, since you will use it regularly as a gas fitter. When any substance is burnt, it releases a certain amount of energy. In order to compare the gross heating value of fuel gases, combustion engineers will burn a standard cubic metre of gas in a special device called a 'bomb calorimeter' and measure the amount of energy released under controlled laboratory conditions.

The simplest definition of heat value for a gas fitter is, 'the amount of heat (MJ) released during the complete combustion of 1 m^3 of gas in air'.

The heat value of each fuel gas can be seen in Figure 4.6. This figure shows that although the same volume of gas is being burnt, different amounts of energy are released.

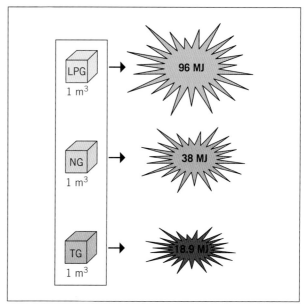

FIGURE 4.6 The heat values of gases

You can see from Figure 4.6 that in each instance 1 m^3 of gas has been completely burnt in air and yet the amount of energy released from this combustion is different in each case. For the same amount of gas, LPG releases approximately two and a half times more energy than NG and almost five times as much as TG. This does not make LPG a 'better' gas, but simply demonstrates that it is a more concentrated source of energy.

In a practical comparison, an NG heater and an identical LPG heater with the same heat input per hour will give out the same amount of heat, but the NG appliance will use two and half times more gas than the LPG appliance to do the same job.

In later sections, you will use your knowledge of heat value to carry out various calculations related to meter reading, gas rate and heat input conversions.

LEARNING TASK 4.2 CHARACTERISTICS OF FUEL GASES

1. If a fuel gas has a relative density greater than 1, would the fuel gas sink or rise when released into the atmosphere?
2. What are the three key elements required for combustion to take place?
3. Explain the terms 'lower explosive limit' and 'upper explosive limit' when referring to fuel gases.
4. Explain the term 'air requirement' when referring to fuel gases.
5. Are NG and LPG regarded as toxic gases?
6. Why is chemical odorant added to fuel gases?
7. What is the unit of measurement for the heat value of fuel gases?

COMPLETE WORKSHEET 1

SUMMARY

In this chapter you were introduced to a range of key constituents that collectively give each of our fuel gases very specific performance characteristics. It is not possible to successfully undertake SAFE gas installation tasks without understanding how these constituents related to all aspects of the operation and performance of a gas appliance. Ensure you have a good understanding of this chapter before moving on to Part B of this text.

GET IT RIGHT

The photo below shows an incorrect practice when using or storing gas cylinders for an LPG appliance. Identify the potential problem and provide reasoning for your answer.

___
___
___
___
___

WORKSHEET 1

To be completed by teachers
Student competent ☐
Student not yet competent ☐

Student name: _____

Enrolment year: _____

Class code: _____

Competency name/Number: _____

Task

Review the sections on *Constituents of fuel gases* and *Characteristics of fuel gases* and answer the following questions.

1 Complete the table below to create an easy reference guide to gas characteristics.

Characteristic	NG	LPG
Relative density		
Flammability range (LEL–UEL per cent)		
Air requirements (m^3)		
Flame speed (mm/s)		
Ignition temperature (°C)		
Toxicity (Y/N)		
Odorant chemical (type)		
Heat value (MJ)		

2 How many volumes of air are required to completely combust the following volumes of fuel gas?

   5.5 m^3 of LPG _____

   21 m^3 of NG _____

3 Describe what would happen if you tried to ignite a NG and air mixture made up from 21 per cent gas and 79 per cent air.

   _____

   _____

   _____

4 Circle the correct answer. A relative density of 1.55 makes LPG:

   more explosive than NG

   heavier than air

   disperse more easily than the same volume of NG

5   Natural gas and LPG are known as slow burning gases. Why?

    _____
    _____
    _____

6   If mercaptans can be detected by smell at 20 per cent of the LEL, what would be the percentage of gas in air at which you would be able to smell a leak of NG?

    _____
    _____
    _____

7   What are the two gases collectively known as LPG?

    _____

8   What key characteristic of LPG makes it particularly dangerous if you are called to a gas leak in a below-ground restaurant?

    _____
    _____
    _____

9   A NG appliance consumes 6 m^3 of gas in one hour. If this appliance were converted to run on LPG, how many cubic metres of gas would it use in one hour?

    _____
    _____
    _____

10 What are the two key hazards associated with carbon monoxide?

    i   _____

    ii   _____

11 In which direction does a gas flame burn? Circle the correct answer.

    Back towards the burner head.

    Away from the burner head.

12 Which gas has a wider flammability range: NG or LPG?

    _____
    _____
    _____

GAS INDUSTRY WORKPLACE SAFETY

Learning objectives

Although this text cannot cover the full range of hazard identification and control processes, the following sections will identify some of the primary safety issues particular to general gas fitting practice.

Areas addressed in this chapter include:
- general safety evaluations
- gas installation compliance
- safe work method statements
- gas leaks
- respiratory hazards
  - asphyxiation
  - carbon monoxide
- basic electrical safety.

This chapter provides fundamental gas industry safety knowledge relevant to all aspects of fuel gas installation practice. You should read and complete all worksheets before commencing any of the units of competency detailed in Part B of this text.

Overview

Workplace safety is the responsibility of all employees and employers. Working in the construction and property services sector means that every day you could be exposed to potential hazards. As a gas fitter, you will also bear the responsibility of dealing with some very serious safety issues specifically associated with working on gas installations and appliances.

General safety evaluations

As a gas fitter safety must be your primary concern in everything you do. Your licence should be seen as a symbol of trust from the community that you will carry out all your duties with safety as your primary concern. Of course, while there are specific gas hazards that you need to be aware of, you also need to take a broader view of hazard identification.

As soon as you walk onto a job site or evaluate a set of plans, you should be mindful of potential hazards of all kinds. General safety evaluations will be covered in more detail as part of your broader training, but it is relevant to briefly review the following considerations.

Safe work method statements

Completing detailed safe work method statements (SWMS) will be an integral part of your ongoing training and eventual activities as a gas fitter. The SWMS is a formal document that enables you to list identified hazards, evaluate the level of risk and determine removal or control measures to mitigate such risk (see Figure 5.1).

General hazards that are commonly encountered on many jobs include:

- working at heights
- working in proximity to electrical services and equipment
- working in and around excavations
- contaminated atmospheres
- noise hazards
- chemical hazards.

Once hazards have been identified, you must determine how each associated risk is controlled so that the work can be carried out safely. Risk control options

High Risk Construction Work Safe Work Method Statement (SWMS) Template

- Work must be performed in accordance with this SWMS.
- This SWMS must be kept and be available for inspection until the high risk construction work to which this SWMS relates is completed.
- If the SWMS is revised, all versions should be kept.
- If a notifiable incident occurs in relation to the high risk construction work in this SWMS, the SWMS must be kept for at least 2 years from the date of the notifiable incident.

PCBU DETAILS		PRINCIPAL CONTRACTOR (PC) DETAILS	
Name:		Name: (*name and contact details*)	
Work supervisor: (*name and contact details*)		Date SWMS provided to PC:	
Work activity:		Workplace and works location:	

HIGH RISK CONTRUCTION WORK DETAILS			
High risk construction work (☑ tick relevant work)	☐ Risk of a person falling more than 3 metres	☐ Work on a telecommunication tower	☐ Work on or near energised electrical installations or services
	☐ Work in an area with movement of powered mobile plant	☐ On or adjacent to a road, railway, shipping lane or other traffic corridor in use by traffic	☐ Temporary load-bearing support for structural alterations or repairs
	☐ Work in or near a shaft or trench deeper than 1.5 m or a tunnel	☐ Work in or near water or other liquid that involves a risk of drowning	☐ Work in an area that may have a contaminated or flammable atmosphere
	☐ Work on or near chemical, fuel or refrigerant lines	☐ Demolition of load-bearing structure	☐ Use of explosives
	☐ Tilt-up or precast concrete elements	☐ Work in or near a confined space	☐ Likely to involve disturbing asbestos
	☐ Work in areas with artificial extremes of temperature	☐ Work on or near pressurised gas mains or piping	☐ Diving work

SWMS COMPLIANCE			
Person responsible for SWMS compliance:		Date SWMS received:	
What measures are in place to ensure SWMS compliance?			
Person responsible for SWMS control measures:		Date SWMS received by reviewer:	
How will the SWMS control measures be reviewed?			
Signature of reviewer:		**Date reviewed:**	

FIGURE 5.1 Example of a safe work method statement

can include one or a combination of the following processes (in order):

1. Eliminate the risk.
2. Substitute the hazard with a safer alternative.
3. Isolate the hazard.
4. Reduce risk through engineering controls.
5. Reduce exposure to the hazard using administrative controls.
6. Use personal protective equipment (PPE).

A SWMS should be completed prior to any job. Some companies will design documents customised to your specific industry needs. Your instructor will ensure that you are provided with the type of SWMS relevant to your area so that you can become familiar with using it during your gas installation training.

General compliance assessment

As a gas fitter, you are required to ensure that all work you undertake complies with relevant codes, standards and legislation. Not only does this include new jobs, but any work associated with existing installations.

If you are about to alter or add to an existing installation, you need to ensure that the rest of the work is up to standard. If you identify a problem, you have a 'duty of care' to do something about it or at least let the owner, site supervisor or technical regulator know.

What is duty of care?
Duty of care can be defined as the implicit responsibility of an individual to adhere to a reasonable standard of care when performing acts that could foreseeably harm others. Every worker has a duty of care to all other people in society.

Before beginning any quotation or work, walk around the site and work through a checklist of possible problem areas. The following examples are common inspection items:

- Consumer-billing meter or LPG cylinders are in correct location with required clearances.
- Materials and components are approved.
- Piping and components are installed correctly.
- Appliance data plates are identified.
- Appliances are installed in correct location and with required clearances.
- Flue system is sized correctly.
- Appliance ventilation is sufficient.

Where something is found to be unsatisfactory you must inform your supervisor and/or the owner before any work commences. Approval and rectification of non-compliant work may need to be added to an existing quote or contract.

Pre-work 'leakage test'

Before starting any work it is a mandatory requirement that you carry out a 'leakage test' of the system in accordance with AS/NZS 5601 (Gas Installations).

You must know that the installation you may be working on is gas-tight. This is a sensible safety practice, as you can then be confident that there is no potential for unburnt gas to be present. It also makes sense to know what you are quoting for. You are unlikely to be paid for additional work that you discover *after* you begin! You will learn how to carry out this form of gas pressure test further into your training.

LEARNING TASK 5.1 GENERAL SAFETY EVALUATIONS

1. What detailed document must be completed before undertaking gas fitting work?
2. What is the purpose of a general compliance assessment?
3. At what stage of the installation is a 'leakage test' to be carried out?

Gas leaks

A gas leak is the escape of unburnt gas from the piping system or appliances. Any leak is dangerous and cannot be disregarded. You should never leave any job without being certain the system you leave behind is 'gas-tight'. Unfortunately, a full-time gas fitter will inevitably encounter gas leaks and must rely upon safe and methodical procedures to control the hazard, find the source of the leak, make repairs and test the system. The presence of unburnt gas creates a hazardous situation where fires, explosions and asphyxiation are possible.

Causes of gas leaks

Causes of leaks are many and may include the following:

- leaks at appliance connections
- kinked and split flexible hose connections
- loose or damaged fittings
- ruptured regulator diaphragms venting to atmosphere
- electrolysis damage to pipes
- physical impact damage to appliances, pipes or meters
- ruptured underground pipes from machinery or poor bedding material
- split pigtail connections on LPG cylinder installations.

Gas leak response

The gas leak itself can range from a barely detectable smell of gas to the obvious sound and smell of a high-pressure rupture. Every instance has its own characteristics and it is impossible provide a complete checklist of what to do. However, the following steps are a useful guide on how to proceed. These points are not necessarily in strict order, as your response will depend upon the circumstances.

1. *Remain calm.* You are the professional to whom people will be looking for direction. Move with a sense of purpose but avoid running frantically about. Try to evaluate the situation quickly, and firmly instruct those near the leak.
2. *Isolate the gas supply.* Turning off the supply of gas at the meter or cylinder should prevent the problem becoming worse; however, always consider the possibility of *another* source with a separate isolation that you may not know about.
3. *Evacuate the building.* This may not be required if you have discovered a simple leak from an appliance connection, but if you are in any doubt it is better to clear the area.
4. *Eliminate all ignition sources.* These include naked flames, cigarettes and any devices that may cause sparking or have high temperature surfaces. Electrical switches usually create a small spark when operated, so *do not* turn lights on or off. Leave them as they are. Be aware that some devices operate on timers or thermostats that may turn on when you least expect it. This is particularly common in commercial catering environments (refrigeration and heating systems).
5. *Ventilate the building.* Open all windows and doors to allow gas to diffuse. Minor escapes can be ventilated easily by flapping a sheet of cardboard up and down. Larger escapes and basement areas may require the use of 'intrinsically safe' extraction fans. Intrinsically safe equipment is designed for use in flammable environments.
6. *Do not enter confined spaces.* Basements and other unventilated areas may contain high concentrations of fuel or inert gases and present a very real asphyxiation threat.

As soon as you feel that, despite your best efforts, the situation is getting beyond your control, do not hesitate to call emergency services. They are trained and equipped to deal with these events.

Emergency services

If you are attending a reported gas leak, you will need to quickly assess whether you are capable of handling the situation. This is not always easy; but if you encounter one of the following problems, then it may be safer to call emergency services:

- You cannot find the gas isolation point.
- There is a major or high-pressure escape.
- There is a gas leak of any size in a confined space or basement.
- The size of the leak warrants evacuation, but you are unable to take action.
- You suspect that gas may have originated from or possibly leaked into adjacent buildings and you are unable to conduct an evacuation.

How to detect gas leaks

In most circumstances the smell of gas will first alert you to the presence of unburnt gas. You will recall that gas is dosed with foul-smelling chemicals called mercaptans and becomes detectable by smell at 20 per cent of the lower explosive limit (LEL). While the smell of gas is evidence of some form of gas leak, it does not normally help you to locate the source of the leak itself. For this you will need to employ other methods, such as those described below.

Soap and water solution

A rich mixture of ammonia-free, non-corrosive soap and water in a spray bottle or other container is the most common method a gas fitter will use to locate a leak. By applying the solution to all suspect joints and fittings, the leak will become evident by the appearance of bubbles around the fault (see Figure 5.2). Minor leaks can be harder to see and you will need to take more time to observe the appearance of very small bubbles, particularly around fitting threads. Commercially available solutions incorporating additives that assist in holding the bubbles around the joint are also available as pump and pressurised sprays.

FIGURE 5.2 Using a soap and water solution to locate a gas leak

To prevent possible ammonia damage to brass fittings, ensure that any soap and water solution is thoroughly rinsed away once testing is complete.

Gas detectors

A number of gas detectors are available, including small units that clip onto your belt or clothing and handheld detectors with extendable probes that allow you to reach into tight corners (see Figure 5.3).

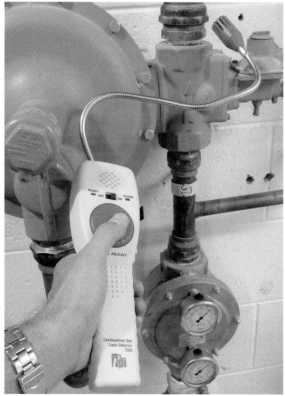

FIGURE 5.3 A handheld gas detector featuring a flexible probe and audible alarm indicator

An audible alarm will normally alert you to the presence of gas. Some devices can determine the amount of gas present as a percentage of LEL and others are designed to take samples as part of confined space entry procedures. Confined space entry is particularly hazardous, and should never be attempted until you have completed all relevant training and using an appropriate detector to monitor the enclosed atmosphere.

REFER TO AS/NZS 5601.1, 'METHODS OF LOCATING GAS LEAKS'

As there are many types of detectors, ensure that the instrument you select is designed to detect the gas you are looking for. A good gas detector is invaluable, but it must be well maintained and calibrated as required.

LEARNING TASK 5.2 GENERAL SAFETY EVALUATIONS

1. What are two common causes of gas leaks?
2. Name two common ways of locating the source of a gas leak.

Respiratory hazards

Gas fitters may potentially be exposed to two key respiratory hazards, and you will need to be able to recognise when these dangers are likely to occur and how to remove or control them. To fully appreciate the nature of respiratory hazards, it helps to have a simple understanding of how respiration occurs in the body.

In basic respiration, your blood contains a molecule called haemoglobin that has the job of transporting oxygen and carbon dioxide around the body. Oxygen is needed to oxidise cellular nutrients required by all body tissues, and carbon dioxide is one of the waste products released from the cells during this process.

When you breathe air in, the gas oxygen diffuses through the walls of the lungs into adjacent blood vessels and is picked up by the passing haemoglobin. Oxygenated blood has a characteristic bright red or crimson colour.

At the same time, the darker, venous blood carrying the carbon dioxide away from the body tissue is released by the haemoglobin into the lungs and exhaled in the next breath.

Your urge to breathe is actually stimulated by a build-up of carbon dioxide in the body, rather than a decrease in oxygen levels. For example, when you hold your breath for a long time, the amount of carbon dioxide in the blood rises as oxygen is consumed, increasing the urge to breathe until you eventually have no choice but to start gasping for air, restoring the correct balance of oxygen and carbon dioxide in the blood.

Respiratory problems start when inhaling other gases disrupts this finely tuned process.

Asphyxiation

Asphyxiation simply means death through the lack of oxygen. An extremely dangerous situation exists when oxygen is displaced from an area by fuel or inert gases. Exposure to concentrated levels of fuel and inert gases can lead to asphyxiation if appropriate precautions are not taken. This may occur during the following circumstances.

- *Gas leaks:* Large volumes of gas can build up in rooms, basements or other enclosed areas, reducing the amount of available oxygen.
- *Purging operations:* Gas fitters will routinely use inert gases such as nitrogen to purge pipelines and large vessels such as gas tanks of fuel gas, so that they will be safe for storage or further work. Although inert gases are generally non-toxic, they are classed as asphyxiates and present a significant hazard if the purging procedure is not conducted correctly. Purging operations are covered in depth in Chapter 11.

When someone enters an enclosed space or an area of high gas concentration, the lack of available oxygen can lead to a condition called hypoxia. This term refers to a lack of oxygen supply to the brain. Symptoms include the following:

- mild hypoxia
  - headaches
  - fatigue
  - shortness of breath
  - a feeling of euphoria
  - nausea
- severe hypoxia
  - cyanosis (blue discoloration of the skin)
  - unconsciousness.

First aid response

- If safe to do so, remove the patient to a well-ventilated, gas-free area.
- Call emergency services if you're not sure how severe the case is, and if necessary apply the latest cardio-pulmonary resuscitation (CPR) procedures.

If sufficient oxygen is not restored to the brain, coma, brain damage and eventual death will occur.

Mild hypoxia can be reversed by restoring a normal oxygen supply, but more severe cases will require immediate medical intervention. If you have any doubt in such a circumstance, you should call for medical assistance.

Caution on the use of nitrogen

Nitrogen is an inert gas commonly used in the gas and plumbing industry. The use of nitrogen in purging operations presents a particular hazard that has claimed many lives. Although nitrogen is not toxic, breathing it in high concentrations will cause a drop in the amount of carbon dioxide in the bloodstream. Without the correct regulating level of carbon dioxide in the body, a person will lose the urge to breathe altogether and will subsequently die of asphyxiation. Do not underestimate the dangers of inert gases.

Never enter a suspected or known gas-affected enclosure or other confined space without the appropriate training, breathing apparatus and assistance.

Carbon monoxide poisoning

Carbon monoxide (CO) gas is a product of incomplete combustion. When a fuel gas is burnt without sufficient oxygen for complete combustion, full oxidation of the fuel to carbon dioxide is prevented and the result is the release of carbon monoxide.

Carbon monoxide is colourless, odourless, tasteless and very toxic. These characteristics make CO a very dangerous gas to deal with and reinforce the point that only a suitably trained and licensed gas fitter should work with gas appliances. Incorrectly installed and serviced appliances have led to the death of many people, and it is your responsibility to ensure that this does not happen.

Recall that during during normal respiration, oxygen binds to receptor sites on the haemoglobin molecule for transfer around the body. However, carbon monoxide has an affinity 240 times greater than oxygen to these receptor sites. Therefore, despite normal oxygen levels in the atmosphere, the CO displaces oxygen on the haemoglobin receptor sites. Worse still, when this happens the presence of the CO will then prevent the release of remaining oxygen still carried by the haemoglobin to the body tissues.

Table 5.1 demonstrates the exposure level as a percentage of CO in air and the associated effects to which such exposure may lead. Blue rows are comparative examples of exposure.

Carbon monoxide is almost impossible to detect with normal senses, and so it is particularly dangerous. Compounding this problem, early symptoms of poisoning also resemble those of other ailments and include:

- dizziness
- headache
- weakness
- shortness of breath
- possible nausea
- possible mental disorientation
- unconsciousness.
  - In some circumstances the skin may turn cherry red in colour.

First aid response

1. If safe to do so, remove the patient from the affected area to open air.
2. Call emergency services.
3. If required, carry out approved CPR procedure until emergency services arrive.
4. If available, administer emergency oxygen.

TABLE 5.1 Exposure level as a percentage of CO in the air

Parts per million (ppm)	Percentage of CO in air	Effects of exposure
0.1	Natural atmospheric background level of CO	
0.5	Average background level in homes	
5–15	Levels found near correctly adjusted gas stoves in homes	
35	0.0035%	Headache and dizziness within six to eight hours of constant exposure.
100	0.01%	Slight headache in two to three hours.
100–200	Levels found in highly polluted cities	
200	0.02%	Slight headache in two to three hours.
400	0.04%	Frontal headache within one to two hours.
800	0.08%	Dizziness, nausea, and convulsions within 45 minutes. Insensible within two hours.
1 600	0.16%	Headache, dizziness, and nausea within 20 minutes. Death in less than two hours.
3 200	0.32%	Headache, dizziness and nausea in five to ten minutes. Death within 30 minutes.
5 000	Home wood fire flue products	
6 400	0.64%	Headache and dizziness in one to two minutes. Death in less than 20 minutes.
12 800	1.28%	Unconsciousness after two or three breaths. Death in less than three minutes.
30 000	Undiluted cigarette smoke	

Portable gas detectors designed to sense the presence of carbon monoxide are readily available. You would be well advised to include one of these instruments in your kit for those times when you have doubts about the safety of a particular installation (see Figure 5.4).

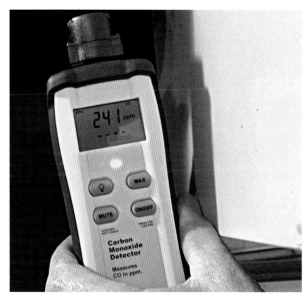

FIGURE 5.4 A handheld carbon monoxide gas detector

COMPLETE WORKSHEET 1

LEARNING TASK 5.3 RESPIRATORY HAZARDS

What are the two key respiratory hazards that gas fitters may be exposed to?

Basic electrical safety

Most gas appliances used today incorporate electrical components of some kind; examples include:
- ignition systems
- control panels
- combustion fans
- solenoid valves.

Even those appliances that do not have any kind of electrical connection are supplied with gas from pipes that may be connected to appliances that do. Although it is illegal for an electrician to fit an earth connection to a gas line, many older homes may still have electrical systems that are earthed via the water supply pipes. If the water pipes are then connected to your gas water heater, any faults in the system could potentially affect the whole installation.

Correctly installed and maintained electrical systems are very safe. However, where poor installation or damage to the system exists, the built-in system safety features may fail to prevent electric shock.

> Electrical hazards are a very real and constant danger to gas fitters. Only correct safety procedures will help protect you and your workmates against possible electrocution, injury and death.

This section will review the key electrical hazards that a gas fitter must be aware of and the equipment and instruments required to safely test for and protect against electrocution. These details are included not only because they are sensible, but also because you are legally required to apply protective electrical precautions in all gas fitting tasks.

> **REFER TO AS/NZS 5601.1 'SAFE PRACTICES AND OHS'**

Electrical terminology

You need to understand the following basic terms to help protect yourself against electrical hazards.

Current
Current is defined as the number or *flow* of electrons past a given point. It is measured in amperes or amps (A).

Voltage
Voltage is defined as the amount of electrical *pressure* applied to get electrons flowing. It is measured as volts (V).

Conductor
A conductor is any material capable of carrying an electric current. Copper is a very good conductor and is used in most wiring. Rubber, on the other hand, does not conduct electricity.

Resistance
Resistance is the measure of opposition in a conductor to the flow of electricity. The greater the diameter of the wire, the lower the resistance to the flow of electricity. Resistance is measured in ohms (Ω).

Alternating current and direct current
Premises connected to the mains electricity supply in Australia are normally provided with alternating current (AC). The electricity flowing in an AC system reverses its direction approximately 50 times per second. In domestic wiring, this flow of electricity comes in via the active wire and returns via the neutral wire.

In comparison, your car and many gas appliance components run on a direct current (DC) system where the flow of electricity is in one direction only.

The symbol for alternating current is '∼'.
The symbol for direct current is ⚌.

Earth
AC electrical supplies incorporate an earth wire system. In the event of an appliance or system fault, potentially dangerous current is provided with a pathway to earth, instead of making pipe and appliance chassis 'live'. Earth wires from appliances and 'bondings' from pipe are taken to ground via an 'earth stake' (see **Figure 5.5**).

In the days when metal pipes were used for all water installations, the earth wire was often clipped to the water pipe itself. With the increasing use of plastic water supply systems, this is no longer suitable and the earth connection will need to be run to an earth stake. Regardless of where it is connected, you must remember one very important rule: *Only a licensed electrician is permitted to remove any form of earth connection*. If you need to remove a pipe that has an

FIGURE 5.5 Earth stake connection

earth or bonding connection clip, you must call an electrician.

Electric shock
Electric shock can occur when the body is placed between two electric conductors or acts as a pathway to earth. Electricity follows two basic laws:
1. It will always try to flow via the path of least resistance.
2. It will always try to return to its source.

If the pathway to earth is your body, then you are likely to suffer an electric shock. Pipe installations and appliances may show no outward sign of electrical danger, but may be carrying lethal electric current. Unless you test the installation, you have no way of knowing that it is safe to touch. It is too late once you do!

How much is too much?

To have a basic grasp of simple electrical theory you need to understand the following three terms:
- voltage (V; measured in volts, V)
- current (I; measured as amperes or amps, A)
- resistance (R; measured as ohms, Ω).

A good way to try to appreciate what each of these terms means is to use the analogy of a plumbing system. Imagine the following relationships, where:
- voltage is equivalent to water pressure
- current is equivalent to flow rate
- resistance is equivalent to the pipe size.

In a plumbing system, any change in either pressure, flow rate and pipe size will have an effect on the other two. The same applies to electrical systems and can be determined with the following formula.

$I = V/R$ Current (flow rate) is equal to voltage (water pressure) divided by resistance (pipe size)

At a given pressure, a certain amount of water can flow out of a 20 mm pipe. If you increase the pressure out of the same pipe, more water will flow. So too in our electrical system: If you increase the voltage without changing the resistance, the current will increase, as shown below:

0.0024 amps = 24 V/10 000 ohms (the approximate resistance of dry human skin)

0.024 amps = 240 V/10 000 ohms

If you changed the pipe to 32 mm, at the same pressure the flow rate would increase. Again, in the electrical system, if you reduce resistance at the same voltage, the current will increase, as shown below:

0.0024 amps = 24 V/10 000 ohms (the approximate resistance of dry human skin)

0.024 amps = 24 V/1000 ohms (the approximate resistance of wet human skin)

Despite it being a simple 24 V system, you can see in the above equation that the simple change between dry skin and wet skin happens to be the difference between a minor and severe shock!

This example is used to demonstrate that injury and death caused by electric shock largely depends upon the amount of current (measured as amperes) forced through the body rather than on the number of volts. Figure 5.6 shows the relationship between different current levels and their effect upon the body.

Work practices to reduce risk

Simple work practices and habits can help to reduce your exposure to electrical hazards. You should treat every job as one that presents some type of electrical danger, and avoid placing yourself in situations that may pose an electrocution risk.

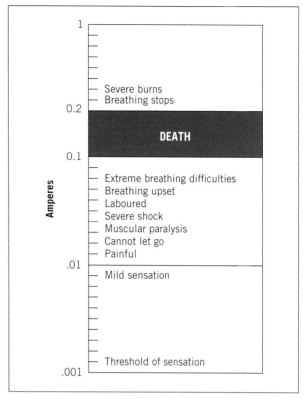

FIGURE 5.6 The relationship between different current levels and the effect upon the human body

Body attitude

The way that you position yourself to undertake a particular task is termed 'body attitude'. Electricity always tries to find its way back to its source or to earth, and therefore you should try to avoid placing yourself in positions that may increase the risk of electric shock.

EXAMPLE 5.1

When moving through a confined space such as a ceiling area or subfloor, try to avoid touching any metal pipe or conduit. If you must do so, test for stray current first. If you are forced to lie on the ground, try to use some form of rubberised ground sheet, and avoid puddles of water.

EXAMPLE 5.2

When disconnecting sections of pipe or components from the system, try to remain on your feet, relying on the insulating properties of your boots to prevent an earth connection.

Move in a deliberate manner

If working around any electrical equipment, you need to work in a steady, deliberate manner. Rushing and moving carelessly will only increase your chance of contact with a live conductor. Ensure that where you do need to touch metal pipe, you test it first and then only touch one section at a time.

Wear boots in good condition

Boots that are full of holes are next to useless in guarding against electric shock. Any insulating value of rubber soles is lost if you are standing in water with wet socks and feet. Keep your footwear in good order.

Avoid working alone

Of course, it is not always possible to work with others. However, when you are planning to work on electrical equipment try to avoid working by yourself. This is particularly important when working in any confined space. Unfortunately, many an electrician or plumber working alone has received an electric shock in a ceiling or an under-floor area only to be found many hours later severely injured or dead.

Electrical hazards – what causes them?

There are many primary causes for electrical hazards, including insulation breakdown, short circuits and component failure. Under normal circumstances, such problems are dealt with safely via the particular wiring system found in most homes in Australia. However, if a fault exists in this system at the same time, or you do not take suitable testing precautions, you are highly likely to receive an electric shock. Figure 5.7 shows the common household wiring layout known as the MEN (multiple earthed neutral) system. As mentioned previously, AC electricity reverses its direction and returns to its source approximately 50 times per second.

In Figure 5.7, follow the path of the electric current into the house via the active wire, through the switchboard and general-purpose outlet (GPO), and on to the appliance. The balance of power not expended in running the appliance then returns from the appliance via the neutral wire to the switchboard. It is here that the electricity passes through a component called the 'neutral bar' before returning to the power lines and its source.

Looking carefully at the diagram, you can see that the earth wire also connects to the neutral bar. The earth wire is connected to earth via the earth stake on a

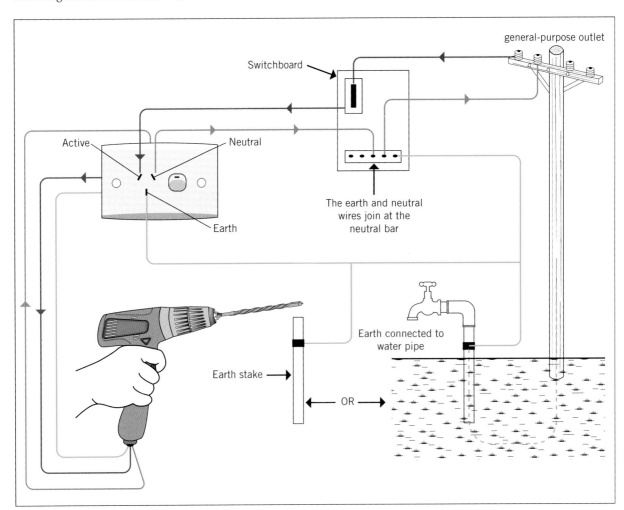

FIGURE 5.7 The multiple earthed neutral (MEN) wiring system

newer building or, in the case of many older premises, via the metallic water pipes.

Figure 5.8 shows what will happen when an active wire comes into contact with a metal part of the appliance. The electricity will be carried away safely from the appliance via the path of least resistance to either the neutral bar or the earth stake, up the nearest earthed power pole and on to its source. On older installations, this condition may be happening while the appliance still seems to be working as normal!

If part of the earth system has fallen off or is being removed from the water pipes or earth stake, anyone touching the appliance at the time of the fault is likely to receive an electric shock (see Figure 5.9).

In Figure 5.10, you can now see that the neutral connection to the house has been damaged and provides no return pathway. In this event, the returning current simply goes via the earth system through the ground and up the nearest earthed power pole.

If you were to disconnect part of the water pipe system or the earth clip itself, your body would then create the alternative pathway to earth and suffer an electric shock. In fact, it is possible that you can receive an electric shock caused by an appliance fault from another house some distance away simply because all of a sudden your body has provided a better return path. This is the danger plumbers and gas fitters face every time they make a break in water and gas piping and do not use bonding straps!

There is often no obvious indication that a fault may exist in an electrical system. You can protect yourself only by carrying out regular and appropriate tests.

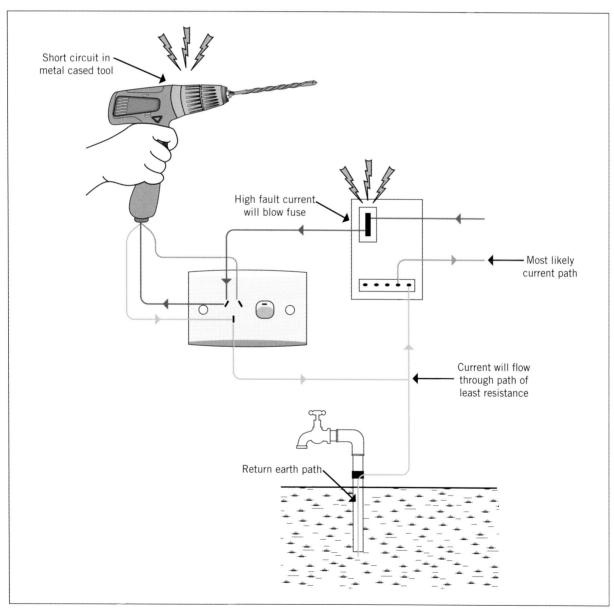

FIGURE 5.8 The MEN system under short circuit conditions

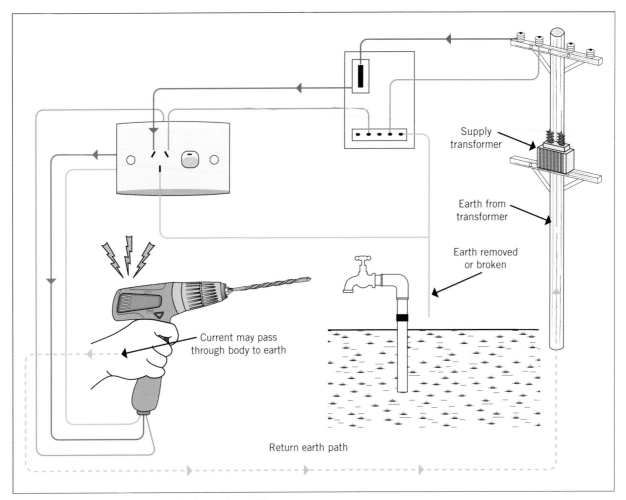

FIGURE 5.9 The MEN system under faulty earth conditions

Electrical test equipment

Fortunately, a wide range of test equipment is now available to the gas fitter for daily use. These items should always be carried on your person or in your main toolbox – they will not save your life sitting in your work vehicle.

Test instruments

There are many types of electrical test instruments, but in this section we will only review those most commonly used by gas fitters.

Non-contact voltage detectors

Perhaps more commonly known as a 'volt stick', this type of instrument provides visual and sometimes audible tones to indicate the presence of electricity. Usually the size of a large marker pen (see Figure 5.11), the volt stick can be carried in your top pocket for quick and easy use as you move around the job.

The volt stick operates by detecting the magnetic field that surrounds a live AC object or cable. Direct contact is generally not necessary, therefore this instrument is ideal for locating electrical cables hidden behind wall cladding or appliance cabinets (see Figure 5.12).

When choosing a volt stick, ensure that the AC voltage detection range is appropriate for the task and that it is designed for the 50/60 Hz electrical systems found in Australia.

You should have this device with you every day and, like any sensitive instrument, keep it safe from damage.

To use a volt stick:
1 Check the device is actually working at a 'known source'. Some devices have a self-proving function to ensure they are always working correctly.
2 Apply the end of the volt stick to the pipe or appliance you wish to check.
3 Check the end bulb. If it lights up, it has detected a source of electricity.

There are a number of advantages of a volt stick:
- It is inexpensive.
- It does not require electricity to pass through the operator to earth.

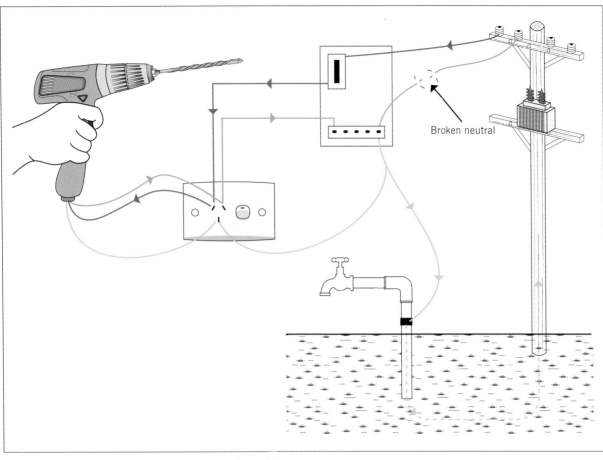

FIGURE 5.10 The MEN system under faulty neutral conditions

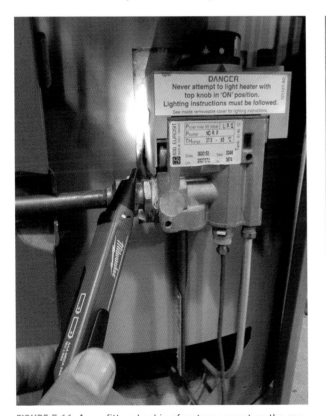

FIGURE 5.11 A gas fitter checking for stray current on the gas supply to a storage water heater

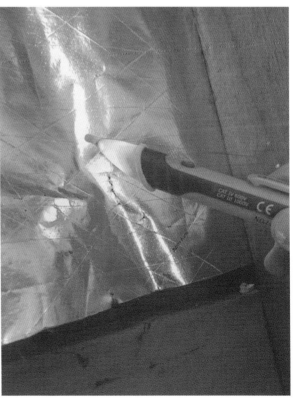

FIGURE 5.12 In this image the gas fitter has detected stray electrical current in old aluminium sheet foil ceiling insulation using a volt stick.

- Most provide a wide operating range, often 50–1000 V AC.
- It is compact.
- It is simple to use.

There are also some disadvantages:

- It runs on batteries, which can go flat. (Although the battery life is normally very good, don't get caught without a spare.)
- It does not detect DC power supplies.
- Most devices will not detect power circuits below 50 V.
- It cannot detect voltage from shielded or armoured cables.
- It may pick up the presence of electricity from nearby components, giving a false indication of malfunction.

Multi-meter

As the name suggests, this is a multi-function device used not only to test for stray current but also to measure the performance of electrical circuits and components. Two probes are applied to whatever you wish to measure and a digital display provides the reading (see Figure 5.13).

It is highly recommended that you learn how to use a multi-meter correctly so that you can safely use its many features in your daily work as a gas fitter. In Figure 5.13, you can see a gas fitter conducting a continuity test on the high-tension ignition leads of a gas cooktop. A high resistance, or 'infinity', reading would indicate a faulty lead that requires replacement.

All gas fitters should become familiar with using multi-meters as they are invaluable for general safety checks and appliance diagnoses. A good quality auto-range, digital device is ideal for plumbers and gas fitters.

The advantages of a multi-meter are that it is:

- a multi-purpose instrument useful to a gas fitter in a wide range of electrical testing tasks for both appliances and consumer pipe installations
- very safe when used correctly.

There are no disadvantages to using a multi-meter, but they are not as simple to use as a volt stick, so it is best to obtain expert instruction before using one.

Power point tester

Most gas appliances come with a plug-in electrical connection; but how do you know that the general-purpose outlet (GPO) on the wall is safe to use? A power point tester is a valuable device that will detect some (but not all) key problems associated with incorrectly installed GPOs.

A power point tester is designed to provide a visual warning of GPO wiring faults. Depending on the model, these instruments provide visual and audible indicators to demonstrate the wiring status of a GPO (see Figure 5.14). A combination of different lights and sounds will indicate polarity problems or missing earth connections. Always read the instructions for each device and check every power socket before use with a power point tester, but be aware that such devices cannot detect every kind of fault.

The type of tester shown in Figure 5.14 is simple and safe to use. You simply plug it in and observe the presence and colour of the lights indicating the safety of the installation. It will detect a 'no earth' problem and incorrect polarity of the active and neutral wires (be aware that some devices may not be designed to detect a neutral–earth polarity fault, high resistance earth fault or a combination of faults).

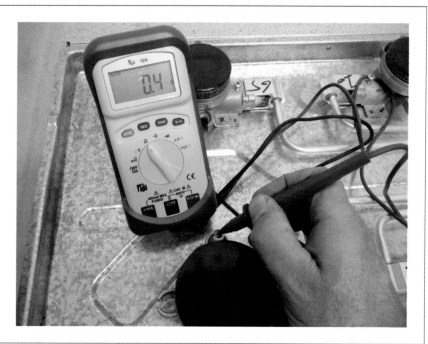

FIGURE 5.13 A gas fitter conducting a continuity test on the high-tension ignition leads of a gas cooktop

FIGURE 5.14 A power point tester detects basic switch wiring problems

What do I do if I find a fault?

If a fault presents itself in a piping system or any part of the building electrical system, carry out the following steps as a minimum:

1. Avoid touching any part of the system.
2. If possible, isolate the electrical supply.
3. If possible, safely apply some form of warning tag.
4. Keep everyone away from the problem area.
5. Notify the owner of the hazard where applicable.
6. Call an electrician or electrical authority if the owner does not provide an electrician.

Bonding straps

A pipe installation may show no sign of stray current unless a break is made in the system and a pathway to earth results. This means that if you remove a valve, section of pipe, regulator or meter from a pipe installation you may actually provide the required path to earth and suffer an electric shock.

The only way to protect yourself is to provide an alternative current path between the two separated halves of the system (see Figure 5.15). By using bonding straps, the gas fitter working on this pipe system has provided an alternative current path for any stray current that may become a hazard after the cut is made. The straps must remain in place until the job is completed. Any potential current will then have no path to earth, allowing you the opportunity to test each side before continuing with the work.

While many plumbers and gas fitters often neglect to use bonding straps, it must be emphasised that the risk of electric shock is very real, and many tradespeople have suffered injury and death as a result of poor work practices. Therefore, the Gas Installation Standards explicitly directs you to use bonding straps. As Australian Standards are specifically referred to ('called up') by legislation, your obligation to use bonding straps is a legal requirement and not a choice.

REFER TO AS/NZS 5601.1, 'ELECTRICAL SAFETY BONDING OR BRIDGING'

In the following procedure, you can see how bonding straps are used to safely disconnect the consumer piping from the outlet of a meter.

1. Connect one end of the bonding strap to the meter upstand (inlet side).
2. Connect the other end to the consumer piping.
3. Ensure you scratch your way through to bare metal in both cases.
4. Make the break in the consumer piping at the outlet of the meter.
5. Without touching any part of the system, remove the end connected to the consumer piping and, using a suitable test instrument, test both the pipe and the bonding strap for any stray current.

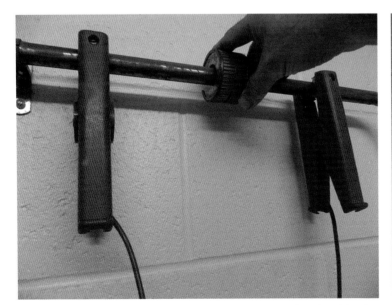

FIGURE 5.15 Bonding straps must be used whenever you make a break in pipework, remove valves or appliances

6. Replace the piping immediately.
7. If you detect current, ensure no one approaches the area and call an electrician or electrical authority.
8. Leave the bonding straps in position until all work is complete.
9. Check and re-check as required.
10. Only remove bonding straps once the break in the pipe has been restored.

 Always wear safety glasses when using bonding straps. Any pipe that is carrying stray current will arc when removing a clamp for testing, possibly causing droplets of molten metal to spray outwards.

LEARNING TASK 5.4 BASIC ELECTRICAL SAFETY

1. What is the purpose of an earthing system on an electrical installation?
2. Why can working on gas appliances be dangerous when an earth system is connected to the pipework?
3. What are two types of electrical test equipment that can be used to detect stray current on an appliance or pipework?

 COMPLETE WORKSHEET 2

SUMMARY

It is a sad fact that many plumbers cut corners when it comes to basic safety requirements, and unnecessary injuries and deaths are a too common occurrence. The use of test equipment (see Figure 5.16) must become part of your normal work practices until using them becomes automatic. As installations become older and electricity is used in almost all gas appliances, these devices are almost inevitably going to protect you from injury or indeed save your life.

It is up to you to understand that you are ultimately responsible for your own safety, and for the safety of those you work with and the general public at large. Unsafe attitudes and behaviours have no place in the gas industry, and you must always strive to ensure that each job you undertake is done so with full consideration given to checks and procedures designed to keep you and others safe from harm. Be very particular about electrical safety and ensure you apply all relevant precautions during your gas installation training and out in the workplace.

Ensure you have a good understanding of this chapter before moving on to Part B of this text.

FIGURE 5.16 A selection of basic testing equipment

GET IT RIGHT

The photo below shows an incorrect practice when disconnecting a gas meter.

Identify the incorrect method and provide reasoning for your answer.

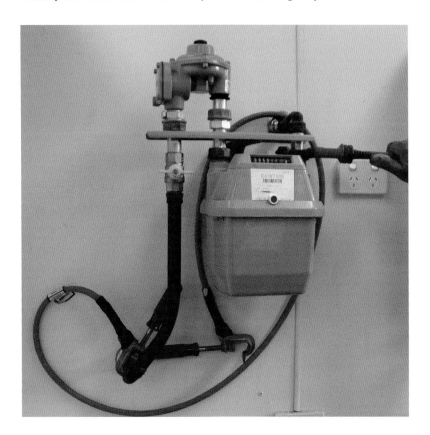

_____

_____

_____

_____

_____

WORKSHEET 1

To be completed by teachers
Student competent ☐
Student not yet competent ☐

Student name: _____

Enrolment year: _____

Class code: _____

Competency name/Number: _____

Task

Review the sections on *General safety evaluations* through to *Respiratory hazards* and answer the following questions.

1 What type of pressure test are you required to apply to an installation before starting work?

   _____
   _____
   _____

2 List at least five compliance-related issues that you would look for in your general inspection of an existing installation.

   i   _____
   ii _____
   iii _____
   iv _____
   v   _____

3 Identify and list the six key actions you may need to take when dealing with a gas-affected building.

   i   _____
   ii _____
   iii _____
   iv _____
   v   _____
   vi _____

4 Imagine you are working on a large job site. You look into a confined space area, and see that someone has collapsed inside. Should you immediately go to the person's assistance and apply first aid procedures? Explain your answer.

   _____
   _____
   _____

5. Briefly describe how carbon monoxide affects the transport of oxygen around the body.

___

6. List eight possible symptoms you might find in a person suffering carbon monoxide poisoning.

   i ___
   ii ___
   iii ___
   iv ___
   v ___
   vi ___
   vii ___
   viii ___

7. If carbon monoxide levels in a room rose to 0.04 per cent of total air content, what symptoms would a person exposed to such a level experience?

___

8. Which gas regulates the urge to breathe in your body?

___

9. What is cyanosis?

___

10. What are two methods you could use to isolate the source of a gas leak?

    i ___
    ii ___

11 What should you do after using a soap and water solution to test for a gas leak around brass fittings? Why?

_____

_____

_____

12 List at least three examples of gas leak situations in which you would need to call emergency services.

　i _____

　ii _____

　iii _____

13 What does the term *hypoxia* describe?

_____

_____

_____

14 What is the primary risk to a gas fitter when using nitrogen?

_____

_____

_____

 **WORKSHEET 2**

To be completed by teachers
Student competent ☐
Student not yet competent ☐

Student name: _____

Enrolment year: _____

Class code: _____

Competency name/Number: _____

Task

Review the sections on *Basic electrical safety* and *Electrical test equipment* and answer the following questions.

1 List two faults that a power point tester might detect.

   i _____

   ii _____

2 A fault is discovered at a GPO. List the steps you would take to control the hazard.

   _____
   _____
   _____

3 You are asked to remove an old gas hot water system and find an earth connection on the adjacent water pipes. Would you be able to remove it so that you could alter the pipe layout?

   _____
   _____
   _____

4 Plastic pipe can build up dangerous levels of static electricity. What procedure does the Gas Installation Standards recommend to safely discharge this electricity?

   _____
   _____
   _____

5 What clause in the AS/NZS 5601.1 directs you to use precautions against electrical hazards?

   _____
   _____
   _____

6 Before using your test instrument, what basic check should you do first?

_____

_____

_____

7 List four instances in which you would need to apply bonding straps.

i _____

ii _____

iii _____

iv _____

8 What path does electricity always prefer to take to earth?

_____

_____

_____

9 If the neutral connection were somehow disconnected from a house, what path would electricity take to return to its source?

_____

_____

_____

10 Does current or voltage have a greater influence on the severity of an electric shock?

_____

_____

_____

11 Using lines, link the following terms with the appropriate symbol:

Volts   ∼

Resistance   ⎓

AC   Ω

Current   V

DC   I

12 Why is the earth system such an important part of the AC electrical system?

_____

_____

_____

13 What are four basic work practices that will reduce your exposure to electric shock?

i _____

ii _____

iii _____

iv _____

14 In the following diagram, insert arrows to indicate the direction of electricity as it travels through the household wiring system.

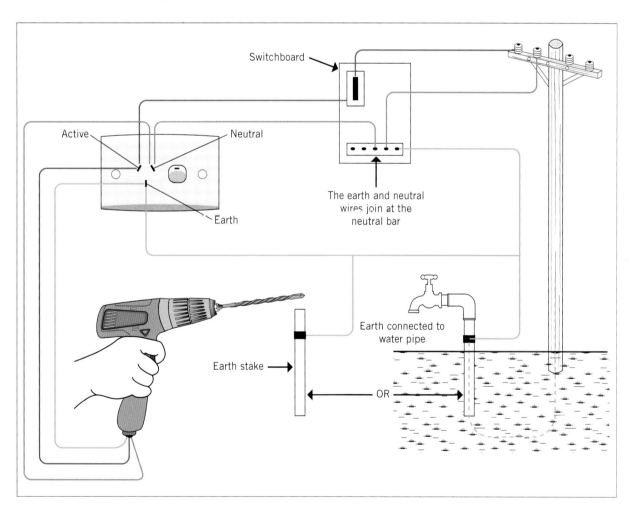

15 What form of current is a volt stick able to detect?

_____

_____

_____

16 This diagram shows damage to the wiring system. Using arrows, show the likely path the current will take back to its source.

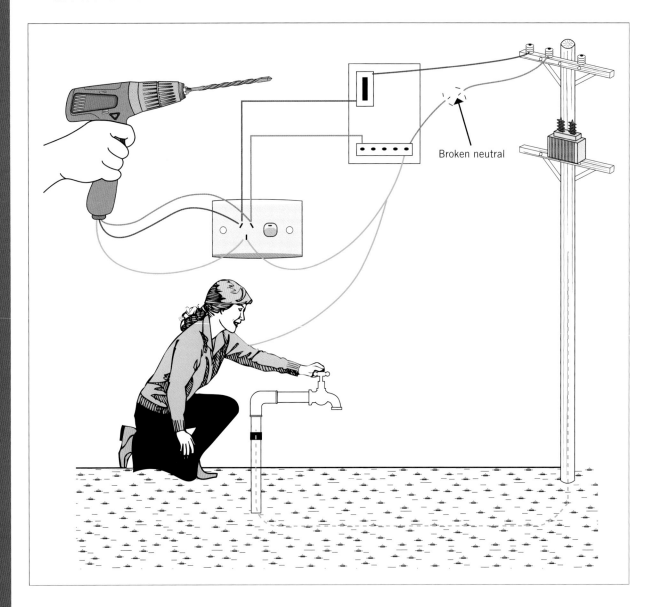

COMBUSTION PRINCIPLES

Learning objectives

Areas addressed in this chapter include:
- basic combustion principles
- burner design and functions
- flame characteristics
- what constitutes a 'good' flame
- how to diagnose a faulty flame.

A sound grasp of basic combustion principles is fundamental to all aspects of gas fitting practice and so this chapter and worksheets should be completed before working through the units of competency contained within Part B of this text.

Overview

One of a gas fitter's primary tasks after installation or servicing is to 'commission' the appliance. To commission an appliance means to bring it into correct working condition.

A gas fitter is required to follow the manufacturer's instructions so that the appliance operates as intended. Part of this process is to ensure that the combustion system of the appliance is adjusted correctly and that complete combustion is taking place. To do so, you will need a strong understanding of combustion principles.

Basic combustion principles

For any form of combustion to take place you must have these three ingredients:

Air + Fuel + Heat

If one of the three is missing, combustion cannot take place. Perhaps more importantly for the gas fitter, when any one of these three ingredients is out of balance, then hazardous, poor or incomplete combustion takes place.

GREEN TIP

Not only is poor combustion a safety issue but it also represents inefficient appliance performance. Gas fitters play a significant role in meeting the latest energy-efficiency standards in new construction and providing cost-effective appliance solutions in existing premises.

When considering how air, fuel and heat interact, you need to know specifically what these terms mean in relation to combustion itself.
- *Air.* The air that surrounds us is made up of a mixture of many gases. Of these, the principal gases that concern you as a gas fitter are oxygen (O_2) and nitrogen (N_2). Characteristically, air is made up from approximately 78 per cent nitrogen and 21 per cent oxygen.
- *Fuel.* The fuel in the combustion process is the gas itself. In the case of LPG, this is propane and/or butane, and in the case of natural gas it is the hydrocarbon methane.
- *Heat.* This not only means the heat required for ignition but also the heat required for continuous, self-sustaining, complete combustion.

Complete combustion itself describes a chemical reaction between a *fuel* (natural gas or LPG) and an *oxidiser* (oxygen), which becomes self-sustaining after the application of the correct amount of *heat* (a match or spark).

You will recall from Chapter 4 that to burn 1 m^3 of natural gas you need approximately 9.5 m^3 of air. This is fine in a laboratory, but gas appliances operate in variable environments and suffer from eventual wear and tear. If an appliance were set up perfectly to combust gas with only the right amount of air, it would only take a minor change in settings or conditions for insufficient air to be available, with a hazardous state resulting.

To avoid such a situation, manufacturers design Type A gas appliances to ensure that, when correctly commissioned, more than sufficient air is available for combustion at all times. This design provision is known as excess air.

Assessing and calculating excess air is a critical task for Type B appliance installations but, fortunately, in order to know that the correct amount of air is available, a Type A gas fitter needs only to ensure that the appliance is correctly ventilated and adjusted according to commissioning instructions.

Products of complete combustion

Your goal as a gas fitter is to commission appliances in a manner that ensures complete combustion of the fuel gas. When a fuel gas is burnt in air under normal atmospheric conditions, the following products of combustion are produced:
- heat
- water vapour
- carbon dioxide.

In Figure 6.1, you can see a cut-away of a hot water cylinder. A well-designed and commissioned appliance will ensure that complete combustion takes place.

What about the nitrogen? You will notice that the 8 m^3 of nitrogen that enters the appliance still exists as 8 m^3 at the flue outlet. Nitrogen is not a product of combustion. Nitrogen is an inert gas and it will not burn; it simply goes in one end and comes out the other with no real change in quantity.

Causes of incomplete combustion

In general gas fitting you will find two main causes of incomplete combustion:
- lack of air
- flame chilling.

These two conditions can occur for a number of different reasons, examples of which are described below:
- *Lack of air.* When the supply of air is restricted, not enough oxygen will be present for the hydrocarbon fuel to be completely oxidised into carbon dioxide (CO_2) and water vapour. Instead, this partial oxidation will result in the production of carbon monoxide (CO). Lack of air can be caused by:
  - poor burner adjustment or ventilation of the appliance
  - too much gas, leading to a proportional reduction in air supply
  - inadequate flueing – a blocked or incorrectly sized flue will cause a build-up of combustion products, reducing the amount of fresh air available for complete combustion.

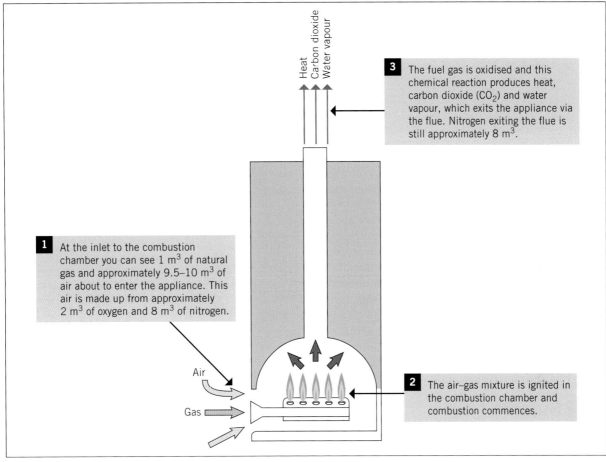

FIGURE 6.1 A cut-away of a storage hot water system, showing how air and fuel combust to create combustion products

- *Flame chilling.* If a solid object is allowed to penetrate the inner cone of the flame, the fuel–air mixture making contact with the object will fall below its ignition temperature, creating a zone of incomplete combustion.

Products of incomplete combustion

Incomplete combustion of a fuel gas will result in the creation of the following products:
- *Carbon monoxide.* As noted in earlier chapters, this odourless, tasteless, explosive and invisible gas is the toxic result of incomplete oxidation of the fuel into CO_2 and water vapour.
- *Soot (carbon).* During combustion, fuel molecules will break down and change in a series of complex chemical reactions. In any flame, there is a 'soot-growth zone', in which particulate solids can form. During complete combustion, these particles normally oxidise completely into CO_2 and water vapour, but where combustion is incomplete, they will stabilise as solid particles in the flame. These particles of hot, glowing soot give any flame its yellow colour. The particles cool as they escape the flame and will deposit as black carbon on any nearby surface. Many authorities regard soot as a carcinogenic (cancer-causing) substance.
- *Aldehydes.* These are a class of organic compounds formed during the partial or incomplete oxidation of hydrocarbon fuels. In high concentrations, aldehydes are poisonous, but this is not generally the case at the low levels released during normal fuel gas incomplete combustion. Of importance to you as a gas fitter is the characteristic sharp, unpleasant odour given off by aldehydes. As such, they provide a warning that incomplete combustion is taking place and that it is highly likely that CO is also present.

The conditions created through incomplete combustion are potentially life threatening. Your understanding of combustion theory is vital to your daily activities as a gas fitter.

COMPLETE WORKSHEET 1

LEARNING TASK 6.1

BASIC COMBUSTION PRINCIPLES

1. When a fuel gas is burnt under normal atmospheric conditions, what are the three products of complete combustion?
2. What are two causes of incomplete combustion?

Burner design

The central component of any Type A gas appliance is the burner itself. You will soon find that there is no such thing as a standard burner. Burners are manufactured in hundreds of different ways in order to suit the size, configuration and application of the appliance. However, all of these individual burners are actually variations on a simple theme and design principle.

In this section, you will learn about some of the different types of burners used in the gas industry, but in particular the design, operation and adjustment of the partial pre-mix natural draught burner.

Burner functions

Any burner can be regarded as having three primary functions:

- mixing – it combines the air and fuel into a homogenous mixture
- delivery – it delivers the air–fuel mix to the point where combustion heat is to be applied
- stability – it provides a stable platform for self-sustaining combustion.

Irrespective of whether a burner is used in domestic, commercial or industrial duties, they all carry out these tasks. The variations on how they actually do so enable burner types to be split into different classifications.

Burner types – classification

Burners can be described according to key functional characteristics. The primary classifications are:

- the point at which combustion air and gas are mixed
- the means of providing combustion air.

Each of these primary classifications is broken down into the actual types of burners and are called secondary classifications. These are shown in Figure 6.2 and are described below.

Primary classification 1: Point at which combustion air and gas are mixed

Under primary classification 1, you will find four main burner types that differ significantly in relation to where the air and gas are mixed.

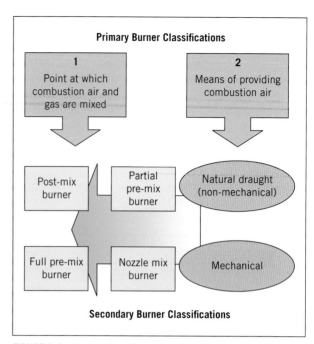

FIGURE 6.2 The relationships between primary and secondary burner classifications

Post-mix burner

The post-mix (post-aerated) burner was designed to introduce combustion air *after* the fuel gas had left the burner. Also known as luminous burners because of their characteristic incandescent flame, post-aerated burners were suited only to the use of gases with high flame speeds, such as towns gas. With oxygen available only after the gas had left the burner, towns gas had to burn very fast to ensure satisfactory combustion.

By comparison, modern gases such as natural gas and LPG are slow-burning gases and therefore unsuitable for post-aerated burners. For this reason, such burners are now found only in some specialist industrial applications and are no longer relevant in relation to modern Type A gas appliances.

Figure 6.3 shows that a natural draught post-mix burner must rely upon the characteristics of a fast-burning gas for complete combustion to be achieved. These images show a very early form of gas continuous flow water heater, traditionally installed above a sink or basin. If you look very carefully at the spiral-shaped burner, you will see very fine burner ports. Raw towns gas would issue from each of these ports and all air for combustion would be supplied from around the flame.

Partial pre-mix burner

The partial pre-mix burner is the main kind of burner found in most modern Type A gas appliances. As the fuel gas is introduced into the burner, it is mixed with part of the required combustion air *before* the gas leaves the burner itself. The remainder of required air is sourced from around the flame within the combustion

FIGURE 6.3 A natural draught post-mix burner

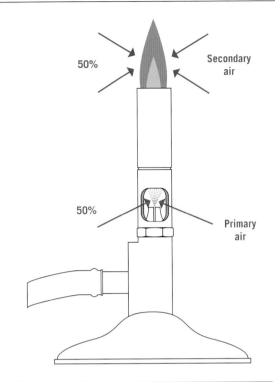

FIGURE 6.4 A partial pre-mix burner

chamber. These burners are also known as 'pre-aerated burners' and 'Bunsen burners'. Figure 6.4 shows that 50 per cent of the air required for combustion is drawn in through the primary air port (primary air), while the remaining 50 per cent of required air is sourced from around the flame itself (secondary air).

Full pre-mix burner

A full pre-mix burner is an industrial type burner for which *all* required combustion air is mechanically mixed with the fuel gas by an aspirator mixer before it leaves the burner. This type of burner normally does not fall into the scope of Type A gas fitting.

Nozzle-mix burner

A nozzle-mix burner is an industrial burner in which the air and fuel gas are kept separate until they exit burner nozzles that are designed to provide a rapid mix of air and gas as they leave the burner. As with a full pre-mix burner, these are not found in normal Type A gas appliances.

Figure 6.5 clearly shows the gas nozzles surrounding the central air outlet in a nozzle-mix burner.

COMBUSTION PRINCIPLES **85 GS**

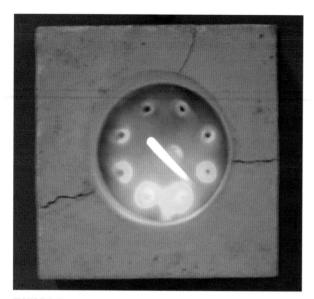

FIGURE 6.5 An industrial nozzle-mix burner mounted within a refractory block

Primary classification 2: Means of providing combustion air

The combustion air supply for the burner can be provided through both natural and mechanical means.

Natural draught burners

Natural draught burners are found in most Type A gas appliances. This type of burner relies on the effect of the gas streaming through a venturi and mixing tube to inspirate the combustion air. The word 'inspirate' means to 'draw in'. Air is also supplied by the natural draught characteristics of the flue system and combustion space. In this sense, a natural draught burner does not rely on any mechanical means to supply combustion air.

The blue arrows in Figure 6.6 indicate the primary air being drawn into the primary air port through the inspirating effect of the gas stream.

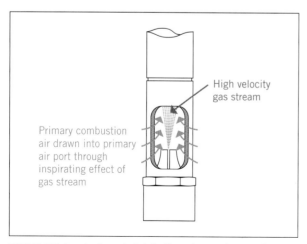

FIGURE 6.6 Inspiration of air into the primary air port of a natural draught burner

Mechanical air supply

A number of Type A gas appliances, including some water heaters and space heaters, use mechanical means to supply combustion air.

An appliance that provides air under pressure into the combustion chamber with a fan or compressor is classed as a forced draught burner.

Where air is drawn into the combustion chamber using fan suction, the appliance is classed as an induced draught burner. The combustion fan in the continuous flow hot water system shown in Figure 6.7 provides air to the sealed combustion chamber mounted above it.

FIGURE 6.7 The air supply to this water heater is provided by a fan below the combustion chamber

Of all these burners, the partial pre-mix is most commonly found in Type A gas appliances and it will be examined in the next section. Although 'partial pre-mix burner' is the correct technical description used by combustion engineers, gas fitters normally associate this name with an industrial burner with mechanical air supply. Now that you are aware of the differences between each classification of burner, for the remainder of this text we shall use the more commonly used term: pre-aerated burner.

The majority of burners that a Type A gas fitter will work on will fit under the classification of pre-aerated natural draught. The pre-aerated burners shown in Figure 6.8 are cooktop, laboratory and kiln burners.

The pre-aerated burner

Robert Bunsen developed the first pre-aerated burner as early as 1852 in Germany. At that time, post-aerated burners were commonly used for street lighting and other gas appliances, but Robert Bunsen needed to outfit his new laboratory with a burner that had a hotter, more compact flame with no luminosity. This new partial pre-mix burner became commonly known as the 'Bunsen burner', a term that can be applied to any pre-aerated burner.

To identify component parts and understand how a pre-aerated burner works, read the text boxes on each of the two burner diagrams in Figures 6.9 and 6.10.

While similar in some aspects, the pre-aerated hot water service burner (Figure 6.10) has some additional features that you need to be aware of. See if you can spot the main differences.

Flame retention design

Compared to towns gas, natural gas and LPG are slow-burning gases. Because they also require greater volumes of air to burn completely, they must also operate at relatively higher pressures in order to inspirate this extra air through the primary air port. These two characteristics mean that the flames of natural gas and LPG can tend to come away from, or 'lift off', the burner. Such a flame is unstable and may easily be extinguished.

With the introduction of modern fuel gases, combustion engineers needed to find a way of holding, or retaining, the flame on the burner. The following flame retention methods are incorporated into many Type A gas appliance burner designs:

- flame retention ports and rings
- recirculation
- retention hardware.

For efficient commissioning and servicing, it is important that you understand what these design features are and can identify them when examining a burner. A more detailed description of each system is provided below. Your teacher should be able to provide many more examples.

Flame retention ports and rings

Look carefully at the laboratory burner in Figure 6.11. You will notice an outer ring surrounding the main burner port. This ring is designed to protect a series of secondary ports drilled through from the main burner head, allowing an air–gas mixture to fill the ring space. In the event of flame instability, the smaller and more stable low-velocity ring of flame will re-ignite the main burner.

Any burner port must be sized to suit the designed mixture volume and large enough to provide a flame of adequate size to do the job. However, if the burner port is made too large for a given supply pressure, combustion stability will decrease as the burner port diameter increases.

The smaller retention flames escape from the burner at a lower velocity than the main burner ports and are therefore more stable. These flames act as a form of

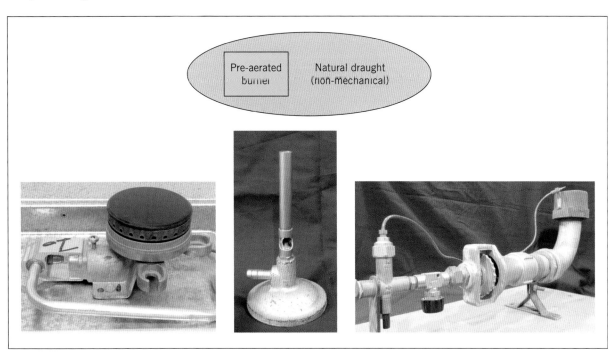

FIGURE 6.8 Examples of pre-aerated burners

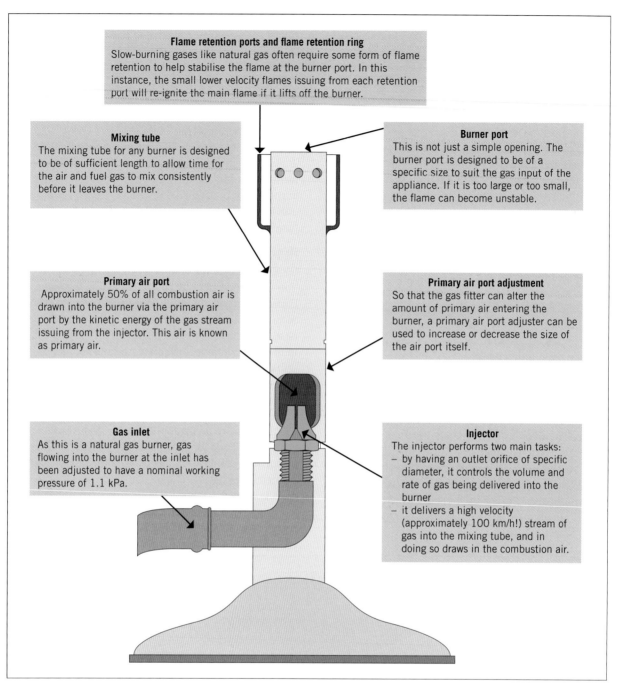

FIGURE 6.9 Laboratory burner

continuous re-ignition, preventing the larger burner flames from being extinguished. Figure 6.12 shows a flame retention burner head commonly used in kiln burners. Such burners may have operating pressures of around 100 kPa and would certainly exhibit combustion instability, causing the flame to lift off the burner entirely, if it were not for the flame retention burner head and the circle of smaller retention flames around the main burner port.

Recirculation

You will often see this form of flame retention used on cast iron burners. The products of combustion around the base of the flame are at a higher temperature than the ignition point of the air–gas mixture. The slightly recessed burner port enables a small amount of these combustion products to be momentarily captured around the base of the flame. If the flame starts to become unstable and lifts off the burner, recirculation of

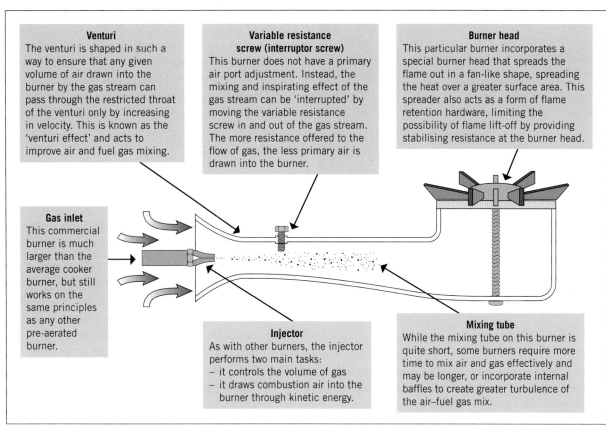

FIGURE 6.10 A commercial storage hot water burner

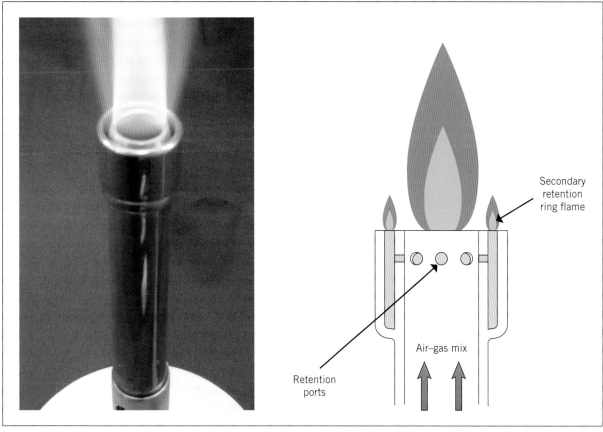

FIGURE 6.11 Laboratory burner flame retention ring and half-section showing the location of retention ports within the outer flame retention ring

FIGURE 6.12 A flame retention burner head in a kiln burner

these very hot combustion products will cause the air–gas mixture to re-ignite. In practice, this would be an almost constant process of re-ignition. Figure 6.13 represents a close-up view of burner ports designed to allow hot products of combustion to accumulate at the base of the flame. Flame instability is reduced by a process of almost constant re-ignition of the air–gas mixture.

While suitable for use in thicker-walled cast iron burners, recessing of the burner ports is generally not possible with many modern pressed metal burners.

Retention hardware

The most common example of the use of flame retention hardware is often found in domestic oven burners. Not only does the oven burner need to be designed to deal with the instability problems normally associated with slow-burning gases, but it is also often subjected to potential flame outage brought about by slamming the oven door.

In Figure 6.14, you will see a metal hood attached over the top of one end of the burner. The turbulence created by this hardware slightly reduces flame velocity and increases flame stability. In the event of a door slam, it also acts to protect that portion of the flame from the sudden rush of air that will act to extinguish the rest of the burner flame. This protected section of flame then acts to re-ignite the main flame.

Slamming an oven door would almost certainly extinguish the small by-pass flame of an oven burner. The flame retention hardware seen here at one end of the burner acts to shield part of the flame from the effects of a door slam.

An extreme example of flame retention can be seen in Figure 6.15. This is known as a 'duckbill burner', often used in commercial wok tables. It is a good example of how a burner can be deliberately designed to operate outside normal performance parameters with flame retention features. This burner is designed for a normal low burner pressure, but is deliberately over-aerated and over-gassed in order to provide a fierce column of heat within a small area.

This would normally be impossible with standard burner design, as the flame would simply lift off and go out, but if you look carefully at each individual burner you will see that the two 'duckbills' are actually bent inwards over the flame, capturing the flame and creating the intense, concentrated heat required for commercial wok cooking. This design feature acts to create enough turbulence to just hold the flame on the

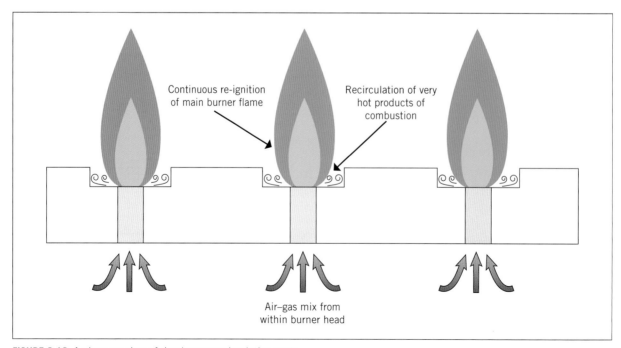

FIGURE 6.13 A close-up view of ring burner recirculation ports

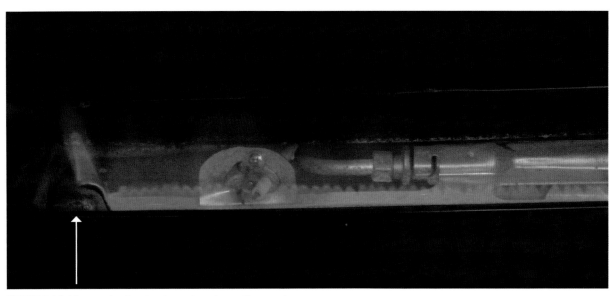

FIGURE 6.14 Flame retention hardware in a domestic oven burner

FIGURE 6.15 The curved shape of a duckbill burner acts as a form of flame retention

burner and still provide more heat than can normally be achieved in a burner area of this size.

The importance of correct injector size

As you have read in the above descriptions of burner components, the injector performs the task of delivering a specific amount of gas and inspirating the primary air into the burner. For these two tasks, it is important to look more closely at the relationship between injector orifice size and burner performance.

For an injector to be able to provide a specific amount of gas to a burner in accordance with the manufacturer's design requirements, it is dependent upon a correctly set and maintained gas burner pressure. If the burner pressure fluctuates excessively, the amount of gas consumed by the burner will change.

If the burner pressure is correct, then the amount of gas coming out of the injector is determined by the size of the orifice itself. At the nominal low burner pressures found in Type A gas appliances, the size of an injector orifice can be measured not only in diameter, but also in the MJ/h it is designed for.

The injectors in **Figure 6.16** are from two different natural gas appliances, each designed to have a nominal

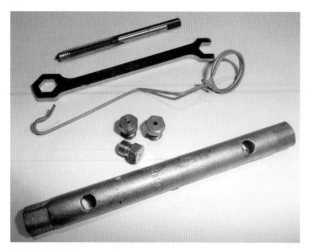

FIGURE 6.16 From the front: tube spanner, three different sized jets, wire jet holder, jet spanner and a thread tap

burner pressure of 1.1 kPa. You can see that the orifice sizes are very different, as are the MJ/h at which each injector is rated. Injectors are carefully sized to measure the exact amount of gas required for the burner. At the same pressure, a larger orifice will deliver more gas.

An injector is sometimes known as a *jet*. Ask your teacher to show you some other examples.

Gas type, pressure and orifice size

A Type A gas appliance designed for a particular gas cannot be fuelled by another gas without conversion. LPG and NG have very different characteristics that restrict their use to those appliances designed specifically for each gas.

For example, examine the two injectors in Figure 6.17. Both serve an 8 MJ/h burner and yet their orifice sizes are entirely different. You can see that the LPG injector is much smaller than the equivalent injector used for NG.

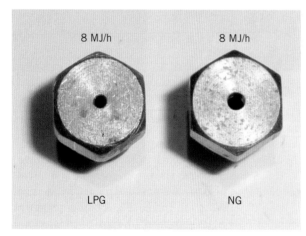

FIGURE 6.17 The difference in orifice size between LPG and NG injectors rated at the same MJ/h

The differences in orifice size are due to the following interrelating factors, as seen in Figure 6.18.

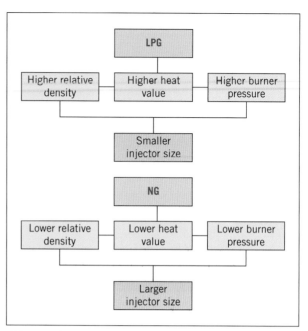

FIGURE 6.18 Differences in orifice size

- *Relative density (RD)*. LPG is heavier than air (RD 1.55) and therefore must have a higher supply and burner pressure (2.75 kPa) so that it can overcome the inertia created by this greater density. As it is running at a higher pressure, its injector orifice does not need to be as large. NG is lighter than air (RD = 0.6) and runs at a much lower pressure (1.1 kPa).
- *Heat value*. LPG (96 MJ/h) has more than twice the heat value of NG (38 MJ/h). Therefore, not as much gas needs to be delivered to the burner to do the same amount of work as NG.

You can now see why it is so important that burner injectors are sized correctly. Operating an appliance designed for NG (with its large injector orifices) on LPG can be disastrous.

LEARNING TASK 6.2 BURNER DESIGN

1. What are the three primary functions of a burner?
2. Why is flame retention designed into natural gas and LPG burners.
3. Why do natural gas and LPG injectors that can serve the same MJ/h burner, have different orifice sizes?

COMPLETE WORKSHEET 2

Flame characteristics

Earlier in this chapter, you were introduced to the principles of fuel gas combustion and reviewed basic burner design as it relates to pre-aerated burners commonly found in Type A gas appliances. This knowledge now allows you to proceed to an understanding of flame characteristics and their relationship to optimum burner and appliance performance.

A gas fitter is responsible for correctly commissioning the gas appliance and in many cases this will involve adjusting the burner to ensure that the flame is burning completely. A poorly adjusted flame burns differently from a well-adjusted flame. The colour and shape of a flame can show a gas fitter how well the appliance is performing and what needs to be adjusted.

The importance of correct gas pressure

During either commissioning or servicing of a Type A gas appliance, you cannot determine anything about flame characteristics until you confirm that the appliance working pressure is set according to the manufacturer's specifications. You will recall the following nominal burner pressures:

Natural gas	1.13 kPa
Liquefied petroleum gas	2.75 kPa

These nominal pressures will apply to many appliances, but remember you *must* always refer to the appliance data plate for confirmation.

There is little point in trying to adjust burner aeration if the gas pressure is too low or too high. You would be wasting your time and, worse still, you may leave the appliance in an unsafe condition.

FROM EXPERIENCE

Correct gas pressure is the fundamental requirement of all gas fitting work. Remember – gas pressure must always be checked before carrying out any other work!

The importance of adequate air supply

Complete combustion is dependent not only on correct gas pressure but also on having an air supply of the correct quality and quantity. If gas pressure is correctly set, most problems associated with flame characteristics relate to air supply. The following are three principal forms of air supply that you need to consider:

- *Ambient air supply*. This is all the air available to a gas appliance. It must be fresh air and not contaminated with chemicals, dust or other combustion products. Appliances installed in an enclosed area must be provided with no less than the minimum amount of air stipulated in the Gas Installation Standards. You will learn later how to carry out such calculations.
- *Primary air supply*. This is the amount of air drawn into the burner itself via the primary air port. Primary air normally represents approximately 50 per cent of all combustion air in a standard natural draught Type A gas appliance.
- *Secondary air supply*. The remaining air required for combustion comes from around the flame itself and is known as secondary air.

What is a 'good' flame?

A correctly adjusted flame from a pre-aerated burner should have the following key characteristics:
- The flame has two visible and well-defined zones of combustion.
- The flame is stable on the burner.

The flame of a pre-aerated burner is actually made up of three distinct zones, though only two of these are visible. Figure 6.19 demonstrates the sort of flame you should be looking for. Flame combustion zones have distinct characteristics that a gas fitter must be able to identify in order to correctly diagnose combustion efficiency.

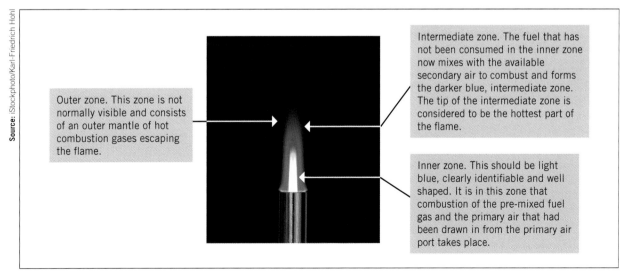

FIGURE 6.19 A laboratory burner detailing flame zones and colours

Exceptions to the rule

Although almost all Type A gas appliances should demonstrate a flame with the characteristics described above, there are some key exceptions. Many decorative flame effect fires are designed to burn with a yellow/orange flame to simulate the visually pleasing flames of a real log fire.

Flame fault diagnosis

Just as a medical doctor is able to diagnose a problem by looking for characteristic symptoms, the gas fitter can also determine a fault and carry out remedial action simply from looking at the burner flame. This skill takes some time to develop and your instructor should be able to provide some examples of each of these characteristics.

> **FROM EXPERIENCE**
>
> As a gas fitter, you need to develop the habit of evaluating every flame you look at to determine its condition. Through repetition, you will become an expert in flame fault diagnosis.

Remember, before proceeding with diagnosis of any flame always satisfy yourself that both the gas supply and the gas working pressures are compliant with the appliance data plate.

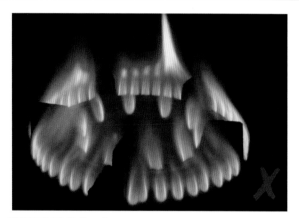 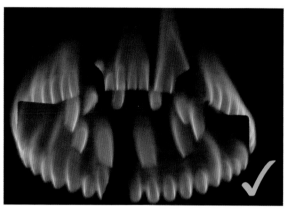

FIGURE 6.20 Yellow tips

Symptom: Yellow tips

Yellow tips (as seen in **Figure 6.20**) are characterised by:
- flame volume – a normal amount of flame
- colour – continuously burning, bright yellow tips on all or most flames
- shape – the flame shape is often too 'soft' and zones may lack distinction.

Consequence: The flame may deposit soot on adjacent surfaces and will produce dangerous CO gas.

Cause: Yellow tips on a normal sized flame indicate *a lack of primary air*.

Remedy:
1. Check that the gas pressure is correct.
2. Adjust the primary air supply until all the yellow tips disappear and the flame regains the correct shape.

Hint: If you experience difficulties in getting rid of yellow tips, check that nothing has blocked the venturi or mixing tube of the burner. A physical blockage (build-up of soot, dust or even a spider's web) may be the cause.

 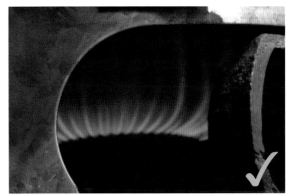

FIGURE 6.21 Streaming flame

Symptom: Streaming flame

The streaming flame (see Figure 6.21) is characterised by:
- flame volume – a normal volume of flame, although this is sometimes difficult to gauge because of its misshapen appearance
- colour – the flame ranges from extensive yellow tips to a darker, dirty yellow flame in severe cases
- shape – flame is long and lazy, streaming upwards, 'hunting' for more secondary air. In severe cases the flame will lack any consistent shape and will show no combustion zones at all.

Consequence: Streaming flames produce large amounts of the poisonous gas CO and will deposit soot throughout the combustion area and flue.

Cause: A streaming flame is caused by *a lack of secondary air*. This is often because of a blocked or inadequately sized flue. This condition of air starvation is often known as 'vitiation' (pronounced *vish-ee-ay-shon*).

Remedy:
1 Check that the gas pressure is correct.
2 Check that the primary aeration is correct.
3 Check for flue blockage.
4 Check that the flue is sized and located correctly, according to the appliance and gas code requirements.

Hint: This problem is often found on commercial catering equipment where the internal flue channels have been blocked by cooking grease. Be aware of CO levels at all times when working on such an appliance in enclosed areas.

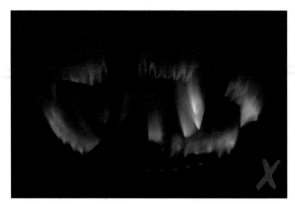 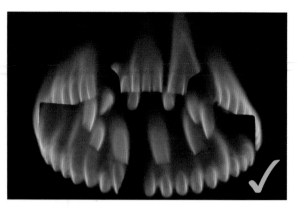

FIGURE 6.22 Flame lift-off

Symptom: Flame lift-off

Flame lift-off (as shown in Figure 6.22) is characterised by:
- flame volume – a normal flame size but unstable on the burner
- colour – generally blue, but the flame may flare with yellow tips because of inconsistent combustion
- shape – a ragged, flickering and noisy flame, often with indistinct combustion zones in severe cases.

Consequences: The flame threatens to lift off the burner and in some cases will do so and go out. Unburnt gas will then enter into the combustion area unless the burner is fitted with some form of flame-failure safety device.

Cause: Flame lift-off is caused by too much primary air.

Remedy:
1. Check that the gas pressure is correct.
2. Reduce the primary air supply until the flame stabilises back on the burner, but not so far as to create yellow tips.
3. Check that any flame retention hardware is positioned where it should be.

Hint: This problem is more commonly found on some natural gas appliances where a full range of adjustment is available at the primary air port. Appliances originally designed for towns gas were particularly susceptible to lift-off after conversion to NG or LPG. It is more unusual to find LPG appliances with a lifting flame because of their overall higher air supply requirements.

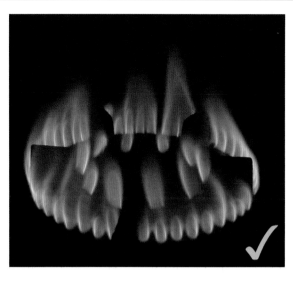

FIGURE 6.23 Excessively large flames

Symptom: Excessively large flames

Excessively large flames (as seen in Figure 6.23) are characterised by:
- flame volume – too much flame for the size of burner
- colour – partial or completely yellow
- shape – no distinct combustion zones, and the flames are more vertical.

Consequences: High CO production and large soot deposits create a high fire hazard.

Cause: Oversized flames can be caused by three main faults:
- excessive gas pressure (normally from a faulty or incorrectly set regulator)
- LPG being used in a natural gas appliance
- an oversized injector.

Remedy:
1. Check that the gas type and appliance are compatible; look at the data plate and double check the gas supply.
2. Check that the gas pressure is correct.
3. Check the injector size against the manufacturer's instructions or data plate.

Hint: The actual amount of gas on the burner is the key symptom that makes this problem different from the normal yellow flames caused by a lack of primary air. Unfortunately, you will occasionally encounter this type of flame on an appliance that has not been correctly converted from NG to LPG.

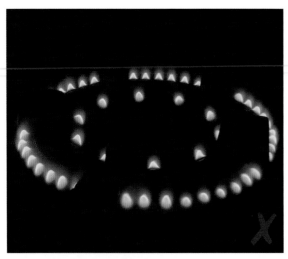 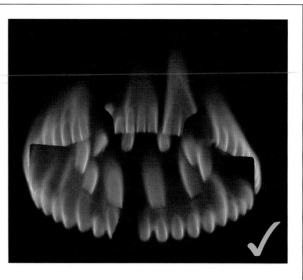

FIGURE 6.24 Undersized flames

Symptom: Undersized flames

Undersized flames (as seen in Figure 6.24) are characterised by:
- flame volume – flames that are too small for the size of burner
- colour – all blue
- shape – small tight flames that may still show very small combustion zones.

Consequences: The flames may be unstable and can 'flash back' to the injector with an obvious roaring sound near the primary air port. This may then present a potential fire hazard or cause internal damage to the appliance through overheating near the injector.

Cause: Undersized flames can be caused by three main faults:
- NG being used in an LPG appliance
- an undersized or blocked injector
- the gas pressure being too low.

Remedy:
1. Check that the gas type and appliance are compatible; look at the data plate and double check the gas supply.
2. Check that the gas pressure is correct.
3. Check the injector sizes against the manufacturer's instructions or data plate.

Hint: Sometimes pipeline contaminants can partially block the injector. If the appliance does flash back in your presence, immediately turn off the appliance. Although the flame will not burn back into the pure gas within the supply line, the considerable heat given off by the flame near the primary air port may cause damage to the appliance's internal components.

FIGURE 6.25 Orange flames

Symptom: Orange flames

Orange flames (as seen in Figure 6.25) are characterised by:
- flame volume – normal
- colour – intermittent orange flares
- shape – generally the same as a well-adjusted flame, but at times it may flare or even sparkle.

Consequences: Depending upon the cause, orange flames generally do not present any problem, but may confuse an inexperienced gas fitter who may mistake them for yellow tips. Orange flames may indicate a hazardous condition where certain chemicals are present in the air.

Cause: Orange flames are caused by *contaminated secondary air*. The presence of airborne oils or chemicals may cause the flame to glow orange, as these contaminants are combusted. This is normally encountered when you turn on the burner of a new appliance for the first time and the production oils on metal parts are burnt away. This will normally cease after approximately 20 minutes.

You may also experience such flames in commercial catering environments, where some appliances may be affected by the cooking vapours of adjacent appliances such as char grills and wok tables.

Although these scenarios do not present any real problems, be aware that secondary air contamination in some environments can be very serious. Examples include dry cleaning and car wash facilities. Certain chemicals can become very corrosive or poisonous during the combustion process. Your job is to ensure that all gas appliances have a constant and correct supply of fresh air.

Diagnosis checklist

To assist with flame characteristics and fault diagnosis, you may find it helpful to copy the following brief notes into the back of the Gas Installation Standards for quick and easy reference whenever you commission an appliance.

Ensure correct gas pressure	☐
Flame should readily ignite	☐
Flame should be entirely blue	☐
Flame zones should be clearly defined	☐
No yellow tips	☐
No lift-off	☐
No streaming	☐

LEARNING TASK 6.3 FLAME CHARACTERISTICS

1. What do yellow tips on a burner with a normal size flame indicate?
2. What is the consequence of flame lift-off, on a burner?
3. What is the cause of orange flames on a burner?

COMPLETE WORKSHEET 3

SUMMARY

Having worked through this chapter you should now have a good appreciation of the direct interrelationship between fuel gas constituents and burner design, and the vital importance of air supply and gas pressure. Complete combustion depends upon each of these different aspects working in correct balance with each other. Being able to observe, diagnose and adjust fuel gas burners will underpin your success as a gas fitter.

Ensure you have a good understanding of this chapter before moving on to Part B of this text.

GET IT RIGHT

The photo below shows an incorrectly burning flame.

Look carefully at the image and identify the condition that is causing such a flame pattern. Provide reasoning for your answer.

_____

_____

_____

_____

WORKSHEET 1

To be completed by teachers	
Student competent	☐
Student not yet competent	☐

Student name: _____

Enrolment year: _____

Class code: _____

Competency name/Number: _____

Task

Review the section on *Basic combustion principles* and answer the following questions.

1. Lack of air is one reason for incomplete combustion. State two factors that would lead to a lack of combustion air.

   i _____

   ii _____

2. This is a list of products of combustion. Circle those products that are specifically related to complete combustion.

   Soot   CO   Water vapour CO_2 Heat   Aldehydes

3. Which of these two gases is poisonous: CO_2 or CO? _____

4. What substance might cause a burner flame to give off a strong, unpleasant odour?

   _____
   _____
   _____

5. What substance causes the yellow glow of an incorrectly adjusted flame?

   _____
   _____
   _____

6. A cooking pot that has been lowered into the inner cone of an otherwise well-adjusted flame may develop black stains on its bottom and sides. What is the name of the condition that causes this? Why does it happen?

   _____
   _____
   _____

7   To burn 1 m³ of natural gas you require approximately 9.5 m³ of air. How much air would you require to burn the same amount of LPG?

    _____

    _____

    _____

8   Why is there no reduction in the volume of nitrogen from the beginning to the end of the combustion process?

    _____

    _____

    _____

9   List three products of incomplete combustion.

    _____

    _____

10 What are three possible causes for lack of air to the burner?

    _____

    _____

WORKSHEET 2

To be completed by teachers
Student competent ☐
Student not yet competent ☐

Student name: _____

Enrolment year: _____

Class code: _____

Competency name/Number: _____

Task

Review the section on *Burner design* and answer the following questions.

1. What characteristic of natural gas makes it unsuitable for use in a post-mix burner?

   _____

   _____

   _____

2. Is a Bunsen burner a pre-aerated or post-aerated burner?

   _____

3. These pictures show parts of two pre-aerated burners. Name the parts of each burner where indicated by an arrow.

COMBUSTION PRINCIPLES

4 What is the more technically correct name for a pre-aerated burner?

5 What is the purpose of a variable resistance screw?

6 Approximately what percentage of all combustion air is drawn in via the primary air port?

7 What are the three primary functions of any gas burner?

   i

   ii

   iii

8 Describe two ways in which a natural draught burner gets its combustion air supply.

   i

   ii

9 Describe how the recirculation of combustion products ensures flame stability.

10 Why was flame retention not generally required on towns gas burners?

11 What are the two key tasks of the injector?

   i

   ii

12. You compare two gas injectors rated at the same MJ/h but one is for NG and the other LPG. Which of the following is true?
    - The LPG injector orifice is larger
    - Both injector orifices are the same diameter.
    - The NG injector orifice is larger than the LPG orifice.

WORKSHEET 3

To be completed by teachers
Student competent ☐
Student not yet competent ☐

Student name: _____

Enrolment year: _____

Class code: _____

Competency name/Number: _____

Task
Review the section on *Flame characteristics* and answer the following questions.

1. Before attempting to adjust a poorly performing flame, what should you check first?

   _____

   _____

   _____

2. An LPG appliance requires around 24 m^3 of air to burn 1 m^3 of gas. How many cubic metres of this would be secondary air in the combustion process itself?

   _____

   _____

   _____

3. What is the primary cause of the condition known as vitiation?

   _____

   _____

   _____

4. Where is the hottest part of the flame?

   _____

   _____

   _____

5. A flame presents itself with the following characteristics:
   - soft-looking flames lacking clear definition
   - yellow tips
   - it deposits soot on adjacent surfaces
   - the volume of flame looks normal.

   What would be the likely cause of these symptoms?

   _____

   _____

   _____

6 What characteristics would a flame exhibit from an appliance that was installed with an undersized flue?

_____
_____
_____

7 A customer phones you to report that one of the burners on her recently installed six-burner cooktop has a very large, yellow flame that leaves soot all over pots and pans. With further questioning, you confirm that all the other burners have normal-looking blue flames. Without having yet seen the job, what do you think might be the likely cause of this problem?

_____
_____
_____
_____

8 Complete the following characteristics of a good flame:

Ensure correct gas _____

Flame should _____ _____

Flame should be entirely _____

Flame _____ should be clearly _____

No _____ tips

No _____ _____

No _____

LPG BASICS

Learning objectives

In this chapter, you will be introduced to some fundamental characteristics of liquid petroleum gases. It is important to have a strong understanding of such characteristics prior to dealing with the more specific requirements of LPG installations covered in Part B of this text.

Areas addressed in this chapter include:
- the effects of pressure and temperature on LPG
- storing and handling LPG
- boiling liquid expanding vapour explosions (BLEVEs)
- use of butane.

Before commencing Part B of this text you should familiarise yourself with the particular basic characteristics of LPG. Knowledge of the key differences between LPG and NG is important for all of the units of competency found within Part B, including the following specific installation units:
- Chapter 16 Install LPG storage of aggregate storage capacity up to 500 litres
- Chapter 17 Install LPG systems in caravans, mobile homes and mobile workplaces
- Chapter 18 Install LPG systems in marine craft
- Chapter 20 Install LPG storage of aggregate storage capacity exceeding 500 litres and less than 8 kL.

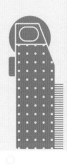

Overview

Liquefied petroleum gas (LPG) is a generic abbreviation used to describe varying mixtures of hydrocarbons. However, the gases propane and butane are of particular relevance to you as a gas fitter. LPG is used around Australia in a wide range of applications, including domestic heating and cooking, industrial processes and manufacturing, auto gas and home barbecues. A Type A gas fitter needs to have an intimate knowledge of the particular characteristics that make LPG a very different substance to natural gas.

In Australia, the principal constituent of commercial LPG is the hydrocarbon propane. Small amounts of butane may also be mixed with the propane in some more consistently warm areas of the country, as discussed later in the chapter.

In previous chapters, you were introduced briefly to the sources and distribution of LPG. This chapter will now assist you to develop a more in-depth knowledge of the properties of LPG and how this relates to your duties as a gas fitter. References to LPG in this section will generally relate specifically to propane (see Table 7.1), but where important differences appear, the properties of butane will also be covered.

TABLE 7.1 Propane key facts (note that some figures are subject to test conditions and temperatures)

Boiling point	−42°C
MJ per kg	49.6
Litres per kg	1.96
Relative density	1.55
Heat value (commercial) MJ	96
Flammability range	2–10 per cent gas in air

Storing LPG

Why does LPG have to be stored in pressurised vessels?

When stored in a vessel under pressure, propane gas actually becomes a liquid. In doing so, LPG becomes a highly concentrated form of energy at normal ambient temperatures. When you turn on a gas appliance, the liquid in the vessel will actually begin to boil and turn into vapour (gas).

How and why this happens requires you to have an understanding of the following LPG characteristics:
- its boiling point
- how it responds to changes in temperature and pressure.

Boiling point

The atmospheric boiling point of any liquid is normally defined as the temperature at which it changes from a liquid to a gas at standard atmospheric pressure (100 kPa). Different liquids boil at different temperatures (see Figure 7.1), but the actual process of boiling remains the same. Propane will boil in much the same way as water, but at the same pressure it will do so at a *different temperature*.

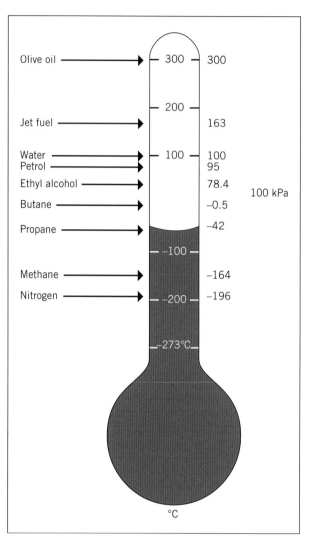

FIGURE 7.1 A graph comparing the boiling points of various common liquids at sea level or approximately 100 kPa (not to scale)

At standard atmospheric pressure (100 kPa):
- the boiling point of water is 100°C
- the boiling point of propane is −42°C
- the boiling point of butane is −0.5°C.

It is common for most people to simply regard heat as being something warm or hot to touch. In fact, heat is a relative term and when you consider that the coldest theoretical temperature is actually −273.15°C, the temperature at which propane boils is comparatively warm!

The boiling point of a liquid is determined not only by its molecular structure but also by atmospheric pressure. Liquid boiling points are generally tested at

1 atmosphere (atm) or 100 kPa. This point will change according to the pressure applied to the surface of the liquid:

- At a pressure lower than 1 atm (100 kPa), a liquid will boil at a lower temperature than normal.
- At a pressure equal to 1 atm, a liquid will boil at its normal boiling point.
- At a pressure of greater than 1 atm, a liquid will boil at a temperature higher than its normal boiling point.

For example, the boiling point of water at sea level, or approximately 100 kPa, is 100°C. If you took a leisurely climb to the top of Mount Everest and put on the kettle, water would boil at 69°C because the atmospheric pressure at that altitude is only around 26 kPa.

The boiling point of water

To understand this better, it is helpful to first grasp what happens when water exposed to standard atmospheric pressure is heated from ambient temperature to a point at which it starts to boil. Follow the text boxes describing the process in Figure 7.2. As more heat is applied to this container of water, the kinetic energy of the water molecules increases until they begin to break free of the surface in the form of vapour or steam.

The boiling point of propane

A village called Oymyakon in the Sakha Republic of Russia is reputed to be the coldest inhabited place on Earth, once recording a winter temperature of −71.2°C. At such a temperature you would be able to walk outside and pour liquid propane into a cup and it would just sit there as a liquid, looking much the same as a cup of water. You would be able to do this because the ambient temperature is well below the boiling point of propane (−42°C). For this reason, any gas appliances in the area would not work because it is too cold for the liquid to vaporise into gas within the cylinder.

However, if the village of Oymyakon was subjected to a sudden heat wave and the temperature rose to a balmy −42°C, the propane in the cup would start to boil, vaporising into gas. At such high temperatures, the locals would at last be able to roll out their gas barbecues and enjoy an afternoon in the sun (however, anyone with a butane cooker would have to wait until the temperature reached at least −0.5°C!).

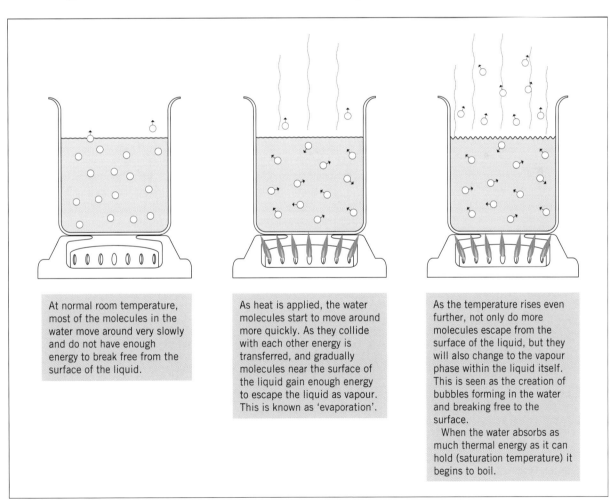

FIGURE 7.2 The process of heating water to boiling point

This somewhat simplistic example seeks only to emphasise the importance of ambient temperature in the use of propane. In the next section, you will see how closely pressure and temperature are interrelated when propane is stored in a pressurised vessel.

Temperature and pressure

Propane is stored, transported and used directly from pressurised vessels known as cylinders. Larger vessels are generally called tanks. When kept under pressure in such vessels, propane exists as both a liquid and a gas. Read the text boxes in Figure 7.3 to see how this happens. Liquid propane will continue to boil at any temperature above −42°C until the pressure of both vapour and pressure in the cylinder equalise.

So that the pressure of liquid and vapour in the cylinder can equalise, a vapour space must always exist. For this reason, LPG cylinders are designed to be filled to no more than approximately 83 per cent of their potential volume. The remaining 17 per cent of cylinder volume is kept as a vapour space and as room for liquid expansion as ambient temperature rises.

What happens if a cylinder is overfilled?

Cylinders must never be filled above the 83 per cent level. Doing so will prevent the vapour and liquid pressures from equalising. As the pressure in the vapour space continues to rise, the relief valve is likely to open and vent the excessive pressure to the atmosphere. In the past, this situation has proven particularly tragic when overfilled barbecue cylinders have vented inside vehicles or houses. Lives have been lost from the resulting explosions and fires.

Cylinder pressure

The temperature surrounding the cylinder will determine how quickly the liquid will boil. While propane will start to boil at −42°C, the actual rate of vaporisation will be very sluggish. As the temperature rises, so too does the vaporisation rate and the amount of pressure required above the liquid to retard boiling.

The actual internal cylinder pressure is entirely determined by the surrounding temperature and not the amount of liquid it holds. To illustrate this fact, consider the diagram in Figure 7.4. These two cylinders are holding different volumes of gas but, because they are exposed to the same ambient temperature, internal cylinder pressures are equal.

Storage and handling of LPG cylinders

The design requirements of cylinders and the physical characteristics of LPG dictate that you must store and handle cylinders carefully. As a gas fitter, you may at times transport supplier cylinders (particularly in country areas), carry small test cylinders on your vehicle and provide advice to customers.

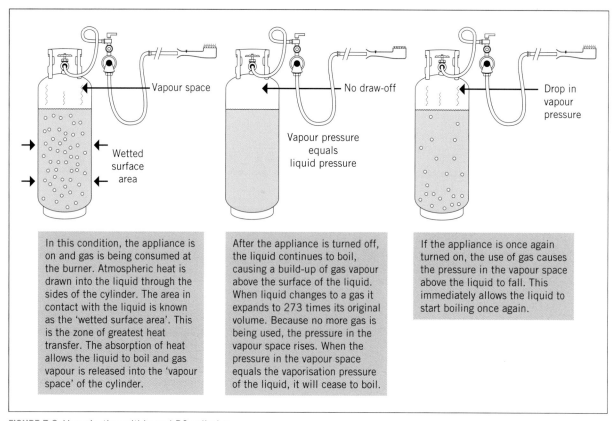

FIGURE 7.3 Vaporisation within an LPG cylinder

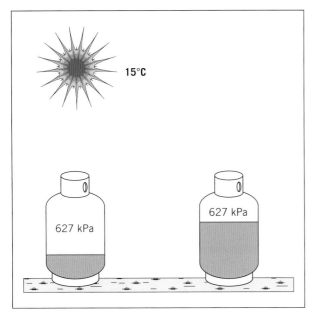

FIGURE 7.4 The cylinder pressure of different LPG volumes exposed to the same ambient temperature

The following information will help you to transport and store LPG safely.

Transport

Important information about transporting LPG cylinders is outlined in Table 7.2.

 Many explosions, injuries and deaths have occurred as a consequence of keeping gas cylinders inside vans and cars. Avoid keeping gas cylinders of any size or type inside enclosed vehicles.

Storage

Important information about storing LPG cylinders is listed in Table 7.3.

LEARNING TASK 7.1 STORING LPG

1. How many litres is contained in 1 kg of liquid LPG?
2. What is the heat value of LPG?
3. To what level is an LPG cylinder filled?
4. In what position should LPG cylinders be stored and transported?

Boiling liquid expanding vapour explosion (BLEVE)

A boiling liquid expanding vapour explosion (BLEVE – pronounced *blev-ee*) results when a vessel containing LPG ruptures because of fire, corrosion or mechanical impact. You will gain a better appreciation of how a BLEVE can unfold by reviewing what would happen if

TABLE 7.2 Transporting LPG cylinders

Consideration	Details
Transport cylinders in open vehicles.	• In an Australian summer, internal vehicle temperatures may rise to 70°C when the vehicle is parked in the sun. • If an LPG cylinder were to vent inside the car, exposure to any form of ignition source, such as electronic switches and door locking systems, could cause an explosion.
Transport cylinders in an upright position.	• Never store cylinders on their side! • The cylinder valve and relief vent are installed and designed to operate from the vapour space of the cylinder. If a cylinder is stored on its side, liquid propane may be released from the valve. • Once liquid propane is exposed to atmospheric pressure, it will immediately vaporise and expand to 273 times its original volume.
Transport cylinders securely.	• Ensure that the cylinder is securely held in place during transport so that it will not fall into or from the vehicle.
Handle with care.	• Even a small 13.5-kg cylinder is heavy when full. Handle cylinders carefully so that you do not injure yourself or the cylinder. • Do not drop cylinders off the side of your ute! Lower them down carefully onto an old rubber tyre or handle them with some mechanical lifting equipment.

TABLE 7.3 Storing LPG cylinders

Consideration	Details
Store cylinders outside.	If a cylinder relief valve were to vent gas inside a building, it could potentially cause an explosion.
Store cylinders upright.	For the same reasons as for transporting cylinders, storing a cylinder in the upright position will prevent the release of liquid propane.
Store with dust caps and plugs inserted.	Ensure the relief valve is fitted with a plastic dust cap and the valve inlet is suitably plugged to prevent insects and dust from blocking the valve.

a cylinder were exposed to an intense fire in the example provided in Figures 7.5 and 7.6.

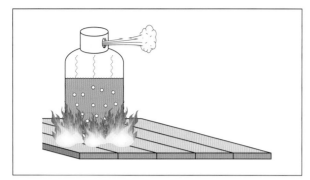

FIGURE 7.5 A cylinder engulfed in flames

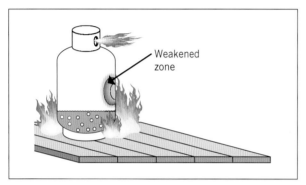

FIGURE 7.6 Cylinder wall weakens

With the cylinder engulfed in flames (Figure 7.5), the pressure within the cylinder begins to rise dramatically as the liquid propane absorbs the heat and boils. As the liquid boils and turns to vapour, the pressure relief valve operates as designed and begins to vent gas vapour in an effort to reduce cylinder pressure.

The vented gas normally ignites as it escapes the relief valve (Figure 7.6) but if the gas fitter has located the cylinder so that the valve points away from the building and other cylinders, this situation is controllable.

While the flames impinge upon the wetted surface area of the cylinder, the liquid acts to absorb this heat, and so the structural integrity of the vessel is maintained. However, as the liquid boils away, more of the vapour space within the cylinder is exposed to the flames. Without the heat-absorbing effect of the liquid, the walls of the cylinder may weaken. It has been reported that just prior to a BLEVE the sound of escaping gas becomes a higher pitched roar, similar to that of a passing freight train.

If the heat of the flames continues to surround the cylinder, the steel may suddenly rupture.

Almost instantly, the remaining liquid is subjected to atmospheric pressure and, in turning to vapour, expands to 273 times its original volume, mixes with the surrounding air and ignites. The blast wave and heat of the explosion is very destructive. Large BLEVE events have recorded destructive blast zones measuring many hundreds of metres in diameter. The cylinder itself may also become an airborne projectile.

For example, the 9-kg cylinder shown in Figure 7.7 was overfilled and subsequently vented into a confined space. During the fire that followed, a BLEVE occurred. You can see that the force of the explosion completely tore off the lower half of the cylinder.

FIGURE 7.7 A 9-kg cylinder ruptured in a BLEVE

A demonstration of the destructive effects of a BLEVE can be viewed online at www.youtube.com > BLEVE (Boiling Liquid Expanding Vapor Explosion) Demonstration – How it Happens Training Video.

A BLEVE can cause a high level of destruction and loss of life. Fortunately, these events are rare. By installing LPG cylinders and tanks correctly, the gas fitter has a significant role to play in ensuring that a BLEVE cannot occur. If handled, stored and installed according to all standards and industry guidelines, LPG cylinder and tank installations are very safe and reliable.

Using butane

With the climatic variations commonly found in Australia, butane use is restricted because of its relatively high boiling point. Butane only begins to boil at −0.5°C; therefore, the ambient temperatures in many regions of Australia would promote only sluggish vaporisation or none at all. You may have noticed how difficult it is to light butane gas lighters or cookers during cold weather.

However, in consistently warmer areas butane is often mixed with propane. The mix may vary, depending whether it is intended for automotive, domestic or commercial applications.

 COMPLETE WORKSHEET 1

LEARNING TASK 7.2

BOILING LIQUID EXPANDING VAPOUR EXPLOSION (BLEVE)

1. When a cylinder is subjected to excessive pressure, where does the gas vapour vent?
2. How many times its original volume does liquid LPG expand when subjected to atmospheric pressure?

SUMMARY

You should now be able to appreciate that LPG is a very different gas from NG. Key characteristics include:
- It is heavier than air.
- It boils and begins to vaporise at −42°C.
- It is stored as a liquid but turns to a vapour when in boils.
- Cylinders are only ever filled to approximately 83% of vessel capacity.
- If installed correctly LPG is very safe, but where cylinders are not sited according to the Standard or are mishandled, a vessel rupture resulting in a BLEVE may occur.

In Part B of this text you will build on this fundamental information and learn about the specific installation requirements related to LPG pressure control, piping systems and appliances.

GET IT RIGHT

The photo below shows an incorrect practice when transporting an LPG cylinder. Identify the incorrect method and provide reasoning for your answer.

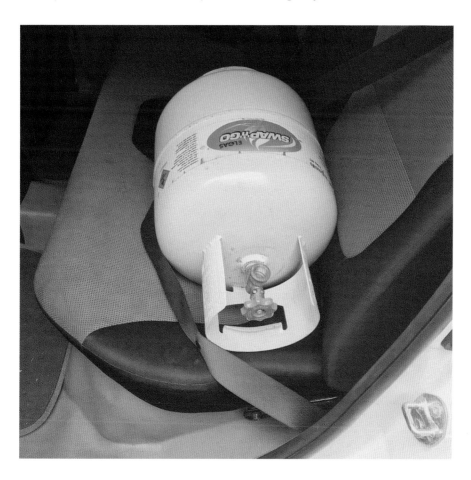

_____
_____
_____
_____

WORKSHEET 1

Student name: _____

Enrolment year: _____

Class code: _____

Competency name/Number: _____

To be completed by teachers
Student competent ☐
Student not yet competent ☐

Task
Review the sections on *Storing LPG* through to *Using butane* and answer the following questions.

1. List three LPG storage considerations.

    i _____

    ii _____

    iii _____

2. By how much does liquid propane expand when it turns into a vapour?

    _____

    _____

    _____

3. Why would butane be unsuitable as a fuel supply for a gas heater in the alpine areas of Victoria?

    _____

    _____

    _____

4. What does the acronym BLEVE stand for?

    _____

    _____

5. List at least three LPG handling considerations.

    i _____

    ii _____

    iii _____

6. What is the heat value of commercial propane?

    _____

    _____

    _____

7   Two LPG cylinders are placed next to each other on a sunny 30°C day. One cylinder is full and the other has only 10 per cent liquid LPG remaining. Which cylinder has the higher internal pressure?

_____

_____

8   Complete the following: Most cylinders can be filled to no greater than _____ per cent of total volume.

9   Why must an LPG cylinder be stored only in the upright position?

_____

_____

10 From the list below, circle the boiling point of liquid propane.

1.55°   −71.2°   −0.5°   −42°   100°

PART B

UNITS OF COMPETENCY

Part B of this text allows you to explore the requirements of the AS/NZS 5601 Gas Installation Standards in greater depth. With interpretation and guidance, you should apply these requirements to your gas installation practice in a methodical and safe manner.

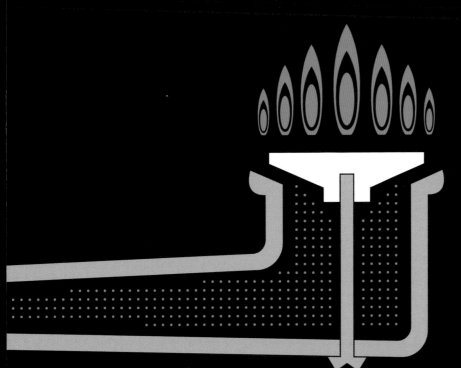

8 SIZE CONSUMER GAS PIPING SYSTEMS

Learning objectives

Areas addressed in this chapter include:
- selecting products using AS/NZS 5601.1
- how to identify a certified or approved product
- how to size single-stage pipe systems
- how to size multistage pipe systems

Before working through this section, it is absolutely important that you complete or revise each of the chapters in Part A 'Gas fundamentals'.

Overview

Gas appliances depend upon the supply of gas at the correct pressure. Pipe sizes that are too small can lead to a significant drop in available gas pressure, causing poor combustion. Incomplete combustion can produce the poisonous gas carbon monoxide, which may injure or kill building occupants. As such, pipe sizing is of fundamental importance to efficient and safe gas appliance operation and must be carried out with due process and accuracy. This chapter will introduce you to the full sizing process in accordance with the Australian Standard.

Note that there is a technical difference between 'pipe' and 'tube', but as is the case in the Australian Standard, these terms will be used interchangeably within this chapter.

Identify job requirements

Prior to undertaking a pipe sizing task you must first know how to choose the correct pipe or tube material for the job. All installation components must be approved or certified for gas use.

Safe gas installations depend upon the correct use of high quality materials, fittings and components. As a gas fitter, it is your responsibility to select the correct items for your installation.

FROM EXPERIENCE

Just because an item may be available for sale from a supplier does not mean that it is suitable for gas fitting work. Develop the habit of checking that every component is fit for purpose.

For example, an isolation valve that is approved for use with water may not necessarily be approved for use with gas. Furthermore, many items that are sold may look very similar to an approved product but may be manufactured to a lower performance standard. Make no assumptions about the materials you use! When it comes to your selection of installation materials, you must remember one important point: *All gas fitting materials, fittings and components must be certified/approved for use.* Every single fitting, component, pipe and appliance you use in gas installations must be approved by a recognised certifying body, or comply with an Australian Standard acceptable to a technical regulator.

REFER TO AS/NZS 5601 'CERTIFYING BODY'

To seek certification, manufacturers must prove that their products meet specified performance guidelines, comply with all associated Australian Standards and satisfy a relevant testing regime. If the application is found to be suitable, the certifying body will list the item as an approved product and issue it with an approval number.

The Australian Gas Association (AGA) and SAI Global are two examples of certifying entities in Australia. Links to these and other bodies can be found via the Gas Technical Regulators Committee (GTRC) website. The GTRC is an association of government departments from across Australia and New Zealand with the primary goal of promoting the safe use of gas. Use the GTRC home page at http://www.gtrc.gov.au to further explore the National Database of Certified Gas Appliances.

Selecting products using AS/NZS 5601

When you are planning your installation, you need to refer to AS/NZS 5601 to select the correct product for your project. Gas Installation Standards provides guidance in the following categories:
- general material conditions
- consumer piping restrictions
- prohibited joints and fittings
- proprietary systems
- pipe selection tables
- flue material selection tables
- component requirements.

REFER TO AS/NZS 5601.1 'MATERIALS AND COMPONENTS'

How to identify a certified or approved product

As it is your responsibility to identify and confirm that you are using approved or certified components, you must form the habit of continually checking each product. There is no single method of identification and if you ever find yourself in doubt, do not hesitate to contact the technical regulator.

The next section will review some of the more common methods of identifying approved or certified materials.

Materials

Check that your copper tube (see Figure 8.1) is manufactured to AS 1432, and is either Type A or B. Each type has a different wall thickness. Some brands may be colour coded:
- Type A – green
- Type B – blue
- Type C – red
- Type D – black.

Proprietary brand piping systems based on cross-linked polyethylene (PEX) or multilayer construction

FIGURE 8.1 Copper tube to AS 1432 Type B

FIGURE 8.3 PEX piping product

FIGURE 8.2 'Nylon' (polyamide) pipe to AS 2944.1

(see Figure 8.3) must also comply with strict certification approvals that you will need to compare with the consumer piping material table in your Standard.

Appliances

Although you are not yet up to the point of appliance installation, as part of your preparation and pipe sizing process you must ensure that the appliances are approved and that you can identify gas input requirements on their data plates. Approved appliances will often display an approval sticker from the certifying body (see Figure 8.4). The approval number can be found on the appliance data plate (see Figure 8.5). Cross check this number with the latest certification approval list. Is the appliance in this photograph still approved for installation?

If in doubt, check it out!

When you are not sure of an appliance approval history, you are able to refer to the certifying body's approval list. This is particularly important when you intend to re-install older second-hand or converted appliances.

FIGURE 8.4 Approval sticker from a certifying body

FIGURE 8.5 Appliance data plate showing its approval number

The *Directory of AGA Certified Products* (see Figure 8.6) is available to all gas fitters as a free internet download. Put together a selection of fittings, materials and appliances then download the current version of the directory and see if you can find each item listed.

In Figure 8.7, you can see how the model number and the AGA approval number of the appliance data plate matches the entry within the *Directory of AGA Certified Products*. However, if the result is unclear or any doubt remains, always contact your technical regulator.

If you are not sure how to interpret data plate information, consult your teacher or supervisor for clarification.

LEARNING TASK 8.1

HOW TO IDENTIFY A CERTIFIED OR APPROVED PRODUCT

Download the latest copy of the Directory of AGA Certified Products. Locate two gas appliances and determine if the appliances are approved by locating the AGA approval number on the appliance and cross referencing with the directory.

Size gas piping systems

As gas travels through a pipe, the friction between the gas and the walls of the pipe will cause a degree of pressure loss. This pressure loss is accumulative and increases as the pipe gets longer. If a pipe is too small, this friction and resulting turbulence may lead to the pressure at the inlet to the appliance being too low for complete combustion and efficient appliance performance. Therefore, pipes must be sized in relation to appliance demand and pipe length, to ensure that the required pressure is available at the inlet to the appliance at all times.

Minimum inlet pressure

Each gas type and appliance has a particular burner pressure at which it is designed to operate. You must ensure each gas appliance is supplied with the minimum required gas pressure so that it can operate as designed. This is known as the minimum inlet pressure and while not directly defined, is still a prescribed minimum within the Gas Installation Standards (Table 8.1).

TABLE 8.1 Minimum inlet pressure

Gas type	Min. pressure at appliance inlet (kPa)
NG and SNG	1.13
LPG	2.75

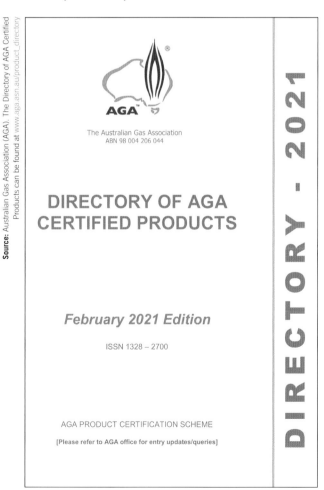

FIGURE 8.6 The *Directory of AGA Certified Products*

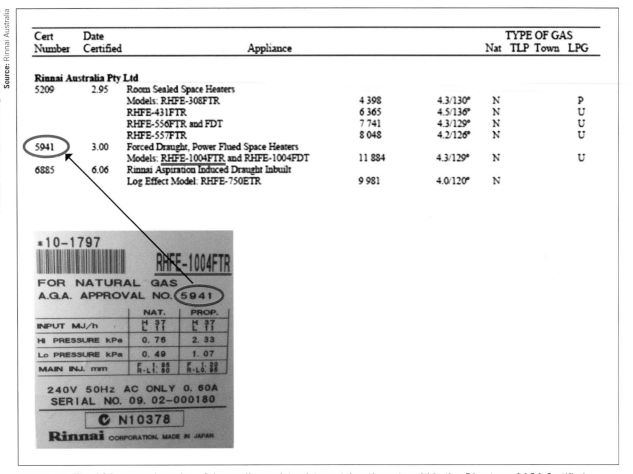

FIGURE 8.7 The AGA approval number of the appliance data plate matches the entry within the *Directory of AGA Certified Products*

REFER TO AS/NZS 5601.1 SECTION 5 'MINIMUM PRESSURE AT APPLIANCE INLET'

Importantly, these pressures must be available at each appliance, with all appliances connected to consumer piping and running at maximum gas consumption. Therefore the overall purpose of pipe sizing is to ensure that regardless of system size the required gas pressure is always available at the inlet to each appliance.

What resources do I need?
Pipe sizing is a simple process and does not require a great deal of sophisticated equipment or resources. The following items will serve as a good indication of what is required:
- AS/NZS 5601.1 (Gas Installations) – you must ensure you are working from the most up-to-date version; this Standard changes regularly
- approved plans – if sizing off the plan, try to work off the latest approved drawings to avoid future unnecessary changes
- a scale rule
- a calculator
- writing materials.

If measuring up on-site, you may also require items such as a measuring tape, ladder and torch.

Methodology
AS/NZS 5601.1:2013 recommends two methods of determining pipe size:
1. flow graphs (F3 to F14) – as often used in New Zealand
2. tables (F6 to F42) – primary method as used in Australia for many years.

You may use either method depending on your preference; however, only the table system will be referred to in this text.

Information required before you start pipe sizing
Before sizing any job, you need to gather certain information. It is important to note at this point that you should make a habit of neatly recording all information for the job. There are two key reasons for this:
- By collating your data in a methodical and neat manner you will lessen the chance of mistakes. Always remember that mistakes are at the least costly and at worst hazardous.

- It may be some weeks or months before you start the job and neat, well-filed records will save you many hours of frustrating repetition.

Using a standard sizing table suitable for most pipe sizing jobs is one way of keeping good records. The example in Figure 8.8 is easily drafted on a computer and can be printed off as and when required. All you need to do is fill in the gaps. Obviously, larger or more complex jobs may require different-sized paper, but the principles of a neat and orderly approach remain the same.

By completing each section within this worksheet, you will have collated all the information necessary to commence pipe sizing. A brief explanation is provided below:

- *Gas type.* As you have already learnt, different gases have different properties including heat value and relative density. Therefore, the type of gas being used has a large influence on which table you consult in AS/NZS 5601. You must choose the table that matches the gas that is being made available to the premises.

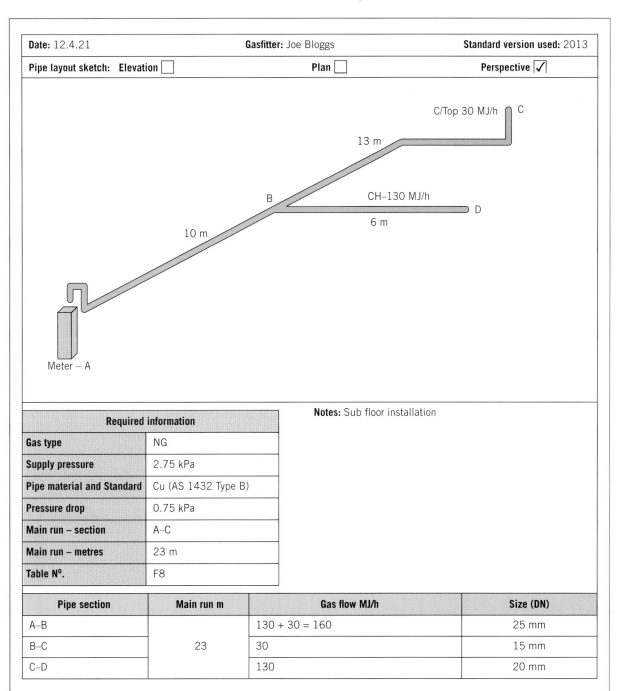

FIGURE 8.8 Example of a pipe sizing worksheet

- *Gas input rating.* From the data plate of each appliance, determine the maximum gas input rating in MJ/h. This information has a direct relationship to the size of each pipe branch connecting each appliance.
- *Supply pressure.* You must confirm the supply pressure that the supplier has or intends to provide to the premises. Knowing this pressure will enable you to determine which sizing table and allowable pressure drop to check in AS/NZS 5601.1.
- *Pipe material.* The choice of pipe material may be determined by personal choice, cost, plan specifications or installation suitability.
- *Pressure drop.* The pressure drop is defined as the maximum allowable drop in pressure between the supply point and the inlet to the appliance. For any given supply pressure, AS/NZS 5601.1 Appendix F quotes a maximum pressure in kPa that may be permitted in the use of a particular table. Before the advent of high-pressure reticulation mains, gas fitters were simply told which pressure drop to use by the gas utility or the technical regulator. However, the sizing tables in AS/NZS 5601.1 Appendix F now allow the gas fitter greater flexibility by more closely matching supply pressures to pressure drop (check with your local technical regulator or distributor first). You must still ensure that the maximum allowable pressure drop is never exceeded. The pipe size tables are designed around certain installation assumptions but do not account for complex pipework with more changes of direction than normal, extra-long pipe runs or special appliance characteristics.
- *Main run.* The sizing calculation revolves around the determination of the main run in any given installation. The main run is simply defined as that section of pipe from the supply point to the *furthest appliance*. It does not matter what size that appliance is. This main run length is used throughout the entire calculation.

Pipe sizing procedure

To assist you in your first pipe sizing exercise, follow the steps below with reference to the natural gas installation in AS/NZS 5601.1 Appendix F. For simplicity, each section is separated from the main sheet and the drawing has been changed to a basic plan view.

Step 1 – Design your layout

Whether you design your pipe layout in elevation, plan or perspective view does not really matter as long as your co-workers and your employer can clearly understand how the installation has been sized and that all pipe sections can be seen. In this example, the perspective view from Figure 8.8 has been changed to a plan view (see Figure 8.9).

1. Note the supply point.
2. Identify each appliance.
3. Record the MJ/h load of each appliance.

Step 2 – Label each section

A 'section' is a length of pipe that carries a certain amount of gas (see Figure 8.10). Once the gas load changes, it becomes a new section of pipe. For example, you can see that the section B–D is carrying 130 MJ/h of gas. Section A–B, however, is carrying the combined load of 130 + 30 = 160 MJ/h.

Starting at the supply point, label each section with capital letters along the main line to the furthest appliance then work back towards the supply point, labelling each branch as you go.

Step 3 – Measure each section length

When measuring your installation, the important thing to remember is that you measure each section from one change of load to the next (see Figure 8.11). Do not measure individual changes of direction around elbows. For example, you can see that Section B–C has a full length of 13 m. The elbow is irrelevant.

Clearly record the length of each section.

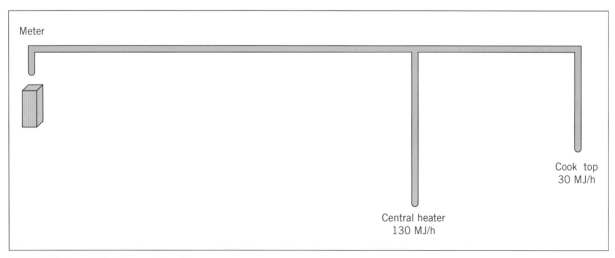

FIGURE 8.9 Step 1: Design your layout

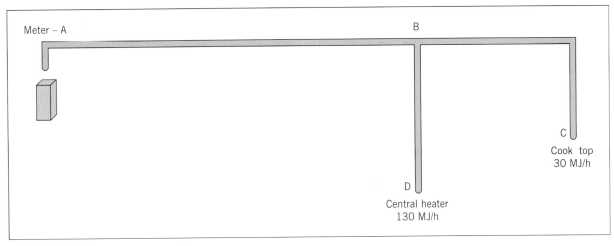

FIGURE 8.10 Step 2: Label each section

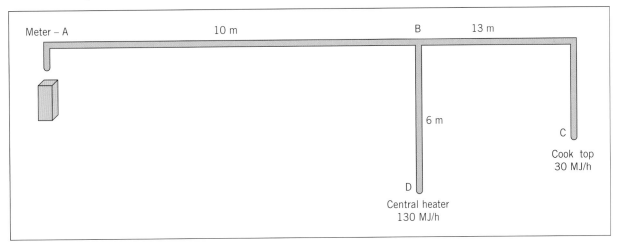

FIGURE 8.11 Step 3: Measure the length of each section

Step 4 – Determine the main run

AS/NZS 5601.1 defines the main run as that section of pipe from the supply point to the *furthest appliance*.

In the example in Figure 8.12 you can see that from the meter (A) to the central heater (D) is 16 m. From the meter (A) to the cooktop (C) is 23 m.

Therefore, the main run for this installation is A–C with a length of 23 m.

Step 5 – Select the pipe material

Select the pipe material that you prefer to use or as stipulated in the approved plans (see Table 8.2).

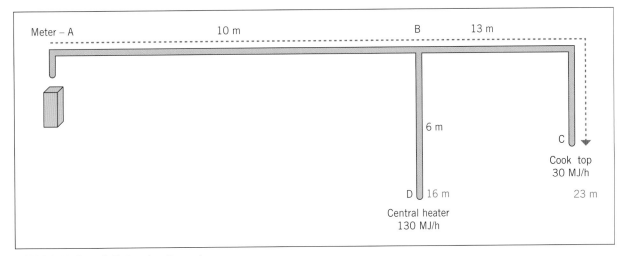

FIGURE 8.12 Step 4: Determine the main run

TABLE 8.2 Step 5: Select the pipe material

Required information	
Gas type	NG
Supply pressure	2.75 kPa
Pipe material and Standard	Cu (AS 1432 Type B)
Pressure drop	0.75 kPa
Main run – section	A–C
Main run – metres	23 m
Table number	F8

Regardless, always ensure the material selected complies with the requirements of AS/NZS 5601.1.

In this example, the gas fitter has selected copper for the job. Steel or composite pipe would also be compliant with AS/NZS 5601.1.

Step 6 – Select the pipe size table

At this point, you will need to refer to AS/NZS 5601.1 Appendix F on the procedures related to choosing a pipe sizing table (see Table 8.3). Having selected the material, to choose the appropriate table you require the following additional information:
- the gas type
- the supply pressure.

TABLE 8.3 Step 6(a): Confirm gas type and supply pressure

Required information	
Gas type	NG
Supply pressure	2.75 kPa
Pipe material and Standard	Cu (AS 1432 Type B)
Pressure drop	0.75 kPa
Main run – section	A–C
Main run – metres	23 m
Table number	F8

Ensure that when you choose the appropriate table you also double check that it is in reference to Australian Standards and OD (outside diameter) measurements. Some tables refer to NZ Standards and an ID (inside diameter) measurement.

Refer to the top of each table and you will find that Table F8 in AS/NZS 5601.1 (see Table 8.4) fulfils these requirements in our example. In this instance, a maximum allowable pressure drop of 0.75 kPa is stipulated. This means that at no time should the gas pressure fall any more than this amount between the supply point and the furthest appliance. Good installation practice that reduces sharp changes in direction and compliance with this sizing table should ensure that this is the case.

Note that although the modern Gas Installation Standards enables the gas fitter to choose a sizing solution based upon supply pressure, some technical regulators and/or gas distributors may still stipulate a

TABLE 8.4 Step 6(b): Select the pipe size table (reference Table F8 in AS/NZS 5601.1)

Required information	
Gas type	NG
Supply pressure	2.75 kPa
Pipe material and Standard	Cu (AS 1432 Type B)
Pressure drop	0.75 kPa
Main run – section	A–C
Main run – metres	23 m
Table number	F8

Pressure drop of 0.75 kPa
Natural gas flow through *copper pipe* (MJ/h)
(This table is suitable for supply pressures within the range 2.75–5 kPa)

specific and mandatory pressure drop requirement in certain areas. Check your local area requirements first!

Step 7 – Determine section gas flow

Now that all the necessary information required to choose the correct table has been collected, it is time to determine the gas flow requirements of each part of the installation.

In relation particularly to individually metered, single-occupancy residences, the following basic requirement should be applied to gas pipe sizing: The minimum *pressure* shall be available at each *appliance* with all *appliances* connected to the *consumer piping* operating at maximum gas consumption.

If you sized your system on the assumption that just one appliance might be working at any one time, it is highly likely that during simultaneous appliance use there would be an unacceptable gas pressure drop. Insufficient pressure to each appliance will lead to incomplete combustion and hazardous conditions. Unless permitted to do so by your technical regulator for particular circumstances, do not size for any diversity in consumption patterns. Always size pipe as if the whole system is designed to run simultaneously, all the time.

Complete your gas flow table as per the example in Table 8.5. Refer back to the design layout to confirm the load requirement of each section.

TABLE 8.5 Example of a gas flow table

Pipe section	Main run (m)	Gas flow (MJ/h)	Size (DN)*
A–B	23	130 + 30 = 160	
B–C		30	
C–D		130	

*DN (diamètre nominal or nominal diameter) refers to the nominal or 'trade' designation of a pipe or tube and not to its actual internal or external diameter.

Source: SAI Global

Step 8 – Find the main run column

Remember the trouble you went to in determining the main run? Well here is where you use that length. Referring to Table F8 once again, identify the row of figures.

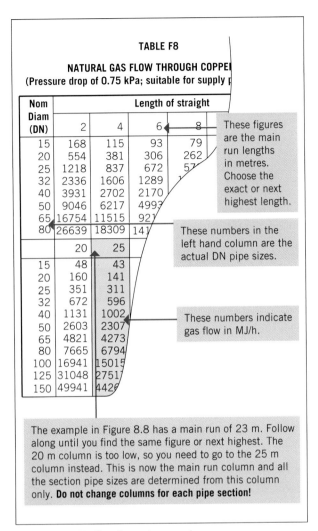

FIGURE 8.13 Finding the main run column

Step 9 – Size each pipe section (see Figure 8.14)
What are the shaded areas in the tables?

You will notice within each of the pipe sizing tables that certain values are lightly shaded. Pipe sizing solutions that fall within these shaded areas are likely to allow excessive gas velocities that may lead to flow noise and possible erosion of pipe and fittings.

REFER TO AS/NZS 5601.1 APPENDIX F 'EXCESSIVE GAS VELOCITY'

If a solution is likely to result in excessive gas velocity, you must go to the next unshaded size up that will carry the gas load. If this is not possible, or you require clarification on any individual situation, you must consult with your technical regulator before proceeding.

COMPLETE WORKSHEET 1

FIGURE 8.14 Sizing each section

Applying higher pressure drop

The Standard permits limited application of a higher pressure drop within a given supply pressure under certain strict circumstances. Where a particular installation is perhaps characterised by a simple straight run or has limited use of fittings and is likely to allow a pipe size reduction while still satisfying minimum inlet pressures, then application of a higher pressure drop can be considered.

REFER TO AS/NZS 5601.1 APPENDIX F 'ALLOWABLE PRESSURE DROP'

The process can be separated into two main steps:
1 Confirm if pressure drop can be increased.
2 Apply an increase to pipe size MJ carrying capacity within the main run column.

Confirm if pressure drop can be increased

The Standard states that the upper limit of an increase to allowable pressure drop is 45 per cent. Raising the pressure drop by this amount would result in a 20 per cent increase in gas flow over a given pipe section. These are the only parameters that you are permitted to apply.

You must first confirm that the supply pressure will permit such an increase.

- Table F8 has a NG supply pressure range of 2.75 to 5 kPa.
  - The minimum inlet pressure of a given NG appliance is 1.13 kPa.
  - Standard allowable pressure drop is 0.75 kPa.
- Increase the allowable pressure drop by 45 per cent:

  1.45 × 0.75 kPa = 1.09 kPa

  This is the new proposed allowable pressure drop.
- Assuming a meter outlet supply pressure of 2.75 kPa, you must ensure that a pressure greater than the minimum 1.13 kPa is still available at the inlet to the appliance:

  2.75 kPa − 1.09 kPa = 1.66 kPa
- As can be seen, 1.66 kPa is available at the inlet to the appliance and therefore for this supply pressure you would be able to proceed with the use of a higher pressure drop.

You must always apply this check first to ensure that the minimum inlet pressure is always available. For example, you cannot apply this process to a supply pressure of 1.25 kPa as the adjusted pressure drop would be too great.

Apply an increase in MJ carrying capacity to the main run column

Having confirmed that a proposed installation supply pressure can accommodate an increase in allowable pressure drop, you now need to apply this to the sizing tables.

Using the NG example in the previous section for a central heater and cooktop installation with a supply pressure of 2.75 kPa, we need to first apply an increase of 20 per cent to the MJ carrying capacity within the main run column:

DN	Gas flow MJ/h in 25 m main run column	
	Standard carrying capacity MJ/h	Modified carrying capacity MJ/h × 1.2 (20%)
15	43	51
20	141	169
25	311	373

Once you have modified the carrying capacity of each pipe size within the main run column you can now re-work the calculation to determine the pipe size of each section.

Pipe section	Main run (m)	Gas flow (MJ/h)	Standard DN	Modified DN
A–B	23	130 + 30 = 160	25	**20**
B–C		30	15	15
C–D		130	20	20

As can be seen in this example, by applying a permitted 45 per cent increase in allowable pressure drop, the resulting 20 per cent increase in flow allows for section A–B to be reduced in size from DN25 to DN20.

GREEN TIP

Applying the strict guidelines within your Standard in reducing installation pipe sizes not only has the potential to save on cost and installation time but, importantly, prevents unnecessary use of valuable materials.

Sizing proprietary brand piping systems

Unlike standard materials such as copper, steel and polyethylene, some pipe products can only be installed with fittings and components specifically produced by the same manufacturer. Pipe sizing solutions to these proprietary brand products are not provided in the Standard, and you are required to use each product's sizing data when using a particular system.

Be aware that some manufacturers will require installers to undergo some form of training prior to sizing and use. Additionally the Technical Regulator may also require installers to be authorised.

There are a number of proprietary brand products available, each with an individual sizing process. For the purpose of clarity these additional processes have not been included in this text. Your teacher may supply an example of product sizing data applicable to systems commonly used in your region.

LEARNING TASK 8.2 SIZE GAS PIPING SYSTEMS

As a class, and in consultation with you class teacher, undertake a pipe sizing exercise based on a four-bedroom home with two bathrooms, laundry and kitchen. Discuss what gas appliances would be installed based on the conditions and gas availability of your local area.

Draw the design and establish the MJ/h load, pressure drop and main run of the installation. Use two different piping materials that are relevant to your local area and complete the tables below. Once completed compare the difference in sizes between the two materials.

Required information		
Gas type		
Supply pressure		
Pipe material and Standard	Material 1:	Material 2:
Pressure drop		
Main run – section		
Main run – metres		
Table number		

Pipe section	Main run, m	Gas flow, MJ/h	Material 1 Size (DN)	Material 2 Size (DN)

How to size multistage gas supply systems

A complex or multistage supply system is defined in the AS/NZS 5601.1 glossary as one in which pressure regulation is controlled in multiple stages. Where the installation demand profile is high or the delivery distance quite long, a higher-pressure first stage allows for the use of relatively smaller pipe sizes. At the point of use, this higher pressure is then reduced to the pressure required at the inlet to the appliance.

REFER TO AS/NZS 5601.1 'GAS PRESSURE REGULATOR'

The regulator nearest the supply point is known as the first-stage regulator, followed by the second-stage regulator and possibly third and fourth stages, depending upon your requirements.

Both natural gas and LPG can be supplied via staged pressure reduction. Although it is fairly common to find a multistage or 'two stage' LPG system used domestically, natural gas multistage supply systems are normally encountered in commercial and industrial applications.

Sizing a multistage system is easy! There are two key things to remember:
- Check the supply pressure requirements of each stage.
- Size each stage separately from the other.

The actual sizing procedures remain the same, although you will need to break up your documentation into different sections.

 COMPLETE WORKSHEET 2

Clean up the job

The design process required for pipe sizing will rarely require much in the way of material clean-up, but the following points may be of importance:

SIZE CONSUMER GAS PIPING SYSTEMS **135 GS**

- Where you have needed to undertake confirmatory measurements on site, ensure that all equipment such as measuring tapes, laser measuring equipment etc. are returned in serviceable order to the required storage area.
- Whether you are using computer software or paper-based calculations, ensure that files, drawings and calculations are completed in a manner that a third party could interpret, and are correctly stored according to company protocols.

SUMMARY

In this chapter you have come to understand the vital importance of correct pipe sizing for the safe and efficient operation of gas appliances. In doing so you have learned how to:
- select pipes and tubes, ensuring that all items are certified for use in gas installations
- interpret the pipe sizing tables to determine pipe sizes that ensure that the maximum allowable pressure drop is not exceeded and that minimum pressures are available at the inlet to all appliances operating at maximum gas consumption
- use the pipe sizing tables to determine pipe sizes for both single stage and multistage piping systems

Now that you know how to size gas piping, you are able to progress to the installation requirements of gas appliances, as covered in the next chapter.

GET IT RIGHT

The photo below shows an incorrect practice when determining the correct pipe sizing.

Identify the problem and provide reasoning for your answer

_____

_____

_____

_____

_____

WORKSHEET 1

To be completed by teachers
Student competent ☐
Student not yet competent ☐

Student name: _____

Enrolment year: _____

Class code: _____

Competency name/Number: _____

Task

Review the section on *How to size pipes* and answer the following questions.

1 How would you check that the PEX pipe you are intending to use is compliant with the Australian Standard as detailed in the Gas Installation Standards?

_____

_____

_____

2 Where on a gas appliance do you look to confirm that it has been issued with an Australian approval or certification number?

_____

_____

_____

3 You now have the opportunity to apply this pipe size procedure to the sizing tasks on the following pages. In each case, half the sheet has been completed. Fill in the remaining information and come up with a sizing solution.

i   Natural gas – low-pressure commercial

Date:		Gasfitter:		
Pipe layout sketch: Elevation ☐		Plan ✓		Perspective ☐

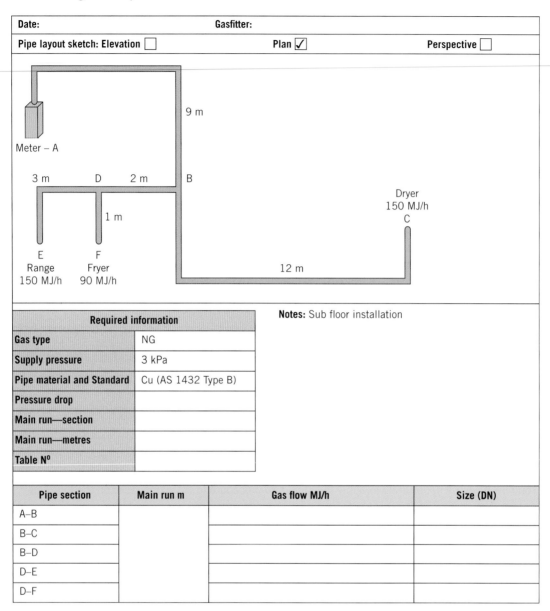

Required information	
Gas type	NG
Supply pressure	3 kPa
Pipe material and Standard	Cu (AS 1432 Type B)
Pressure drop	
Main run—section	
Main run—metres	
Table N°	

Notes: Sub floor installation

Pipe section	Main run m	Gas flow MJ/h	Size (DN)
A–B			
B–C			
B–D			
D–E			
D–F			

ii   LPG – low-pressure domestic

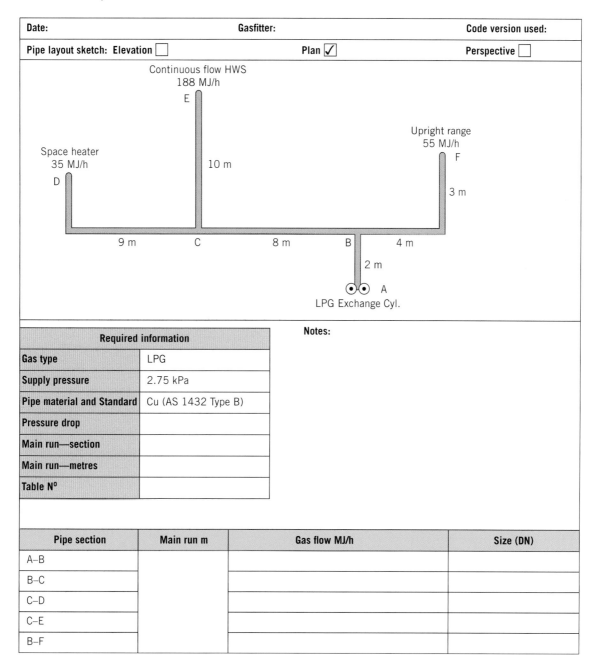

Required information	
Gas type	LPG
Supply pressure	2.75 kPa
Pipe material and Standard	Cu (AS 1432 Type B)
Pressure drop	
Main run—section	
Main run—metres	
Table Nº	

Notes:

Pipe section	Main run m	Gas flow MJ/h	Size (DN)
A–B			
B–C			
C–D			
C–E			
B–F			

WORKSHEET 2

Student name: _____

Enrolment year: _____

Class code: _____

Competency name/Number: _____

To be completed by teachers

Student competent ☐

Student not yet competent ☐

Task

Review the section on *How to size multistage gas supply systems* and answer the following question.

1 You now have the opportunity to size two multistage supply installations. In each case, half the sheet has been completed. Fill in the remaining information and come up with a sizing solution.

   i Natural gas – commercial car wash facility – 1st stage

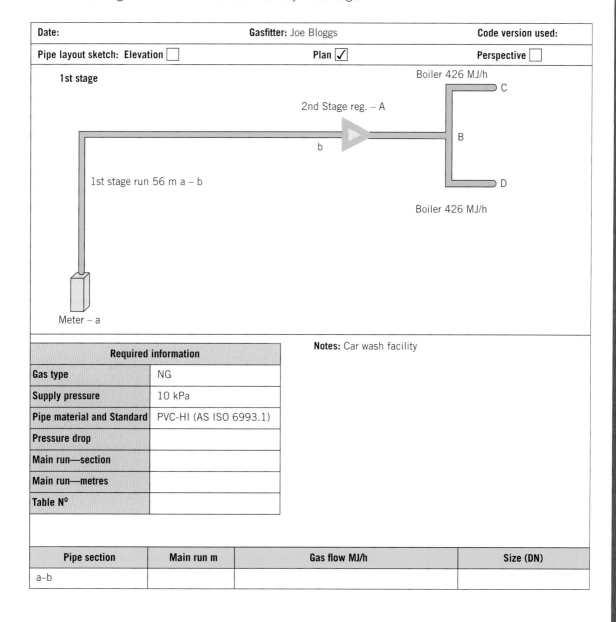

Date:	Gasfitter: Joe Bloggs	Code version used:
Pipe layout sketch: Elevation ☐	Plan ✓	Perspective ☐

1st stage
- 1st stage run 56 m a – b
- Meter – a
- 2nd Stage reg. – A
- b
- Boiler 426 MJ/h → C
- B
- D
- Boiler 426 MJ/h

Required information	
Gas type	NG
Supply pressure	10 kPa
Pipe material and Standard	PVC-HI (AS ISO 6993.1)
Pressure drop	
Main run—section	
Main run—metres	
Table N°	

Notes: Car wash facility

Pipe section	Main run m	Gas flow MJ/h	Size (DN)
a–b			

SIZE CONSUMER GAS PIPING SYSTEMS 143 GS

ii   Natural gas – commercial car wash facility – 2nd stage

Date:	Gasfitter: Joe Bloggs	Code version used:
Pipe layout sketch: Elevation ☐	Plan ☑	Perspective ☐

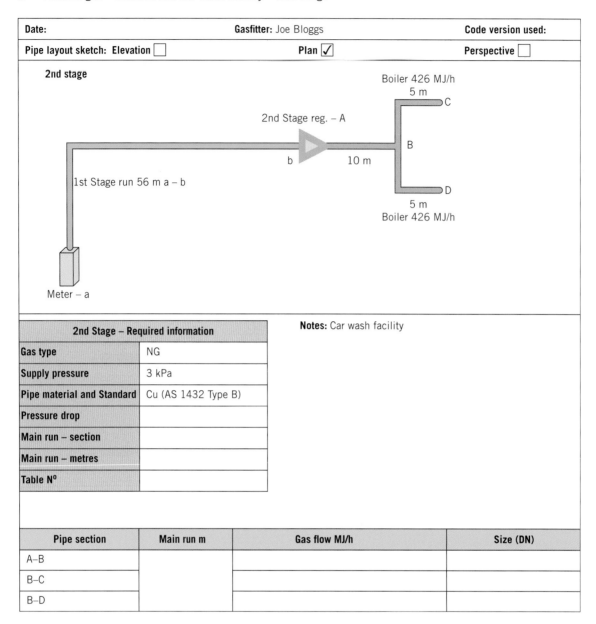

2nd Stage – Required information	
Gas type	NG
Supply pressure	3 kPa
Pipe material and Standard	Cu (AS 1432 Type B)
Pressure drop	
Main run – section	
Main run – metres	
Table Nº	

Notes: Car wash facility

Pipe section	Main run m	Gas flow MJ/h	Size (DN)
A–B			
B–C			
B–D			

iii LPG – football change room showers hot water service

Date:	Gasfitter:	Code version used:

Pipe layout sketch: Elevation ☐   Plan ✓   Perspective ☐

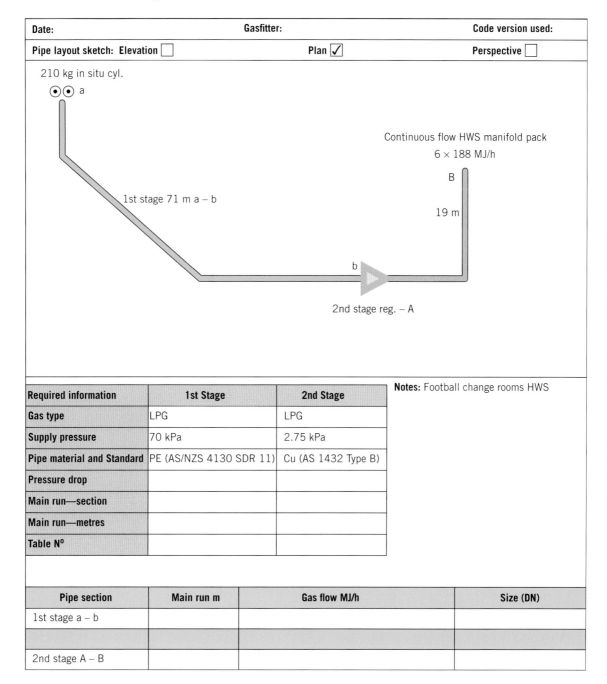

Required information	1st Stage	2nd Stage
Gas type	LPG	LPG
Supply pressure	70 kPa	2.75 kPa
Pipe material and Standard	PE (AS/NZS 4130 SDR 11)	Cu (AS 1432 Type B)
Pressure drop		
Main run—section		
Main run—metres		
Table N°		

Notes: Football change rooms HWS

Pipe section	Main run m	Gas flow MJ/h	Size (DN)
1st stage a – b			
2nd stage A – B			

INSTALL GAS PIPING SYSTEMS

Learning objectives

Areas addressed in this chapter include:
- general requirements for installing pipes
- piping in concealed locations
- requirements for installing underground pipes
- piping support
- quick-connect devices
- how to pressure test a gas installation
- purging (<30 L).

Before working through this section, it is absolutely important that you complete or revise each of the chapters in Part A 'Gas fundamentals', with particular emphasis on the electrical safety section within Chapter 5 'Gas industry workplace safety', as well as the pipe sizing processes within Chapter 8 'Size consumer gas piping systems'.

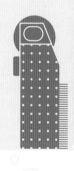

Overview

For most Type A gas fitters, installing consumer piping is a large part of the job. The safe and efficient functioning of a gas installation depends upon a well-installed, correctly sized consumer piping system.

In this section, you will be introduced to some of the key requirements for pipe and component installation, as stipulated within AS/NZS 5601. This text does not cover every individual clause, but it will assist you in becoming familiar with some of the main categories covering pipe installation specifics. References to the Australian Standard are included in each subsection, enabling you to cross-reference as you progress. As you do so, read more than the single reference and, before you know it, you will have covered most of the AS/NZS 5601 requirements.

However, you must understand that it is impossible to learn about pipe installation entirely from a textbook, and you will need to work closely with your supervisor and teacher in order to blend basic theoretical knowledge with practical installation experience.

Identify gas piping system requirements

Prior to commencing any pipe installation task it is vital that you correctly identify the particular needs related to the job. A successful outcome is dependent upon applying a methodical approach in reference to Australian Standards, industry and site requirements.

Obtain building plans and specifications

All new installations are planned for and carried out in reference to approved plans and specifications. Your first task when asked to undertake new work will be to examine the installation plans in detail. Plans will be made available by the builder, owner or your employer. The following points apply:
- Always ensure the plans and specifications you are using are the latest approved versions.
  - It is normally a legal requirement that a master set of plans and specifications be held on site at all times, enabling you to confirm the currency of your copies as required.
  - On smaller jobs the specifications are often included in the construction plans, but for larger projects are normally in a separate document.
- Identify any additional directions or instructions that may apply in addition to the construction plans. These may be issued from:
  - local council
  - water authorities
  - gas technical regulator
  - gas distributor
  - engineers
  - environmental protection authority.

Work on existing properties may or may not include plans, depending upon the scale of work. Where there are no plans, you and/or your employer will need to undertake a detailed site inspection as part of the quotation and subsequent job commencement process. In doing so, you must make detailed notes and measurements to ensure all requirements are considered. This documentation must be completed in a clear and methodical manner, so that any employee within your business is able to interpret and apply.

Work, health and safety considerations

Working as a gas fitter means that safety must be your primary concern in everything you do. Your licence should be seen as a symbol of trust from the community that you will carry out all your duties with both their and your own safety in mind. Of course, although there are specific gas hazards that you need to be aware of, you also need to take a broader view of hazard identification.

As soon as you walk on to a job site or evaluate a set of plans, you should be mindful of potential hazards of all kinds. As a bare minimum you must have on hand basic personal protective equipment (PPE) necessary for most jobs. This includes:
- safety boots
- safety glasses
- hearing protection
- dust masks and/or respirators
- gloves
- overalls or relevant protective clothing
- hard hats.

A gas fitter will install piping systems for new construction projects but will also alter, maintain and extend existing gas pipe installations. In addition to the normal work, health and safety considerations that cover all jobs, work on existing installations exposes you to significant electrical hazards that you must be aware of.

Electrical hazards

Electrical faults are an ever-present hazard for gas fitters and it is vital that you apply appropriate risk assessment processes in all your work. Ensure that you revise the electrical safety section within Chapter 5 'Gas industry workplace safety' before proceeding further.

Type of pipe materials

When preparing for a gas installation, one of your primary tasks is to confirm the type of piping materials to be used. On many jobs this may be left to the discretion of the gas fitter, but always ensure you read through any plans and specifications before assuming this is the case. Often a particular pipe material is pre-specified for certain commercial and industrial installations. Common pipe materials include:
- copper
- PEX – cross-linked polyethylene

- PE – polyethylene (Figure 9.1)
- PVC-HI – high impact PVC
- multilayer.

FIGURE 9.1 Certified polyethylene gas pipe complying with AS/NZS 4130

The materials selected must not only be an economical choice but, importantly, be 'fit for purpose' on the basis of the installation characteristics. For example, the installation of black and galvanised steel pipe is much slower than for other systems, but may be pre-specified for certain projects where high impact resistance or pipe rigidity is required. The full list of approved pipe systems, jointing requirements and limitations is detailed in your Standard.

REFER TO AS/NZS 5601.1 'CONSUMER PIPING MATERIALS'

Importantly, all pipe materials used for gas fitting purposes must be 'certified' for use. This means that the product has met the performance requirements necessary for gas. The Australian Gas Association (AGA) and SAI Global are examples of two entities that provide certification services. Lists of certified products can be downloaded or can be searched for from the Gas Technical Regulators Committee (GTRC) website: www.gtrc.gov.au.

Fittings and components

All components must meet the requirements of the certifying body or technical regulator. Check every valve or fitting you purchase for compliance. If no details are written on the component do not use it until you can confirm its certification history. Be aware that items that have previously been approved sometimes lose their certification. Do not assume that items are permanently listed.

AS/NZS 5601.1 states that all copper alloy flared compression fittings (see Figures 9.3 and 9.4) must meet the requirements of AS 3688. Approved fittings will have this stamped somewhere on the actual fitting

FIGURE 9.2 The *Directory of AGA Certified Products*

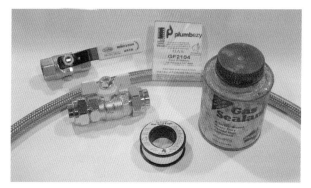

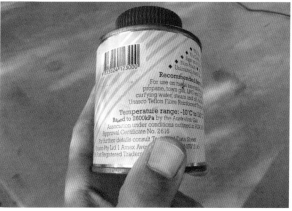

FIGURE 9.3 Always check for current product certification

FIGURE 9.4 Certified copper alloy flared compression fittings

itself. Most will also indicate that the fitting is dezincification resistant with the abbreviation 'DR'. Be aware that in recent years, some imported fittings and components have been found to have fake approval marks. The only way you can be sure is to confirm such claims against the relevant list of currently approved products.

 Never use pipe, fittings or components that are not certified for use. Using non-compliant materials is not only a breach of the Standard and your legal obligations, but in many instances will make your installation unsafe.

 COMPLETE WORKSHEET 1

Confirm installation design and gas load requirements

From the plans and/or a site inspection, your next step is to confirm the pipe installation design and gas appliance load requirements.

Installation design

Although some plans may detail gas pipe installation locations, in most cases you will need to determine this yourself. The following points may apply:
- Identify the gas supply point. This may be a consumer billing meter, LPG cylinder, sub-meter or point of extension from existing pipe.
- Identify all gas appliance locations.
- Determine a proposed route for the pipe to take from the supply point to the appliances, ensuring that the locations meet Australian Standard requirements.
- Ensure your design keeps changes of direction to a minimum to reduce frictional pressure losses.
- Ensure your design takes the most direct and efficient route possible to reduce costs and satisfy a need for the sustainable use of materials.

- Confirm that the proposed pipe material type meets the requirements of the design application and location.

Piping in concealed locations

This section deals with the installation of consumer piping in concealed locations other than underground or under buildings. A number of restrictions apply to the installation of gas piping within and around building structures. These restrictions are related to three considerations:
- *Gas pressure in the pipe.* For the purposes of discussing concealed pipe, pressures are broken into pressures below 7 kPa and pressures above 7 kPa.
- *Accessibility of the pipe.* 'Accessibility' is a term defined in AS/NZS 5601.1 as meaning, 'Access can be gained without hazard or undue difficulty for inspection, repair, testing, renewal or operational purposes'. You need to assess and determine the accessibility of every installation.
- *Ventilation of the pipe.* If the cavity the pipe is installed within has some form of cross-ventilation, it is deemed to be ventilated to the satisfaction of the standard. If not, additional ventilation may need to be provided if the pipe is to remain in that location.

 REFER TO AS/NZS 5601.1 'PIPING IN A CONCEALED LOCATION OTHER THAN UNDERGROUND OR EMBEDDED IN CONCRETE'

The permissible jointing system and pipe material selection is dependent upon the location of each pipe installation. There are a number of combinations of these restrictions, demonstrated in the following examples.

In Figure 9.5, you can see that the piping is made accessible via the personnel access door. Ventilation is provided from subfloor vents.

Is installation permitted? *Yes*. Pipes and joints as per AS/NZS 5601.1.

Running a gas consumer piping line in the cavity of a brick veneer dwelling makes the pipe inaccessible, as shown in Figure 9.6. However, it receives ventilation through the cavity area from above the top plate and below the bottom plate of the wall.

Is installation permitted? *Yes*. Pipes and joints as per AS/NZS 5601.1, but the joints are to be kept to a minimum.

Where a gas line must be installed through the studs of a chamfer board dwelling, the pipe is sealed up

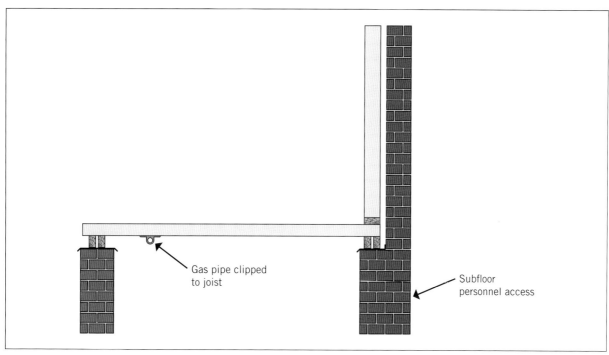

FIGURE 9.5 Less than 7 kPa, accessible and ventilated

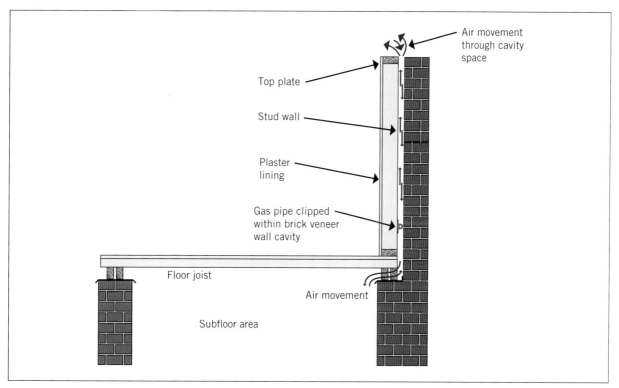

FIGURE 9.6 Less than 7 kPa, inaccessible and ventilated

between each stud and is classed as inaccessible and not ventilated (see Figure 9.7).

Is installation permitted? *Yes*. Pipes are as permitted in AS/NZS 5601.1, but all joints are to be kept to a minimum and must be *permanent joints*. This is a defined term (joints such as flared compression are not permitted).

A permanent joint is one that cannot be readily disassembled. Examples include brazed, welded, crimped and hydraulically pressed fittings. The proprietary brand composite pipe system in Figure 9.8 is designed for use with permanent crimped fittings. Notice the witness hole on the left showing full engagement of the pipe.

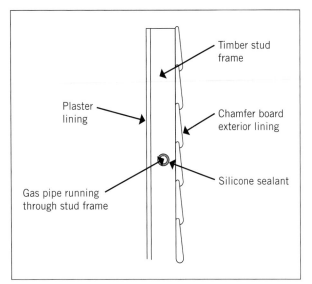

FIGURE 9.7 Less than 7 kPa, accessible or inaccessible and not ventilated

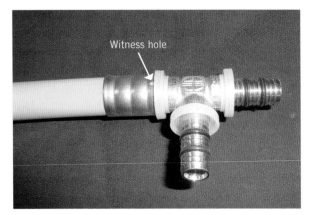

FIGURE 9.8 Permanent crimped fittings designed for use with a proprietary brand composite pipe system

The installation in Figure 9.9 is accessible and ventilated along its entire length.

Is installation permitted? *Yes*. Pipes are permitted in AS/NZS 5601.1 but all joints are to be kept to a minimum and must be *permanent* joints.

The installation in 9.10 is very similar to the one in Figure 9.9, but you will note that part of the consumer piping is now concealed in an inaccessible location.

Is installation permitted? *No!* Consumer piping above 7 kPa in an inaccessible location *cannot* be installed.

Provision of ventilation

Where ventilation of concealed piping does not exist and will need to be provided to ensure compliance with the standard, you will need to install ventilation of the correct size and location.

REFER TO AS/NZS 5601.1 'VENTILATION OF CONCEALED PIPING'

Underground piping requirements

Consumer gas piping can be installed underground subject to depth and clearance requirements. Examples of underground installation include the need to run a line from a boundary meter to the building, from remote LPG cylinders to a commercial restaurant or perhaps between detached dwellings in a caravan park or retirement village.

REFER TO AS/NZS 5601.1 'INSTALLATION OF CONSUMER PIPING UNDERGROUND'

Depth of cover and bedding

Minimum depth of cover refers to the distance between the top of the pipe and the ground surface, not the depth of the trench. An allowance needs to be made for both the diameter of the pipe and the bedding material around the pipe when planning your trench depth. The minimum depth of cover is determined by the location and ground material in which the trench is dug (see Figure 9.11).

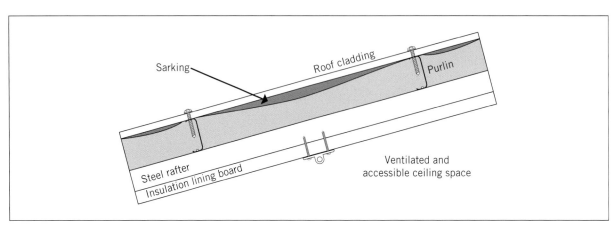

FIGURE 9.9 Greater than 7 kPa, accessible and ventilated

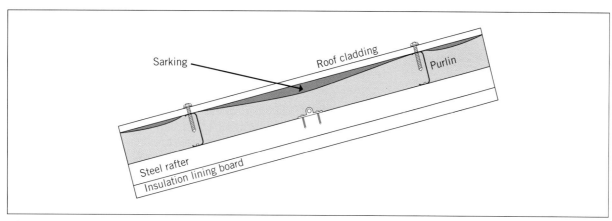

FIGURE 9.10 Greater than 7 kPa, inaccessible

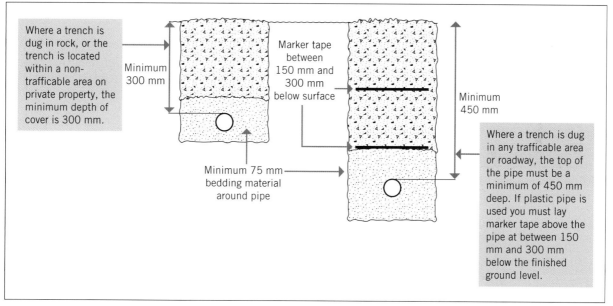

FIGURE 9.11 Minimum depth of cover for an underground gas pipe (not to scale)

Notice the use of fine bedding material around the pipe in Figure 9.11, and the required marker tape between 150 mm and 300 mm above the plastic installation.

Bedding materials for all types of pipe must be of a grade that will not be abrasive, and they must be clear of any stones or sharp objects within 75 mm of the pipe itself. If you excavated a sandy loam free from any stones, then bedding material would not be required. For all other instances, you will need to lay additional material and consolidate within the trench to ensure the pipe is supported upon a stable surface.

Underground use of plastic pipe

For long installations, it is often more cost effective to use a plastic pipe for all underground sections. In using any of the approved plastic materials such as PE, nylon or PVC-HI, you must be mindful of the following considerations:

- *Traceability*. Plastic gas pipe is normally untraceable by above-ground detectors. This would present a very dangerous situation if any subsequent excavation were carried out in the area. When laying plastic pipe, make sure you include a single core copper or stainless steel trace wire running along the length of the pipe and terminating above ground at the riser (see Figure 9.12). The wire can then be located with a pipe detector at a later date. Some marker tape products that are designed for use above plastic gas pipes are manufactured with an integral trace line.
- *Transition to metallic pipe*. Underground plastic consumer gas pipes must transition to metal pipe before coming out of the ground. Therefore, at the point at which you wish your gas line to come out of the ground you need to ensure the transition is compliant with the standard. All plastic pipes must

FIGURE 9.12 The termination of underground trace wire from an underground plastic pipe to the metallic riser of a consumer billing meter

FIGURE 9.13 An electrofusion adaptor fitting to enable transition of an underground PE pipe to a copper riser

terminate horizontally at least 300 mm below ground with an approved adaptor fitting connecting to the metallic pipe riser. The riser itself must be secured from undue movement.

Clearances from underground electrical cables and other services

Some installations may include the use of a common trench for incoming services to the property.

A common trench with gas and electrical services is shown in Figure 9.14. A common trench may offer some efficiencies, but you must be particularly mindful of ensuring that your gas line is located in the trench with all required clearances and bedding, and that it is backfilled carefully with appropriate material. Do not trust other contractors to do this job properly! You must supervise this yourself, as you are the licensed and responsible person.

In the example shown in Figure 9.15, a separation of 300 mm must be observed from the 100 mm house drain, while the 25 mm water service must be at least 100 mm away from the gas pipe.

Gas load

The input of gas appliances is measured in megajoules per hour (MJ/h) and relates directly to the amount of gas consumed by each appliance and the effect this gas load has on the pipe sizes within the system.

You can determine the system gas load from the following sources:
- the approved plans and specifications detailing the MJ/h for each appliance
- knowing the type of appliances required, and accessing specific MJ/h input data from manufacturers' website or product literature
- the MJ/h input on the data plate of each appliance.

Collate all appliance MJ/h input gas load for use in the pipe sizing process.

Pipe sizing

Review Chapter 8 'Size consumer gas piping systems' to confirm your understanding of the pipe sizing process.

Material quantities

Having confirmed the suitability and size of the selected piping system, you need to conduct a quantity take-off from the approved plans or during a site inspection. A list of job quantities must be detailed, neat and written in a methodical manner so that other people within your business or at the suppliers can easily determine all details. This is a basic quality assurance requirement of any gas fitting business. Always ensure you use appropriate and accurate Système International (SI) units of measurement to avoid confusion when ordering or during installation. Some SI units and other abbreviations related particularly to pipe installations are included in the table (overleaf).

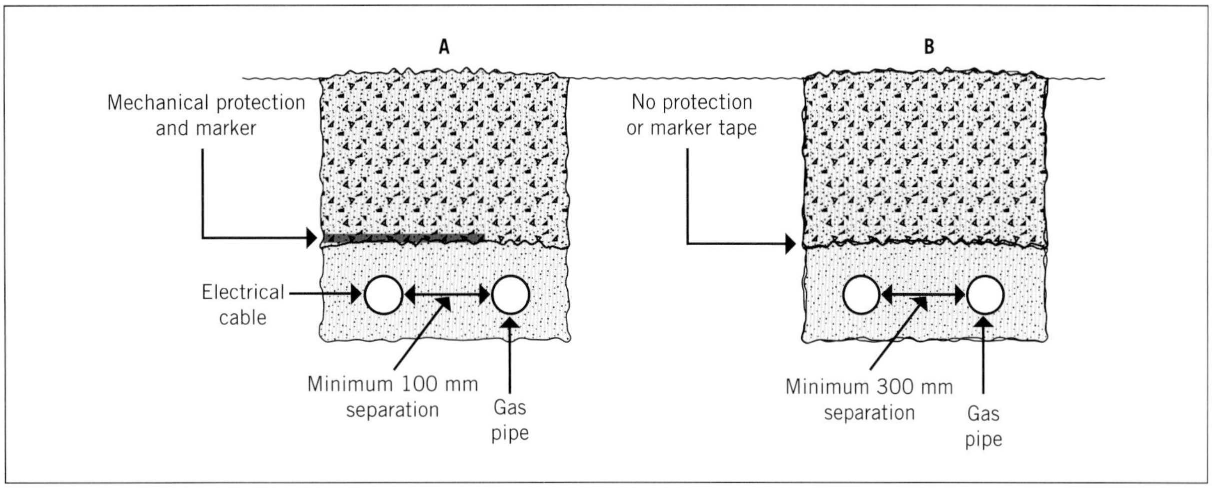

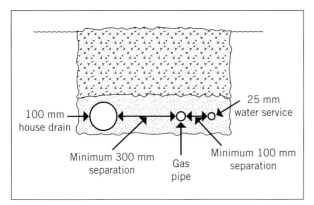

FIGURE 9.14 The required clearances between an underground gas pipe and an electrical conduit (a) where electrical cable is marked and protected and (b) where electrical cable is not marked and protected

FIGURE 9.15 The required clearances between a gas pipe and other services

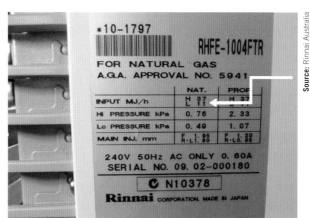

FIGURE 9.16 The maximum MJ/h gas load of an existing appliance can be determined directly from the data plate

SI unit or abbreviation	Meaning and application
mm	• Millimetres – all pipe sizes are quoted in millimetres
m	• Metres – used for the overall measurement of pipe lengths and associated installation dimensions
L	• Litres – pipe volume is measured in litres per metre
m^3	• Cubic metres – pipe volume can also be measured in cubic metres
kPa	• Kilopascals – the most common expression of pressure in Type A gas fitting; certain pressure limitations apply to pipework and fittings
DN	• Nominal diameter – the standard expression of pipe and tube diameter (mm) used during the sizing process and when ordering materials

Quantity details will include:
- material type, DN size and length of each pipe section
- DN size and number of all fittings
- DN size and number of all fixings
- any special fittings such as those required for meter connections or reversion fittings where proprietary pipe systems are specified.

Once all details are collated, materials can be ordered through your supplier, making particular note of any lead time necessary for items that must be ordered in. Delivery delays must be accommodated in your installation schedule as early as possible.

General compliance assessment

As a gas fitter, you are required to ensure that all work you undertake complies with relevant codes, standards and legislation. Not only does this include new jobs, but any work associated with existing installations.

If you are about to alter or add to an existing installation you need to ensure that the rest of the work is up to standard. If you identify a problem, you have a 'duty of care' to do something about it or at least let the owner, site supervisor or technical regulator know.

What is duty of care?
Duty of care can be defined as the implicit responsibility of an individual to adhere to a reasonable standard of care when performing acts that could foreseeably harm others. Every worker has a duty of care to all other people in society.

Before beginning any quotation or work, walk around the site and work through a checklist of possible problem areas. The following is an example:
- Consumer-billing meter or LPG cylinders are in correct location with required clearances.
- Materials and components are approved.
- Piping and components are installed correctly.
- Appliance data plates are identified.
- Appliances are installed in correct location and with required clearances.
- Flue system is sized correctly.
- Appliance ventilation is sufficient.

LEARNING TASK 9.1 WRONG

A compliance check of this existing installation reveals the incorrect use of multilayer pipe (PE/AL/PE). Refer to your Australian Standard and answer the following:
- What is at fault?
- How could this be rectified?
- What are the Australian Standard reference/s?

Where existing and proposed installation locations do not meet job set-out requirements or Australian Standards, report full details to your supervisor immediately.

COMPLETE WORKSHEET 2

LEARNING TASK 9.2
UNDERGROUND PIPING REQUIREMENTS

1. What is the purpose of trace wire and why is it used?
2. At what distance should a plastic pipe terminate horizontally with an approved adaptor fitting when transitioning to a metallic riser?

Prepare for installation

All materials used for consumer piping must be checked for serviceability to ensure that manufacturing faults or pre-purchase damage has not occurred. If you find a fault in a pipe or fitting (such as a crack in a brass casting), you are not permitted to repair it. It must be replaced. Only the best quality materials should be used in gas installations.

REFER TO AS/NZS 5601.1 'MEANS OF COMPLIANCE – MATERIALS, FITTINGS AND COMPONENTS'

Always make an extra effort to protect your materials during storage and handling. Here is a list of things you must take into consideration.
- Keep dust and water out of pipes with caps or plugs.
- Keep fittings in an orderly and clean manner, ensuring that they are not contaminated with dirt and other foreign substances.
- Ensure that plastic and multilayer pipe is not stored in direct sunlight as this will degrade the pipe.
- Avoid rough handling of your pipe, fittings and materials.
- Ensure that brass flared fittings are always stored with the nut screwed on, to prevent damage to the mating faces of the flare.
- Ensure press-fit copper connectors are kept dirt free.

Appropriate tools and equipment

Apart from general hand tools used on almost all jobs, tools and equipment used for pipe installation will to a large extent be dictated by the type of pipe material and jointing method being used. This is particularly important to consider where proprietary piping systems have been specified. Having identified the type of pipe from plans and specifications, you must access relevant websites and installation guides to ensure you know what tools and equipment may be required. The table overleaf shows just a selection of items that may be required.

Other considerations related to handling of the pipe and the proposed installation location may include:
- elevated work platforms
- ladders
- hand trolleys, forklifts and hoists

Material	Tools and equipment
Copper	• Oxy/acetylene brazing equipment • Press-fit tool and matching jaws • Flaring tools • Copper pipe cutters and reamer • Benders • Internal and external bending springs • Expanders • Branch formers • Shifters
Galvanised steel	• Threading stocks and dies • Threading machine • Steel pipe cutting tool and reamer • Steel pipe bender • 'Stillsons' and 'Footprint' grips
Polyethylene	• Electrofusion welding equipment • Peeler • Cleaning solution and wipes • Poly pipe cutter • Butt welder
Nylon, Multilayer and PEX	• Proprietary crimping/pressing tools • Cutting tool • Solvent cement and primer

- trenching equipment – manual and machine
- laser levels.

FIGURE 9.17 Both oxy/acetylene and press fit equipment may be required for copper pipe installations

Equipment required for pressure testing

In addition to general hand tools, you will require some specific equipment and instruments for pressure testing. The following items will serve as a good indication of what is required:

- *AS/NZS 5601.1 (Gas installations)*. Make sure you have the *latest* copy of the Standard, as it changes on a regular basis. You are required to use only the most up-to-date version.
- *Test instruments*. AS/NZS 5601.1 provides a list of approved test instruments and their limitations of use. Refer to this table first before choosing the instrument you prefer for the job.

REFER TO AS/NZS 5601.1 APPENDIX E 'TEST INSTRUMENTS'

For standard testing procedures, the use of a water gauge manometer or digital manometer is recommended.

Test tee

A test tee enables you to connect both a test instrument and some form of pressure source (hand pump or inert gas cylinder) to a pipe installation. The two valves on the tee enable you to isolate the pressure source or instrument as required. A low pressure test tee, such as the one seen at the top of Figure 9.18, can be purchased from most plumbing hardware suppliers.

FIGURE 9.18 Low and medium-pressure test tees

The tee incorporates an O-ring seal and a DN 25 union. This is designed for easy connection to a standard meter-adaptor fitting and will also fit some (but not all) DN 25 tube bushes and adaptors. Medium pressure test tees can be easily fabricated from standard fittings, valves and gauge of the correct pressure range.

Hand pump

Aspirator bulbs can be purchased from your plumbing hardware supplier. The bulb is fitted with a special one-way valve and a short length of hose, and is fitted to one side of your test tee. You could also use a simple bicycle pump.

Silicone hose

For many years, gas fitters used traditional natural rubber manometer hose to connect a manometer to test fittings or appliances. However, such hose does not last very long, and it perishes particularly quickly when used with LPG. It is recommended that you use clear silicone hose instead as it is very flexible, can withstand the high temperatures sometimes encountered when testing working pressures, and will not split or degrade.

Adaptor fittings

So that you can connect your test equipment to different types and sizes of pipe, you will need to put together a box full of adaptors and other fittings useful for testing purposes (see Figure 9.19).

FIGURE 9.19 A selection of basic pressure testing equipment

These include different sized adaptors, caps and plugs in both BSP and SAE threads, and soapy water. Also shown in Figure 9.19 are some specially fabricated adaptors that enable connections in tight locations or where pipe movement is limited. You should think about keeping your test equipment in a special toolbox so that such items are always on hand.

Soapy water solution

Testing for leaks requires the use of a non-corrosive soap and water solution. A rich mix of dishwashing detergent and plain water is all you need. This can be brushed on or squirted from a garden-type trigger spray. Some proprietary-brand spray-pack products also include additives in the soapy solution that will hold the bubbles around the joint for an extended time. The method you choose is up to you.

Electrical test instrument

You must also have some form of test instrument available so that a test for stray current can be carried out whenever required during pressure testing procedures.

Choice of pressure test instrument

Your choice of test instruments should primarily be dictated by AS/NZS 5601.1, as described earlier. In addition to these basic requirements, you should consider the following points in relation to your choice of instrument.

U-tube manometer

There are a number of advantages to using a U-tube manometer:
- It is always accurate and regular calibration is not required.
- It is simple to use.
- It is inexpensive to purchase or you can even make one at home or on the job
- An experienced gas fitter can learn to interpret what is happening in an installation from the way the water meniscus sits in the tube.
- It is very sensitive to pressure changes.

There are also some disadvantages:
- It has a limited pressure range, which is dictated by the length of the instrument (for example, a 1 m long manometer can only effectively measure pressures around 8 kPa).
- There is nuisance water loss.
- Larger instruments can be cumbersome in some situations.

FIGURE 9.20 Two forms of U-tube manometers

Digital manometer

The advantages of a digital manometer are that:
- it is compact
- it is accurate
- some instruments feature multifunction pressure measurement
- it has wide pressure range
- it is particularly good for appliance testing.

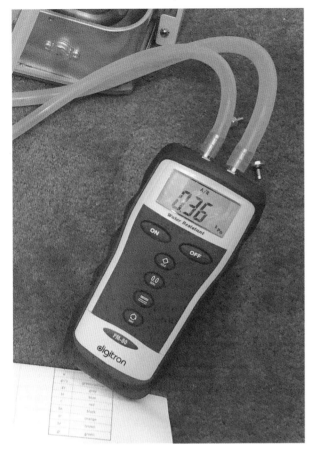

FIGURE 9.21 A digital manometer

The disadvantages are:
- it requires frequent calibration checks to ensure accuracy (some technical regulators will not accept digital manometer results without a calibration certificate)
- occasional battery failure.

Bourdon gauge

Low pressure bourdon gauges are no longer recommended for pressure testing purposes as per AS/NZS 5601.1 Appendix E. Such gauges are not as accurate as other forms of measurement and are subject to internal mechanical damage from sudden variations in pressure. Bourdon gauges should be replaced by water and digital manometers. If in doubt, contact your technical regulator.

LEARNING TASK 9.3 PRESSURE TESTING

1. What is the purpose of a test tee?
2. Why is an electrical test used when carrying out test procedures?
3. What are two types of manometer?

 COMPLETE WORKSHEET 3

Install and test piping system

This section will detail specific requirements relating to the installation, jointing and testing of gas consumer piping systems.

Piping support

Gas consumer piping must be secured to structures correctly using the appropriate materials, systems and spacing.

Pipe support fixings include:
- saddles
- clips
- threaded rods and clip heads
- brackets.

In selecting the pipe support devices for your installation, you should ensure that they are compatible with the pipe being used. For example, the use of steel clips may be compatible with steel pipe, but if used with copper tube, the two incompatible materials will react with each other in a condition called 'electrolysis'. This has been known to lead to pinholes developing in the tube where the clip and the pipe are in contact.

The spacing of support devices is based upon the rigidity of the pipe material and is the same in both the vertical and horizontal plane.

 REFER TO AS/NZS 5601.1 'SUPPORT OF CONSUMER PIPING'

Hose assemblies

Where approved for use, a hose assembly connection is advantageous for both the gas fitter and the owner/operator. These advantages include:
- a flexible form of appliance connection free from vibration
- allowing movement of the appliance by the owner/operator
- in conjunction with the use of a quick-connect device, allows the owner/operator to disconnect and connect the appliance safely.

 **REFER TO AS/NZS 5601.1 'USE OF HOSE ASSEMBLIES'**

Hose assemblies must be considered an extension of the pipe installation itself and, as such, should be sized accordingly. Additionally, you need to ensure that hose assemblies are approved for use in gas installations. All hose assemblies must be manufactured according to AS/NZS 1869 and be of Class A, B, C or D, depending on working pressure and temperature requirements (see Table 9.1). You must choose the right hose for the job.

For example, the lower temperature working limit of the Class C hose in Figure 9.22 makes it unsuitable for

TABLE 9.1 Gas hose assemblies compliant with AS/NZS 1869

Class	Max working pressure at 23°C ± 2°C	Max working temperature
A	7 kPa	−20° to 65°C
B	7 kPa	−20° to 125°C
C	2600 kPa	−20° to 65°C
D	2600 kPa	−20° to 125°C

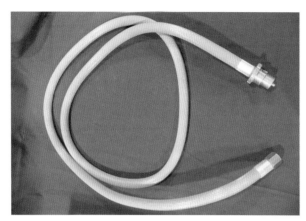

FIGURE 9.22 A Class C hose as often used for flueless space heaters

FIGURE 9.23 Bayonet fittings

use with cooking appliances and is not permitted for use in such installations (refer to the standard).

Additional requirements apply to the use of hose assemblies depending upon the appliance type and manufacturer's instructions. If an appliance is not approved for use with a hose assembly then you cannot use one. In addition, you are only permitted to install a hose assembly connection point in certain locations.

Quick-connect devices

Quick-connect devices are also known as bayonet fittings (see Figure 9.23). They are a two-part fitting between the consumer piping and a hose assembly that allows for the connection and disconnection of the appliance from the gas supply without the use of special tools. The device itself consists of a spring-loaded valve that seals off the gas supply when the hose end is withdrawn.

REFER TO AS/NZS 5601.1 'QUICK-CONNECT DEVICES'

Quick-connect devices come in chrome-plated finish and stainless steel flush wall fittings for internal installation and plain brass for external or commercial applications. There are a number of restrictions placed upon the use of these devices including, of course, the same location restrictions as for hose assemblies.

It is important that dust, water and any other contaminants are not permitted to settle in the valve, and for this reason valves must point downwards when installed outside. Devices installed indoors may be installed facing either downwards or sideways.

Jointing

Jointing of pipes using fittings and welding techniques should always be done according to Australian Standards and to manufacturer's requirements, and noting the following considerations:
- Use certified thread sealant sparingly on male threads only, wiping paste into threads from the back of the thread to the end of the fitting/pipe. This avoids a build-up of paste in the bore of the pipe that would obstruct the flow of gas.
- Certified PTFE (Teflon) tape should be wrapped carefully from the second thread back from the end of the fitting/pipe. This is to avoid 'stringing' of the tape into the bore of the pipe as it is tightened. Do not use too much tape.
- When gas-welding capillary fittings, ensure you have full engagement and penetration of the pipe into the fitting and ensure you are using a neutral flame at the correct pressure.
- Double check that you have full pipe engagement before compressing press-fit copper or multilayer fittings.

Restrictions on the use of thread sealant paste

Certified thread sealant paste should be used only in an approved manner and be restricted to the sealing of threaded pipe and fittings.

Note that it is *not* permissible to use thread sealant on any Prest-O-Lite (POL) fitting or any compression unions or flares.

Flared compression fitting

Both 45° and 60° flared compression fittings are used extensively throughout the gas fitting industry. It is not permissible to use thread paste on any surface of the flare itself or the nut, for the following two reasons:

- Excess thread paste will be squeezed into the bore of the pipe and obstruct the flow of gas and/or break away and end up blocking injectors and other gas appliance components.
- Thread paste is not designed for use with compression flares and, while it may provide an initial seal, many joints end up leaking because the paste has dried out. This is exacerbated because the fittings were never as tight as they would have been if thread paste had not been used.

If a flare is made correctly, you will have no need to add any other form of sealant! The following procedure is a recommended method of flaring that will provide consistently gas-tight joints with no use of sealant on the flare.

HOW TO

PREPARE GAS-TIGHT FLARED JOINTS

1 Check that both faces of the flare are in good condition and are not scratched or dented.

2 Ream out the burr on the inside of the cut pipe to ensure the flare is formed correctly.

3 Anneal the very end of the pipe to be flared. Point the flame towards the end of the pipe and heat no more than is necessary.

4 Place the nut on the pipe and use an appropriate flaring tool to form the flare.

5 The copper is now work-hardened and should be softened again by gently reheating the flare. This softened copper face will create an excellent seal with the brass fitting. Again, only heat as much as you need to.

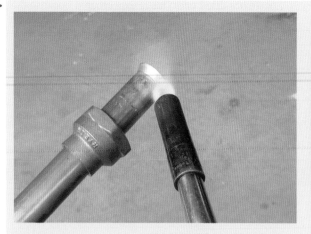

6   Sometimes annealed copper pipe has a roughened surface. To prevent binding of the pipe to the fitting as you tighten it, apply some ammonia-free soapy water to the back of the flare for lubrication. This allows the fitting to be made gas-tight without applying any paste and avoids twisting and deforming the pipe.

7   Tighten and test joint for leaks with soapy water. While this method takes a little more time to create the flare than some other shortcut procedures, you will find that leaks will be very rare, saving you time in the long term and making your work safer.

Pipe identification

In all buildings other than single-occupancy residential premises, you must use various forms of pipe identification to indicate the following:
- type of gas
- direction of flow
- pressure of gas.

REFER TO AS/NZS 5601.1 'IDENTIFICATION OF PIPEWORK'

FIGURE 9.24 Examples of pipe identification for natural gas and LPG

Even within normal single-occupancy residential premises, if you determine that a pipe is located in such a way that its purpose is not clear, it is worth using some form of identification to ensure that it is not mistaken for a water supply line.

Stickers and adhesive tape are commonly available to suit different gases and pipe diameters. Pipes can also be painted in recognised Australian Standards colour schemes, so where this is to be done in commercial/industrial applications, ensure you specify the correct paint code as per AS/NZS 5601.1.

Figure 9.24 shows some examples of pipe identification for both natural gas and LPG. Stickers are normally manufactured with arrows pointing in both directions, so remember to cut one off to make it clear in which direction the gas is flowing. Poor pipe identification is confusing, wastes time and can be very dangerous.

Reversion fittings

There are now many manufacturers of approved proprietary multilayer piping systems, giving the gas fitter a wide range of installation choices. Each manufacturer has specialised fittings and jointing tools that are generally not compatible with competitor systems. This is fine as long as the manufacturer remains in business and continues to provide the same fittings and tools for extension and/or repair of the installation.

However, gas installations can remain in service for decades and although it is perfectly possible to maintain, repair and extend existing DN copper systems, this may not be possible when multilayer system-specialised fittings and pipes become obsolete or unavailable for any reason.

To avoid the inevitable problems where proprietary multilayer systems might become unsupported at some future time, the Gas Installation Standards now requires the use of reversion fittings for any system in a Class 1 building that is longer than 10 metres. Such fittings are designed to allow 'reversion' back to a standard pipe and thread system.

FROM EXPERIENCE

An important attitude that you must develop as a professional gas fitter is to plan for and carry out your installations with consideration to future maintenance and extension needs. After all, it is entirely probable that you will be the one to come back in 10 years to do extra work!

You must provide some form of standard thread (e.g. BSP) or annealed copper tube at specified accessible locations to enable simple connection if required in the future, as shown in Figure 9.25. This is a mandatory requirement and the Standard provides examples of how this is to be done.

FIGURE 9.25 Provision of a reversion assembly in an accessible sub-floor space

REFER TO AS/NZS 5601.1 'PROPRIETARY MULTILAYER PIPING'

LEARNING TASK 9.4
INSTALL AND TEST PIPING SYSTEM

1 When installing a freestanding gas appliance, which 'class' hose assembly would be suitable for use? Provide a reference from AS/NZS 5601 to support your answer.
2 Is thread sealant paste permitted to be used on compression unions or flare joints?
3 What is the purpose of 'reversion fittings' when installing approved proprietary multilayer piping systems?

Testing gas piping systems

All gas consumer piping installations must be pressure tested to ensure that they are gas-tight. No leaks. No exceptions. It is one of your most important duties as a gas fitter.

Note: Appendix E, 'Testing for gastightness' of the Standards, is a *normative* appendix. This means you *must* follow its requirements! Whenever a new standard is introduced, you are obliged to follow the latest procedures. Refer to your teacher if new standards are introduced during the life of this edition.

REFER TO AS/NZS 5601.1 APPENDIX E 'TESTING FOR GASTIGHTNESS'

There are four main types of pressure test:
- *pipework test* – for testing consumer piping without appliances connected
- *installation test* – for testing consumer piping with appliances connected
- *leakage test* – for testing the existing consumer piping before starting work
- *final connection test* – for testing any connection after a pressure test has been completed.

Limitations on test procedures

The AS/NZS 5601.1 only provides specific test procedures for pressure testing of a gas installation up to 30 L (0.03 m^3) of pipe volume. The Standard provides no direct guidance on test time for installations that exceed 30 L. Where pipe volume is greater than 30 L, you must provide a written test procedure to the technical regulator. However, *Informative Table E2* does provide some test-time guidance when testing above 30 L.

The next section will show you how to determine pipe volume. It will also provide a *potential* test-time solution for installations above 30 L.

Note: This test-time solution is a suggested guide only and you must always consult with your technical regulator as to what test requirements will be necessary for installations above 30 L in volume. Your teacher should be able to advise you on local requirements.

Determining installation pipe volume

As you read further into this section, you will see that certain test procedures will require you to determine the volume of installed pipe. These test procedures will state that an installation up to 30 L in pipe volume must show no loss of pressure over a period of *five minutes*.

The volume of any installation can be quickly determined by referring to Table D1 in AS/NZS 5601.1 (see Figure 9.26). To do this you will need to know the pipe material, pipe diameter and the pipe length.

To interpret the table, follow the simple steps as seen in Figure 9.26 and in reference to AS/NZS 5601.1.

Determining test time – greater than 30 L volume

The test procedure below is one option for the testing of pipe volumes greater than 30 L. However, before undertaking such work you must confirm the requirements of your local State or Territory technical regulator.

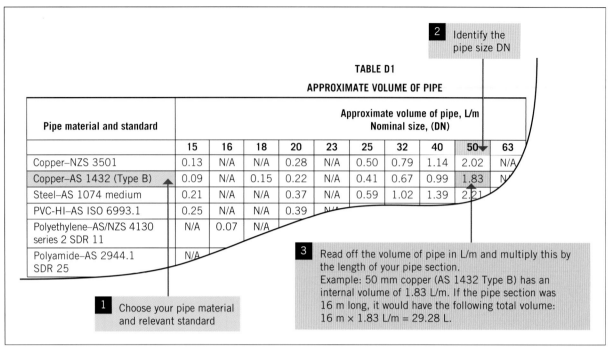

FIGURE 9.26 Using Table D1 from AS/NZS 5601.1 to determine installation pipe volume

Test time – single run installation

Once you have worked out the actual volume of installed pipe, you can then determine the required minimum test time for the installation, remembering the following directions from AS/NZS 5601.1:

For consumer piping with a volume not exceeding 30 L there is to be no loss of pressure during a test period of 5 minutes.

and:

Where the pipe volume exceeds 30 L, the test period is to be extended by 5 minutes for every additional 30 L or part thereof.

Refer to Table D1 and confirm the following test times for single pipe-run installations, for both less than and exceeding 30 L of volume:
- 28 m of DN 40 copper pipe (Type B): 27.72 L, five-minute test time
- 15 m of DN 25 steel pipe: 8.85 L, five-minute test time
- 18 m of DN 50 copper pipe (Type B): 32.94 L, 10-minute test time (five minutes for every 30 L part of the installation)
- 18 m of DN 80 copper pipe (Type B): 75.06 L, 15-minute test time (five minutes for every 30 L part of the installation).

Test time – multi-branch installations

Most installations will include multiple branches and different pipe diameters. To determine the test time of the whole installation, you need to work out the required test time for each section then add these times together.

Using the example in Figure 9.27, itemise the details of each pipe section (Table 9.2).

You can now see that the requirements of each section add up to a total test time of 25 minutes for the entire installation. This is slightly in excess of the minimum requirements and it is possible to work out

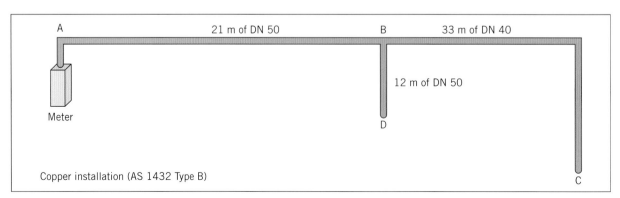

FIGURE 9.27 Documenting the diameter and length of each section

TABLE 9.2 Test-time solution for installation: greater than 30 L (check with your technical regulator first!)

Section	Length and diameter	Length m × Vol L/m	Test time
A–B	21 m of DN 50 copper pipe	21 × 1.83 = 38.43 L	10 minutes
B–C	33 m of DN 40 copper pipe	33 × 0.99 = 32.67 L	10 minutes
B–D	12 m of DN 50 copper pipe	12 × 1.83 = 21.96 L	5 minutes
Total installation test time			25 minutes

the minutes and seconds for every metre over 30 L. However, this is unnecessarily complicated; it is far better simply to apply the gross test time to the full installation.

Standard test procedures – less than 30 L pipe volume

Now that you have gained an understanding of pipe volume and test time requirements, you can move on to a description of each test procedure in more detail.

Pipework test (no appliances connected)

You must carry out a pipework test on any newly installed pipe immediately after fitting and prior to connecting appliances. You must also carry out this test after any alterations or additions to existing pipework. This is often known as the 'rough-in test'.

REFER TO AS/NZS 5601.1 APPENDIX E 'E4 PIPEWORK TEST PROCEDURE'

Note that appliances must never be exposed to a pipework test. You must isolate or disconnect all appliances to ensure that internal components are not damaged by the higher than normal pressure of the test.

Step 1: Carry out an electrical safety test and use bonding straps

a Never touch any piping or appliances unless you have carried out a test for stray current first.
b You must also ensure that you use bonding straps across any intended break in the pipework or disconnection of any component or appliance.

REFER TO AS/NZS 5601.1 'ELECTRICAL SAFETY BONDING OR BRIDGING'

You have no choice about whether or not to complete these tests: safety is paramount and you are required by AS/NZS 5601.1 to comply.

Step 2: Assemble the test equipment

Set up your test equipment as shown in Figure 9.28. You can undertake the test at any part of the installation as long as all sections are capped off. This could be at the meter–cylinder connection or at the end of one of the branches.

FIGURE 9.28 Setting up the test equipment

In Figure 9.28 the gas fitter is about to undertake a pipework test at the cylinder connection point of an LPG installation. A new branch has been installed in the house and all existing appliances have been disconnected in preparation for the test.

Step 3: Prepare to pressurise the pipework

a Ensure that the meter or cylinder is disconnected from the pipework.
b Seal all open ends with a cap, a plug or a temporary pipe crimp and solder.
c Ensure that any appliance connected to the system is fully isolated.

Step 4: Pressurise the system

a Use the hand pump to raise the pipe pressure to 7 kPa or 1.5 times the *operating pressure* of the system, whichever is the greater. For testing purposes, operating pressure is defined as the pressure that any part of the installation will be subjected to under *normal operating conditions*. For testing purposes, operating pressure should be regarded as the lock-up pressure of the next upstream regulator. This is normally slightly higher than the nominal supply pressure.

 REFER TO AS/NZS 5601.1 'OPERATING PRESSURE'

EXAMPLE 9.1

If your supply pressure is 3 kPa out of the meter with a lock-up of 3.2 kPa, then 3.2 kPa is deemed to be the operating pressure. 1.5 × 3.2 kPa = 4.8 kPa, so as 7 kPa is the greater figure, you will use this pressure for the test.

b Once you have pumped up the system to the correct pressure, apply some soap and water to the hose connection between the manometer and the test tee only. This is just to ensure that you do not chase a leak elsewhere on the system when it is at the test instrument itself.

Step 5: Allow for temperature stabilisation

a You need to allow at least two minutes for the pressurised air within the pipework to stabilise with the ambient temperature. In practice, this can take some considerable time, depending upon the weather. Be aware that cold winds, draughts and alternating cloudy and sunny periods can extend this time.

b If the pressure drops a little during this period, it usually means that colder outside air is having an effect upon the pressure. Add a little more pressure and you will find it will normally equalise. (Hint: Avoid using any cold soapy water unless you know you have a leak, as this may cause the pressure to drop.)

If the pressure seems to rise during the stabilisation period, it is likely that some form of heat is being applied to the pipework. *Do not* start your test until this rise stops, as it may mask a leak. Bleed off a little pressure and then pump back up to correct pressure. For this reason, never apply a test immediately after working on the consumer piping with oxy-acetylene.

c Do not start your test time until you are satisfied that the system has stabilised.

Step 6: Start the test time

a The test time for an installation of up to 30 L of volume is five minutes. There must be no loss of pressure during this period.

b If the pressure drops during this period, then you must find the source of the pressure loss. Use your soapy water to find the leak. Look for the following:
   – Leaking flares. Very small leaks may amount to only one small bubble per minute. You must look very carefully.
   – Silver-soldered fittings in tight corners that may not have been fully soldered. Check the backs of each fitting with a mirror.
   – Threaded fittings that could be leaking.
   – Any damage or pinholes in your pipe.
   – Leaking test tee connections. Double-check all connections around your test equipment.

Leaks can be very time consuming and difficult to find. However, you must *never* complete a test unless you are absolutely satisfied that what you have tested is gas-tight. There is no such thing as a permissible leak in modern gas fitting.

Once you have found the leak, repeat the full test procedure from the beginning.

Step 7: The system passes the test

Once you are sure the system has passed the test, remove your equipment and seal the open end.

Installation test (appliances connected – no gas connected)

You must carry out an installation test on any newly installed pipe system after the connection of appliances but before connection of the gas supply and commissioning of the appliance. You must also carry out this test after any alterations or additions to existing pipework. This is often known as a 'lock-up test'.

The importance of an installation test is that you are testing not only the pipework but also the appliance connections right through the internals of the appliance itself, up to the control valves. In this way, leaking connections and appliances with internal damage or faults are identified before gas is connected to the system. You can therefore be confident when the gas is connected that the system is already gas-tight.

 **REFER TO AS/NZS 5601.1 APPENDIX E 'E5 INSTALLATION TEST PROCEDURE'**

Step 1: Carry out an electrical safety test and use bonding straps

a Never touch any piping or appliances unless you have carried out a test for stray current first. Refer to the electrical safety section within Chapter 5 'Gas industry workplace safety' for guidance.

b You must also ensure that you use bonding straps across any intended break in the pipework or disconnection of any component or appliance.

You have no choice about whether or not to complete these tests: safety is paramount and you are required by AS/NZS 5601.1 to comply.

Step 2: Assemble the test equipment

Set up your test equipment in the manner shown in Figure 9.28.

Step 3: Prepare to pressurise the pipework and appliances

a Ensure that the meter or cylinder is disconnected from the pipework.

b   Seal all open ends with a cap or plug.
c   Ensure that the appliance control valves (i.e. the individual burner valves) are turned off.
d   Check that all system manual shut-off valves, including appliance isolation valves, are turned on. This is so that the test pressure can be applied right through to the appliance internals.

Step 4: Pressurise the system

a   Use the hand pump to raise the pipe pressure to the operating pressure or 2 kPa, whichever is the greater. For testing purposes, operating pressure should be regarded as the lock-up pressure of the next upstream regulator.
b   Once you have pumped up the system to the correct pressure, apply some soapy water to the hose connection between the manometer and the test tee only. This is just to ensure that you do not chase a leak elsewhere on the system when it is at the test instrument itself.

Step 5: Allow for temperature stabilisation

You now need to wait for the temperature to stabilise, as described in the pipework test procedure above. Do not start your test time until you are satisfied that the system has stabilised.

Step 6: Start the test time

a   The test time for an installation up to 30 L of volume is five minutes. There must be no loss of pressure during this period.
b   If the pressure drops during this period then you must find the source of the pressure loss. Use your soapy water to find the leak. Look for leaks in the following:
   – *Flares.* Very small leaks may amount to only one small bubble per minute. You must look very carefully.
   – *Isolation valves.* Check all appliance isolation valves, particularly flared gas valves.
   – *Regulators.* Check the connections around the regulators, and look for any escapes around breather holes.
   – *Test tee connections.* Double-check all connections around your test equipment.

If you are confident that you have no leak in your pipe installation itself, isolate each appliance in turn in order to identify the one at fault. Once you have done so, you will need to contact the manufacturer for warranty instructions, or service the appliance if it is second-hand.

Once you have found the leak, repeat the full test procedure from the beginning.

Step 7: The system passes the test

Once you are sure the system has passed the test, remove your equipment and seal the open end.

Leakage test (for existing installations)

The leakage test is a form of lock-up test that confirms an existing installation is gas-tight before and after work. As noted in Chapter 5, you must carry out this test *before* commencing work on any existing installation, as well as after any work on the installation is completed.

REFER TO AS/NZS 5601.1 APPENDIX E 'E6 LEAKAGE TEST FOR EXISTING INSTALLATIONS'

Step 1: Carry out an electrical safety test and use bonding straps

a   Never touch any piping or appliances unless you have carried out a test for stray current first.
b   You must also ensure that you use bonding straps across any intended break in the pipework or disconnection of any component or appliance.

REFER TO AS/NZS 5601.1 'ELECTRICAL SAFETY BONDING OR BRIDGING'

You have no choice about whether or not to complete these tests: Safety is paramount and you are required by AS/NZS 5601.1 to comply.

Step 2: Depressurise system and prepare for test

- Isolate gas supply and depressurise system.
- Disconnect installation from meter or cylinder.
- Ensure all appliance pilots are off.
- Check that all appliance isolation valves are open.
- Ensure that the appliance's internal burner valves are off.

Step 3: Connect a suitable test instrument

a   Attach a manometer to any suitable test point upstream of a regulator. A meter outlet nipple is ideal or, in some cases, a cooker test nipple.
b   By watching the manometer for any sign of a pressure rise over a period of five minutes, ensure that the gas supply test point is not passing gas.

Step 4: Pressurise the system

Turn on the gas supply, so that it locks up at its *operating pressure* or 2 kPa, whichever is the greater.

REFER TO AS/NZS 5601.1 'OPERATING PRESSURE'

EXAMPLE 9.2

If your supply pressure is 1.5 kPa out of the meter with a lock-up of 1.6 kPa, then 1.6 kPa is deemed to be the *operating pressure*. As this is lower than 2 kPa, you would need to use a hand pump to raise the system pressure to the prescribed amount. If the normal system lock-up pressure is greater than 2 kPa then this higher pressure is deemed to be the system operating pressure.

Step 5: Isolate the pressure source and apply the test
Turn off the pressure supply valve. This locks up the system.

Step 6: Start the test time
a Observe the manometer for a period of five minutes and ensure that there is no pressure loss.
b On older systems where the technical regulator permits a pressure drop, ensure that it does not exceed the amount allowable. (Note that this allowance does not normally apply to modern gas installations. You should always strive to achieve a no-loss test for these. If in doubt, contact your technical regulator.)
c If the pressure drops during this period then you must find the source of the pressure loss. Use your soapy water to find the leak. You will need to look at every joint in the pipework and at each connection point for appliances. Remember that older appliances may also be leaking internally.

Once you have found the leak, repeat the full test procedure from the beginning.

Step 7: The system passes the test
a Once you are sure the system has passed the test, remove your equipment and seal the open end.
b Apply a final connection test as outlined in AS/NZS 5601.1 and summarised below.

Final connection test
The final connection test is required for any connection made after the removal of test equipment. This normally relates to the connection point to which the test gear was connected. It must be tested at the *operating pressure* (lock-up pressure) of the system. Use soapy water to ensure that the joint is gas-tight.

Once testing is complete, all open ends must be fully sealed. Record any test data as required by company or gas regulator protocols.

Depending on the job requirements, it may be some time before you come back to install and commission appliances. When you return to the site to complete the job, you will need to conduct a further leakage test to ensure the installation has not been damaged by other trades during the interval.

REFER TO AS/NZS 5601.1 APPENDIX E 'TESTING A CONNECTION MADE AFTER A TEST PROCEDURE'

COMPLETE WORKSHEET 4

Purging procedures

Upon completion of a successful leakage test process on an existing rough-in, you will need to conduct a purge of the pipework. If you attempted to light an appliance without having purged the line, an air–gas mix within the system may ignite. This may result in damage to the appliance or an explosion at the gas meter.

Never try to light an appliance unless all air has been purged from the line!

Purging a small installation through an appliance
The following section details the particular requirements related to the purging of a small installation. Large installation purging is covered in more detail within Chapter 11.

Piping of less than 30 L (0.03 m^3) is classed as a small installation. AS/NZS 5601.1 sets out three detailed appliance purge procedures:
- purging through an appliance fitted with an open burner (an open burner is a burner that has no flame-failure device fitted)
- purging through an appliance fitted with an electronic flame safeguard
- purging through an appliance fitted with a thermoelectric flame safeguard.

REFER TO AS/NZS 5601.1 APPENDIX D 'PURGING'

Before commencing an appliance purge, consider the following:
- Plan the purge in a logical sequence so that the most pipe is purged in one go and no pockets of air are left in the pipe.
- Try to purge through an open burner first, as this is the quickest procedure.
- Avoid purging into a confined space, particularly on larger installations. Use a plastic hose to vent purged gas to the atmosphere.
- Isolate any ignition device that may inadvertently activate when you turn on a gas control during the purge.

Purging through an open burner
Purging through an open burner is the easiest procedure of these three and will be covered here in detail. Be aware that few new appliances will now have an open burner, but the process is still relevant for appliance purging of existing installations.

HOW TO

1. Turn on one burner gas control until the presence of gas is detected. Put your ear to the burner and listen to the escaping air. As the gas comes through, you will hear a change in pitch due to the difference in density between gas and air. Use your sense of smell too, although you will find this is not as reliable as you might think.
2. Let the gas flow for a few seconds longer then turn off and allow sufficient time for any accumulated gas to disperse before proceeding. Be particularly careful when purging LPG, as it is heavier than air and will tend to sit in appliance recesses. Fan with a piece of cardboard if necessary.
3. Turn on one gas control valve and keep a continuously burning flame at the burner until the gas is alight and the flame is stable.

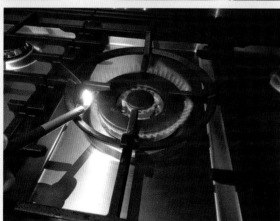

4. Continue to purge until gas is available at each appliance. Do this at each appliance before you adjust the pressure or aeration. You must ensure that all air is out of the system before continuing with the commissioning process.

Purging through an appliance with a flame safeguard device

Because flame safeguard devices prevent the release of any gas from a burner without the presence of a flame, you cannot efficiently purge the appliance from a burner. The only option is to purge at the appliance inlet connection. The amount of air left in the appliance internals is negligible and does not pose a hazard. The procedure is described below:

HOW TO

1. Isolate any electrical supply to the appliance.
2. Fit a bridging device (bonding strap) across the appliance inlet union connection when purging from the closest union to the appliance inlet.

3. Slacken the union to allow gas to flow out, but do not fully disconnect it. This procedure maintains a level of controlled gas release.
4. Turn on the appliance manual shut-off valve until the presence of gas is detected. Put your ear to the fitting and listen to the escaping air. As the gas comes through, you will hear a change in pitch due to the difference in density between gas and air. Use your sense of smell too, although you will find this is not as reliable as you might think.
5. As soon as the presence of gas is detected, tighten the union and test it with soap and water solution.
6. Allow sufficient time for any gas to disperse.
7. Remove the bonding straps.
8. Turn on the power supply and activate the ignition source.
9. Remember, ignition may not be successful immediately and lock-out may occur a number of times before combustion is satisfactory.
10. Allow sufficient time for any unburnt gas to disperse before resetting the system.

The procedure for purging an appliance fitted with a thermoelectric safety device is similar to that outlined above, with additional steps needed in order to light the pilot.

LEARNING TASK 9.5 TESTING GAS PIPING SYSTEMS

1. What are the four main types of pressure tests for gas piping systems?
2. At what stage is purging of gas pipework carried out on an existing rough in?

Clean up the job

The job is never over until your work area is cleaned up and tools and equipment accounted for and returned to their required location in a serviceable condition.

Clean up your area

The following points are important considerations when finishing the job:
- If not already covered in your site induction, consult with supervisors in relation to the site refuse requirements.
- Cleaning up as you go not only keeps your work area safe, it also lessens the amount of work required at the end of the job.
- Dispose of rubbish, off-cuts and old materials in accordance with site-disposal requirements or your own company policy.
- Ensure all materials that can be recycled are separated from general refuse and placed in the appropriate location for collection. On some sites, specific bins may be allocated for metals, timber etc. or, if not, you may need to take materials back to your business premises.

Check your tools and equipment

Tools and equipment need to be checked for serviceability to ensure that they are ready for use on the job. On a daily basis and at the end of the job, check that:
- all items are working as they should with no obvious damage
- high-wear and consumable items such as cutting discs, drill bits and sealants are checked for current condition and replaced as required
- items such as laser levels, crimping tools, electrofusion equipment etc. are working within calibration requirements
- where any item requires repair or replacement, your supervisor is informed and if necessary mark any items before returning to the company store or vehicles.

Complete all necessary job documentation

Most jobs require some level of documentation to be completed. Avoid putting it off until later, as you may forget important points or times.
- Ensure all SWMS and/or JSAs (Job Safety Analysis) are accounted for and filed as per company policy.
- Complete your timesheet accurately.
- Reconcile any paper-based material receipts and ensure electronic receipts have been accurately recorded against the job site and/or number.

 COMPLETE WORKSHEET 5

SUMMARY

In this unit you have examined the requirements related to the installation of gas piping systems including:
- revision of general and electrical hazards associated with pipe installation
- preparation requirements necessary for successful installation
- specific installation requirements related to gas pipe installations
- pressure testing requirements
- purging requirements for small volume installations (<30 L)
- clean-up of job and return of equipment.

GET IT RIGHT

The photo below shows an incorrect practice when using a bubble leak detector on a gas meter. Identify the problem and provide reasoning for your answer.

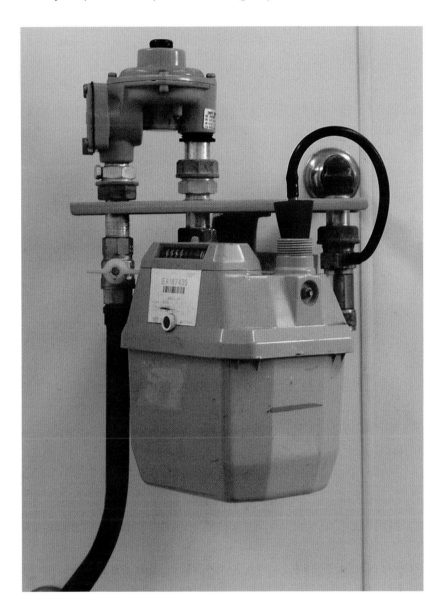

_____

_____

_____

_____

WORKSHEET 1

Student name: _____

Enrolment year: _____

Class code: _____

Competency name/Number: _____

To be completed by teachers
Student competent ☐
Student not yet competent ☐

Task

Review the sections on *Selecting products using AS/NZS 5601* and *How to identify a certified or approved product* and answer the following questions.

1. When selecting consumer piping for your installation, which table in AS/NZS 5601.1 do you refer to?

   _____

2. Are you permitted to use soft-soldered capillary fittings in gas installations? Provide a reference from AS/NZS 5601 to support your answer.

   _____

   _____

   _____

3. Define the term *push-on connector*. If you encountered a commercial rice cooker with this type of connection at the gas inlet, what would you do? Provide a reference from AS/NZS 5601 to support your answer.

   _____

   _____

   _____

4. Are compression fittings with nylon olives permitted for use with gas consumer piping? Provide a reference from AS/NZS 5601 to support your answer.

   _____

5. If you were installing a galvanised steel consumer piping system supplying gas at 10 kPa, would you be able to use malleable cast iron fittings to AS 3673? Provide a reference from AS/NZS 5601 to support your answer.

   _____

   _____

   _____

6. Copper tube can be used as consumer piping up to a pressure of 200 kPa. What two types of copper tube are permitted for use? Provide a reference from AS/NZS 5601 to support your answer.

   i _____

   ii _____

7. Is plastic-coated copper tube permitted for use underground and beneath a building? Provide a reference from AS/NZS 5601 to support your answer.

   _____

   _____

   _____

8. Are you permitted to apply jointing compound certified to AS 4623 to flared compression joints? Provide a reference from AS/NZS 5601 to support your answer.

   _____

   _____

   _____

9. Does AS/NZS 5601.1 permit branches to be formed in copper tube? Provide a reference from AS/NZS 5601 to support your answer.

   _____

   _____

   _____

10. Copper press-fit fittings rely upon an HNBR (hydrogenated nitrile butadiene rubber) O-ring to ensure a gas-tight seal. What colour is the O-ring that identifies such fittings for use with gas installations? Provide a reference from AS/NZS 5601 to support your answer.

    _____

    _____

    _____

11. When using some form of proprietary brand piping system such as copper press-fit or multilayer composite pipes, you are required to identify the system type used in the premises. How does the Standard require this to be done? Provide a reference from AS/NZS 5601 to support your answer.

    _____

    _____

    _____

12. Are press-fit copper end connectors permitted for use as the final connection to an appliance? Provide a reference from AS/NZS 5601 to support your answer.

   _____

   _____

   _____

13. Can cross-linked polyethylene multilayer pipe be installed above ground and exposed to sunlight?

   _____

   _____

   _____

WORKSHEET 2

Student name: _____

Enrolment year: _____

Class code: _____

Competency name/Number: _____

To be completed by teachers
Student competent ☐
Student not yet competent ☐

Task
Review the section *Identify gas piping system requirements* and answer the following questions.

1 Is it permissible to fix consumer piping to a fence?

_____
_____
_____

2 Are screwed fittings and compression flare joints permitted for use in inaccessible locations? Provide a reference from AS/NZS 5601 to support your answer.

_____

3 Is it permissible to run an LPG gas line operating at 70 kPa within the concealed cavity of a double-brick commercial construction? Provide a reference from AS/NZS 5601 to support your answer.

_____
_____
_____

4 What is the minimum vertical separation between underground consumer piping and any other service? Provide a reference from AS/NZS 5601 to support your answer.

_____

5 During an installation, you form a bend in copper pipe with your pipe benders. Upon inspection you notice that the inside radius of the bend has a series of buckles along its length. Is this a problem? Provide a reference from AS/NZS 5601 to support your answer.

_____
_____
_____

WORKSHEET 3

Student name: _____

Enrolment year: _____

Class code: _____

Competency name/Number: _____

To be completed by teachers
Student competent ☐
Student not yet competent ☐

Task

Review the section *Prepare for installation* and answer the following questions.

1 List at least four ways in which you can protect your gas installation materials during storage and handling.

_____

_____

_____

_____

2 Does AS/NZS 5601.1 still recommend the use of low pressure Bourdon gauges for installation testing? Provide a reference from AS/NZS 5601 to support your answer.

_____

_____

_____

3 List at least three tools and pieces of equipment that may be required when installing above-ground multilayer pipe.

_____

4 What are least three items of pressure testing equipment you will require when conducting a test procedure from the meter connection point?

_____

WORKSHEET 4

To be completed by teachers
Student competent ☐
Student not yet competent ☐

Student name: _____

Enrolment year: _____

Class code: _____

Competency name/Number: _____

Task
Review the section *Install and test piping system* and answer the following questions.

1. Where required, at what maximum interval should pipe identification stickers be placed? Provide a reference from AS/NZS 5601 to support your answer.

   _____

2. Horizontal DN 25 copper pipe fixed to a wall requires support clips to be fitted at a maximum interval of 2.5 m. True or false? _____

3. You are installing a multilayer composite pipe system through a ceiling space of a Class 1 building with an approximate run of around 16 m. What are you required to fit into this line immediately before the first branch and immediately after the last branch take off? Provide a code reference to support your answer.

   _____

   _____

   _____

4. A flueless space heater is to be installed in a room. What is the closest permissible distance between the hose assembly connection point and a doorway? Provide a reference from AS/NZS 5601 to support your answer.

   _____

5. List the four key pressure tests carried out on gas consumer piping.

   _____

   _____

   _____

   _____

6. To what pressure do you subject consumer piping when undertaking a pipework test? Provide a reference from AS/NZS 5601 to support your answer.

   _____

   _____

   _____

7 What kind of test would you undertake before starting work on an installation to ensure that it is gas-tight?

___

8 After pressurising your installation, heavy clouds obscure the sun and it becomes noticeably colder. What is likely to happen to your manometer reading? What would you do if this happens?

___

___

___

9 You are conducting an installation test with all appliances connected. Should manual shut-off valves be left open or closed? Why?

___

___

___

10 Is the temperature stabilisation period part of the five-minute test or in addition to it?

___

11 You have successfully completed an installation test so you remove your test instrument from the connection point. What do you need to do now? Provide a reference from AS/NZS 5601 to support your answer.

___

___

___

12 Complete the test time details in the table below.

Length and diameter	Test time
76 m of DN 20 UPVC pipe	
35 m of DN 25 copper pipe	
37 m of DN 50 steel pipe	
120 m of DN 100 copper pipe	
8 m of DN 80 steel pipe	

13 In relation to test procedures, what is the definition of the term 'operating pressure'?

___

___

___

WORKSHEET 5

Student name: _____

Enrolment year: _____

Class code: _____

Competency name/Number: _____

To be completed by teachers

Student competent ☐

Student not yet competent ☐

Task

Review the section *Purging procedures* and answer the following questions.

1 What is the definition of an *open burner*?

_____

_____

2 Why must bonding straps be used when purging through the appliance inlet union?

_____

_____

_____

3 Name two methods of detecting the presence of gas during a purge.

_____

_____

4 Why is appliance purging a required part of the commissioning procedure?

_____

_____

_____

5 In relation to the purging process, what maximum volume of pipe constitutes a small installation?

_____

_____

6 Where an appliance is fitted with a thermoelectric flame failure device, at what location do you carry out the pipework purge?

_____

_____

INSTALL GAS PIPING SYSTEMS **183 GS**

7 What should be done with an electrofusion machine that is found to be out of calibration?

_____

_____

INSTALL GAS PRESSURE CONTROL EQUIPMENT

10

Learning objectives

In this chapter, you will have the opportunity to investigate each of the following aspects related to gas pressure control.

Areas addressed in this chapter include:
- principles of regulator operation
- adjustment of regulators
- application of regulators based upon gas type, location and pressures
- venting of regulators.

Completing the review questions at the end of the chapter will also prepare you for the practical application of this knowledge further into your training.

Before working through this section, it is absolutely important that you complete or revise each of the chapters in Part A 'Gas fundamentals' with particular emphasis on the section 'Units of pressure' in Chapter 2 'Units of measurement and gas industry terms', as well as the section 'Basic electrical safety' in Chapter 5 'Gas industry workplace safety'.

Overview

All gas appliances have pressure control systems in some form. As a gas fitter, you cannot simply install regulators and associated pressure control equipment without understanding how they operate. You must develop a strong understanding of the various ways in which pressure is regulated to enable you to safely undertake standard commissioning and maintenance tasks. The first part of this chapter is specifically dedicated to the fundamental operation of regulators.

Pressure control fundamentals – regulators

Regulators are devices designed to control pressure, and are regarded as being the heart of all gas systems. Some are attached directly to the appliance, while other types are used in the consumer piping or reticulation systems. Regardless of where regulators are installed, they all have the following primary tasks:

- They reduce higher supply pressures to a lower outlet pressure.
- They are designed to maintain a consistent outlet pressure despite changes in supply pressure and appliance demand.

In addition to basic pressure control, some regulators may be designed with more features that enable them to:

- relieve excess pressure
- shut off the supply of gas when no appliances are being used so that unacceptably high pressures do not build up in the system.

It is important to note at the outset that regulators cannot raise insufficient supply pressure; they can only reduce pressure to suit design requirements. For this reason, appliance consumption load or demand must always be within the capacity of the supply system, to deliver the correct amount of gas at the right pressure.

If supply and demand pressures always remained constant, then a simple restriction device would suffice to reduce pressure to an appliance. Figure 10.1 shows a cross-section of a pre-set needle valve restricting inlet pressure to a desired outlet pressure.

However, such a device cannot compensate for changes in either supply or demand. If supply pressure rises, as in Figure 10.2, the outlet pressure also rises in direct proportion, which then provides too much gas to the burner.

If the inlet pressure were to rise, the needle valve would still reduce the outlet pressure to some extent, but the maximum pressure permitted for correct combustion to take place would be exceeded. For example, although the needle valve setting in Figure 10.2

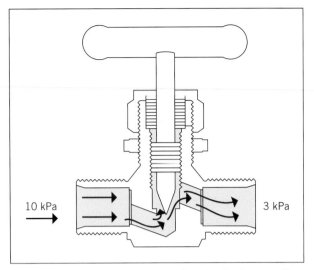

FIGURE 10.1 Needle valve cross-section showing how gas flow is restricted

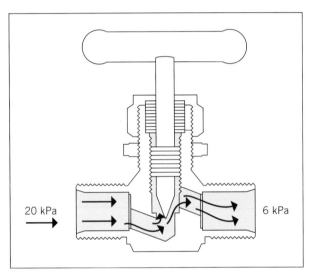

FIGURE 10.2 A simple needle valve cannot maintain a constant outlet pressure

has not changed, it cannot compensate for the alteration in inlet pressure, and therefore the outlet pressure rises as well (the figures are indicative only).

Regulators are primarily required because both supply and demand pressures are continually changing. There are a number of reasons for variations in pressure, including:

- changes in appliance demand, such as turning the cooker burners off and on
- the operation of other high-demand or additional appliances on the same consumer line
- fluctuations in reticulation supply pressures.

Regulators are designed to accommodate these changes and still provide a specific and constant pressure to the burner.

You will now review the design, operation and purpose of the following six regulator types:
- constant pressure appliance regulators
- low-pressure compensating regulators
- dual stage LPG regulators
- first and second stage LPG regulators
- industrial LPG regulators
- service regulators.

The AS/NZS 5601 Gas Installation Standards defines regulator use in more generic categories and, as a result, you may find some variations in terminology that may be confusing at first. Work closely with your teacher so that you can reconcile any minor variations between industry and Australian Standard terminology.

REFER TO AS/NZS 5601 SECTION 1 'GAS PRESSURE REGULATOR'

So that you have an appreciation of how and where these regulators are used, refer to the schematics in Figure 10.3. Also refer to the Standard and familiarise yourself with the basic symbols often used in gas installation and valve train schematics.

REFER TO AS/NZS 5601.1 APPENDIX P 'SYMBOLS USED IN GAS CONTROL SYSTEM DIAGRAMS'

In addition to the main six types of regulator listed above, other types of regulator include Zero Regulators and District Service Regulators, but the use and operation of these devices is outside the scope of normal Type A gas fitting. If necessary, ask your trainer for further information on these and any other regulator types. Now review each of the sections below before proceeding with the rest of the chapter.

Constant pressure appliance regulators

All natural gas appliances in Australia are fitted with some form of appliance regulator. While some are built into the appliance, others are fitted as separate components to the appliance inlet (see Figure 10.4).

In this section, operating the standard constant pressure appliance regulator is examined in detail, the basics of which can be applied to most other types of regulator.

Figure 10.5 shows a cross-section of a constant pressure appliance regulator with all component parts named.

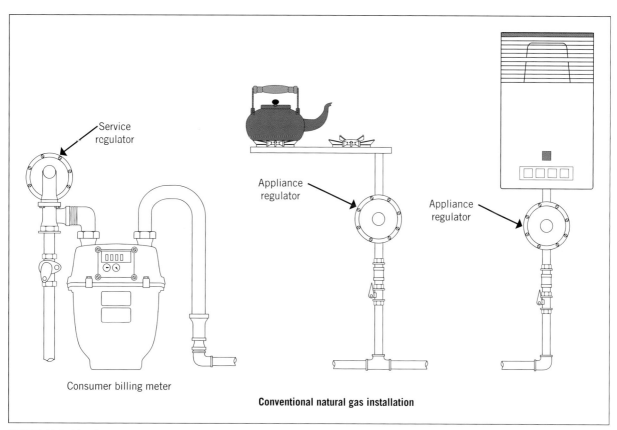

FIGURE 10.3 Simple system layouts showing where regulators are applied in normal Type A gas fitting practice   *continues*

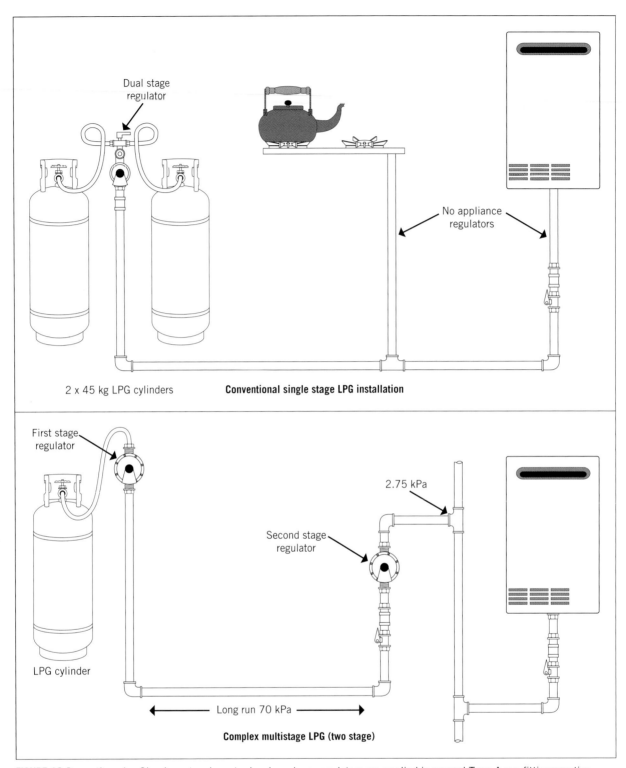

FIGURE 10.3 *continued* Simple system layouts showing where regulators are applied in normal Type A gas fitting practice

FIGURE 10.4 Appliance regulators (a) installed as a separate item and (b) designed as part of the appliance itself

1. *Compression spring.* The compression spring provides the resistant pressure against that of the gas below the diaphragm. Compression springs are specifically designed for each type of regulator. Never stretch or cut springs. If the regulator does not perform to standard, replace it or contact the manufacturer.
2. *Bearing plate.* The bearing plate provides a degree of rigidity to the diaphragm, preventing it from ballooning inside the regulator. It also acts as a wear-resistant base for the adjustment spring.
3. *Diaphragm.* This is a neoprene membrane that separates the two halves of the regulator. It acts as a sensing element subject to changes in inlet and outlet pressures.
4. *Gas inlet.* Meter outlet pressures vary from region to region and therefore inlet pressures for standard Type A installations will also vary from approximately 1.5–3 kPa. Most Type A appliance regulators are designed to have an inlet pressure no greater than 3.5 kPa.
5. *Valve.* Attached to and subject to the movement of the diaphragm, the valve provides a form of variable restriction to the flow of gas moving between it and the valve seat.
6. *Adjustment cap and lock nut.* Adjustment of the regulator is achieved by loosening the locknut and screwing the adjustment cap in or out to change the gas outlet pressure. Screwing the cap in will increase outlet pressure. Screwing the cap out will decrease outlet pressure.
7. *Breather hole.* The breather hole is designed to allow for displacement of air above the diaphragm as it moves up and down. The breather hole is also sized specifically to provide some degree of resistance to airflow, cushioning or slightly retarding such movement to prevent a condition, called 'chatter', where the excessive diaphragm movement will create a noisy vibration.
8. *Gas outlet.* The nominal burner pressure of a Type A natural gas appliance is 1.1 kPa, but this will vary according to manufacturer specifications. You will adjust the regulator outlet pressure to achieve the required pressure at the burners.

Test point (not seen). All modern appliance regulators will incorporate an integral test point to allow for easy testing of regulator outlet pressure.

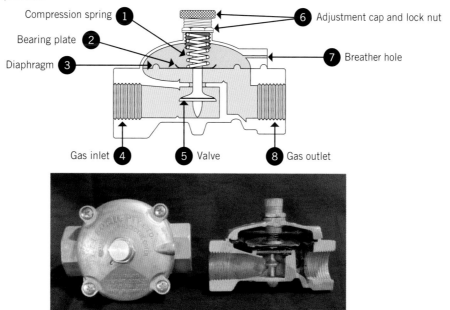

FIGURE 10.5 Constant pressure appliance regulator – component identification

Operation of the constant pressure appliance regulator

In order to gain an understanding of how a constant pressure appliance regulator works, each of the diagrams in **Figures 10.6** to **10.10** demonstrates a condition of gas flow, with explanatory notes describing the actions of the regulator at each stage.

1. In condition A in **Figure 10.6**, you can see that no gas is flowing and that the pressure at both the inlet and outlet is approximately equal. These regulators are not designed to lock up or shut off supply gas, and therefore in a no-flow condition pressure will equalise across the device.
2. Turning on a burner will cause a momentary drop in pressure at the outlet of the regulator as gas exits the burner (**Figure 10.7**). With this drop in pressure, the diaphragm is forced down by the pre-adjusted compression spring, in turn moving the valve away from the valve seat. This allows more gas to flow through the valve orifice, restoring outlet pressure to the specified level.
3. In condition C in **Figure 10.8**, the demand for gas from the use of the additional burner will once again create a momentary drop in pressure at the outlet of the regulator. The diaphragm will fall in response, allowing more gas to flow through the valve orifice.
4. With the regulator providing gas to run two burners, the sudden reduction in demand caused by turning off one burner will cause a brief increase in pressure on the outlet side of the regulator (**Figure 10.9**). This will cause the diaphragm to be pushed upwards against the compression spring until the gas pressure and opposing resistance from the spring equalise. In doing so, the valve has moved upwards, creating greater resistance to the flow of gas through the valve orifice, and thus ensuring the outlet pressure remains constant.
5. The simultaneous use of another gas appliance is likely to cause a momentary reduction of supply pressure at the inlet to the regulator (see **Figure 10.10**). In this instance, the resistance of the compression spring will be greater than that of the opposing gas pressure and the diaphragm will be forced down, allowing more gas to flow past the valve and valve seat. Outlet pressure is thus maintained despite the change in supply pressure.

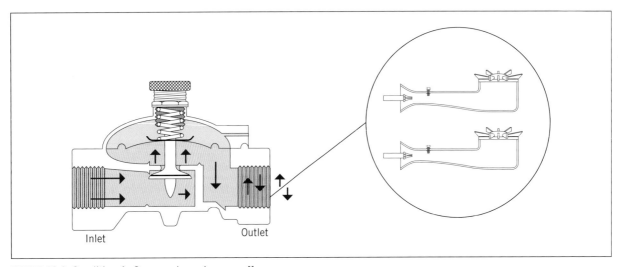

FIGURE 10.6 Condition A: Gas supply on, burners off

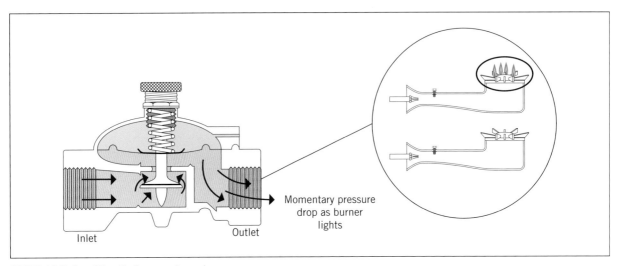

FIGURE 10.7 Condition B: Gas supply on, burner on

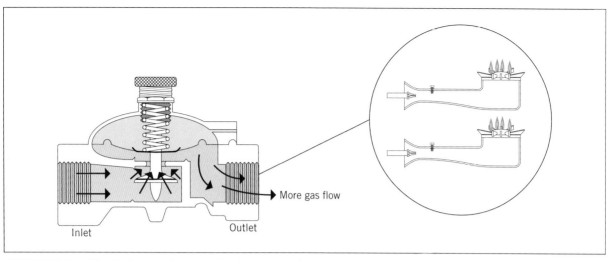

FIGURE 10.8 Condition C: Gas supply on, another burner turned on

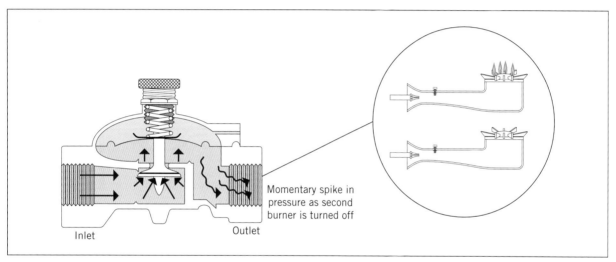

FIGURE 10.9 Condition D: Gas supply on, second burner turned off

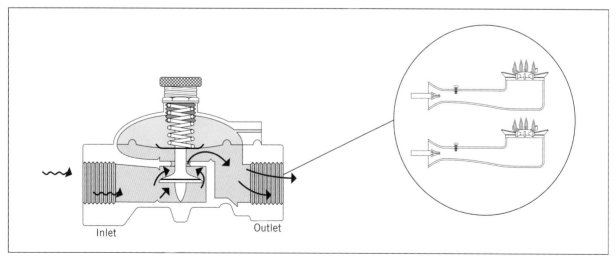

FIGURE 10.10 Condition E: Gas supply pressure changes, burners on

It is worth noting that these changes in regulator function happen very quickly, with slight changes in outlet pressure almost imperceptible at the burner.

Increasing regulator outlet pressure

Adjusting regulator outlet pressure requires that you attach a manometer and screw the adjusting cap in or out, observing the pressure as you do so. By screwing the cap in you are compressing the spring more, which means it will impart greater resistance, and therefore a longer response time, to the gas pressure below the diaphragm. This allows more gas to flow through the regulator before the pressure rises sufficiently to cause the diaphragm and valve to rise and once again restrict the flow of gas.

To increase pressure, screw the cap inwards.

Decreasing regulator outlet pressure

Screwing the adjusting cap out reduces the compression pressure of the spring against the diaphragm and incoming gas pressure. This means that it will be more sensitive to changes in gas pressure, move more quickly in response and cause the valve to restrict the flow of gas sooner. This results in a lower outlet pressure.

To decrease pressure, screw the cap outwards.

Valve effect

You will notice from the cross-sectional diagram in Figure 10.7 that the gas enters the constant pressure appliance regulator from underneath the valve, applying pressure to both the bottom of the valve and the diaphragm against the downward force of the adjustment spring. Therefore, when inlet pressure increases, the diaphragm and valve move upwards, reducing outlet pressure. When the inlet pressure decreases, the opposite occurs. This is fine on small, low-input installations, but this force imbalance causes problems in larger applications.

Installations such as commercial kitchens that feature multiple, high-input appliances may experience momentary but significant changes in supply pressures as appliances are turned on and off. The simple design of the constant pressure appliance regulator cannot effectively deal with such changes, leading to unacceptable fluctuations in outlet pressure. What is needed is a regulator designed so that it can be set for a certain outlet pressure and still maintain this pressure despite changes in supply.

Larger appliances and installations require the use of a slightly different regulator designed to deal with such demands. This regulator is called the low-pressure compensating regulator.

Low-pressure compensating regulators

The basic principles of regulator function are the same for most regulator types. However, the low-pressure compensating regulator has a different internal layout that is designed to overcome problems caused by the condition known as 'valve effect'. You will find this kind of regulator on larger domestic appliances, commercial catering appliances and on an industrial valve train (see Figure 10.11).

Figure 10.12 shows a cross-section of a low-pressure compensating regulator. You can see that internally this

FIGURE 10.11 Low-pressure compensating regulators used as part of an industrial boiler valve train

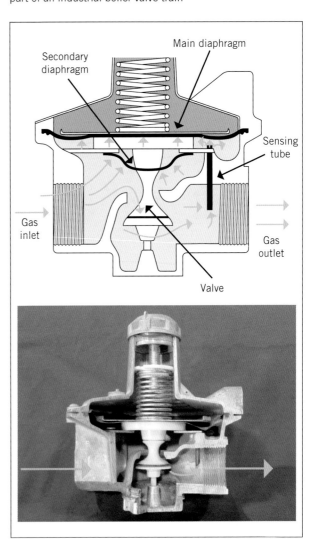

FIGURE 10.12 A cross-section of a low-pressure compensating regulator

regulator differs considerably from the constant pressure appliance regulator. The regulator incorporates a second subsidiary diaphragm of the same surface area as the valve. Inlet gas pressure pushes down on the valve but also up against the subsidiary diaphragm. Because they have the same surface area, this two-way application of pressure would cancel out any movement of the valve, so valve effect cannot occur.

As gas passes through the valve into the bottom chamber of the regulator, it is able to apply pressure to the larger main diaphragm via a sensing tube or channel. Adjustments to set pressure and movement of the diaphragm to maintain constant outlet pressure can occur only via pressure applied to this main diaphragm with its larger surface area.

Dual stage LPG regulators

There are some considerable differences between LPG and natural gas installations. In Figure 10.13, you can see that LPG systems do not generally use appliance regulators, as each appliance is designed to operate at a minimum inlet pressure of 2.75 kPa. Instead, it is common for one regulator to control pressure for the whole installation.

LPG cylinder and tank pressure will rise and fall in a direct relationship to changes in ambient temperature, and it is quite common for these storage vessels to have internal pressures in excess of 1000 kPa. In simple

1. *Compression spring.* The compression spring provides the resistant pressure against that of the gas below the diaphragm. Compression springs are specifically designed for each type of regulator. Never stretch or cut springs. If the regulator does not perform to standard, replace it or contact the manufacturer.
2. *Secondary diaphragm.* This is a neoprene membrane that separates the two halves of the main regulator body. It acts as a sensing element subject to changes in inlet and outlet pressures.
3. *Fulcrum arm.* This provides the linkage between the valve and the diaphragm.
4. *Primary regulator and diaphragm.* Cylinder pressure is reduced to approximately 40 kPa by the primary regulator and diaphragm.
5. *Gas inlet.* The inlet is subject to actual cylinder/tank pressure.
6. *Valve.* Attached to the fulcrum arm and subject to the movement of the diaphragm, the valve provides a form of variable restriction to the flow of gas moving between it and the valve orifice seat. It also enables the LPG regulator to entirely close the inlet when no gas is being used, preventing high pressures entering the system.
7. *Relief vent.* If the inlet valve of the LPG regulator was to become faulty or obstructed by pipe debris, high pressure gas would continue to leak through the valve and may rise to dangerous levels. To prevent this, a relief vent is incorporated into the centre of the diaphragm and bearing plate and is designed to release this pressure to atmosphere via the vent outlet.
8. *Test point.* Most larger-sized LPG regulators will incorporate a tapping to allow the fitting of a test point or gauge.
9. *Bearing plate.* The bearing plate provides a degree of rigidity to the diaphragm, preventing it from ballooning inside the regulator. It also acts as a wear-resistant base for the adjustment spring.
10. *Gas outlet.* The nominal working pressure of a Type A LPG appliance is 2.75 kPa, but this will vary according to manufacturer specifications. You will adjust the regulator outlet pressure to achieve this minimum appliance inlet pressure at all appliances.
11. *Vent outlet.* The vent outlet is designed to allow for displacement of air above the diaphragm as it moves up and down. The breather hole is also sized specifically to provide some degree of resistance to airflow, cushioning or slightly retarding such movement to prevent a condition called 'chatter', where the excessive diaphragm movement will create a noisy vibration. Some LPG regulators use a neoprene one-way flap to act as a restricted breather outlet that, in the event of a diaphragm rupture, is designed to open fully to relieve pressure.
12. *Adjustment disc.* Adjustment of the regulator is achieved by removing the dust cap and, using a suitable tool, screwing the adjustment disc in or out to change the gas outlet pressure. Screwing the disc in will increase outlet pressure. Screwing the disc out will decrease outlet pressure.

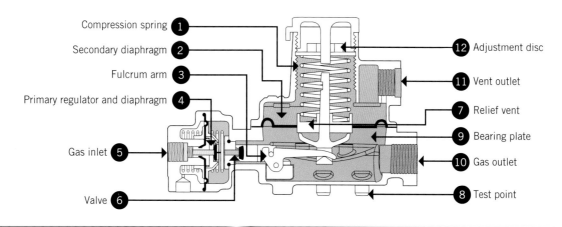

FIGURE 10.13 Dual stage LPG regulator

terms, the dual stage regulator must reduce this high, fluctuating pressure to a minimum inlet pressure of 2.75 kPa at the inlet of each appliance. Importantly, all LPG regulators (except some industrial types) include an over-pressure relief valve to ensure that any excess pressure is vented safely to the atmosphere in the event of regulator failure.

The dual stage regulator is actually two regulators in one and is able to reduce the pressure in two steps, hence the name 'dual stage' or 'integral two stage' regulator. Figure 10.13 shows a cross-section of a dual stage LPG regulator. You can see that it is mechanically very different from the constant pressure appliance regulator.

Operation of the dual stage LPG regulator

In order to gain an understanding of how a dual stage LPG regulator works, each of the following diagrams demonstrates a condition of gas flow. The explanatory notes describing the actions of the regulator at each stage are listed.

1. With the gas turned off (Figure 10.14), the spring pressure forces the diaphragm down and, as a result, the valve is in the open condition.
2. As full cylinder pressure is applied to the regulator (Figure 10.15), the gas makes its way up to the primary regulator, where it passes through the small inlet and applies pressure to a primary diaphragm and spring. In doing so it will cause the high cylinder pressure to be reduced to approximately 45 kPa.
3. Having been substantially reduced in pressure, the gas now passes into the main regulator chamber below the secondary diaphragm, which is then forced up against spring resistance. As it does so, the valve, acting via fulcrum arm movement, draws closer to the valve seat and begins to restrict the flow of gas. The diaphragm will continue to rise until the valve sits firmly against the valve seat, 'locking up' the supply of gas. At this point, static or 'lock-up' pressure measured at the outlet of a standard domestic regulator should not exceed 3.5 kPa.

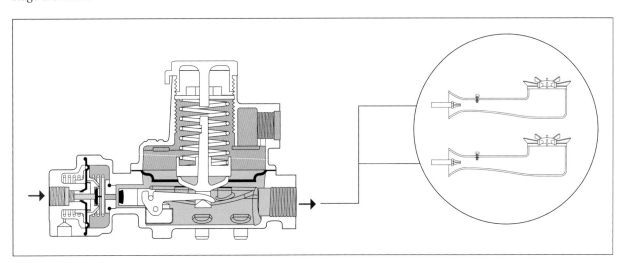

FIGURE 10.14 Condition A: No gas on

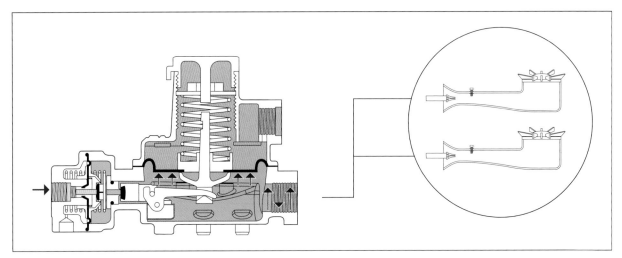

FIGURE 10.15 Condition B: Gas supply on, burners off

4. Turning on a burner (Figure 10.16) will cause a momentary drop in pressure at the outlet of the regulator as gas exits the burner. With this drop in pressure under the diaphragm, spring pressure will cause it to lower and, via the fulcrum arm connection, open the valve, allowing gas into the regulator. The incoming gas pressure will in turn force the diaphragm up until equilibrium is achieved between the downward spring pressure and the dynamic gas pressure.

5. Once again, the demand for gas from the use of an additional burner (Figure 10.17) will create a momentary drop in pressure at the outlet of the regulator. The diaphragm will fall in response, allowing more gas to flow through the valve orifice. The incoming gas pressure will, in turn, force the diaphragm up until equilibrium is achieved between the downward spring pressure and the dynamic gas pressure.

6. With the regulator providing gas to run two burners, the sudden reduction in demand caused by turning off one burner (Figure 10.18) will cause a brief increase in pressure on the outlet side of the regulator. This build-up of pressure will cause the diaphragm to be pushed upwards against the compression spring resistance until the gas pressure and opposing resistance from the spring equalise. In doing so, the fulcrum arm transfers this movement to the valve, moving it closer to the valve seat and so providing greater restriction to the flow of gas.

7. Changes in ambient temperatures will cause a wide variation of supply pressures from the cylinder (Figure 10.19). As the compression spring has been adjusted to provide a set resistance and consistent outlet pressure, the diaphragm and valve assembly will continue to move up and down to accommodate such inlet pressure fluctuations. Outlet pressure is thus maintained despite the change in supply pressure.

Increasing regulator outlet pressure

Having attached a manometer to an appropriate test point with the aim of adjusting the pressure, remove the regulator dust cap and screw the adjusting disc in a clockwise (inwards) direction. By screwing the adjusting disc in, you are compressing the spring more, with the

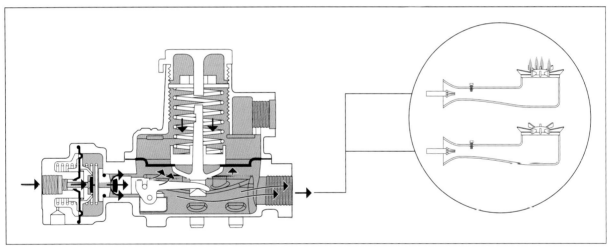

FIGURE 10.16 Condition C: Gas supply on, burner on

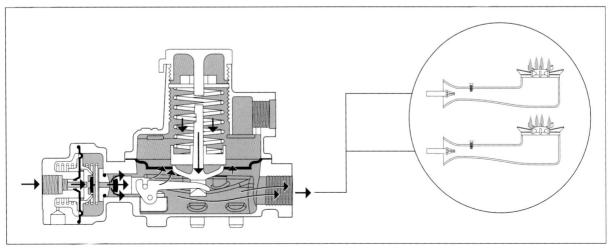

FIGURE 10.17 Condition D: Gas supply on, another burner turned on

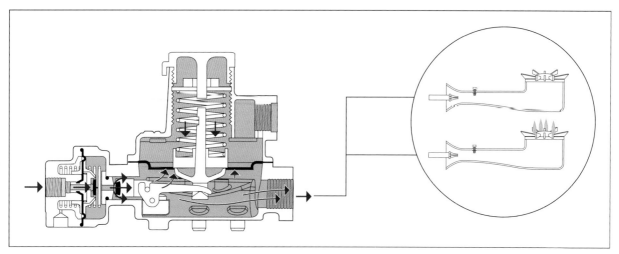

FIGURE 10.18 Condition E: Gas supply on, second burner turned off

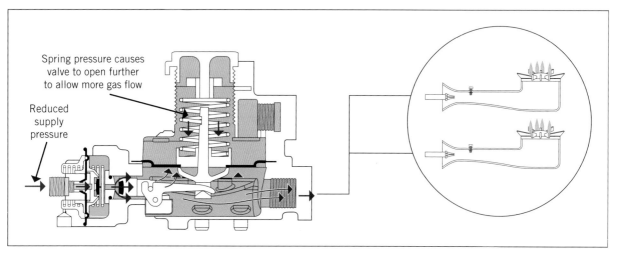

FIGURE 10.19 Condition F: Gas supply pressure changes, burners on

result that it imparts greater resistance and therefore a longer response time to the gas pressure below the diaphragm. This allows more gas to flow through the regulator before the pressure rises sufficiently to cause the diaphragm to rise and once again restrict the flow of gas.

Decreasing regulator outlet pressure

Screwing the adjusting cap out or anticlockwise reduces the compression pressure of the spring against the diaphragm. This means that it will be more sensitive to changes in gas pressure, move more quickly in response and cause the valve to restrict the flow of gas sooner. This results in a lower outlet pressure.

Operation of the over-pressure relief valve

As stated earlier in this section, LPG regulators have the job of reducing the very high pressures found in cylinders (often over 1000 kPa) down to the much lower pressures used by appliances (generally around 2.75 kPa). If the main valve becomes impregnated with contaminants such as copper swarf or dirt, it will fail to completely seal off. As a result, gas will continue to flow slowly into the regulator and the pressure inside the regulator and consumer piping may rise to dangerous levels. This could be catastrophic. You will encounter this on occasion when conducting a 'lock-up' test. Your manometer will display an unacceptable increase in pressure, even though the appliances are all off. This is often known as 'regulator creep' and can often result in downstream pipework and appliances being exposed to dangerous cylinder pressures. While some larger regulators can be serviced, in many cases it means that the upstream regulator must be replaced.

In order to prevent dangerously high gas pressures from building up in the consumer piping and appliances, LPG regulators include a device called an 'over-pressure relief valve' that is designed to vent excessive gas pressure to the atmosphere. Look carefully at Figure 10.20 and you will identify the relief valve and relief spring located at the centre of the diaphragm.

How does it work?

When the pressure rises, the secondary diaphragm will rise against the main compressing spring until the inlet valve closes on the valve seat and the diaphragm cannot

FIGURE 10.20 LPG regulator over-pressure relief valve and spring

rise any more. If the pressure continues to rise and exceeds the relief spring resistance of approximately two to three times the set outlet pressure of the regulator, the valve will start to lift away from the diaphragm (see Figure 10.21). This allows gas to escape from below the diaphragm and vent to atmosphere via the relief vent terminal.

First and second stage LPG regulators

If you review Figure 10.3, you will see two different LPG systems in use. A 'single stage' LPG system relies upon one regulator at the cylinder providing a low pressure supply sufficient to provide a minimum inlet pressure of 2.75 kPa for the whole installation. This is normal installation practice for standard domestic and small commercial installations.

However, at such low pressure long pipe runs or appliances with high gas input demand would require larger pipe diameters to be used, in order to counteract frictional pressure drop in the pipe system. At a certain point, this may become uneconomical and a multistage LPG system becomes a viable option. In a multistage system, gas is supplied at much higher pressures that can overcome frictional pressure losses, and is reduced to the required pressure in stages much closer to the point of use. Such a system is commonly called a 'two stage system' but many modern LPG (and some NG) installations can comprise additional third and fourth stages. Therefore, the term 'multistage system' is a more accurate description.

A multistage LPG system allows the gas fitter to use smaller pipe sizes over longer distances by reducing the supply pressure in two or more separate stages (Figure 10.22).

1 *Stage 1.* In this example, you can see that cylinder pressure is reduced to approximately 70 kPa at the outlet of the first stage regulator. The use of a higher pressure significantly reduces the impact of frictional pressure losses, and allows the gas fitter to use the less expensive, smaller-diameter piping right up to the inlet of the second stage regulator.
2 *Stage 2.* The second stage regulator is normally sited close to the appliance installation itself. The inlet pressure of approximately 70 kPa is reduced to provide the required nominal 2.75 kPa for each LPG appliance.

First and second stage LPG regulators operate on the same principles as other LPG regulators, incorporating both dual stage pressure reduction and relief venting.

Industrial LPG regulators

Industrial-type LPG regulators are designed specifically to control pressure for industrial-type gas appliances or systems. Common applications for industrial LPG regulators include direct-fired air heaters, pottery kilns and automotive powder-coat ovens. Firing glazed clay vessels and other objects in a pottery kiln requires the creation of variable combustion atmospheres within the kiln. The potter adjusts supply pressures in order to achieve different effects in the glazing.

An industrial regulator is often characterised by the lack of an integral relief valve and relief vent terminal.

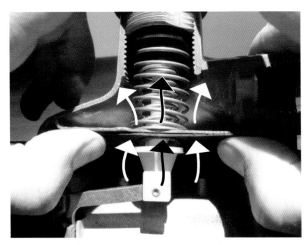

FIGURE 10.21 Relief valve opening at approximately two to three times the set outlet pressure

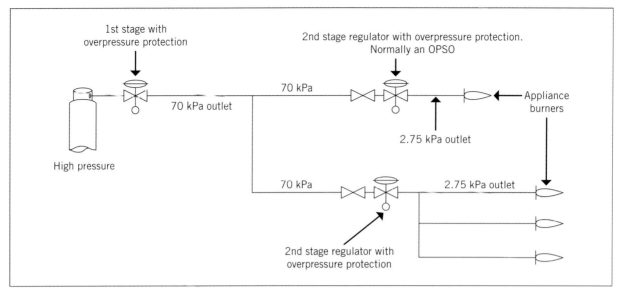

FIGURE 10.22 Detailed schematic of a multistage system

This means that you are not permitted to install an industrial regulator in anything but industrial applications. Note that instead of a large relief vent, an industrial regulator incorporates only a simple and small breather vent and, as a result, certain applications will require installing a separate relief valve. Figure 10.23 shows an example of a commonly found industrial regulator. If you look carefully, you can see the small breather vent. Remember, they are NOT permitted for use in normal Type A consumer piping systems.

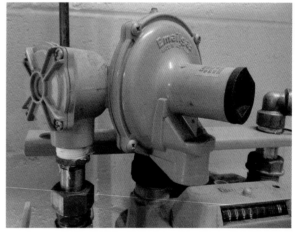

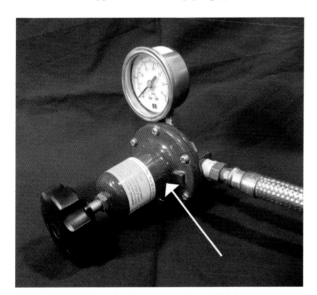

FIGURE 10.23 Example of a commonly used high-pressure industrial regulator (with arrow showing breather vent)

Service regulators

A service regulator is installed by the gas distributor at the inlet of the consumer billing meter and is used to reduce reticulation mains supply pressures (see Figure 10.24).

The internal design and operation of most modern service regulators is quite similar to that of an LPG

FIGURE 10.24 Two types of service regulator

regulator. However, one notable difference is the inclusion of a filter, which you can see in **Figure 10.25**.

FIGURE 10.25 Cut-away view of a service regulator filter

It is important to note that the service regulator is the property and responsibility of the gas distributor.

> **FROM EXPERIENCE**
>
> Unless specifically permitted by the gas distributor and technical regulator, a gas fitter is not allowed to adjust a service regulator.

Single stage LPG regulators

You will often come across LPG installations fitted with an old single stage regulator, similar to the one shown in the foreground in **Figure 10.26(a)**. It is important to upgrade old single stage LPG installations with a new and safer dual stage regulator.

In **Figure 10.26(a)**, you can see the obvious physical differences between the single stage regulator in the foreground and the modern dual stage regulator to the rear. Of particular note is the extreme difference between the relief vents of each regulator, as seen in **Figure 10.26(b)**. The dual stage regulator features a very large, screened vent terminal; compare this to the inadequately sized slot beneath the dust cap of the single stage regulator.

Many old-style single stage regulators had inadequate relief valves and vents, and were not designed to prevent cylinder-pressure gas from entering the building in the event of a malfunction. It is not uncommon to test these regulators with a manometer and find them to be in an over-pressure condition. These regulators are now becoming quite old and, in most areas, if you are going to work on such an installation, the technical regulator will require you to replace single stage regulators with the much safer dual stage regulators.

Regulator service life

There is no overall rule that states when a regulator should be replaced. However, most regulator manufacturers recommend a maximum service life of 15 years. You can determine regulator age from the date of manufacture that is stamped into the body of most regulators. Service life depends to some extent on the type of regulator and its duty requirements. If you were changing a 10-year-old cooker, you would be well advised to change the appliance regulator too. However, larger consumer piping regulators that might be found in a commercial installation could easily be serviced with the replacement of springs, valves and diaphragms. Any regulator that has been subjected to flooding should always be replaced. In all cases if you are in any doubt, contact the manufacturer for advice.

Definitions used in the Standard

You have just been introduced to a range of regulator types commonly used in NG and LPG installations. It is important to also understand how these regulators are categorised and defined within the Australian Standard, so you can accurately apply each of their specific

(a)

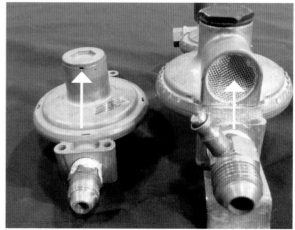

(b)

FIGURE 10.26 A single stage regulator and a modern dual-stage regulator, showing the differences in their breather or relief vents

INSTALL GAS PRESSURE CONTROL EQUIPMENT **199 GS**

requirements. Of these, the definition of 'consumer piping gas pressure regulator' is particularly important to understand.

You should be familiar with the definitions within the Standard for the following terms:
- Gas pressure regulator
- Automatic change-over regulator
- Consumer piping gas pressure regulator
- Cylinder regulator
- Gas appliance regulator.

REFER TO AS/NZS 5601.1 AND REVIEW 'DEFINITIONS'

COMPLETE WORKSHEET 1

LEARNING TASK 10.1
PRESSURE CONTROL FUNDAMENTALS – REGULATORS

1. What type of regulator is installed upstream of the consumer billing meter?
2. Are constant-pressure appliance regulators only built into natural gas appliances?
3. What is the 'valve effect' and how can it be reduced?
4. What is required if you come across a single stage LPG regulator?

Identify requirements for gas pressure control equipment

Having gained an understanding of basic regulator types and principles of operation, it is now time to consider requirements related to the installation of pressure control equipment on the job.

Work, health and safety considerations

For all new work, work, health and safety considerations relating to the installation of pressure control equipment varies little from most plumbing work. You still need to adhere to relevant site SWMS, and apply industry-approved risk assessment and mitigation protocols through site or company JSAs.

Of particular note, however, is any work on existing installations and the associated repair and replacement of regulators. This work is particularly hazardous, and the following must be incorporated into your job task JSA.
- Gas isolation requirements: confirm isolation options and ensure the line can be kept isolated during all work.
- Electrical safety: making a break in any piping system or removing a component is always hazardous. You are at risk of electrocution if your body creates an alternative path to earth once pipework continuity is broken. You must always apply bonding straps across any proposed break in pipework and ensure they stay in place until all work is complete. This is not a negotiable procedure and is stated clearly in the Standard (review Chapter 5 once again if necessary).
- Gas purging requirements: where the line must be opened to replace a component, you must have a clear purging plan in place (see Chapter 11 for more details on purging).

REFER TO AS/NZS 5601.1 AND REVIEW 'ELECTRICAL SAFETY BONDING OR BRIDGING'

Review the plans and specifications

It is generally the case that for most larger installations and commercial or industrial applications you will be provided with well-detailed plans and specifications showing the layout and requirements of the proposed gas installation. However, you need to consider the following:
- Never assume the designer is always right. Mistakes can be made and site conditions can change during the period of construction. Despite what it may say on a set of plans, it is the licensed gas fitter who takes responsibility for the installation; therefore, never simply install to plan without checking it complies with the Standard.

For smaller projects and general residential work, it is actually not common to find specific layout requirements in the plans. The designer will normally provide locations of appliances, but then state that all work must be in accordance with the relevant Standard. It is up to you to design the system either on the day or prior to going on site.

Confirm gas type, load and supply pressures

Gas type. On all jobs you must confirm the type of gas that is available. It is not uncommon for NG to be available down the street while some properties in the same area are on LPG. Always check.

Appliance load. Determine appliance MJ/h rating to calculate probable peak input requirements.

Supply pressure. If LPG is to be used, you need to determine if a single stage system or multistage system is proposed. You also must check with the LPG supplier and confirm the sort of cylinders they can get on site.
- Exchange – exchange cylinders can normally be supplied for most sites, although in hilly areas or sites with lots of steps, the supplier may refuse to supply 45 kg cylinders and restrict you to more portable 13.5 kg cylinders.

- In situ fill – do not assume in situ fill cylinders are available in all areas. The supplier may not run tanker trucks in that location, the site may be unsafe or too hilly. You also need to confirm the length of hose run, and verify that the truck driver has a direct line of sight between the cylinders and the truck itself.

All of these factors may have an influence on what type of system and pressure control you apply.

Confirm design compliance

You need to know the requirements related to the installation of consumer piping gas pressure regulators as defined in the Standard. Review the relevant section in the Standard and, in particular, each of the following main points and apply them where necessary in your installation.

- Location – you must always consider location requirements whenever siting regulators. Of these, accessibility and protection from physical damage are two important considerations. It is also worth noting that the Standard makes note that a manual isolation valve should be installed upstream of a consumer piping regulator to facilitate servicing. While noted as discretionary, in this instance it should be regarded as good trade practice to always use isolation valves upstream of all consumer piping regulators (see Figure 10.27).
- Prohibited locations – when reviewing plans and specifications always remined yourself of these prohibited locations and ensure the proposed design complies.
- Requirement for regulators – the need for consumer piping regulators is specifically related to the pressure intended in each section of the installation. The Standard details three specific requirements.
- Outlet pressure notice – wherever the outlet pressure exceeds 1.5 kPa for NG and 3.5 kPa for LPG, the outlet pressure of the regulator must be noted on a sticker or notice near the regulator itself.
- Requirement for pressure test points – the installation of pressure test points is a commonly missed requirement in many installations. The Standard is quite explicit on their location and use. You must ensure a test point is installed at the outlet of any gas pressure regulator. Some regulators have an integral outlet test point that will satisfy this requirement. Where a regulator test point is blanked off, you can replace with a test point suitable for the outlet pressure. You must also install a test point at the inlet to an appliance where no other option exists. Lastly, it is important to note that where pressures exceed 7 kPa you need to use a self-sealing test point (see Figure 10.28).

REFER TO AS/NZS 5601.1 AND REVIEW 'INSTALLING GAS EQUIPMENT'

Determining regulator venting requirements

Wherever possible you should try to avoid installing regulators inside a building. However, many appliances and installations require the use of various regulators indoors. Where this is the case, you need to determine whether such regulators need venting to the outside atmosphere.

If a regulator diaphragm ruptures, or the internal relief valve opens, gas will escape from the regulator into the room via the breather hole. Some gas pressure regulators may not require venting to outside because of their small size and low gas pressure. AS/NZS 5601.1 contains specific information on how to determine venting requirements, and this is dealt with in the first part of this section.

Once you have confirmed the need for regulator venting, you need to work out the size of the vent itself. An incorrectly sized vent line could cause back pressure resistance upon the top of the diaphragm, altering the outlet pressure of the regulator. Correct vent sizing is therefore vital to the correct performance of the regulator.

FIGURE 10.27 Isolation valves should be installed upstream of consumer piping regulators wherever possible

FIGURE 10.28 Pressure test points must be installed at the outlet of all gas pressure regulators

Is it a 'breather vent' or a 'relief device'?

Whenever a regulator is sited indoors and you consider its need for venting, you must first determine whether it has a breather vent or a relief device. It is important you establish this immediately because of the following points:

- *All* relief devices *must* be vented to the outside atmosphere.
- Venting of regulator breather holes is dependent upon gas supply pressure, room volume and breather hole orifice diameter and may, or may not, be required accordingly.

Breather vents

All standard regulators have some form of breather vent (see Figure 10.29). The breather hole or vent allows atmospheric pressure to act upon the topside of the regulator diaphragm. The movement of the diaphragm during normal use displaces air in and out of the top regulator chamber. This air flows through the breather vent orifice.

FIGURE 10.29 A breather vent from a low-pressure compensating regulator (the threaded terminal allows for the use of an adaptor if venting to atmosphere is required)

If in doubt, contact the manufacturer of the regulator or your technical regulator to confirm this.

Breather vents may need to be vented to the outside atmosphere, but this will depend upon the relationship between room volume, gas supply pressure and orifice size.

REFER TO AS/NZS 5601.1 AND DEFINE 'BREATHER VENT'

Where such an orifice above the regulator diaphragm has this sole purpose and no internal relief device exists, it is classed as a breather vent. Most natural gas appliance regulators fall into this category.

REFER TO AS/NZS 5601.1 'VENTING'

Relief vents

A regulator relief vent is designed to allow gas to escape from the regulator in the event of an over-pressure condition or diaphragm rupture. It also acts as a breather hole, but is normally of a larger size to enable the passing of large volumes of gas.

As all LPG regulators are fitted with an internal relief valve, they must always be vented to the outdoors. There are no exceptions (subject to manufacturer advice).

In Figure 10.30, you can see the central relief valve and the large relief vent terminal outlet. The plastic insert in the vent terminal features a one-way neoprene flap with a small breather orifice that will open up fully in the event of a full relief discharge. This restricted orifice prevents regulator 'chatter' or humming during normal operation.

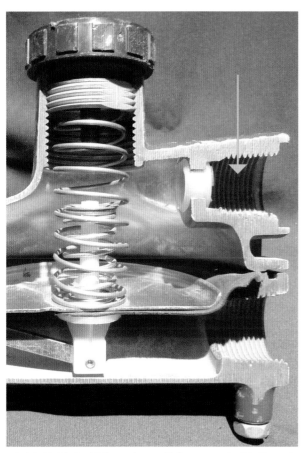

FIGURE 10.30 An LPG regulator relief vent terminal indicated by the arrow

Determining breather venting requirements

AS/NZS 5601.1 provides a solution to breather venting requirements based upon the worst-case scenario of a diaphragm rupture allowing gas into a room or enclosure. Where devices with a breather vent are used in non-domestic installations, you must refer to the tables provided in Appendix G of the Standard or, where these are out of range, use the formulas provided in Section 5 of the Standard.

REFER TO AS/NZS 5601.1 'DISCHARGE FROM BREATHER VENTS' AND APPENDIX G

The tables used in Appendix G can determine minimum effective room size, orifice diameter or supply pressure as long as you know two of the following three values:
- room effective volume (m^3)
- breather vent orifice diameter (mm)
- service inlet pressure (kPa).

The effective room volume takes into account the relative density of each family of gases and is determined as follows:

NG – area of the room that is above the device
LPG – area of the room that is below the device.

AS/NZS 5601.1 provides a worked example on how to use the tables in Appendix G.

EXAMPLE 10.1

You are to install a boiler in an enclosure and the constant pressure appliance regulator has a breather vent orifice of 0.9 mm and an inlet pressure of 3 kPa. To determine the minimum effective room volume permitted for this breather vent to remain unvented, follow these steps:
1. For a device inlet pressure of 3 kPa, use Table G1.
2. Look down the column showing inlet pressures until you find 3 kPa.
3. Move right along the row until it lines up with the column showing a breather vent diameter of 0.9 mm.

Where these two values intersect, you can see that the minimum effective room size must not be less than 12.4 m^3.

Domestic installations

A breather vent from a device located in a domestic Class 1 or 2 building must not exceed 0.7 mm in diameter.

REFER TO AS/NZS 5601.1 'DOMESTIC APPLICATIONS'

Vent line materials

A wide selection of materials is approved for vent line installation. Although the use of copper tube is generally most common, other materials can be used subject to suitability and limitations.

REFER TO AS/NZS 5601.1 SECTION 4 'VENT LINE MATERIAL'

Vent sizing

Vent lines from a regulator must be sized to ensure that any back pressure does not prevent the regulator from operating correctly or, in the case of a diaphragm rupture, impede the flow of gas escaping to atmosphere.

REFER TO AS/NZS 5601.1 'SIZE OF VENT LINE FOR A CONSUMER PIPING GAS PRESSURE REGULATOR BREATHER VENT OR RELIEF DEVICE'

Where the distance from the regulator vent connection to the proposed termination of the vent line is less than 10 m, the size of your vent line must be the same nominal size as the connection itself.

EXAMPLE 10.2

If the vent connection size is 20 mm and the proposed length of run for the vent line is 7 m, you would use DN 20 pipe for the vent.

Where the distance from the regulator vent connection to the proposed termination of the vent line is greater than 10 m but less than 30 m you must use one standard size larger pipe.

EXAMPLE 10.3

If the vent connection size is 10 mm and the proposed length of run is 13 m, you would use DN 15 pipe for the vent.

Vent sizing for safety shut-off systems

When determining the vent size of a safety shut-off system, you need to use a different calculation. The most commonly encountered example of a safety shut-off system is an over-pressure shut-off (OPSO) device.

In Figure 10.31, you can see a 50 mm vent going through the outside wall to atmosphere. The data plate of this OPSO device states that the vent valve orifice has a diameter of 12.1 mm. You use this figure in your calculation of vent size. If the data plate is missing or you are still unsure, contact the regulator manufacturer or supplier to confirm.

To determine the venting requirements of such a device, you need to refer again to AS/NZS 5601.1.

REFER TO AS/NZS 5601.1 'SIZE OF VENT LINE FOR A VENTED SAFETY SHUT-OFF SYSTEM'

FIGURE 10.31 A vented OPSO regulator

HOW TO

To size the vent, follow these steps.
1 Determine the vent valve nominal size (in this case, DN 20) and find this in the left-hand column of the sizing table.
2 Look horizontally across to the first number in the sizing table. The figure in the middle of the table represents the maximum distance you can run the sized vent line, as indicated in the top row. In this example, you can run a DN 20 vent a maximum distance of 10 m. Alternatively you could run a DN 25 vent a maximum of 25 m or a DN 32 vent a maximum distance of 64 m.

LEARNING TASK 10.2

DETERMINING REGULATOR VENTING REQUIREMENTS

1 When installing a 'relief device' indoors, where must it terminate?
2 When installing breather vents, must they always be vented to the outside atmosphere?

Determining over-pressure protection requirements

Modern gas systems often feature supply pressures ranging from standard low-pressure requirements for domestic situations to multistage installations with variable pressures branching off a high-pressure supply line. Gas components, piping and devices are rated to allow only a certain maximum inlet pressure. If this is exceeded, the component, piping or device may be damaged or malfunction.

AS/NZS 5601.1 requires that piping and components be protected from excessive inlet pressures. You will normally encounter these requirements in large residential or commercial/industrial applications. In this section, you will gain a basic understanding of over-pressure protection systems and how to employ them in gas piping installations.

Key definitions and symbols

Before going any further, you must understand some key terms. Confirm these definitions in AS/NZS 5601.1.

- *Rated working pressure*. This is defined as the maximum allowable inlet pressure specified by the manufacturer of the component, pipe or device. You can normally determine this by looking at the data plate or sticker. Some overseas-made products may require unit conversion to kPa. If you cannot find any information, you will need to refer to the latest certification list and/or the technical regulator.
- *Maximum over-pressure*. This is the maximum pressure at which any portion of an installation remains safe. If not stated on a component, then this is deemed to be 1.5 times the rated working pressure.
- *Operating pressure*. This is defined as the gas pressure that any part of an installation will be subjected to under normal operating conditions.
- *Over-pressure protection*. This is defined as a system that prevents the pressure in piping or an appliance from exceeding a predetermined point. Having determined the rated working pressure and the maximum over-pressure, you employ an over-pressure protection system to provide the necessary protection.

A number of schematic drawings are used to graphically represent gas installations and over-pressure protection systems. Standard gas fitting symbols are used. Refer to Figure 10.32 and the Standard as you progress through the section.

REFER TO AS/NZS 5601.1 APPENDIX P

Over-pressure protection – what's it all about?

An over-pressure condition can occur when a fault in a regulator allows the downstream pressure to rise to a point where components and appliances might be damaged or caused to operate outside of their design parameters. This situation could lead to very hazardous conditions and must be prevented.

For every component, pipe, device and appliance you install, you need to ask yourself this question: What is the rated working pressure and maximum over-pressure of the gas system?

If the supply pressure is greater than the maximum over-pressure of components downstream from the regulator, then you will need to employ some form of over-pressure protection.

REFER TO AS/NZS 5601.1 'OVER-PRESSURE PROTECTION'

Two separate requirements for over-pressure protection are detailed in AS/NZS 5601.1. These relate to differences between natural gas and LP gas.

- *Natural gas*. If the operating pressure at the inlet of the gas pressure regulator exceeds 7 kPa and also exceeds the maximum over-pressure of the system downstream of the gas pressure regulator, then over-pressure protection is required.
- *LPG*. If the operating pressure at the inlet of the gas pressure regulator exceeds 14 kPa and also exceeds the maximum over-pressure of the system downstream of the gas pressure regulator, then over-pressure protection is required.

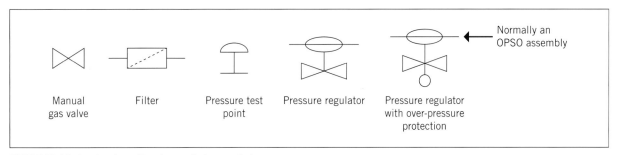

FIGURE 10.32 A selection of basic gas fitting symbols

EXAMPLE 10.4

To help you understand these requirements, the example below shows a schematic layout of an industrial installation that requires over-pressure protection.

You can see at point A that the supply pressure of this NG system is 70 kPa. The maximum over-pressure of the downstream system B is only 7.5 kPa. Because the inlet pressure to the consumer piping regulator exceeds 7 kPa *and* the maximum over-pressure of the downstream system, over-pressure protection is required. If the standard consumer piping regulator C were to fail, downstream components would probably be damaged or malfunction. Components, piping and devices downstream of point C require protection from an over-pressure protection device. This industrial installation requires an over-pressure protection solution at point C.

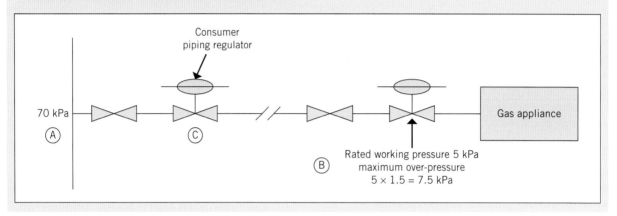

Types of over-pressure protection devices

Three types of over-pressure protection will be detailed here:
1 relief valves
2 monitor regulators
3 over-pressure shut-off devices.

Relief valves

Relief valves may be an integral part of a regulator, as they are in every modern LPG regulator, or they may be installed as separate components in the consumer piping. Relief valves are normally installed in the line between the upstream and downstream regulator. If the pressure from the upstream regulator were to exceed design requirements, the relief valve would open and vent gas to the atmosphere until the gas supply returned to the correct pressure.

The advantages of relief valves are that:
- the gas system remains operational for the customer
- they are economical.

The disadvantages are that:
- a vent line is required to ensure that any gas released is vented safely
- because the system continues to run, a pressure problem may not be discovered for some time.

There are many types of relief valve. The two images in Figure 10.33 show the cross-section of an inline relief valve and a photograph of a valve in an offset configuration. Manufacturers provide comprehensive sizing and product literature to assist you in choosing the correct device.

Monitor regulators

A monitor regulator assembly describes the installation of two regulators installed to work together in order to protect the downstream system. Follow the text relating to Figure 10.34 to gain an understanding of how this form of over-pressure protection works. The colours denote the different pressures within the assembly.

The 'blue' pressure will operate the monitor regulator if the controlling regulator malfunctions or is set incorrectly. The monitor regulator can only alter 'yellow' pressure if the 'blue' pressure is too high.

The upstream regulator 'A' is known as the monitor regulator. Its diaphragm is isolated from incoming gas pressure and does not react to it. Under normal conditions it is wide open.

The downstream regulator 'B' is known as the controlling regulator. The diaphragm of this regulator is in direct contact with the incoming gas and will move up and down to regulate pressure.

These two regulators are joined via a control line, 'C'. This control line allows the assembly outlet pressure to be applied under the monitor regulator diaphragm.

Under normal conditions the monitor regulator 'A' does nothing and allows gas to flow right through it, as its diaphragm is not affected by incoming gas pressure. It is the controlling regulator 'B' that reduces system pressure.

If the outlet pressure of the controlling regulator 'B' rises to the set point of the monitor regulator 'A' then this pressure is transferred via the control line 'C' to the underside of the monitor regulator's diaphragm. The diaphragm and valve in 'A' will rise and reduce

FIGURE 10.33 Cross-section of an inline relief and photograph of a valve in an offset configuration

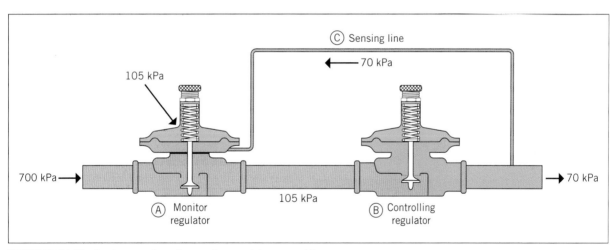

FIGURE 10.34 The basic operation of a monitor regulator assembly

incoming gas pressure in proportion. The controlling regulator outlet pressure is therefore also reduced.

The advantages of a monitor regulator assembly are that:
- the gas system remains operational for the customer
- there is no release of vented gas.

The disadvantages are that:
- there is a higher component cost
- frequent testing is required to check the serviceability of the monitor regulator and to check for blockages.

Over-pressure shut-off devices

Over-pressure shut-off devices can be sourced as separate inline devices, or more commonly as an integral part of a combined regulator–OPSO assembly.

If outlet pressure rises above a predetermined set point, the device will activate and shut off the supply of gas. The device must be manually reset, so bringing attention to the fact that a fault exists.

You can see from the cross-section of an integral regulator–OPSO shown in Figure 10.35 that the 'blue' outlet pressure, as controlled by the regulator, is also applied under the diaphragm of the OPSO via a sensing tube. If this 'blue' pressure were to rise to a point above that of the OPSO trip setting, the diaphragm pin would disengage from the spring-loaded slam shut valve, allowing it to fly forwards and shut off the supply of the 'yellow' inlet gas.

The advantages of over-pressure shut-off devices are that:
- downstream components are completely protected from excessive pressure

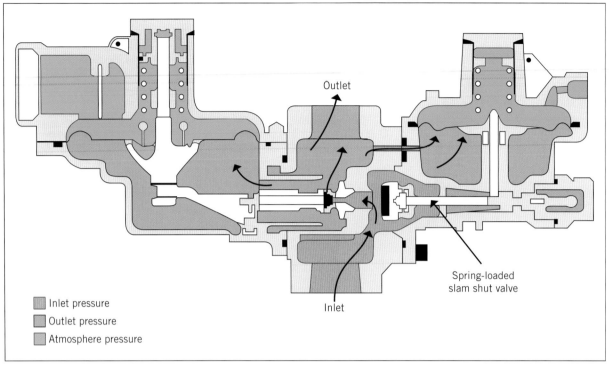

FIGURE 10.35 Cross-section of an integral regulator–OPSO

- the shut-down ensures that someone must reset the device and take the opportunity to find out what is wrong.

The disadvantages are that:
- the customer is left without gas until a gas fitter is available
- they can be subject to nuisance shut-downs if pressures fluctuate or if downstream fast-acting shut-down valves are used.

LEARNING TASK 10.3

DETERMINING OVER-PRESSURE PROTECTION REQUIREMENTS

When is an over-pressure device required for the following installations?
- Natural gas
- LPG

Prepare for installation

Not only must you ensure that all gas pipe and components are certified for use, you must also check that all materials have been delivered in a serviceable condition to ensure that manufacturing faults or pre-purchase damage has not occurred. If you find a fault in a pipe or fitting (such as a crack in a brass casting), you are not permitted to repair it. It must be replaced. Only the best quality materials should be used in gas installations.

Always make an extra effort to protect your materials during storage and handling.
- Keep dust and water out of pipes with caps or plugs.
- Keep fittings in an orderly and clean manner, ensuring that they are not contaminated with dirt and other foreign substances.
- Ensure that plastic and multilayer pipe is not stored in direct sunlight as this will degrade the pipe.
- Avoid rough handling of your pipe, fittings and materials.
- Ensure that brass flared fittings are always stored with the nut screwed on, to prevent damage to the mating faces of the flare.
- Ensure press-fit copper connectors are kept dirt free.

Appropriate tools and equipment

Apart from the general hand tools used on almost all jobs, tools and equipment used for pipe installation will, to a large extent, be dictated by the type of pipe material and jointing method being used. This is particularly important to consider where proprietary piping systems have been specified. Having identified the type of pipe from plans and specifications, you must access relevant websites and installation guides to ensure you know what tools and equipment may be required. The table below shows just a selection of items that may be required.

Material	Tools and equipment
Copper	• Oxy/acetylene brazing equipment • Press-fit tool and matching jaws • Flaring tools • Copper pipe cutters and reamer • Benders • Internal and external bending springs • Expanders • Branch formers • Shifters
Galvanised steel	• Threading stocks and dies • Threading machine • Steel pipe cutting tool and reamer • Steel pipe bender • 'Stillsons' and 'Footprint' grips
Polyethylene	• Electrofusion welding equipment • Rotary peeler • Cleaning solution and wipes • Poly pipe cutter • Butt welder
Nylon, multilayer and PEX	• Proprietary crimping/pressing tools • Cutting tool • Solvent cement and primer

Other equipment related to handling of the pipe and the proposed installation location may include:
- elevated work platforms
- ladders
- hand trolleys, forklifts and hoists
- trenching equipment – manual and machine
- laser levels
- bonding straps – vital for existing installations.

Instruments required for pressure testing

In addition to general hand tools, you will require some specific equipment and instruments for pressure testing. Your test instrument must match the pressure being tested and may be digital or water gauge as appropriate. The following items as listed in the Standard will serve as a good indication of what is required:
- *AS/NZS 5601.1 (Gas Installations)*. Make sure you have the *latest* copy of the standard, as it changes on a regular basis. You are required to use only the most up-to-date version.
- *Test instruments*. AS/NZS 5601.1 provides a list of approved test instruments and their limitations of use. Refer to this list first before choosing the instrument you prefer for the job.

REFER TO AS/NZS 5601.1 APPENDIX E 'TEST INSTRUMENTS'

FIGURE 10.36 Subject to job requirements, oxygen acetylene/LPG brazing and press-fit equipment may be needed for copper pipe installations

Install and commission control and regulating equipment

Having previously confirmed the planned location of regulators complies with the Standards, you also need to apply some specific requirements during the installation process.

Vent interconnection

The standards do not permit you to interconnect vents unless they are breather vents from the same gas appliance, or they are vents from the same safety shut-off system fitted to the same appliance.

REFER TO AS/NZS 5601.1 'LIMITATION ON INTERCONNECTION OF VENT LINES'

If vent lines from two different appliances were connected, the operation or malfunction of one appliance regulator would affect the performance of the correctly operating regulator on the other appliance. This is why the interconnection is prohibited.

However, where vents are from the same appliance or device on the same appliance, the danger is minimal, as it is likely the appliance would be shut down. When you do need to install a common vent line on a single appliance, you need to ensure that the main vent is equal to or greater than the combined cross-sectional areas of the two interconnecting vents.

REFER TO AS/NZS 5601.1 'SIZE OF A COMMON VENT LINE'

Vent terminal location

It is very important that vent lines are terminated in a position that will not allow the ignition of the gas or allow such gas to enter into a building.

REFER TO AS/NZS 5601.1 'VENT TERMINAL LOCATION'

When installing a vent line, the end of the line is called the 'terminal'. All restricted zones are measured from the outlet of this terminal. A regulator breather outlet or relief outlet without a vent line is also called a terminal and the same measurements apply.

Vent terminal exclusion zone

All vent terminals must be sited to ensure that no sources of ignition exist within an exclusion zone detailed in the Standard. Before looking at these requirements in more detail, you must first fully understand the scope of the Standards that deal with LPG installations.

- *AS/NZS 5601.* The provisions within the Gas Installation Standards relate to everything downstream of the outlet of the LPG cylinder regulator. The vent terminal exclusion zones detailed within the Standard cannot be applied to the cylinder regulator.

REFER TO AS/NZS 5601.1 SECTION 1 'APPLICATION'

- *AS/NZS 1596.* This Standard deals only with LPG installations upstream of the cylinder regulator outlet and has nothing to do with downstream installation requirements. It also does not relate to LPG installations in caravans, catering vehicles or marine craft.

REFER TO AS/NZS 1596:2014 'SCOPE AND GENERAL'

Having confirmed the scope of each Standard in relation to these requirements, you can now review the specific requirements related to the exclusion zone around vent terminals. The exclusion zones shown in Figure 10.37 apply to LPG installations from the outlet of the cylinder regulator and NG installations from the outlet of the consumer billing meter.

In Figure 10.38, you can see how the exclusion zone is projected from the vent terminal in the direction of discharge. No ignition sources, objects or openings into buildings can be placed in this zone.

REFER TO AS/NZS 5601.1 'VENT TERMINAL LOCATION'

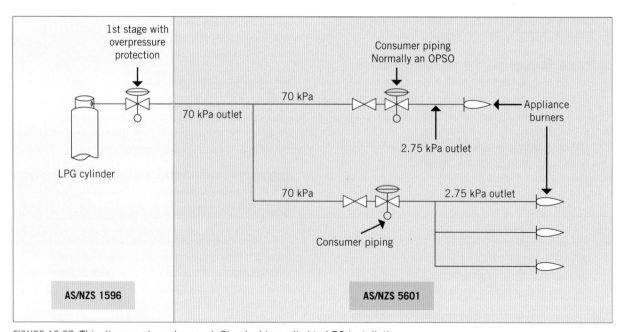

FIGURE 10.37 This diagram shows how each Standard is applied to LPG installations

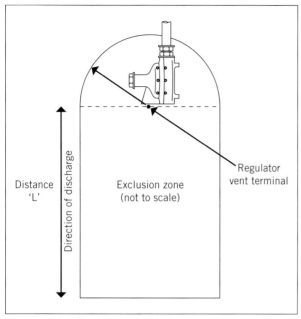

FIGURE 10.38 Vent terminal exclusion zone

As shown in Figure 10.39, when a consumer piping regulator or vent line is mounted on a wall, the exclusion zone is extended to the distance of L all around the terminal. So instead of a cylinder-shaped exclusion zone projected in the direction of discharge, the zone becomes a much wider hemispherical exclusion zone. Be aware of this requirement if you ever provide a second stage regulator adjacent to appliances on a wall. You will need to maintain this clearance to all ignition sources such as flue terminals, power outlets and air intakes.

Protect against bugs

To work correctly, a vent terminal must remain clear of all obstructions. It must be turned down to prevent ingress of water and, in particular, you need to ensure that the terminal has a wire or plastic mesh inserted to prevent insects such as mud wasps from building nests at the outlet that could block the pipe. A blocked terminal will retard movement of the diaphragm and lead to poor combustion.

The easiest way to do this is to remove the mesh that comes with the regulator and insert it in a flare nut at the end of the line.

In Figure 10.40, the vent lines are turned down to prevent ingress of water. The wire mesh from each regulator terminal is removed and pushed into the flare nut at the end of each vent line to prevent the line from being blocked by insects.

OPSO installation requirements

Once you have determined where an OPSO device needs to be installed and the pressure at which you need to set its shut-off device, you need to ensure that the device is installed according to AS/NZS 5601.1 requirements.

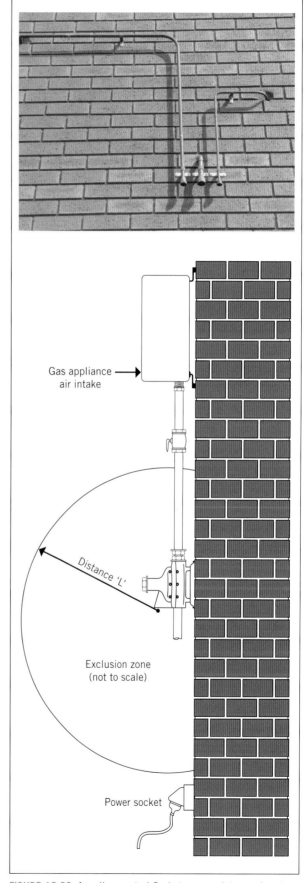

FIGURE 10.39 A wall-mounted 2nd stage regulator and vent terminal exclusion zone

FIGURE 10.40 Vent lines turned down to prevent water entering them

REFER TO AS/NZS 5601.1 'REQUIREMENTS WHERE OVER-PRESSURE PROTECTION SYSTEM SHUTS OFF THE GAS SUPPLY'

All OPSO devices must be installed with a filter within 5 m of the inlet to the device. This is to ensure that pipe contaminants do not damage or block the operation of the device.

The filter is designed to stop the passage of foreign objects and, as such, may eventually become blocked. Always fit your line filters with suitable pressure test points or gauges to ensure that you can measure inlet and outlet pressures across the filter. For example, the OPSO device in **Figure 10.41** is protected by a filter immediately upstream of its inlet. Notice the use of self-sealing test point fittings (often known as Pete's Plugs) on the front face of the filter. The pressure can be read by inserting the long probe of the test gauge through the internal one-way valves of the Pete's Plug fitting.

The filter in **Figure 10.42** has an oil-filled gauge fitted at the inlet and outlet to ensure that any pressure loss through the filter is observed. Gauges filled with oil are protected from spikes in pressure that can occur in gas systems using fast-closing solenoid valves.

OPSO trip setting

OPSO devices do not come factory-set, ready to go. Although you are able to order these devices with different pressure ranges, springs and orifice sizes, they must still be adjusted and set to suit the installation requirements.

You can pre-set an OPSO device prior to installation or carry out the adjustment in situ. Although you should

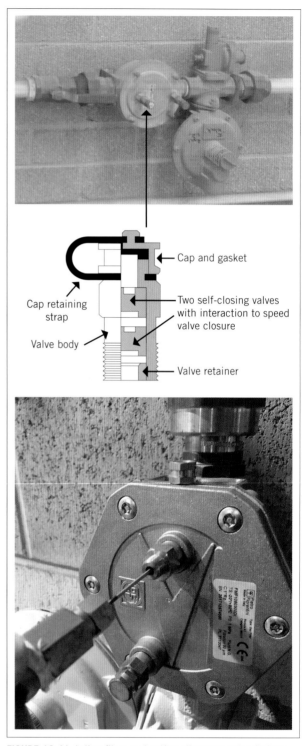

FIGURE 10.41 Inline filter and self-sealing test point fittings

always refer to the manufacturer's instructions on how to set a particular device, the two methods described below may act as a general guide.

FIGURE 10.42 A filter with an oil-filled gauge fitted at the inlet and outlet

HOW TO

PRE-INSTALLATION TRIP SETTING

1. Plug the outlet of the device.
2. Connect a 25 mm meter adaptor fitting to the inlet (you may require some additional adaptor fittings to do so).
3. Connect a standard low-pressure test tee, hand pump and 7 kPa manometer to the inlet (for this exercise you must ensure that your manometer water levels are 'zeroed').
4. Remove the dust cap from the OPSO device and pull out the reset stem firmly until an audible click is heard.
5. Replace the dust cap.
6. Pressurise the regulator slowly and note the manometer reading when you hear the valve slam shut. This sound represents the set shut-off pressure.
7. Adjust the OPSO trip pressure:
    a. Turn the OPSO adjustment screw clockwise to increase the trip pressure and anticlockwise to decrease it, until you are satisfied that the device will operate before system maximum over-pressure is reached.
    b. When setting an OPSO that serves a Type B appliance, you must ask the Type B gas fitter what the trip setting should be set at. AS 3814 requires that the trip pressure of a Type B appliance must not exceed 35 per cent of regulator outlet pressure.
8. Reset and double check the settings – the device is now ready for installation and adjustment of the regulator during commissioning.

If carrying out the adjustment in situ, the following 'How to' offers a basic guide.

HOW TO

IN-SITU TESTING OF AN OPSO

1. Ensure that the installation is safe to touch – use an appropriate instrument to test for stray current.
2. Turn off all downstream appliances.
3. Attach a manometer to the nearest downstream pressure test point.
4. Remove the OPSO regulator dust cap and slowly increase the pressure by turning the adjustment disc clockwise – note the manometer reading when the shut-off valve operates.

    If the device requires adjustment:
5. Reduce the regulator pressure, remove the reset stem dust cap, and pull on the stem to re-engage the device.
6. Replace the dust cap.
7. Screw the OPSO adjustment in or out to alter the trip setting pressure.
8. Slowly increase regulator pressure once again and note the manometer reading when the shut-off valve operates. Repeat this procedure until the device is operating correctly.

There is one other important point to make in relation to OPSO settings: If you are called out to a job to restore gas supply that has been shut down by an OPSO, *always investigate why the OPSO tripped in the first place!* Do not simply reset it and then walk away. Check the system out and satisfy yourself that everything is performing to specification before leaving the job.

LEARNING TASK 10.4

INSTALL AND COMMISSION CONTROL AND REGULATING EQUIPMENT

1. What must be installed within 5 m of an over-pressure shut-off device inlet?
2. Do over-pressure shut-off devices require additional adjustment after installation?

Clean up the job

The job is never over until your work area is cleaned up and tools and equipment accounted for and returned to their required location in a serviceable condition.

Clean up your area

The following points are important considerations when finishing the job:
- If not already covered in your site induction, consult with supervisors in relation to the site refuse requirements.

- Cleaning up as you go not only keeps your work area safe it also lessens the amount of work required at the end of the job.
- Dispose of rubbish, off-cuts and old materials in accordance with site-disposal requirements or your own company policy.
- Ensure all materials that can be recycled are separated from general refuse and placed in the appropriate location for collection. On some sites specific bins may be allocated for metals, timber etc. If not, you may need to take materials back to your business premises.

Check your tools and equipment

Tools and equipment need to be checked for serviceability to ensure that they are ready for use on the job. On a daily basis and at the end of the job check that:

- all items are working as they should with no obvious damage
- high-wear and consumable items such as cutting discs, drill bits and sealants should be checked for current condition and replaced as required
- items such as laser levels, crimping tools, electrofusion equipment etc. are working within calibration requirements
- where any item requires repair or replacement, your supervisor is informed and if necessary mark any items before returning to the company store or vehicles.

Complete all necessary job documentation

Most jobs require some level of documentation to be completed. Avoid putting it off till later as you may forget important points or times.

- Ensure all SWMS and/or JSAs are accounted for and filed as per company policy.
- Complete your timesheet accurately.
- Reconcile any paper-based material receipts and ensure electronic receipts have been accurately recorded against the job site and/or number.

 COMPLETE WORKSHEET 3

SUMMARY

In this unit you have examined the requirements related to the installation of gas pressure control systems including:

- regulator types:
  - constant pressure appliance regulators
  - low pressure compensating regulators
  - dual and single, and second stage LPG regulators
  - industrial LPG regulators
  - service regulators
- regulator principle of operation and application
- relief and breather vents
- regulator venting requirements
- vent terminal locations
- over-pressure protection devices and application
- clean-up of job and return of equipment

GET IT RIGHT

The photo below shows an incorrect practice when installing a constant pressure appliance regulator. Identify the problem and provide reasoning for your answer.

_____
_____
_____
_____
_____

WORKSHEET 1

To be completed by teachers
Student competent ☐
Student not yet competent ☐

Student name: _____

Enrolment year: _____

Class code: _____

Competency name/Number: _____

Task

Review the sections *Pressure control fundamentals – regulators* through to *Identify requirements for gas pressure control equipment* and answer the following questions.

1   On the diagram below, name the parts of a constant pressure appliance regulator.

2   Are LPG appliances generally installed with an appliance regulator?

_____

_____

_____

3   Does a regulator allow you to increase the outlet pressure above that of the inlet pressure?

_____

_____

_____

4   If you found that the pressure available at the outlet of a consumer billing meter was inadequate, would you be able to adjust the service regulator?

_____

_____

_____

INSTALL GAS PRESSURE CONTROL EQUIPMENT

5. In your own words, describe the difference between a breather hole and a relief vent.

6. Does an NG appliance regulator shut off the supply of gas when no appliances are being used?

7. What design difference means that you are not permitted to use an industrial regulator in a domestic installation?

8. Identify the type of regulator and name its parts.

9. An NG upright cooker is being used with only one burner in operation. If the second burner is turned on, how will the regulator react to maintain correct burner pressure?

10. Having tested an LPG dual stage regulator with a manometer, you find that the burner pressure at the appliance is too low. In which direction would you turn the adjustment disc to increase the pressure to the correct amount – clockwise or anti-clockwise?

11 What is the name given to the pressure fluctuation condition that low-pressure compensating regulators are designed to prevent?

_____

_____

12 The diagram below shows a low-pressure compensating regulator. Indicate the gas inlet and outlet with arrows.

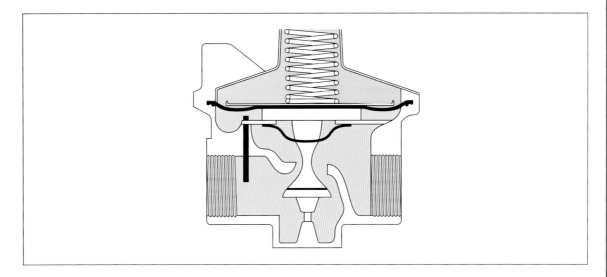

13 At what point above set outlet pressure will an LPG relief valve operate?

_____

_____

14 What is the maximum permissible lock-up pressure for a domestic dual stage LPG regulator?

_____

_____

15 How does the Gas Installation Standards define 'consumer piping gas pressure regulator'? Provide a reference from AS/NZS 5601.

_____

_____

16 From the Standards, identify the following gas fitting symbols. Use a line to link each symbol with its name.

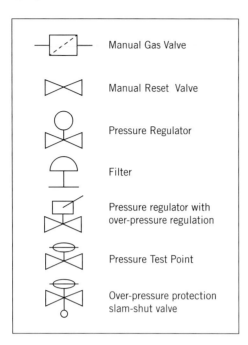

WORKSHEET 2

To be completed by teachers
Student competent ☐
Student not yet competent ☐

Student name: _____

Enrolment year: _____

Class code: _____

Competency name/Number: _____

Task
Review the section *Determining regulator venting requirements* and answer the following questions.

1 Why must all regulators fitted with a relief device be vented to atmosphere?

_____

_____

_____

2 What is the purpose of a breather vent?

_____

_____

_____

3 Determine the maximum breather vent orifice diameter permitted in a room of 900 m^3 effective volume and where a natural gas device has an inlet pressure of 5 kPa.

_____

4 Is a union required between a regulator vent outlet and the vent line? Provide a reference from AS/NZS 5601 to support your answer.

_____

5 What size vent line would you install where the regulator outlet is DN 32 and the distance between the outlet and the vent line terminal is 16 m?

_____

6 Determine the maximum breather vent orifice diameter permitted in a plant room of 60 m^3 effective volume and where a natural gas device has an inlet pressure of 1.5 kPa.

_____

7 What are three forms of over-pressure protection device?

_____

_____

_____

8 Under what circumstances is an over-pressure protection device required on a natural gas installation? Provide a reference from AS/NZS 5601 to support your answer.

9 Define these two terms:

Rated working pressure

Maximum over-pressure

WORKSHEET 3

To be completed by teachers

Student competent ☐

Student not yet competent ☐

Student name: _____

Enrolment year: _____

Class code: _____

Competency name/Number: _____

Task

Review the section *Determining over-pressure protection requirements* and answer the following questions.

1. Within what distance must a filter be installed upstream of an OPSO? Provide a reference from AS/NZS 5601 to support your answer.

   _____

2. Complete this installation design showing locations of control valves, regulators and over-pressure protection devices where required. Use industry standard symbols.

   Kiln
   30 kPa
   Rated working pressure

   40 kPa ⊳⊲

   Commercial kitchen
   5 kPa
   Rated working pressure

3. What type of test point must be fitted to a regulator with an outlet pressure of 70 kPa? Provide a reference from AS/NZS 5601 to support your answer.

   _____

4. True or False. The AS/NZS 1596 is applied to all parts of a gas installation from the outlet of the cylinder regulator.

   _____

5. What is the ignition source exclusion zone radius above a DN 25 vent terminal where no obstructions in the direction of discharge exist? Provide a reference from AS/NZS 5601 to support your answer.

_____

6. What must you use on pipework when removing a faulty regulator?

_____

7. Are the vent lines of safety shut-off systems on two separate appliances permitted to be interconnected so that only one vent line is taken to outside? Provide a reference from AS/NZS 5601 to support your answer.

_____

8. What must be installed as part of or immediately downstream of every second stage regulator? Provide a reference from AS/NZS 5601 to support your answer.

_____

9. Does the AS/NZS 1596 apply to the cylinder regulator vent terminal exclusion zone of a caravan cylinder installation? Provide a reference from AS/NZS 5601 to support your answer.

_____

10. Where fixed gauges are installed in a gas line or component, what must also be fitted in case the gauge is suspected of being inaccurate or faulty? Provide a reference from AS/NZS 5601 to support your answer.

_____

PURGE CONSUMER PIPING

Learning objectives

Areas addressed in this chapter include:
- the three main types of large volume purge
- an outline of the equipment required to purge an installation
- how to purge a sub-meter
- purging calculations and planning
- applying your calculations to the job
- clean up the job.

Where applicable, the equipment and procedure options are detailed in each section of the chapter. A simple example installation is provided for each section so that you are able to follow each step in reference to an actual task. Further examples are provided in the worksheet at the end of the chapter so that you can assess your knowledge.

Before working through this section, it is absolutely important that you complete or revise each of the chapters in Part A 'Gas fundamentals'. In particular work through Chapter 5 'Gas industry workplace safety' and Chapter 6 'Combustion principles'.

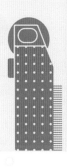

Overview

The purpose of purging a consumer gas system is to ensure that a flammable mixture of gas and air does not exist in the pipe or meter. Before work is commenced you must ensure that all gas is removed safely and replaced with either air or an inert gas. Gases that are known as 'inert' cannot support combustion and include nitrogen and carbon dioxide. Purging enables you to safely undertake all necessary installation or maintenance tasks.

After all work is completed, you must ensure that all air is removed entirely from the system and replaced with pure fuel gas so that there is no chance of an explosive condition existing in the pipework or meter. Ensure you have revised Chapter 6 'Combustion principles' before proceeding.

Purge requirements for large volume gas installations

The procedures in this section relate specifically to the requirements for purging a large volume gas installation. AS/NZS 5601.1 provides basic considerations and requirements for large volume purging.

REFER TO AS/NZS 5601.1 APPENDICES D, D5 AND D6

A large volume purge is generally considered to be one in which the total installed pipe volume exceeds 0.03 m^3 (30 L) but less than 1.0 m^3 (1000 L). Any purge requirements above this amount must be referred to the technical regulator and fall outside the scope of this chapter.

The approximate volume of any installation can be quickly determined by referring to Table D1 in AS/NZS 5601.1. This chapter provides tables to help you work out exactly how much purge medium you will require on a per metre basis. Note, however, that although the additional tables, calculations and procedures in this book have been considered standard industry practice for many years, it is imperative that you personally check with both your local technical regulator and the gas distributor prior to commencing any large volume purge.

References in this chapter to meter purging principally relate to sub-meters. In most distribution areas a gas fitter is not permitted to touch a consumer billing meter, and although the process is the same you must always check with the gas distributor first.

It is also important to note that the procedures described in this chapter principally apply to purging natural gas, which is lighter than air. While the same procedures can be used to purge LPG, you must check with your technical regulator if this is permissible in your area. As LPG is heavier than air, it is vital that an LPG purge process eliminates any chance of the gas accumulating. For this reason, it is often preferable to purge LPG using a flare stack, incorporating a special anti-flashback flame trap. The use of such specialised equipment and processes is deemed outside the scope of this text.

Finally, please remember that the equipment shown in this section must *never* be used for flare purging. Contact your technical regulator if in doubt.

Purging types

There are three main types of large volume purging:
1 fuel gas purge – to purge air or inert gas with fuel gas
2 inert gas purge – to purge fuel gas with inert gas
3 air purge – to purge/vent fuel gas with air.

The calculations required for such tasks mainly relate to the determination of pipe volume and measurement of purge medium. These calculations are essentially the same for both a fuel gas and an inert gas purge; you simply use the information specific to the need. An air purge shares some of the planning and safety requirement, but is only applied to some installations less than 30 L and does not require the detailed volume calculations. These differences are illustrated in Figure 11.1.

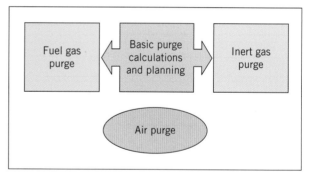

FIGURE 11.1 The three main types of purging

Prepare for purging

By its very nature purging will result in the controlled release of fuel gas at some point, and when working with gas pipework you will also be exposed to potential electrical hazards. As such, it is imperative that you fully evaluate the safety requirements of the task before commencing work. Basic safety equipment required for purging includes the following items:
- warning signs
- barrier mesh and witches hats to create a cordon area
- fire extinguisher

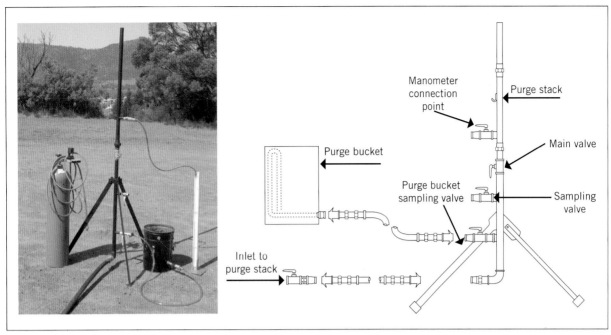

FIGURE 11.2 Basic equipment required for large volume purging

- electrical test instruments such as a volt stick and multi-meter
- bonding straps
- restraint harnesses when working at heights
- standard PPE including safety glasses, safety boots and protective clothing.

Tools and equipment required for purging

To carry out a large volume purge you require some special tools and equipment (see Figure 11.2). The items discussed below are the minimum requirements for most large volume purging tasks.

Purge stack
The purge stack is fabricated from 25 mm steel pipe (see Figure 11.3) and allows lighter-than-air gases to be safely vented at a height that permits easy dispersion. You could also use it with LPG, as long as the purge zone is large enough to allow full dispersion of the gas. A flame trap could also be adapted to the design if required for flare purging; however, this is not covered in this unit and falls outside the scope of the Standard. You must consult with the technical regulator in this instance. Never undertake flare purging without full approval.

Purge bucket
The U-shaped internal piping of the purge bucket allows a sample of gas to be released and lit without the chance of any flashback while the sample is being taken (see Figure 11.4). It also allows gas to blow through the soapy water without flowing back up the purge line.

A purge bucket is simply fabricated from a 20 L steel bucket and copper pipe. It should be filled with a 1:10

FIGURE 11.3 A purge stack fabricated from mild steel pipe and bar

FIGURE 11.4 A purge bucket

FIGURE 11.5 'Camloc'-style fittings

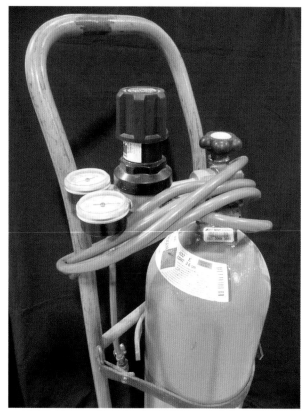

FIGURE 11.6 'E'-size nitrogen cylinder, regulator and connection hose

mixture of detergent and water. The water level should be filled to a maximum of 50 mm above the outlet. This is deep enough to create bubbles but not so deep as to retard the flow of gas.

Note: Appendix D of the Standard now recommends the use of a calibrated gas detector instead of a purge bucket to determine when a purge is complete. However, Appendix D is an 'Informative' appendix and you should contact your technical regulator for clarification of equipment use.

Purge hose and quick release fittings

A long length of hose enables the purge stack to be located well away from sources of ignition and openings to buildings. Quick-release 'Camloc'-type fittings (see Figure 11.5) make connection and disconnection of purge hoses to both the stack and consumer piping fast, secure and efficient.

Nitrogen cylinder/s and regulator

Most purging tasks can be carried out with a simple 'E'-size cylinder (see Figure 11.6). Higher-volume jobs may require multiple or larger cylinders. You will also require a specific nitrogen regulator.

In addition to the nitrogen cylinder you will require a supplementary regulator that is adjusted to ensure that the installation is subjected to no more than 3 kPa flow pressure. A spare LPG second stage regulator is useful for this application. The purge regulator assembly is easily made up from standard components (see Figure 11.7).

Flow meter

The flow meter is attached to the outlet of the nitrogen regulator, and can be adjusted so that gas is released in

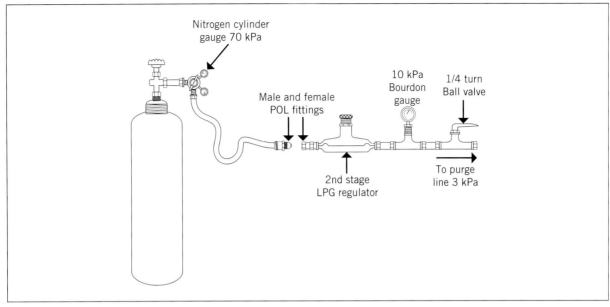

FIGURE 11.7 Purge regulator assembly

litres per minute. This is particularly useful where you are purging a system without a meter. Rather than measuring your purge gas requirements in cubic metres, you simply convert to litres by multiplying by 1000. You can then set your flow meter to the required volume and measure the gas flow in litres per minute.

Barrier mesh, signage and fire extinguisher

You must prevent access of unauthorised people to the purge zone area by using barriers and signs that clearly cordon off the purge zone. Having a fire extinguisher on hand may assist if something does go wrong.

Gas detector

A gas detector that can measure the relative percentages of gas and air in the air–gas mixture (the lower explosive limit, or LEL) can be very useful for taking quick and simple samples from the purge stack. The gas detector should be capable of taking readings as low as 10 per cent of the LEL.

Manometer

A manometer can be attached to the purge stack so that you can determine when gas is being purged. This is particularly useful where there is some distance between the piping being purged and the stack itself. A water gauge manometer is generally more useful than a digital unit in this instance, as fluctuating movement of the water level quickly demonstrates that gas is flowing through the purge stack.

Two-way radio

Purging requires considerable coordination, often between the ends of pipe installations that are not in sight of each other and sometimes over a considerable distance. While it is possible to carry out the work with the use of a person acting as a 'runner' between the two points, it is far more efficient to use two-way radio communication.

Which inert gas should I use?

The recommended inert gas is nitrogen, which is available in a range of cylinder sizes to suit different volume requirements. Carbon dioxide used to be regarded as an alternative; however, it is no longer mentioned in the Standard.

REFER TO AS/NZS 5601.1 APPENDIX D

Nitrogen and carbon dioxide can displace oxygen and you risk asphyxiation if they are handled incorrectly. Always follow recommended procedures when using inert gases.

LEARNING TASK 11.1 LPG

1. What is the flammability range of natural gas in air?
2. What is the LEL of LPG?
3. Is LPG heavier or lighter than air?

Identify purge requirements

Safe work within the gas industry is all about following detailed procedures. Taking shortcuts can lead to injury and death for yourself or other people. Successful purging operations depend upon fully evaluating the job requirements before commencing work.

Check installation compliance

When first evaluating the purge requirements of any job, you need to confirm that the installation complies with the Gas Installation Standards and the relevant job specifications. There is no point in going any further on an installation that does not meet minimum requirements. Common compliance issues related immediately to purging include the following:

- Is the meter or cylinder installation in the correct location and compliant with Standards and gas distributor requirements?
- Are pipe materials certified and fit for purpose?
- Has the pipe installation been installed in accordance with material limitations and locations?
- Do fittings match the selected pipe material?
- Are minimum isolation requirements and means of disconnection applied to relevant sections of the installation?

Perform a leakage test

Before undertaking any purge you must know that the pipe installation is gas-tight. The goal of the purging process is to ensure that fuel gas and inert gas are released in a safe location and in a controlled manner. Failure to check that the installation is sound may mean that while purging is underway, gas may actually be leaking somewhere in the installation and creating a potential hazard.

Prior to commencing any purging work you must carry out a leakage test to ensure that the installation is safe to work on. Note that the Gas Installation Standards no longer provides testing solutions for installations above 30 L of volume. Contact the technical regulator in your State or Territory to confirm local requirements.

Purging a sub-meter

A sub-meter must be purged with at least five times its cyclic capacity or volume. This is to ensure that no trace of gas, air or inert gas is left inside the meter body. You will find the cyclic capacity of a meter on its badge plate. Sometimes, however, this may be abbreviated and you may be unsure of which number you should be looking for. In such cases, you must contact your distributor or meter manufacturer for clarification.

In the example shown in Figure 11.8, you can see that the cyclic capacity of the meter is 2 litres (2 dm³). Therefore, to comply with AS/NZS 5601.1 you would need to pass at least 10 litres of purge medium through the meter to be confident that the purge was complete.

REFER TO AS/NZS 5601.1 APPENDIX D

Where the meter is part of a larger pipe system purge, you will add this volume to the overall pipe volume.

FIGURE 11.8 A meter with a cyclic volume of 2.0 dm³ or 2 litres

Purging calculations and planning

Standard reference tables

You will need to refer to Tables 11.1, 11.2 and 11.3 throughout your purging calculations.

TABLE 11.1 Volume of full nitrogen cylinders in m³

Nitrogen cylinder volume (full at approx. 14 000 kPa)	
Cylinder size	Cylinder volume, m³
D	1.3
E	3.2
G	6.4

TABLE 11.2 Relationship of nitrogen volume to cylinder pressure drop

Cylinder pressure drop (kPa)	Equivalent nitrogen volume in m³ (normal temperature and pressure)		
	D-size cylinder	E-size cylinder	G-size cylinder
1 000	0.09	0.23	0.46
2 000	0.18	0.46	0.92
3 000	0.27	0.69	1.38
4 000	0.36	0.92	1.84
5 000	0.45	1.15	2.30
6 000	0.54	1.38	2.76
7 000	0.63	1.61	3.22
8 000	0.72	1.84	3.68
9 000	0.81	2.07	4.14
10 000	0.90	2.30	4.60
11 000	0.99	2.53	5.06
12 000	1.08	2.76	5.52
13 000	1.17	2.99	5.98
14 000	1.30	3.20	6.40

TABLE 11.3 Pipe volume in cubic metres per metre

Pipe diameter (DN)	Pipe volume (m³ per metre of length)		
	Copper (Type B) to AS 1432	Steel (medium) to AS 1074	Polyethylene (SDR 11) to AS 4130
15	0.0001	0.0002	0.0001
20	0.0002	0.0003	0.0002
25	0.0004	0.0006	0.0003
32	0.0006	0.0010	0.0005
40	0.0010	0.0014	0.0009
50	0.0018	0.0022	0.0012
65 (63 PE)	0.0030	0.0037	0.0022
80 (90 PE)	0.0040	0.0052	0.0043
100 (110 PE)	0.0075	0.0088	0.0064
125	0.0110	0.0130	0.0083
150 (160 PE)	0.0167	0.0190	0.0130
200	0.0310		0.0210

Purge calculation worksheets

Purge calculations require quite a number of steps. Developing a purge calculation sheet will make this process easier. Use the pipe installation and worksheet shown in Figure 11.9 as an example to refer to while working through each purge type.

GREEN TIP

Natural gas is principally made up of methane, which, when released to atmosphere is a potent greenhouse gas. It is important that your purge calculations are accurate so that only the minimum required amount of methane is release during the purge.

The purge calculation sheet is designed to provide structure to your information. By entering the data and following the prompts, you will arrive at consistently accurate answers for any purge task.

The example in Figure 11.9 will be used throughout this section.

The main steps for purge calculations and planning

Listed below are the main steps required to calculate and plan for a fuel gas and inert gas purge:
1. Identify the purge zone and your worker requirements.
2. Determine the cyclic capacity of the meter in cubic metres.
3. Label and measure the main run and branches.
4. Calculate the volume of the main run and branches.
5. Multiply the pipe volume by a safety factor of 1.5.

Step 1: Identify the purge zone and your worker requirements

a Ensure that your purge zone is located outside and at least 6 m away from any potential ignition source or opening into a building. When purging raw LPG, ensure that there is sufficient area for effective ground-level dispersion.
b Purging very large volumes of natural gas may require clearance checks where any potential flight path may exist over the purge zone. Gas fitters seldom encounter such large purges but, if you are unsure, do not hesitate to contact your technical regulator.
c Determine how many people you need for the purge:
   – Number 1. This person controls the gas inlet and reads the meter.
   – Number 2. This person controls the purge stack and purge zone, and takes the completion test sample.
   – Runner. Where either end of the installation is out of sight, a runner is often required. This person also assists the Number 2 person with the completion test sample.

Step 2: Determine the cyclic capacity of the meter in cubic metres

a Look for the meter's cyclic capacity on its badge plate. Enter this onto the worksheet sketch in cubic metres.
b Multiply the cyclic capacity of the meter by 5. Enter the result into the worksheet table.

Step 3: Label and measure the main run and branches

a The main run is from the meter to the furthest point. Label, measure and record its length and DN size on your sketch.
b All of the remaining sections are branches. Label, measure and record their length and DN size on your sketch.

Step 4: Calculate the volume of the main run and branches

Follow the steps in the worksheet table. Refer to Table 11.3 and then multiply the cubic metres per metre of length (m³/m) figure by the length of the section. Enter this into the worksheet.

Step 5: Multiply the pipe volume by 1.5 safety factor

A standard safety factor of 1.5 applies to all purging. This ensures that there is more than ample purge gas to ensure that the process is carried out completely. Simply multiply your pipe volumes by 1.5 and enter this into the worksheet.

At this point, you have completed all the required calculations. You now need to apply these figures to the actual job. Apply your figures to either a fuel gas or inert gas purge.

Purge Calculation Worksheet

Date: 12.10.21 **Gas fitter:** Joe Bloggs **Standard version used:**

Pipe layout sketch:

Elevation ☐   Plan ☑   Perspective ☐

Material: Steel ☐   Copper ☑   UPVC ☐   PE ☐   Polyamide ☐

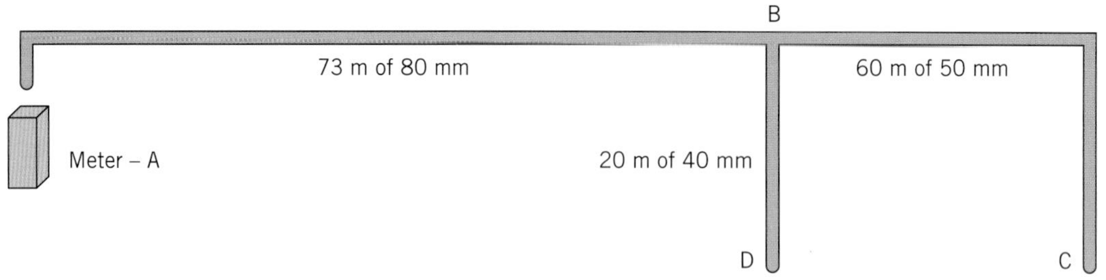

```
                              B
        ┌────── 73 m of 80 mm ──────┐     ┌── 60 m of 50 mm ──┐
     [Meter – A]                    │     │                   │
                                    │ 20 m of 40 mm           │
                                    D                         C
```

Meter cyclic capacity? 0.005 m^3

Pipe volume data

Main Run Section (meter outlet to furthest point)	DN	Length (m)	Pipe vol/m^3 (ref. pipe vol table)	Vol m^3 (m × vol/m^3)	Safety factor (vol m^3 × 1.5)
A to B	80	73	0.0040	0.292	0.438
B to C	50	60	0.0018	0.108	0.162
Main run pipe sections volume (all sections from meter outlet to furthest point)					0.06
Meter volume m^3 (meter cyclic capacity in m^3 x 5)					0.025
Total main run purge volume m^3 (add all sections from meter to furthest point)					MR = 0.625

Branch Sections	DN	Length (m)	Pipe vol/m^3 (ref. pipe vol table)	Vol (m × vol/m^3)	Safety factor (vol m^3 × 1.5)
B to D	40	20	0.0010	0.02	0.03
Total branch purge volumes (add all branch sub-totals together)					B = 0.03
Total required purge gas m^3 = total main run (MR) + branches (B)					MR + B = 0.655

FIGURE 11.9 A simple purge calculation worksheet

LEARNING TASK 11.2 PURGING CALCULATIONS AND PLANNING

1 In relation to the purging of LPG, what must be considered in relation to the size of the purge zone?
2 Why is it not desirable to release large amounts of raw methane gas into the atmosphere?
3 What volume of purge medium must be passed through a sub-meter to be sure it is fully purged of gas?

Carry out and test purge operation

Now that all job planning and calculations are completed, you are ready to undertake the work. Purging is a process that demands strict adherence to a safe and set procedure. The steps detailed in this section should be read in conjunction with your Standard and any requirements of the technical regulator in your State or Territory.

HOW TO

FUEL GAS PURGE

Listed below are the main steps required to purge air or inert gas with fuel gas:
1. Carry out an electrical safety test and a pipe leakage test.
2. Assemble the equipment.
3. Purge the meter and the main run.
4. Take a completion test sample.
5. Purge all branches; repeat the procedure.
6. Carry out a final installation test.

Step 1: Carry out an electrical safety test and a pipe leakage test

a As with any practical aspects of gas fitting, never touch any piping or appliances unless you carry out a test for stray current first.
b You must also undertake a leakage test of the system before commencing any work. This requirement is to ensure that any work is carried out on a gas-tight system.

Step 2: Assemble the equipment

a Set up barriers and signage around your purge zone.
b Connect the purge hose from the end of the main run to the purge stack.
c Ensure that all system isolation valves are off. These include:
   – meter inlet valve
   – appliance isolation valves
   – purge hose inlet isolation
   – purge stack main control valve
d Ensure that the purge test bucket is ready and adjacent to the purge zone.

Step 3: Purge the meter and the main run

a Take note of the current meter dial reading and add the amount you want to run through the system. Refer to your worksheet for the total main run figure in m^3.
b Slowly turn on the meter inlet valve.
c Slowly turn on the purge hose inlet valve and check the purge equipment for any gas escapes.
d Open the purge stack main valve. You will begin to hear the sound of flowing air/inert gas. As the fuel gas makes its way through, you may also hear a change in pitch due to the density differences between the gases.
e When the meter indicates that the correct amount of fuel gas has passed through, the Number 1 should let the Number 2 know to turn off the purge stack main control valve.

Step 4: Take a completion test sample

If using a gas detector:
a Insert the sampling probe or hose into the purge stack sampling valve.
b Open the valve and draw a sample into the device. The minimum acceptable reading is 95 per cent gas.

If using a purge bucket:
a Connect the purge bucket hose to the appropriate sampling valve on the purge stack. Open the valve and watch as detergent bubbles begin to fill the bucket.
b Remove the bucket from the purge zone and, using a long taper, attempt to light the bubbles. The status of the purge is determined by the following result:
   – No ignition indicates that the air or inert gas has still not been removed. The purge must be continued.
   – A sharp, crackling ignition indicates that an air–fuel mixture is present. The purge must be continued.
   – Quiet ignition with a soft flame indicates that there is pure gas in the line and the purge is complete.
c If the result is satisfactory, purge for one minute more and then isolate all valves in the following order:
   i purge stack main control valve
   ii meter inlet valve
   iii purge hose inlet isolation
d Seal all open ends quickly.

Step 5: Purge all branches; repeat the procedure

a Commencing at the branch closest to the supply point, repeat the purging procedure on all branches moving away from the supply point. Refer to your worksheet for each branch purge volume.
b Seal all open ends quickly.

Step 6: Carry out a final installation test

a Open all appliance isolation valves.
b Commission all appliances.
c Conduct a full system installation test.

HOW TO

INERT GAS PURGE

Listed below are the main steps required to undertake a purge of fuel gas with inert gas:

1. Carry out an electrical safety test and a pipe leakage test.
2. Assemble the equipment.
3. Purge the meter and the main run.
4. Take a completion test sample.
5. Purge all branches; repeat the procedure.
6. Carry out a final installation test.

Step 1: Carry out an electrical safety test and a pipe leakage test

a   As with any practical aspect of gas fitting, never touch any piping or appliance unless you have carried out a test for stray current first.
b   You must also undertake a leakage test of the system before commencing any work. This requirement is to ensure that any work is carried out on a gas-tight system.

Step 2: Assemble the equipment

a   Set up barriers and signage around your purge zone.
b   Connect the purge hose from the end of the main run to the purge stack.
c   Ensure that all system isolation valves are off. These include:
   - meter inlet valve
   - appliance isolation valves
   - purge hose inlet isolation
   - purge stack main control valve.
d   Ensure that the purge test bucket is ready and adjacent to the purge zone.
e   Set up a nitrogen cylinder and set the cylinder regulator to 70 kPa (or the maximum inlet pressure of your purge assembly regulator).
f   Connect the purge regulator assembly at the gas supply point, pre-set to a maximum of 3 kPa working pressure.

Step 3: Purge the meter and the main run

a   Take note of the current meter dial reading and add the amount you want to run through the system. Refer to your worksheet for the total main run figure in m^3.
b   If using a flow meter, convert all m^3 quantities to litres. Set the flow meter to approx. 30 L/min and calculate how long each section will take to be purged.
c   Slowly turn on the purge regulator inlet valve.
d   Slowly turn on the purge hose inlet valve and check the purge equipment for any gas escapes.
e   Open the purge stack main valve. You will begin to hear the sound of flowing air/inert gas. As the fuel gas makes its way through, you may also hear a change in pitch due to the density differences between the gases. If necessary, refer to your manometer to be satisfied that gas is flowing.
f   When the meter (or flow meter) indicates that the correct volume of nitrogen has passed through, the Number 1 should let the Number 2 know to turn off the purge stack main control valve.

Step 4: Take a completion test sample

a   If using a gas detector, insert the sampling probe or hose into the purge stack sampling valve. Open the valve and draw a sample into the device. The minimum acceptable reading is no greater than 10 per cent of the LEL of the fuel gas being used.
b   If using a purge bucket, connect the purge bucket hose to the appropriate sampling valve on the purge stack. Open the valve and watch as detergent bubbles begin to fill the bucket. Remove the bucket from the purge zone and, using a long taper, attempt to light the bubbles. The status of the purge is determined by the following result:
   - No ignition indicates that all fuel gas has been removed. The purge is complete.
   - A sharp, crackling ignition indicates that an air–fuel mixture is present. The purge must be continued.
   - A quiet ignition with soft flame indicates that pure gas is in the line. The purge must be continued.
c   If the result is satisfactory, purge for one minute further and then isolate all valves in the following order:
   i   purge stack main control valve
   ii   nitrogen cylinder regulator valve
   iii purge regulator assemble isolation valve
   iv   purge hose inlet isolation.
d   Seal all open ends quickly.

Step 5: Purge all branches; repeat the procedure

a   Beginning at the branch closest to the supply point, repeat the purging procedure on all branches moving away from the supply point. Refer to your worksheet for each branch purge volume.
b   Seal all open ends quickly.

Step 6: Carry out a final installation test

Conduct a full system installation test.

Installations of less than 30 L of volume – or larger installations where circumstances permit – may be vented to atmosphere to effect a purge. In such situations, the process of purging fuel gas with air is simple – follow the steps below to complete this.

HOW TO

AIR PURGE

1. Check the installation size and type of work.
2. Identify the purge zone.
3. Carry out an electrical safety test and a pipe leakage test.
4. Isolate the system from the gas supply point.
5. Open all pipe ends and vent to outside atmosphere.
6. Seal all open ends once the purge is completed.
7. Conduct a final installation test.

Step 1: Check the installation size and type of work

a Installations of a certain size *must* be purged with inert gas. The table below shows the pipe length above which an inert gas purge is required.

REFER TO AS/NZS 5601.1 TABLE D2 'PIPE LENGTH ABOVE WHICH AN INERT GAS PURGE IS REQUIRED'

Nominal pipe size (mm)	Approximate pipe length (m)
50	25
65	15
80	10
100	5
150	3
Over 150	All

b Regardless of pipe size, if any hazardous work such as welding or electric cutting is to be carried out on the installation, or if the pipe system is to be decommissioned or incorporates any vessel such as a surge tank, the installation *must* be purged using an inert gas.

c Where these conditions are not evident, a simple venting of the system to atmosphere can be regarded as being sufficient.

Step 2: Identify the purge zone

Ensure that all purge outlets are at least 6 m away from any ignition sources or openings into buildings. A length of clear plastic hose may be useful for this task.

Step 3: Carry out an electrical safety check and a pipe leakage test

a As with any practical aspect of gas fitting, never touch any piping or appliances unless you have carried out a test for stray current first.

b You must also undertake a leakage test of the system before commencing any work. This requirement is to ensure that any work is carried out on a gas-tight system.

Step 4: Isolate the system from the gas supply point

a Although you are able to isolate the supply by turning off the valve at the supply point, it is possible for a hazardous situation to be created if someone turns on the supply while you have all ends open for venting. Therefore, it is advisable that you apply electrical bonding straps to the meter outlet and fully disconnect the piping from the supply.

b Ensure that you plug or cap the meter outlet.

Step 5: Open all pipe ends and vent to outside atmosphere

Open all pipe ends to allow the system to vent to outside atmosphere. A length of clear plastic hose may assist in venting internal appliances to the outside. Lighter-than-air gases naturally disperse when released, and by opening the pipe ends you will find the gas will self-purge.

Remember: This only applies to smaller installations using natural gas!

Simple venting of any size of installation is rarely sufficient for LPG installations because LPG is heavier than air and will tend to sit in any low sections of the pipe. It will not self-purge. Air or inert gas under pressure is therefore required to purge all LPG from piping.

Step 6: Seal all open ends once the purge is completed

One of the golden rules of gas fitting is to never leave an open end. Even after purging, ensure you plug or cap each pipe end so that at no time can gas be released.

Step 7: Conduct a final installation test

Even if you are going to work on the installation in the near future, it is wise to carry out an installation test to ensure that all parts of the system are gas-tight. Never leave a job unless you are completely certain that the installation is sound.

LEARNING TASK 11.3 CARRYING OUT A PURGE

1. What is the minimum distance required between any ignition source and the purge hose outlet?
2. What type of pressure test must be performed prior to carrying out a purge?

Clean up the job

The job is never over until your work area is cleaned up and tools and equipment accounted for and returned to their required location in a serviceable condition.

Clean up your area

The following points are important considerations when finishing the job:
- If not already covered in your site induction, consult with supervisors in relation to the site refuse requirements.
- Cleaning up as you go not only keeps your work area safe, it also lessens the amount of work required at the end of the job.
- Dispose of rubbish, off-cuts and old materials in accordance with site-disposal requirements or your own company policy.
- Ensure all materials that can be recycled are separated from general refuse and placed in the appropriate location for collection. On some sites, specific bins may be allocated for metals, timber etc. or, if not, you may need to take materials back to your business premises.

Check your tools and equipment

Tools and equipment need to be checked for serviceability to ensure that they are ready for use on the job. On a daily basis and at the end of the job, check that:
- all items are working as they should with no obvious damage
- high-wear and consumable items such as cutting discs, drill bits and sealants are checked for current condition and replaced as required
- items such as laser levels, crimping tools, electrofusion equipment etc. are working within calibration requirements
- where any item requires repair or replacement, your supervisor is informed and if necessary mark any items before returning to the company store or vehicles.

Complete all necessary job documentation

Most jobs require some level of documentation to be completed. Avoid putting it off until later as you may forget important points or times.
- Ensure all SWMS and/or JSAs are accounted for and filed as per company policy.
- Complete your timesheet accurately.
- Reconcile any paper-based material receipts and ensure electronic receipts have been accurately recorded against the job site and/or number.

> **LEARNING TASK 11.4 PURGE REGULATOR**
>
> During the purging operation, you notice that the Bourdon gauge on the purging regulator assembly is not returning to the zero mark after use. What action should you take?

  COMPLETE WORKSHEET 1

SUMMARY

In this unit you have examined the requirements related to the purging of consumer gas pipe installations with a volume greater than 30 L. This included:
- revision of gas UEL and LEL ranges
- revision of safety considerations including electrical testing and the potential hazards associated with the use of inert gases
- job preparation and calculation of purge volumes
- the steps involved in undertaking the actual purge including:
  - purging fuel gas with an inert gas
  - purging inert gas with fuel gas
- clean-up of job and return of equipment

GET IT RIGHT

The photo below shows an incorrect practice that can be performed when using a purge bucket. Identify the incorrect method and provide reasoning for your answer.

_____
_____
_____
_____
_____

WORKSHEET 1

To be completed by teachers
Student competent ☐
Student not yet competent ☐

Student name: _____

Enrolment year: _____

Class code: _____

Competency name/Number: _____

Task
Review the sections *Purging types* through to *Purging calculations and planning* and answer the following question.

1 You now have the opportunity to apply your knowledge to this exercise. Half the sheet has been completed. Fill in the remaining information and come up with a purging solution.

Purge Calculation Worksheet				
Date: 12.10.21		**Gas fitter:** Joe Bloggs		**Standard version used:**
Pipe layout sketch: Elevation ☐		Plan ☑		Perspective ☐
Material:	Steel ☑	Copper ☐	UPVC ☐	PE ☐ Polyamide ☐
Meter cyclic capacity? 0.010 m³				

```
D
  79 m of 32 mm    C    71 m of 50 mm    B    54 m of 40 mm
                   |                     |                     |
                   |                     |                     |
              20 m of 25 mm  E           |                     |
                                   30 m of 80 mm              F
                                         |
                                      Meter: A
```

PURGE CONSUMER PIPING 239 GS

Pipe volume data

Main run section (meter outlet to furthest point)	DN	Length (m)	Pipe vol/m³ (ref. Table 3)	Vol m³ (m x vol/m³)	Safety factor (vol m³ x 1.5)
A to B					
B to C					
C to D					
Main run pipe sections volume (all sections from meter outlet to furthest point)					
Meter volume m³ (meter cyclic capacity in m³ x 5)					
Total main run volume m³ (add all sections from meter to furthest point)					MR =

Pipe volume data

Branch sections	DN	Length (m)	Pipe vol/m³ (ref. Table 3)	Vol (m x vol/m³)	Safety factor (vol m³ x 1.5)
B to F					
C to E					
Total branch volumes (add all branch sub-totals together)					B =
Total required purge gas m³ = total main run (MR) + branches (B)					MR + B =

	Size	Full volume m³	Number of cylinders
Nitrogen cylinders required for inert gas purge	D	1.3	
	E	3.2	
	G	6.4	

CALCULATE AND INSTALL NATURAL VENTILATION FOR TYPE A GAS APPLIANCES

12

Learning objectives

This chapter covers the skills and knowledge that a licensed gas fitter must apply when determining the natural ventilation requirements of a Type A gas appliance. A Type A appliance is one for which a certification scheme exists and are generally almost exclusively found in domestic and commercial applications.

Areas addressed in this chapter include:
- the basic requirements and need for Type A gas appliance ventilation
- general ventilation calculations – these are the primary natural ventilation calculations you need to apply in almost every domestic, commercial and industrial installation to ensure adequate air supply for combustion
- specific ventilation calculations – certain appliances and appliance locations require additional ventilation consideration
- appliance and flue testing requirements – the vital checks necessary after all ventilation installations
- clean up the job.

Before working through this section, ensure you have completed Part A of this text and if necessary revise Chapter 6 'Combustion principles'.

Overview

For complete and clean combustion to take place, you must ensure that during installation and maintenance each appliance receives satisfactory amounts of fresh air. The AS/NZS 5601 Gas Installation Standards provides explicit detail on how to achieve this.

Inadequate ventilation can have many consequences, including damage to appliances, blocked flues and poor appliance efficiency.

Type A gas appliance ventilation basics

All gas appliances require a good air supply to ensure complete and safe combustion. However, appliances designed for installation indoors require additional ventilation consideration.

Outdoor appliances

These appliances are only designed to be used outdoors. The appliance will draw in air for combustion from its surroundings. Despite being outside a building, you must still ensure that a clean cross-flow of air passes by each appliance. The air must also be free from any contaminants, such as chemical and fuel vapour, dust and products of combustion from other appliances.

The storage water heater in **Figure 12.1** is designed to be installed outdoors. Note the use of a balanced flue separating the combustion air supply and the escaping flue gases.

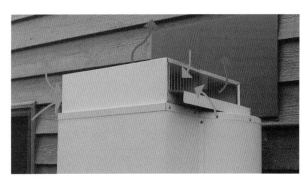

FIGURE 12.1 A balanced flue storage water heater designed for installation outdoors

Room-sealed appliances

Some appliances are installed inside buildings, but are designed in such a way that all the air for combustion is drawn in, and all products of combustion are exhausted, outside the building. This is done via a *natural draught or fan-assisted, balanced flue* (see **Figure 12.2**). A room-sealed appliance is effectively sealed from the air inside the building. Such an appliance does not generally require any further internal ventilation as long as

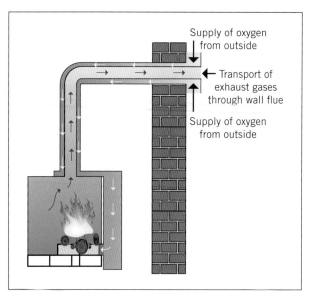

FIGURE 12.2 Operation of a room-sealed appliance

ambient temperatures around the appliance are maintained in line with the manufacturer's recommendations.

Internal appliances

An appliance designed for internal installation and which is not classed as a room-sealed appliance, will derive all its ventilation air from the room in which it is installed. You must ensure that there will be enough combustion air available from the room itself or by providing additional ventilation where determined by AS/NZS 5601.

The heater in **Figure 12.3** draws its combustion air from within the living area. This room needs to be ventilated correctly to ensure an adequate supply of air for both the appliance combustion process and *any inhabitants* in the room.

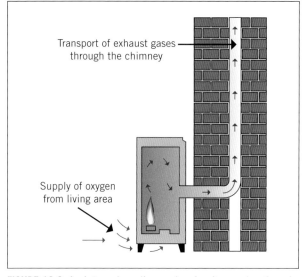

FIGURE 12.3 An internal appliance drawing its combustion air from within the living area

Air requirements and heat value

Oxygen is required for any combustion process to take place. This simple statement is true for all forms of combustion. The air around us is made up of the following key constituents:

- 20.95 per cent oxygen
- 0.038 per cent carbon dioxide
- 78 per cent nitrogen
- minor trace gases.

When combustion takes place, it is the oxygen that is consumed. The nitrogen passes through the combustion process largely unchanged. For any combustion process to take place, certain quantities of air are required to deliver the necessary amount of vital oxygen.

Type A gas appliances are designed to burn hydrocarbon-based fuel gases. For the appliance to do this safely and efficiently, there is a minimum volume of air required to ensure the gas is completely consumed during combustion.

In order to completely burn 1 m^3 of natural gas (NG), approximately 9.5 m^3 of air is required. In comparison, liquefied petroleum gas (LPG) requires a minimum of 24 m^3 of air for 1 m^3 of gas to be completely burnt. These figures can be expressed as simple ratios:

NG: 9.5:1
LPG: 24:1

You will recall from Chapter 4 that fuel gases have a specific heat value. This is the amount of energy released when 1 m^3 of gas in air is combusted under standard test conditions. The heat value for our two principal fuel gases are:

NG: 38 MJ
LPG: 96 MJ

This is the amount of heat released when the gas is burnt in correct proportion to the air supply.

As a licensed gas fitter, it is your responsibility to ensure that these minimum requirements are met, through strict adherence to the manufacturer's instructions and the AS/NZS 5601 Gas Installation Standards.

Gas rate

A gas appliance is rated according to the amount of gas it consumes in one hour and this information is noted on all appliance data plates. Approximate gas rates for some common appliances are:

- upright domestic cooker – 55 MJ/h
- decorative flame-effect fire – 30 MJ/h
- continuous flow water heater – 240 MJ/h
- cooktop – 27 MJ/h

This information is vital to your ventilation calculations, as the total gas consumption of all appliances in the room space has a direct relationship to the volume of air required and size of any additional ventilation. The example below provides some sense of how these gas characteristics interrelate and the importance of correct ventilation.

EXAMPLE 12.1

A small commercial kitchen connected to LPG has a total gas rate of 1056 MJ/h. You know the following facts:
- 1 m^3 of LPG has a heat value of 96 MJ.

Therefore:

$$\frac{1056 \, MJ/h}{96 \, MJ} = 11 \, m^3/h \text{ of gas}$$

You also know:
- 1 m^3 of LPG requires 24 m^3 of air to enable complete combustion.

Therefore, at full gas rate this installation will require the following amount of air:

$$11 \, m^3 \times 24 \, m^3 = 64 \, m^3 \text{ of air } every \ hour$$

Indications and effects of poor ventilation

Inadequate ventilation may be characterised by some or all of the following:

- formation of soot (carbon)
- streaming, yellow flames searching for air within the combustion chamber
- an unstable flame
- spillage of flue products
- production of the poisonous gas carbon monoxide (CO, which can be fatal)
- complaints by building occupants of headaches and irritated eyes (a warning that CO is likely to be present)
- noxious smells (often caused by aldehydes).

Incorrect appliance ventilation can lead to tragic circumstances in which deaths result from carbon monoxide poisoning and asphyxiation. You must never take appliance ventilation for granted and must always apply a methodical approach to such calculations for *every* installation.

Ventilation choices

Ventilation for internal appliances can be provided in two different ways, depending upon the nature of the installation and the requirements of AS/NZS 5601.1.

Natural ventilation

When gas streams out of the injector, it passes the primary air port and has the effect of drawing into the burner approximately 50 per cent of the air required for combustion. This is known as air inspiration. This air must come from the room in which the appliance is installed. Where calculations from AS/NZS 5601.1 indicate that this is sufficient air, it is known as 'adventitious air' or, more simply, air that makes its way in around cracks in doors and windows.

Sometimes the gas input and air requirements of the appliance are too much for a given room volume, and an additional air supply must be sought. Where this happens, you must then calculate the size of ventilation openings needed into the room in order to gain the additional air necessary for both complete combustion and the building inhabitants.

Sources of natural ventilation air

Where your calculations indicate that an appliance installation will require additional ventilation, you must decide whether the air is to be drawn from one of the following two sources:
- Direct to outside – this is the ideal solution to ventilation needs, but must be planned in respect to the very specific descriptions provided within the Standards. Options include:
  - directly through the wall to outside – this is always the preferred solution
  - through to the outside wall but offset, ensuring that the path of air is not blocked by insulation or any other obstruction
  - into a cavity that is in turn vented to outside air
  - into an under-floor space that is vented to outside
  - into a roof space that is vented to outside
- Air from an adjacent room – where additional ventilation cannot be achieved directly to outside, the Standard provides calculations that may permit air to be drawn from adjacent room space. This room may then be vented to outside or in some circumstances may provide sufficient air in itself.

Mechanical ventilation

Sometimes the only means of ensuring adequate ventilation is to provide additional air supply by mechanical means. A below-ground commercial kitchen is one instance where this may be required, as it would be very difficult to source additional natural ventilation direct from outside.

The required ventilation is achieved by the provision of a fan-driven air inlet and either a natural or a fan-driven outlet (not to be confused with cooking vapour extraction fans and canopies). In such cases, the gas supply to the installation must also be interlocked with the air supply system to ensure that if the fan fails, the gas supply is cut off.

Quality of ventilation air

To ensure safe and complete combustion, gas appliances must have a constant supply of fresh air that does not contain contaminants that might affect the combustion process, cause damage to the appliance or create noxious fumes. In situations where vapours, dust or lint may be present at any time, you may need to ensure that all combustion air is ducted from an alternative location.

REFER TO AS/NZS 5601.1 'QUALITY OF AIR SUPPLY'

LEARNING TASK 12.1

1. How many volumes of air are required to completely combust the following volumes of fuel gas?
   5.5 m³ of LPG _____

   21 m³ of NG _____

2. Circle the correct answer – a room-sealed appliance sources its combustion air from:
   a inside the room space
   b through the outer tube of a balanced flue
   c through the inner tube of a balanced flue

3. How many cubic metres of air does a natural gas appliance with a gas rate of 190 MJ/h require per hour?

Prepare for work

All tasks must be assessed and planned before they begin. A gas fitter must ensure that a thorough and methodical approach is taken to documenting ventilation calculations so that mistakes and oversights can be eliminated. Getting it wrong can lead to tragic results.

Important terms to understand

Part of the preparation process is evaluating the site so that you are able to match the design with a particular ventilation solution. Before commencing your ventilation calculations, it is important to clarify some key terms that are used throughout the process. Terms that apply to ventilation of appliances specifically are as follows:
- *Room* – while not directly defined in the Standards, this term is used in the context of any normally habitable room space. For example, this could be a kitchen, lounge area, bedroom, office or other workspace.
- *Enclosure* – an enclosure is defined in the Standards as a compartment or space primarily used for the installation of an appliance and associated equipment.
- *Plant room* – despite what is often regarded as a plant room, this is defined as a room *designed* to accommodate one or more *appliances* in which the *appliances* can be fully maintained and is not *normally occupied or frequented for extended periods*.

- *Free ventilated area* – this is defined as the cross-sectional area in mm² of the actual unobstructed free gaps within a vent component. To prevent blockage by dust and lint, the minimum size of any free ventilated opening is 6 mm.

REFER TO AS/NZS 5601.1 SECTION 1 'DEFINITIONS'

Sources of planning information

Sometimes the planning process for ventilation will take place on-site during a quote or site inspection. At other times, calculations will be based on building plans prior to or during construction. As with all forms of gas installation planning, ventilation details should be recorded on simple documents for clear and concise reference. Regardless of which stage of ventilation is being assessed, sourcing and recording all relevant information using a table format and headings is a simple and clear way to document all necessary data. Table 12.1 is an example of how you might set out a simple planning table.

Safety considerations

All jobs require some form of safety evaluation that is formalised through the application of Job Safety Analysis (JSA) and Safe Work Method Statements (SWMS). You must always follow your company protocols and reporting policies, to ensure such documents are used in the correct manner.

Installation of gas appliance ventilation involves a number of tasks that can prove to be hazardous. Some considerations are discussed below:

TABLE 12.1 Planning for ventilation

Sources of ventilation planning information	Specific detail	Notes
Plans and specifications	<ul><li>Approved plans<ul><li>Is the building a new construction?</li><li>If an existing building, when was construction completed?</li></ul></li><li>Appliance type, rating and number</li><li>Flue requirements</li><li>Room dimensions</li><li>Intended purpose of room/building</li><li>Cross-ventilation</li><li>Ancillary air extraction systems – will it affect appliance ventilation?</li></ul>	
Site inspections	<ul><li>Confirm actual construction with original plans</li><li>Confirm source of air supply as intended on plans is actually possible<ul><li>Direct from outside</li><li>Adjacent room</li></ul></li></ul>	
AS/NZS 5601.1	<ul><li>General ventilation options</li><li>Specific appliance requirements</li><li>Testing requirements</li></ul>	
Manufacturer's appliance information	<ul><li>Data plate:<ul><li>Confirm appliance is Certified</li><li>MJ/h</li></ul></li><li>Specific installation and ventilation instructions</li></ul>	

Through building a permanent list of sub-headings that you can 'tick off', it is less likely that you will miss vital considerations

In this column, write in all relevant information for the calculation

Potential hazards associated with installation of ventilation solutions	
Working at heights	Many ventilation installations will require the installation of a high level vent. This may vary from a simple domestic room to locations many metres above the floor in large commercial and industrial buildings. This work may require the use of: • platform ladders • elevated work platforms (EWP) • scaffolding – fixed or mobile.
Asbestos	Asbestos is a highly dangerous Class 1 carcinogenic (cancer-causing) substance that gas fitters must be on the lookout for on a daily basis. The walls of many older buildings may be clad with asbestos-containing materials, and strict laws and guidelines exist on how this must be dealt with. If in any doubt you must contact your supervisor before commencing work.
Electrical hazards	Electrical hazards may exist behind any wall cladding. Always check before making large penetrations in walls with tools that may cut power cables and possibly lead to electrocution. Items that may help include: • non-contact voltage detectors • inspection cameras with flexible probes.
Dust, noise and flying debris	Where hazards cannot be eliminated or modified, the use of appropriate PPE is necessary. This includes safety glasses/shields, hearing protection, and breathing mask appropriate to the type of dust being generated. In some instances and job sites you will be required to ensure that dust and debris is fully captured with industrial vacuums fitted with rated HEPA (High Efficiency Particulate Arresting) filters.

What other hazards might you anticipate?

Tools and equipment

Installation of ventilation for gas appliances will be largely determined by factors that include type of wall material, wall thickness and room height. The table below shows some examples.

Grinders
When fitted with diamond blades, grinders are a useful tool for cutting out vent holes through masonry products.

Plaster saw (jab saw)
Cutting through plaster sheeting is made much easier with the use of a short plaster saw.

Source: Imagery courtesy Milwaukee Tool

>>

Stud finder
Before making any hole in plaster sheet walls you must determine the location of studs and noggins. A stud finder locates obstructions before you penetrate the wall.

Inspection camera
Cutting into certain wall, ceiling and floor cavities without knowing the location of other services such as electricity and gas can be very dangerous. An inspection camera enables you to view cavity spaces prior to making larger cuts and penetrations.

Source: Imagery courtesy Milwaukee Tool

Rotary hammer drill
These tools enable quick drilling of holes in masonry material for various fasteners.

Source: Imagery courtesy Milwaukee Tool

Dust extraction equipment
On many construction sites, the uncontrolled generation of dust is no longer permitted and contractors are required to ensure that all dust is captured with HEPA-rated vacuum equipment.

Elevated work platform
For the high walls found in larger commercial and industrial applications, some form of elevated work platform enables safe and efficient installation of upper wall vents.

What other items can you think of?

LEARNING TASK 12.2

1. A domestic gas dryer and gas storage water heater are to be installed in a laundry. In reference to the designation of spaces as used in the Standards, what type of space would this be?
   - plant room
   - room
   - enclosure
2. Where would you find the gas rate and certification details of an appliance?

Identify natural ventilation requirements

Having collected all the necessary information, you are now ready to use AS/NZS 5601.1 to undertake a ventilation calculation. In this section, we will cover the steps required to work out the appliance ventilation needs for most general installations. Details for more specific appliance installations will follow later in the chapter.

To meet the latest mandated energy ratings, new buildings and renovations are now constructed in a way that has significantly reduced the amount of adventitious ventilation available to gas appliances. While older buildings tended to 'leak' more air, modern buildings are more airtight. As a result, the ventilation calculations in the AS/NZS 5601 have been split into two key methods:
- Method 1 – ventilation of spaces for buildings approved for construction *before* adoption of the AS/NZS 5601 – 2013
- Method 2 – ventilation of spaces for buildings approved for construction *after* adoption of the AS/NZS 5601 – 2013.

The emphasised words are important. In each State or Territory, the Standard was adopted or 'called-up' by technical regulators at different times. You need to find out when they were adopted for your own area, as the calculations are very different for each situation.

Ventilation Method 1 – before adoption of AS/NZS 5601 – 2013

To help you learn how to calculate ventilation of appliances installed in a room or enclosure, follow each step of this installation example with reference to AS/NZS 5601.1.

EXAMPLE 12.2

A customer has asked you to install a non-room-sealed natural draught continuous flow water heater in a residential laundry. Upon inspection, you record the following information:

Room volume	3 m × 2.5 m × 2.4 m high = 18 m³
MJ/h input	127 MJ/h
Flue system	Natural draught – not room sealed
Potential air access	Direct from outside, if required
Does room have door isolation?	Yes
Room/enclosure or plant room?	Room – this determines what factor is applied.

Having gathered all the information, it is now time to come up with a solution. Use the following steps as a guide.

Step 1 – Is additional ventilation required?
To determine whether additional ventilation is required, the Standard states that the total input of all appliances in the room must not exceed 3 MJ/h for each cubic metre of the room volume.
- If gas input is 3 MJ/h/m³ or less, it is considered that there is enough air for combustion provided via adventitious openings (gaps around doors and windows).
- If the input is greater than 3 MJ/h/m³ then you are required to provide additional ventilation.

To work this out, divide the MJ/h input of all appliances in the room by the volume of the room in cubic metres.

In this example, you can see that the input of the continuous flow hot water service is well in excess of the maximum 3 MJ/h/m³.

$$\text{Room volume} = L \times W \times H$$
$$= 3\,m \times 2.5\,m \times 2.4\,m$$
$$= 18\,m^3$$

Less than or greater than 3 MJ/h/m³?

$$3\,MJ/h/m^3 = \frac{\text{heater input}}{\text{room volume}}$$
$$= \frac{127\,MJ/h}{18\,m^3}$$
$$= 7.05\,MJ/h/m^3$$

Therefore, you must provide additional ventilation so that there is enough air for this appliance to carry out complete and safe combustion.

Step 2 – What size are the vents?
Now that you have determined that additional ventilation is required for this laundry installation, the next step is to work out the minimum size of each vent.

To work out the actual size of each vent providing air direct from outside, you need to apply the formula $A = F \times T$ where:

A = the minimum free ventilation area (mm²)
F = the factor as given in Table 6.1 in the Standard
T = the total gas consumption of all appliances (MJ/h).

The size of each vent as determined by the formula $A = F \times T$, relates specifically to *the opening itself* and not the physical size of the vent. You will need to calculate the actual free-ventilated area of any commercially made vent and ensure that it complies.

Each opening in the 'egg-crate' grille-style vent shown measures 1 cm² (100 mm²). These vents come in a wide range of sizes and provide an off-the-shelf solution for most internal ventilation requirements.

For the laundry installation, we can calculate the size of each vent using the formula $A = F \times T$.

Vent size calculation:

$A = F \times T$
$A = 300 \times 127\,MJ/h$
$A = 38\,100\,mm^2$ (actual opening for each vent)

Hint: The square root of 38 100 mm² gives you a vent of approximately 196 mm × 196 mm. A rectangular vent might be approximately 392 mm × 98 mm for the same area.

Step 3 – Where are the vents located?
AS/NZS 5601.1 provides two solutions for vent locations:
- For flued appliances, the lower vent must be located at or below the level of the burner while the upper vent must be sited at or higher than the appliance draught diverter.
- For flueless appliances, the distance between the top of the upper opening and ceiling and the distance between the bottom of the lower opening and floor does not exceed 5 per cent of room height. The diagram below shows the cross-section of a room indicating how vents may be positioned to achieve good cross-ventilation for flueless appliances.

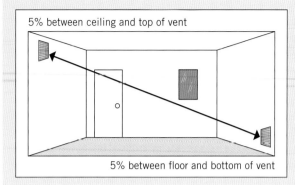

5% between ceiling and top of vent

5% between floor and bottom of vent

Using a percentage to determine the location means that regardless of the particular dimensions of the room, the vents will always be sited in proportion to the room height.

Typically, and where possible, these vents should be positioned so that a good cross-flow of air is achieved across the room. Where the room height is low enough, these two vents may be combined into one, as long as the top and bottom of each opening comply with the limits as stated in AS/NZS 5601.1.

Permanent ventilation

Occasionally a client will suggest that rather than provide additional ventilation into the room, they will promise to keep the door or window ajar. This is not permitted. Despite the best intentions on the part of the owner, you have no guarantee that this will always be the case, particularly when the property changes ownership. If your calculations show that additional ventilation is required, then this *must* be permanent.

Ventilation of a room or enclosure via an adjacent room

At times, you will encounter a job or plan that would be impossible to ventilate directly to the outside. A split-level home where part of the building is below ground level might be one instance. Where this happens, you will need to assess whether air might be available via an adjacent room. To help you learn how to calculate ventilation of appliances installed in a room or enclosure via an adjacent room, follow each step of Example 12.3, with reference to AS/NZS 5601.1.

EXAMPLE 12.3

A customer has asked you to install two internal commercial storage water heater cylinders in a storeroom adjoining a workshop. You inspect the site and record the following information:

Room 1 volume	3.5 m × 4 m × 2.4 m high = 33.6 m³
Room 2 volume	6 m × 4 m × 2.4 m high = 57.6 m³
Total MJ/h input	200 MJ/h
Flue system	Natural draught – not room sealed
Potential air access	Via an adjacent room only
Does room have door isolation?	Yes
Room/enclosure or plant room?	Room – this determines what factor is applied.

Step 1 – Is additional ventilation required?

You will recall from Example 12.2 that the first step is to determine whether there is sufficient air in the room to carry out combustion. Work through the following:

Room volume = $L \times W \times H$
= 3.5 m × 4 m × 2.4 m
= 33.6 m³

Less than or greater than 3 MJ/h/m³?

$$3 \text{ MJ/h/m}^3 = \frac{\text{total input}}{\text{room volume}}$$
$$= \frac{200 \text{ MJ/h}}{33.6 \text{ m}^3}$$
$$= 5.95 \text{ MJ/h/m}^3$$

The answer from the above calculation is greater than 3 MJ/h/m³. With two storage water heater systems in such a small room you can see that additional air will need to be drawn from the adjacent room.

Step 2 – What size are the vents?

To determine the size of each vent, refer to the following calculation, with reference to the AS/NZS 5601.1.

REFER TO AS/NZS 5601.1 'NATURAL VENTILATION VIA AN ADJACENT ROOM'

Vent size calculation

$A = F \times T$
$A = 600 \times 200$ MJ/h
$A = 120\,000$ mm² (actual opening for each vent)

Hint: The square root of 120 000 mm² gives you a vent 347 mm × 347 mm. A rectangular vent might be 600 mm × 200 mm.

Step 3 – Where are the vents located?
As in Example 12.2, the vents must be installed at or below the lowest burner and at or above the draught diverter, and placed in such a way as to provide as much cross-ventilation as possible.

Air from the adjacent room
Air from an adjacent room is not deemed to be as 'fresh' as that drawn from outside; therefore, internal venting is designed to be larger to improve the flow of air between rooms.

The storeroom has now been ventilated via an adjacent room. For the purposes of ventilation, once vents have been installed, rooms 1 and 2 are now regarded as one room volume, with all the air available for combustion. This is important to understand, as you now need to check whether further ventilation is required based upon this new combined room volume.

Is additional ventilation required?
You now need to recalculate the relationship of MJ/h to room volume. If you arrive at an answer of 3 MJ/h/m³ or less, you would not need to provide any additional ventilation. If it is still greater than 3 MJ/h/m³, you would have to continue to provide more ventilation, either through each subsequent adjacent room according to $A = 600(F) \times T$ or, if possible, direct to outside according to $A = 300(F) \times T$.

Combined room volume = room 1 + room 2
$$= 33.6\,m^3 + 56.6\,m^3$$
$$= 90.2\,m^3$$

Less than or greater than 3 MJ/h/m³?

$$3\,MJ/h/m^3 = \frac{\text{total input}}{\text{total volume of rooms}}$$
$$= \frac{200\,MJ/h}{33.6\,m^3}$$
$$= 5.95\,MJ/h/m^3$$

Applying this to our example, you can see that with ventilation drawn from two rooms, there is now sufficient air for combustion as there are only 2.22 MJ/h/m³.

Ventilation of a plant room
Where the room is identified as a plant room, you apply the same steps to the calculation, but ensure that you choose the matching factor for a plant room location from Table 6.1 in the Standard. Because a plant room is not normally occupied, the vent sizes are half the area of those used for a room or enclosure.

 COMPLETE WORKSHEET 1

Ventilation Method 2 – after adoption of AS/NZS 5601 – 2013
Although it has some similarities to Method 1, the overall approach to determining ventilation requirements for spaces approved for construction after adoption of AS/NZS 5601 – 2013 is very different.

Ventilation Method 2 can be regarded as having three primary categories:
1 Ventilation for flued appliances only
2 Ventilation for flueless appliances only (other than flueless space heaters)
3 Ventilation for both flued and flueless appliances (other than flueless space heaters).

The processes for each of these categories all relate to specific factors and appliance locations including a room, an enclosure, a residential garage and a plant room. It is important to note that where appliances are installed in a residential garage, there is no allowance for adventitious ventilation and as such these spaces must always be ventilated. Furthermore, a residential garage is not permitted to be ventilated via an adjacent room.

REFER TO AS/NZS 5601.1 TABLE 6.2 'NATURAL VENTILATION FOR GAS APPLIANCES' AND ATTACHED NOTES

In addition to examples provided within the Standard, we have provided a flowchart to demonstrate how the calculations are carried out (see Figure 12.4). This example is based on a calculation for flued appliances, but a similar process is used for each of the remaining two categories. As a considerable number of steps are involved, you should follow the sequence detailed in the Standard every time you carry out a calculation.

GREEN TIP
Where possible always try to locate vents in a position so that air movement will not be a nuisance to room occupants, particularly in residential applications. Draughts will not only upset occupants but in colder climates will also potentially increase fuel consumption and heating costs to compensate.

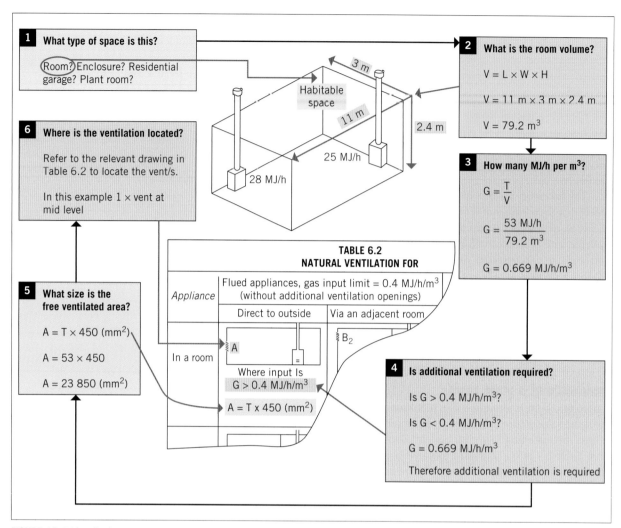

FIGURE 12.4 Ventilation – spaces containing only flued appliances

Mechanical ventilation

Where the provision of a natural air supply is not feasible, you may need to consider the use of mechanical ventilation. This is normally found in commercial and industrial applications and the Gas Installation Standards provide simple guidance on the necessary requirements.

Some basic rules apply to mechanical ventilation installations:
- Air must be drawn directly from outside.
- An interlock mechanism must be fitted to ensure that if the mechanical air supply stops for any reason then the gas supply will be shut down immediately. This would generally be some form of sail-switch or pressure switch linked to a solenoid valve in the gas supply line.
- A low-level natural air supply and high-level mechanical exhaust is not permitted; air supply must always be provided by mechanical means.
- A choice of high-level natural or mechanical exhaust is permitted.

REFER TO AS/NZS 5601.1 SECTION 6 'MECHANICAL VENTILATION'

Follow the example provided and the requirements in the Gas Installation Standards to see how a mechanical ventilation solution can be provided.

It is important to note that where mechanical ventilation is used, you must ensure that no flue spillage or disturbance to the combustion process happens as a result of any negative pressure effects. The use of a smoke candle near the draught diverter while following the flue spillage test procedures in Appendix R of the Standard, as described in Chapter 14, will help you confirm this.

EXAMPLE 12.4

You are asked to install 2 × 500 MJ/h flued natural draught appliances in a plant room. You decide to calculate the requirements for a mechanical ventilation solution.

Step 1 – Low-level mechanical supply
Type of appliance burner – atmospheric natural draught burner

$$\text{Mechanical supply fan air flow in L/s} = (\text{MJ/h}) \times 0.5$$
$$\text{L/s} = 1000 \times 0.5$$
$$\text{L/s} = 500$$

Location: As the appliances are flued, this mechanical air supply would be located at or below the level of the burners.

Step 2a – High-level exhaust (mechanical)
The exhaust is to be sized to between one quarter and one third of the required air inlet rate.

Mechanical inlet L/s = 500

Therefore, a mechanical exhaust must have an extraction rate of between approximately 125 L/s and 166 L/s.

Step 2b – High-level exhaust (natural)

$$\text{Natural exhaust size (mm}^2) = (\text{MJ/h}) \times 150$$
$$= 1000 \times 150$$
$$= 150\,000 \text{ mm}^2$$

Location: Both of these exhaust solutions would be located above the level of the draught diverters and in a position to provide as much cross-ventilation as possible.

Specific ventilation calculations

The ventilation of a room, enclosure, residential garage and plant room are regarded as general ventilation calculations. Knowing how these ventilation solutions are achieved will allow you to provide sufficient air supply for the majority of gas appliance installations.

However, certain other gas installations have additional and/or different ventilation needs that you need to be aware of. These are broadly termed 'specific ventilation calculations'.

Additional considerations
The following installations have particular needs and should be discussed separately when considering ventilation:
- appliances in roof spaces and under floors
- decorative flame effect fires
- commercial kitchens
- overhead radiant tube heaters.

Appliances in roof spaces and under floors
The provision of ventilation for appliances in roof spaces and under floors is actually no different than when using general calculations. The installations may sometimes be unusual, but you will still need to provide sufficient ventilation to ensure complete combustion.

Commonly used as the location of central heaters and hot water systems, these spaces, if not used for any other purpose, are generally classed as enclosures (though this may be open to interpretation by your technical regulator). One difference relates to calculating the volume of a pitched roof space to determine the need to ventilate. To do this you would need to use the following formula:

Volume of a roof space = height to ridge × ½ the width × length of roof

Of course, where composite roof profiles intersect, you will need to combine roof space volumes in a number of calculations.

Once you have confirmed a need to provide additional ventilation, vents are positioned according to the standard requirements discussed earlier. Eaves and gables are ideal places to install vents, which should be sized according to the relevant factor to gain air direct from outside.

Decorative flame effect fires – additional ventilation needs
Before reviewing this section, you should check your understanding of what a decorative flame effect fire is.

REFER TO AS/NZS 5601.1 GLOSSARY 'DECORATIVE FLAME EFFECT FIRE'

Where the installation of either a Type 1 or Type 2 decorative flame effect fire is planned, special ventilation needs must be considered. This specific ventilation is in addition to any general ventilation requirements. Reliance upon adventitious ventilation is not permitted and vent size is calculated in a different manner.

In particular, you should note the following:

One or more openings with a combined free ventilation area of not less than the equivalent cross-sectional area of the *flue cowl* shall be provided for each *decorative flame effect fire*.

Therefore, your first task is to determine what type of appliance you are installing. This information can be found in the manufacturer's literature.

EXAMPLE 12.5

A Type 1 decorative flame effect fire is designed primarily to be installed in an open fireplace, more for its aesthetic qualities than for heating. The burner itself is not enclosed. A Type 1 decorative flame effect fire requires a chimney outlet of at least 40 000 mm^2 or as specified by the manufacturer. In this case, to comply with the ventilation requirements you would need to install one or more openings in the room with an area at least equal to the cross-sectional area of the flue cowl. These openings would need to go direct to outside by one of the methods described in AS/NZS 5601.1.

EXAMPLE 12.6

A Type 2 decorative flame effect fire is an appliance that is fully enclosed and, depending upon the model, may be installed in a fireplace with its own separate chimney liner or as a freestanding appliance. The flue liners for these appliances are generally 100 mm in diameter. You will note that AS/NZS 5601.1 does not specify which types of decorative flame effect fires should be ventilated in this manner. It simply states that each appliance must be ventilated in accordance with the cross-sectional area of the flue cowl. Therefore, to ventilate a Type 2 decorative flame effect fire, one or more openings of an equivalent area to the flue cowl must be installed direct to outside.

Commercial kitchens

Commercial kitchens provide the gas fitter with some unique challenges when it comes to calculating ventilation needs. Depending upon their size, these kitchens are often characterised by multiple appliances that each or collectively have much greater MJ/h inputs than domestic installations. Larger kitchens can commonly have a total gas input well in excess of 1500 MJ/h.

Therefore, it is vital that you ensure a complete ventilation calculation is carried out for every commercial kitchen installation and maintenance job. Do not assume there is enough room volume.

One particular feature of a commercial kitchen that needs to be taken into account is the effect of commercial range hoods on normal ventilation and air movement. Such canopies are fitted with large extraction fans and are installed over all the cooking appliances to remove cooking vapours and airborne fats and oils from the kitchen. Without a supply of relief air, the strong airflow created by these fans would create a negative pressure atmosphere in the room and adversely affect gas appliance combustion air supply.

Relief air is a supply of air in addition to the air required for combustion.

Modern commercial kitchen exhaust hoods must be designed in accordance with AS 1668.2 'Mechanical ventilation in buildings', and will normally incorporate either an integral relief plenum or remote relief air inlet system. This means that the same amount of air removed by the exhaust hood is brought back into the kitchen through the relief. Figure 12.5 provides an example of how this relief air is designed to replace the same amount of air extracted from the cooking area, thereby maintaining positive pressure within the kitchen. Exhaust systems designed in this manner should have no ill effect on the gas combustion process.

However, be very wary of older installations where there is no relief air or it is not the correct volume/flow rate. Even if you have correctly calculated and installed ventilation according to AS/NZS 5601.1, appliances may still be starved of air if overpowered by these extraction fans. Therefore, you must ensure that each commercial kitchen is checked for any combustion flame disturbance and, where necessary, is fitted with an additional relief vent that is sized to provide *the same volume of air that is being mechanically extracted*. Remember, this relief vent is *in addition* to standard ventilation requirements.

Where you are in any doubt as to the design of the kitchen extraction system, contact a reputable manufacturer of compliant exhaust hoods for an inspection report and/or advice.

Flueless space heaters

There is no single solution to determine the ventilation needs for a flueless space heater. This is because many states have different minimum requirements, and in some situations these appliances cannot be installed at all.

A flueless space heater installation must only be used in a ventilated room as stipulated by the manufacturer. You also need to refer to AS/NZS 5601.1 for any additional state-specific requirements.

REFER TO AS/NZS 5601.1 'RESTRICTIONS ON INSTALLATION OF A FLUELESS SPACE HEATER'

While standard ventilation requirements apply, you also need to ensure that the installation of a flueless space heater does not exceed the maximum permitted gas consumption per cubic metre of room volume. This is to ensure that the consumption of oxygen by the appliance is not greater than that available in the room at any time.

If you refer to AS/NZS 5601.1, you will see that you need to consider two possible calculations based on whether or not the appliance is thermostatically controlled. Refer to Chapter 13 for more specific information relating to flueless space heaters. Some basic examples are given here.

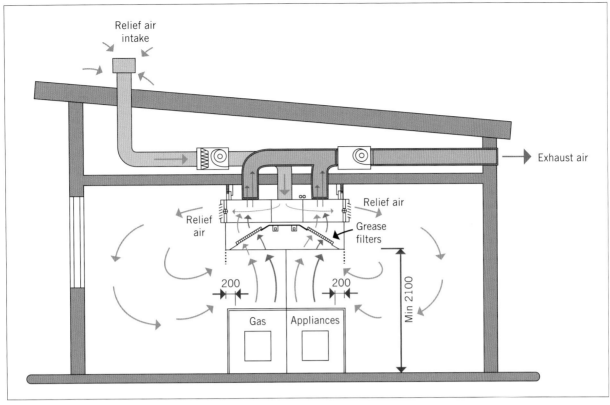

FIGURE 12.5 A commercial exhaust hood with an integral relief air inlet system

EXAMPLE 12.7

A thermostatically controlled flueless space heater is to be installed in a room 6 m × 4 m × 2.4 m high. To determine the maximum permitted gas consumption, you need to apply the following calculation:

$$\begin{aligned}
\text{Maximum gas consumption} &= \text{Volume of room (m}^3) \times 0.4 \\
&= (6\,\text{m} \times 4\,\text{m} \times 2.4\,\text{m}) \times 0.4 \\
&= 57.6 \times 0.4 \\
&= 23.04\,\text{MJ/h}
\end{aligned}$$

This means that the thermostatically controlled flueless space heater that you wish to install must not have a greater MJ/h input greater than 23.04 MJ/h.

EXAMPLE 12.8

To install a non-thermostatically controlled flueless space heater in a the room of the same dimensions as in Example 12.7, you would need to apply the following calculation.

Because a non-thermostatically controlled space heater does not modulate on and off, it requires a much greater safety margin. Therefore, the maximum MJ/h input is reduced accordingly.

$$\begin{aligned}
\text{Maximum gas consumption} &= \text{Volume of room (m}^3) \times 0.2 \\
&= (6\,\text{m} \times 4\,\text{m} \times 2.4\,\text{m}) \times 0.2 \\
&= 57.6 \times 0.2 \\
&= 11.52\,\text{MJ/h}
\end{aligned}$$

Overhead radiant tube heaters

Overhead radiant tube heaters are a common heating solution for large commercial and industrial applications (see Figure 12.6), particularly in open draughty areas where convective heating would be very inefficient. Under normal circumstances you will calculate ventilation requirements for these heaters in the same manner as for other gas appliances.

If a radiant tube heater is to be used in any area where there is the chance of a contaminated atmosphere being created (dust, vapours or lint), then you must ensure that combustion air is ducted directly from outside and that combustion products are flued to outside. This requirement would effectively negate the requirement to provide normal ventilation solutions. In such instances the Standard advises that you should consult with the technical regulator where necessary.

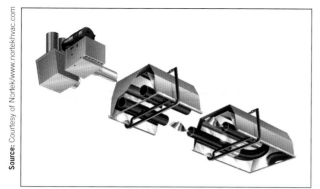

FIGURE 12.6 Cut-away example of an overhead radiant tube heater, which may require specific inlet air and combustion exhaust solutions

 REFER TO AS/NZS 5601.1 'OVERHEAD RADIANT TUBE HEATERS'

LEARNING TASK 12.3

1 For the purposes of determining which ventilation method to use based upon age of the building, what is the adoption date of AS/NZS 5601:2013 in your State or Territory?
2 As detailed in AS/NZS 5601, what specific requirements exist in your State or Territory in relation to the installation of flueless space heaters?
3 You have been asked to install a gas appliance at the back of a residential garage built in 1980. Are you permitted to determine if adventitious ventilation will provide sufficient air supply for the installation? Provide a Standards reference to support your answer.

COMPLETE WORKSHEET 2

Install ventilation and test appliance

Requirements related to the installation of vents will be dependent upon the materials of the building wall and the size of vents being fitted. While the installation of vents during new construction is basically a matter of coordinating with other trades, working on existing buildings can present some unique challenges.

It is impossible to cover all situations, but some common considerations are provided below.

Electrical tracing and stud finding

One you have chosen a likely location for your wall or ceiling vents, do not simply start cutting holes. While wall studs are generally found at 450 mm or 600 mm centres, they can just as easily be anywhere in between, particularly in older buildings. Failure to investigate the suitability of the vent location will inevitably lead to unnecessary damage and costly repairs.

- Always use a non-contact voltage detector (see Figure 12.7) to confirm the possible presence of power cables and understand that even if the instrument does not pick up any voltage it could simply be that the cables are too far back for the device to detect or the wall material is too thick.

FIGURE 12.7 Always use a volt stick and stud finder before making large cuts for vents

FIGURE 12.8 Vent cut-outs in plaster and lath walls and ceilings must be carried out very carefully

- For standard plaster walls, carefully glide a stud finder (see Figure 12.7) across the area and discretely mark the location of any studs, noggins or ceiling joists. Rather than using pen or pencil marks that can be difficult to remove from wall paint, use small pieces of painter's masking tape to show where studs and ceiling joists are located.
- Once you have narrowed down the location, square off the cut-out to suit the vent size and if you have one on hand, insert a flexible camera probe into a small hole to detect any other services. If something is found, it is far easier to repair small inspection holes than larger cut-outs.

Working in older buildings

This section is particularly important for those of you who may end up working in heritage listed buildings.

Before the introduction of paper-lined sheets of gypsum-based dry plaster, internal walls were wet plastered by hand. Many older buildings constructed up into the 1950s will commonly have internal walls that featured a lining system known as 'plaster and lath'. This was made up from multiple strips of thin-section timber nailed to the studs or ceiling joists leaving a small gap between each strip. The wet plaster was then trowelled by hand over these strips with the gaps helping the plaster mix stay in place until dry.

Plaster and lath walls can be quite fragile once disturbed, and if cut incorrectly can result in extensive cracking and breakage of plaster, particularly along the run of the lath strips.

- Do not use reciprocating power saws or course-toothed jab saws. The lath strips will tend to jam the blades suddenly and this normally results in significant cracking of the plaster.
- Having checked for power cables, other services and studs/ceiling joists, use a 3 mm drill bit to create enough space to insert a 24 TPI (teeth per inch) hacksaw blade fitted to a jab handle and keeping the saw at a flat angle, patiently use even strokes to cut your way across the lath. It will take some time but will generally avoid any plaster damage.

Removal of masonry

Installation of vents will sometimes require the removal of clay or masonry bricks. For small vent dimensions the removal of one or two bricks is generally straightforward, but for larger vent areas you need to consider the impact of the vent cut-out on the structural integrity of the wall.

- Larger penetrations in brick walls will often require the installation of a steel lintel to prevent cracking and/or collapse of bricks above the cut-out area (see Figure 12.9).
- Where possible, use a small grinder fitted with a diamond blade and/or raking chisel to remove the mortar from the brick course immediately above the area to be removed and insert the lintel before

FIGURE 12.9 A steel lintel used to prevent cracking and/or collapse of bricks

cutting out lower bricks. Where this cannot be done, you may need to support the upper bricks while lower bricks are being removed.
- Where a cut is proposed through concrete panels or possible locations where structural steel reinforcement may be present, always seek professional advice before proceeding.
- If in any doubt and to avoid unnecessary damage, it is advisable to have the proposed opening inspected by a bricklayer, builder or building surveyor to ensure the best approach is applied.

Appliance and flue testing

Following vent installation and as part of the appliance commissioning process you need to test the ventilation system for adequacy.

- Where flueless appliances only are installed and following standard commissioning, operate all appliances and check for correct combustion characteristics. Fundamental flame evaluation should always follow the list below:
  - Flames should ignite easily.
  - Flames should be entirely blue.
  - Flame zones should be clearly defined.
  - There are no yellow tips.
  - There is no lift-off or instability.
  - There is no streaming.

- Where flued appliances only or where both flued and flueless appliances share the same space with mechanical extraction systems, you must strictly follow the flue-spillage testing guidelines in the Standards. For some installations, the provision of standard fixed ventilation may not be sufficient to ensure the flues draw correctly under all conditions.

Flued appliances must be designed to safely convey potentially hazardous products of combustion away from the appliance through a flue and terminal. In order for these combustion products to be displaced outside to the atmosphere, the same amount of air must enter the building. Under favourable conditions, sufficient combustion and flue dilution air is made available through various forms of adventitious or fixed ventilation. However, problems can occur when some form of air extraction is used in the same zone as the flued appliance. This mechanical extraction can actually be more powerful than the aeromotive force or fan system in a flue system.

Flue spillage

Driven by modern construction materials and energy efficiency requirements, many buildings are now quite airtight and do not permit a great deal of adventitious air to flow inside. In such conditions, the use of kitchen range hoods, exhaust fans and air transfer systems operating in the proximity of a flued appliance may cause combustion gases to be sucked out of the flue system into the room. Any subsequent rise in carbon monoxide levels may be lethal. Be aware that these same hazards apply equally to caravan and marine craft installations.

To mitigate flue-spillage hazards, the Gas Installation Standards details a very specific procedure to test for spillage and enables you to apply a prescribed solution.

REFER TO AS/NZS 5601.1 APPENDIX R 'SPILLAGE TESTS FOR FLUED APPLIANCES'

This test procedure can be broken up into three main sections as follows:
1. Baseline test. Determine if spillage is occurring with neither appliances or air extraction fans operating.
2. Smoke test. Appliances are still not operating. Determine if spillage is occurring with air extraction fans operating at their highest setting and determine whether additional ventilation area is required to prevent flue gases being drawn away from the appliance/s.
3. Spillage test. This test procedure requires that the installation be tested with appliances operating, having first used combustion-detection equipment to confirm background air quality readings.

It is important to note that for your own safety and that of others in the vicinity, you must only undertake flue spillage testing in the strict order detailed in your Standard. To assist you the Standard provides simple flow charts for each stage of this procedure. Work through the Standard in partnership with your teacher and/or supervisor to attain a firm understanding of this vital test procedure. Also access this link to test for spillage using CO detection equipment: www.esv.vic.gov.au. > Gas professionals > Technical information sheets >Sheet 38.

Safety considerations during test

You must use some form of CO testing equipment (see Figure 12.10) to determine background air quality prior to commencing the spillage test and then while appliances are operating to ensure that the presence of CO is detected well before it reaches hazardous levels.

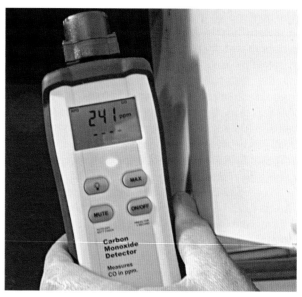

FIGURE 12.10 A hand-held CO gas detector used to monitor gas levels near the downdraught diverter during a spillage test

Smoke generation options

The spillage test requires an efficient and simple form of smoke generation that creates enough smoke for the test without creating a nuisance within the building. Three main options exist for the gas fitter in this regard.
- Smoke match – smoke matches are simple to use and generate enough smoke for quick tests of draught diverters and appliance surrounds. However, depending upon the brand, their burn time is only 20–30 seconds and they require that you be present to strike and hold the match in position (see Figure 12.11).
- Smoke candles – smoke candles have the advantage that they can be lit and left in a tray adjacent to a draught diverter while you carry out the test. Some can produce a considerable amount of smoke for up to 2–5 minutes, so you must be careful that smoke alarms are not set off unnecessarily, or that building occupants are not affected

FIGURE 12.11 Use of a smoke match at an appliance draught diverter shows that the flue is drawing correctly

- Incense sticks – simple incense sticks available from chemists and homeware stores are perfectly suitable for most tests. They can be lit and left in position, do not create too much smoke and have a long burn time. These are recommended by the author for most testing requirements.

Regardless of what type of smoke generator you use, always ensure that any combustible surfaces are protected while placing them in position. Also check if any building occupants have any respiratory conditions that may be affected by the presence of smoke and discuss how the test may be carried out without risk to their health.

Clean up the job

Your appliance ventilation installation task is not over until your work area is cleaned up and tools and equipment accounted for and returned to their required location in a serviceable condition.

Clean up your area

The following points are important considerations when finishing the job:
- If not already covered in your site induction, consult with supervisors in relation to the site refuse requirements.
- Cleaning up as you go not only keeps your work area safe it also lessens the amount of work required at the end of the job.
- Ensure any dust or debris left around your wall ventilation penetration is swept up and vacuumed if required.
- Dispose of rubbish, duct offcuts and old materials in accordance with site-disposal requirements or your own company policy.
- Ensure all materials that can be recycled are separated from general refuse and placed in the appropriate location for collection. On some sites specific bins may be allocated for metals, timber etc. or if not, you may need to take materials back to your business premises.
- Once all equipment is taken away, vacuum carpets in the area of your work to ensure that all dust is removed.

FROM EXPERIENCE

Many ventilation tasks may take place in residential premises. In such instances it is a sign of professional respect for the occupants that you keep the site clean and ensure that all dust and debris is removed as you go. Cleanliness is often more remarked upon by clients than the actual job and has a significant influence on your reputation and chance of repeat work. Clean up and grow your business!

Check your tools and equipment

Tools and equipment need to be checked for serviceability to ensure that they are ready for use on the job. On a daily basis and at the end of the job check that:
- all items are working as they should with no obvious damage
- any vacuum extraction equipment is emptied of dust and rubbish and that the HEPA filter is still serviceable and if collection bags are used that these are in good condition or replaced
- high-wear and consumable items such as cutting discs, drill bits and sealants are checked for current condition and replaced as required
- items such as stud finders, volt sticks and inspection cameras are in working order and batteries in good condition
- you check dates on any height-restraint harnesses, lanyards etc., and if getting close to replacement and/or service date report this to your supervisor
- where any item requires repair or replacement, your supervisor is informed and if necessary mark any items before returning to the company store or vehicles.

Complete all necessary job documentation

Most jobs require some level of documentation to be completed. Avoid putting it off till later as you may forget important points or times.

- Ensure all SWMS and/or JSAs are accounted for, completed and filed as per company policy.
- Complete your timesheet accurately.
- On some jobs it is advisable that photographs are taken of completed work as evidence that all was done to specification and left in neat order.
- Reconcile any paper-based material receipts and ensure electronic receipts have been accurately recorded against the job site and/or number.

COMPLETE WORKSHEET 3

LEARNING TASK 12.4

1. What is the name of the component that may need to be fitted to support brickwork above a ventilation cut-out?
2. List the fundamental flame characteristics that indicate good combustion.
3. What type of filtration system is now often required on many job sites for industrial vacuum cleaners?
4. List three options that may be used to generate smoke suitable for flue spillage testing.

SUMMARY

In this unit you have examined the requirements related to the sizing, installation and testing of gas appliance ventilation including:
- basic combustion principles:
  - air requirements
  - heat value
  - gas rate
- job planning and safety considerations
- natural ventilation calculation methods
- mechanical ventilation
- specific appliance requirements
- ventilation testing requirements
- clean-up of job and return of equipment.

GET IT RIGHT

The photo below shows an incorrect practice when cutting out a section to make allowance for a ventilation grille.

Identify the incorrect method and provide reasoning for your answer.

_____

_____

_____

_____

_____

WORKSHEET 1

To be completed by teachers
Student competent ☐
Student not yet competent ☐

Student name: _____

Enrolment year: _____

Class code: _____

Competency name/Number: _____

Task

Review the section *Identify natural ventilation requirements* and answer the following questions.

1 What formula do you apply in order to determine the vent size from a room containing flueless appliances into an adjacent room within a building constructed in 2004? Provide a reference to AS/NZS 5601 to support your answer.

_____

_____

_____

2 You have been asked to install a 130 MJ/h central heating furnace at the rear of a residential garage beneath a split-level home at least 20 years old. An existing storage water heater with a gas input of 27 MJ/h is also installed in the garage. Two vents already exist and each measures 22 000 mm^2. Will this be sufficient ventilation for the two appliances?

_____

_____

_____

_____

_____

_____

3 What is the minimum vertical dimension of any free-ventilated opening? Provide a reference to AS/NZS 5601 to support your answer.

_____

4 Provide at least four examples of what the AS 5601 standard means by 'direct to outside'.

i _____

ii _____

iii _____

iv _____

_____

_____

5 Would it be possible to use mosquito netting on your vents? Justify your answer.

_____

_____

_____

6 A customer asks you to install a flued 127 MJ/h continuous flow water heater in a laundry within a building constructed in 1962. It is not a room-sealed appliance. The room measures 3 m × 2 m × 2.4 m high. The laundry is isolated from the rest of the house with a door. Possible ventilation is direct to outside. Work out a ventilation solution for this installation.

_____

_____

_____

_____

_____

_____

_____

_____

WORKSHEET 2

To be completed by teachers
Student competent ☐
Student not yet competent ☐

Student name: _____

Enrolment year: _____

Class code: _____

Competency name/Number: _____

Task

Review the sections *Identify natural ventilation requirements* and *Specific ventilation calculations* and answer the following questions.

1a A large commercial car wash business provides its hot water with four natural draught flued 427 MJ/h boilers. The building was constructed after the adoption of the AS/NZS 5601:2013. These appliances are installed in their own space to avoid possible contamination of the combustion air by cleaning chemicals. The space is only frequented for routine appliance maintenance. The room size is 3 m × 4 m × 4 m high. You inspect the premises and determine that the existing ventilation is inadequate. Work out a new natural ventilation solution direct to outside.

1b Provide a mechanical ventilation solution for the scenario in Question 1a. Make the exhaust ventilation natural.

2. For a flueless cooker installation, where would two vents be located on a wall that measures 4.5 m high?

_____

_____

3. An enclosure containing flued appliances in a new construction needs to be ventilated into an adjacent room. It then requires ventilation direct to outside. What formulas would you use to size these vents?

_____

_____

4. What is the largest gas input permitted for a flueless non-thermostatically controlled space heater in a room measuring 5.5 m × 7 m × 2.7 m?

_____

_____

5. In addition to the requirements of Clause 6.4.4 and 6.4.5, what is the minimum size of additional ventilation required for the installation of a Type 1 decorative flame effect fire fitted with a 225 mm flue cowl? Provide a reference from AS/NZS 5601 to support your answer.

_____

_____

6a. A flued natural draught central heater of 90 MJ/h input is to be installed in a ceiling space under a simple gable roof used for no other purpose. The building was constructed in 1980 and the roof measures 2.1 m to the ridge, 8 m wide and 16 m long. Determine a natural ventilation solution for this installation. Provide a reference to AS/NZS 5601 to support your answer.

_____

_____

_____

_____

_____

6b Assume the building from Question 6a was constructed after the adoption of the AS/NZS 5601:2013. If all other factors are the same, what is the ventilation solution? Provide a reference from AS/NZS 5601 to support your answer.

_____

_____

_____

_____

_____

_____

_____

7 A client asks you to modify the currently inadequate natural ventilation system for a bank of laundry dryers with a new high-level mechanical exhaust system while retaining the current low-level natural ventilation inlet. Is this a workable solution? Provide a reference from AS/NZS 5601 to back up your answer.

_____

_____

_____

WORKSHEET 3

To be completed by teachers
Student competent ☐
Student not yet competent ☐

Student name: _____

Enrolment year: _____

Class code: _____

Competency name/Number: _____

Task

Review the section *Install ventilation and test appliance* and answer the following questions.

1. What device should always be used to assist in finding power cables hidden behind walls?

   _____
   _____
   _____
   _____

2. You are installing ventilation in an old heritage-listed dwelling. List two tools that should not be used to cut an old 'plaster and lath' wall surface.

   _____
   _____
   _____
   _____

3. In your own words define your understanding of the term 'flue spillage'.

   _____
   _____
   _____
   _____

4 In what condition are gas appliances during the test procedure described in Appendix R3.1? Circle the correct answer.

Appliance burner is ON

Appliance burner is OFF

_____

5 For how long must a Type 1 decorative gas flame effect appliance be operated before conducting the spillage test?

_____

_____

6 For additional safety reasons, what instrument is recommended for use during the flue spillage testing process?

_____

_____

7 Prior to returning height-restraint harnesses and associated equipment to the store, what do you need to check on the product tags?

_____

_____

INSTALL AND COMMISSION TYPE A GAS APPLIANCES

13

Learning objectives

In this chapter, you will be introduced to many of the considerations and requirements related to the installation and commissioning of Type A gas appliances. Once piping and ventilation has been installed, the safe installation and commissioning of a gas appliance requires strict adherence to the detailed procedures of both the manufacturer and the Australian Standards.

Areas addressed in this chapter include:
- General appliance installation considerations
- Appliance-specific installation considerations
- Protection of combustible surfaces
- Appliance commissioning
- Clean up the job.

Before working through this section, it is very important that you complete or revise each of the chapters in Part A 'Gas fundamentals' with particular emphasis on the electrical safety section within Chapter 5 'Gas industry workplace safety', and you must also have completed Chapter 12 'Calculate and install natural ventilation for Type A gas appliances'.

Overview

The chapter will focus on four key areas:
- General installation requirements and preparation. These are general rules that must be applied to all appliance installations where applicable.
- Specific appliance requirements. Each appliance type has specific and very individual needs. To assist you in how you must use the standards to determine Type A appliance requirements, a selection of appliances have been reviewed with accompanying activities relating to AS/NZS 5601.1.
- How to protect a combustible surface. Combustible surfaces must be protected from the heat given off by gas appliances and flue systems. This section will help you identify combustible surfaces and to understand what the Standards require you to do to ensure such surfaces are protected.
- Appliance commissioning procedures. There is no such thing as one standard commissioning procedure. Appliances vary between types and models – so much so that you must always rely on manufacturers' requirements. In this section, the basic commissioning procedure is described, with sub-sections detailing the differences in appliance purging.

Identify gas appliance installation requirements

First and foremost, all appliances to be installed must meet current specifications, performance standards and certification. You are not permitted to install any appliance that is not a listed and currently certified Type A appliance. Be very wary of any appliance that may have been sourced directly from overseas. Some clients are ignorant or at times dismissive of the strict certification requirements for gas appliances. Do not allow yourself to be pressured into installing a non-approved appliance.

REFER TO AS/NZS 5601.1 SECTION 6 'REQUIREMENTS FOR GAS APPLIANCES'

The following sections will detail both general and specific requirements related to the installation of Type A gas appliances.

General installation requirements

The Gas Installation Standards lists a number of general rules relating to installation practice that the gas fitter must apply to all appliances where relevant. These rules may not be applicable to every installation, but you will need to be sufficiently familiar with this section so that when particular circumstances dictate, you will be able to recognise where they need to be applied.

Some (there are many more) of these requirements are covered in the following sections.

REFER TO AS/NZS 5601.1 'GENERAL INSTALLATION REQUIREMENTS'

Conversion

Recall that the two principal fuel gases currently found in use around Australia are NG and LPG. When a new gas is introduced into an area, or if appliances are moved to a new site, it will be necessary for you to consider converting appliances from one gas type to another. The Gas Installation Standards permits appliance conversion in principle, with two key provisos:
- The appliance must be suitable for conversion to a particular gas. For instance, many years ago some models of towns gas cookers with old fashioned flash-tube ignition were found to be unsuitable for conversion to LPG because of the differences between the relative densities of the two gases; they needed to be replaced entirely.
- The conversion procedure is acceptable to the manufacturer. You need to contact the manufacturer to confirm requirements, determine if conversion kits are available and, if not, seek their approval for your conversion method, which may need to be submitted to the technical regulator as a formal procedure.

Some authorities require gas fitters to hold a licence endorsement for conversions, and others may require other forms of specific approvals to be sought prior to commencing the job. As always, assume nothing. Give them a call!

Manufacturer's instructions

At all times, certified appliances must be installed according to the manufacturer's instructions. As part of the certification process, the instructions for every appliance are endorsed, and therefore they must be adhered to. However, if at any time you believe that there is a conflict between the manufacturer's instructions and the Gas Installation Standards, you must contact the technical regulator. Do not make up your own interpretation!

Following practical instruction on appliance use, the manufacturer's instructions must either be left with the customer, or left in a prominent position if the customer is not present. This is part of your broader obligation to ensure the customer knows how to operate the appliance.

Indoor and outdoor appliances

Every certified appliance will clearly state whether it is designed for indoor or outdoor installation. You must adhere to these requirements. For example, you would not be permitted to install a space heater that is certified for indoors on a deck or veranda.

While some variations are permitted, you must always seek approval from the technical regulator before commencing the job. Some commercial catering installations and appliance locations within multi-level car parks are instances where you will need to obtain clarification.

Temperature limitations

Wherever appliances are installed, adjacent combustible surfaces must be protected from a temperature that may exceed 65°C above ambient. Normally where appliances are installed according to manufacturer's instructions, pre-determined clearances will ensure this is the case. However, if you determine that a surface needs protection, you will need to apply some form of compliant protection in the form of additional clearances and/or fire-resistant material. More information on the protection of combustible surfaces can be found later in this chapter.

Prohibited locations for flueless appliances

Flueless appliances are prohibited for use in some rooms, because of the risk of asphyxiation. These include a:
- bedroom
- bathroom
- toilet
- sauna
- spa room.

Appliances on wheels or castors

Any appliance fitted with wheels or castors must be fitted with a hose assembly. Where such an appliance also weighs in excess of 20 kg, you will need to restrain that appliance so that it cannot be withdrawn by more than 80 per cent of the length of the hose assembly. This ensures the hose is not subjected to any forces that may kink or damage the connection.

You will need to apply these requirements most commonly when installing commercial catering equipment. These appliances are regularly moved for cleaning and servicing purposes and hose assemblies with bayonet type quick-connection ends are particularly useful.

Appliance location

Many general requirements apply to the location of appliance installations, and these need to be considered by the gas fitter. For example, some relate to conditions that apply where appliances are intended for installation in a residential garage, under a floor, in a roof space or elevated structure.

REFER TO AS/NZS 5601.1 'GAS APPLIANCE LOCATION'

Appliance isolation and connection

All appliances have very specific isolation and connection requirements. The Gas Installation Standards sets out some basic requirements, but at all times you must also comply with the manufacturer's specifications as well.

REFER TO AS/NZS 5601.1 'GAS APPLIANCE CONNECTION'

Determining appliance isolation requirements is very simple:
- If the installation is for a commercial application, then *all* appliances must have some form of isolation at the inlet. There are no exceptions.
- If the installation is for single residential premises, then isolation for space heaters, cooking appliances and gaslights is optional. All other appliances must have a means of isolation.

When you connect an appliance, you must ensure that it can be easily disconnected for future servicing and replacement without having to make significant changes to existing pipework. You are not permitted to screw an isolation valve or tube bush directly onto the appliance. A means of disconnection must be included on the outlet of any valve or pipe connected to an appliance.

In **Figure 13.1(a)**, a continuous flow water heater is connected with isolation valves incorporating union connections on the outlet side of the valve. This complies with the AS/NZS 5601.1 requirement. **Figure 13.1(b)** shows another water heater where flared adaptors are used to provide a means of disconnection.

Figure 13.2, on the other hand, is a good example of what not to do. Look carefully and you will see that both the water and gas supply are screwed directly to the appliance with no means of disconnection. Also, you will note the use of a water isolation valve with an integral non-return valve – which is not permitted or required on these appliances, according to clear manufacturer's instructions. Last, though not obvious in this image, the gas valve is not certified. (The pipework is certainly not something to be proud of either!)

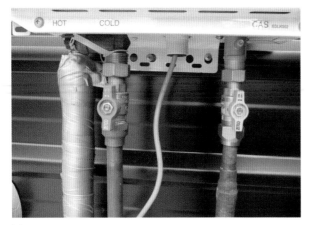

(a)

(b)

FIGURE 13.1 Examples of water heaters that can be disconnected (a) using isolation valves incorporating union connections on the outlet side of the valve, and (b) using flared adaptors

COMPLETE WORKSHEET 1

Specific installation requirements

In this section, a selection of appliance groups will be introduced, some key installation points will be discussed where applicable, and you will then have the

FIGURE 13.2 Example of poor valve selection and appliance connection

opportunity to learn more about the particular appliance installation requirements through specific questioning.

REFER TO AS/NZS 5601.1 'ADDITIONAL REQUIREMENTS FOR INSTALLATION OF SPECIFIC GAS APPLIANCES'

Domestic cookers

The term 'domestic cooker' covers a number of appliance types that are principally used to prepare food for personal consumption. They are generally smaller and to some extent less robust than commercial equipment used in professional catering kitchens. Appliance types include the following.

Cooktops

A gas cooktop is normally found in a four- or six-burner configuration and is generally regarded as an appliance that is installed into a kitchen bench top, entirely separate from an oven or grill. Burners vary in size and function and may include simmer, rapid, wok and fish burners depending upon the model.

You need to be familiar with these commonly used terms:

- *Trivet* – a trivet is the frame upon which cooking vessels or pans are placed. The trivet provides a stable base and maintains the correct clearance from the burner and flame. Trivets are made from either cast iron or enamelled steel.

- *Hob* – the hob is a defined term and means the cooking surface that supports the trivets.
- *Cooking surface area* – this is used in the Standard in the context of the part of the appliance where cooking actually takes place. It would not include a splashback or side panel where control knobs are located.

The four-burner cooktop in Figure 13.3 features a 12 MJ/h wok burner, cast iron trivets and thermoelectric flame failure valves. You will notice that the back wall is within 200 mm of the periphery of the rear burners and as it is a timber-framed, plaster-sheeted wall it has been protected with ceramic tiles in accordance with Appendix C of the Standard.

FIGURE 13.3 A four-burner cooktop with wok burner, cast iron trivets and thermoelectric flame failure valves

FIGURE 13.4 An enamel-finished upright cooker with separate griller and a gas connection point under the appliance (behind the removable kick-panel)

Installation requirements

Clearances from combustible surfaces make up the majority of installation requirements for gas cooktops. Horizontal clearances to vertical surfaces are measured from the periphery (edge) of the burner. Vertical clearances are measured from the highest part of the highest burner to the nearest overhead surface.

The majority of cooktop installations require a hard connection. Remember that hose assemblies must be used only where approved for use by the manufacturer and AS/NZS 5601.1.

Upright cookers

Traditional upright cookers are freestanding appliances that generally combine an oven, separate griller and four-burner cooktop into the one unit. Models range from simple, base-level cookers to those incorporating fan-forced ovens and electronic ignition.

Installation requirements

Freestanding cookers are divided into two groups according to the installation standard:
- freestanding cookers with a high-level gas connection point
- freestanding cookers with an under-cooker gas connection point (see Figure 13.4).

The location of a freestanding cooker connection point determines whether you can use a flexible hose assembly or not (see Figure 13.5). Hose assemblies are

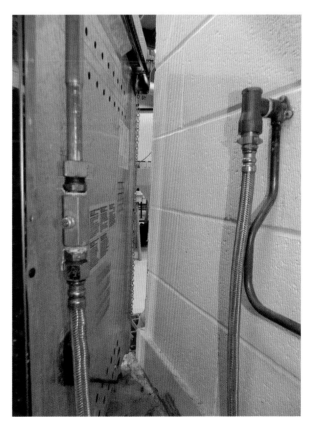

FIGURE 13.5 An upright cooker with a high-level connection point and which has been approved for use with a hose assembly

not permitted where under-cooker connections are required.

The appliance in Figure 13.6 has therefore been installed with a copper loop to enable the unit to be brought forward from its recess when required without disconnection. Notice the restraint plate on the floor; all upright cookers require some form of restraint to prevent tilting.

FIGURE 13.6 An upright cooker installed with a copper loop so that it can be moved forward without disconnection

Domestic ranges

A domestic range is normally regarded as a wider and larger version of a standard upright cooker. While still a freestanding appliance, they are often designed and built to a heavier, semi-commercial standard of finish and may feature two or more ovens and six cooktop burners, including wok and fish burners (see Figure 13.7).

Installation requirements

As a freestanding cooker, domestic range installation must comply with the same requirements as standard upright cookers. Most will feature a high-level gas connection, making them suitable for connection with a flexible hose assembly of between 1 m and 1.2 m in length, where approved by the manufacturer. Alternatively, an annealed copper loop may be used.

Elevated cookers

Elevated cookers were originally designed to make access to the oven much easier by placing it alongside and at a similar height to the cooktop and benches. The griller is situated under the cooktop.

With the advent of underbench gas ovens, elevated cookers are not as common as they once were, though new models are still available.

Installation requirements

An elevated cooker has particular installation requirements related to oven flue clearance and gas connection method. Where older units like the model shown in Figure 13.8 are to be replaced, ensure the

FIGURE 13.7 A range-style cooker with a large single oven

installation complies with the current standard. Rigid supply pipe connections are not permitted for elevated cookers, though you may still find them on some older installations. You must now use a certified hose assembly or an annealed copper loop to enable the appliance to be manoeuvred for connection and disconnection.

Inbuilt ovens

Inbuilt ovens come in two basic forms: underbench and wall.

- *Underbench ovens*. Gas underbench ovens are designed to be installed beneath a standard-height kitchen bench or, if desired, they can be installed at a mid-level height, much like a wall oven (Figure 13.9). Unlike a wall oven, however, the oven is flued out through the front of the unit into the room itself. Most underbench ovens incorporate an integral gas or electric griller.
- *Wall ovens*. Wall ovens are available in a number of configurations, with one option being the oven and separate griller arrangement seen in Figure 13.10. The oven is normally installed at a height so that it can be accessed while the operator is standing. Many models include multifunction and programmable ovens and often feature a hot-surface ignition system to make lighting as simple as pressing a button.

FIGURE 13.8 Elevated cookers are still available to replace older models as shown here

FIGURE 13.10 A fan-forced gas oven with a separate griller situated below the main oven cabinet

FIGURE 13.9 A standard inbuilt underbench oven

Installation requirements

When installing inbuilt ovens, you should always try to work closely with the builder and/or cabinetmaker to ensure that the surrounding benches are dimensioned according to the oven manufacturer's requirements. The majority of kitchen cupboards and bench tops are made from some form of combustible fibreboard. Therefore, you must ensure that all clearances are maintained to ensure surrounding surfaces are not exposed to more than 65°C. If in doubt, protection with complying fire-resistant material should be considered.

Most wall ovens will require a conventional flue system to be installed. This may present a problem where this has not been identified by the designer and/or builder, and it may be impossible to install in multi-level premises. Always check manufacturer's specifications before finalising your quote or installing the appliance. Wall ovens in particular generate considerable cabinet and flue heat, which can easily result in a fire if not installed correctly.

A note on using annealed copper loops

Many appliances are not approved for use with a certified hose assembly. Where this is the case, the only method of connection that still provides some level of flexibility is a connection made with an annealed copper loop. Normally a maximum of one and a half or two

INSTALL AND COMMISSION TYPE A GAS APPLIANCES

loops behind the appliance is sufficient to allow the appliance to be manoeuvred for testing purposes or to enable the operator to clean around the unit.

Caution! If you have ever created a loop from a roll of annealed copper, you will know that to push it in and out takes considerable effort. When this loop is connected to an appliance, the act of pushing the appliance in and out may cause torsional twisting forces to be transferred to the connection point. In this situation, flare nuts and, in particular, appliance connection fittings using fibre washers may come loose and start to leak. Even more concerning, this may not happen straight away, with the connection point gradually becoming loose over time. This problem mainly occurs where copper loops are fitted to wall ovens and freestanding cookers with a high-level connection point. You must ensure that the loop is fitted is such a way that the final torsional stress point is removed from the actual connection by creating an offset in the pipe and securing it at this point to the chassis of the appliance. In this way, the stress point is transferred from the connection to the securing clip.

Figure 13.11 shows how a copper loop is installed behind an upright cooker. By securing the loop after the offset, the connection point is relieved of torsional stresses and remains gas-tight.

Note: Always ensure that you are permitted to fit a securing clip on the back of the appliance. Contact the manufacturer if in doubt. You must be able to see that the area where you intend to fit any clip is entirely free of any components and that the clip will not damage the appliance in any way. Always ask the manufacturer first.

LEARNING TASK 13.1

IDENTIFY GAS APPLIANCE INSTALLATION REQUIREMENTS

1 When installing a gas appliance with wheels or castors and weighing over 20 kg, a restraint chain is required. What is the maximum length of the restraint chain as a percentage of the length of the hose assembly?
2 Are you permitted to screw an isolation valve directly to an appliance?
3 How could 'torsional twisting' occur when installing an annealed copper loop?

COMPLETE WORKSHEET 2

Commercial catering equipment

There is a wide range of Type A commercial catering appliances (see Figure 13.12). Some examples include:
- char grill
- salamander
- deep fryer
- pasta cooker
- range
- wok (Chinese cooking table)
- tandoori oven
- brat pan

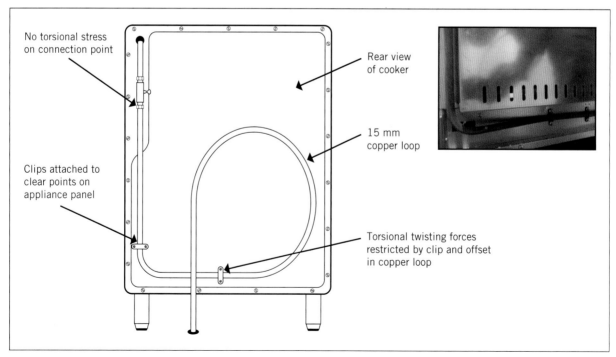

FIGURE 13.11 How to install a copper loop behind an upright cooker

FIGURE 13.12 New char grills and wok tables ready for installation

- combi steamer
- pizza oven
- solid grill plate.

Installation requirements

The Gas Installation Standards has specific requirements related to the installation of commercial catering equipment that you must be aware of. A number of requirements for installing commercial catering equipment particularly relate to gas supply and clearances to other appliances and combustible material. In addition, certain restrictions apply when installing what is known as a combination cooking range.

What is a combination cooking range?

A combination cooking range is one in which a number of appliances (no more than six) have been connected together behind one purpose-built fascia. They are then collectively considered as one appliance in relation to additional gas supply requirements. However, you must be aware that some technical regulators also interpret a combination cooking range to simply mean six appliances in one kitchen. This is a big difference, and you will need to check with your local authority for their interpretation.

Regardless of the interpretation, where a combination cooking range is used it must comply with the following:

- There must be no more than six appliances (additional appliances will need to be branched off before the isolation valve to the combination cooking range).
- There must be one gas supply source.
- A common gas supply pipe is to be sized to ensure that all appliances can operate simultaneously.
- A manual shut off-valve that can isolate all appliances together must be installed in an accessible location.

Although the appliances in Figure 13.13 are not technically a combination cooking range, the gas fitter

FIGURE 13.13 Commercial appliances featuring a main gas supply isolation valve installed in an accessible location

has still ensured that the main gas supply isolation valve has been brought up to an accessible location. This means that in the event of an emergency (such as a fat fire), the chef will be able to quickly and easily isolate the gas supply to all appliances.

It is strongly recommended that individual isolation is always provided for each appliance in addition to the main manual shut-off valve. This is to ensure that essential, and often unforeseen, maintenance work can be carried out while the rest of the kitchen remains in operation.

COMPLETE WORKSHEET 3

Continuous flow water heaters (instantaneous water heaters)

Traditionally called instantaneous water heaters, it is now thought more accurate to call this form of water heater by the modern name, continuous flow water heaters. Whether they are simple, mechanically operated systems or more sophisticated, electronically controlled systems, continuous flow water heaters operate on a heat exchange principle. These appliances do not store water in any form of tank; instead, they heat water with a relatively large burner as it passes through a copper heat exchanger. The burner operates only when a tap is turned on and the water flows. When the tap is turned off the burner shuts down and the appliance returns to a standby mode, consuming little if any energy at all. For this primary reason, such water heaters are very energy efficient, as they heat water only when it is actually required for use.

Installation requirements

Continuous flow water heaters are available in both internal and external models (see Figure 13.14). Basic requirements relate to the following:

- Unflued water heaters must not be installed indoors. Many years ago, unflued instantaneous water heaters were installed above baths, showers and sinks. These appliances have been prohibited for many years now, but you may still find them around. Unflued appliances installed indoors can create an extremely dangerous asphyxiation hazard, and if you encounter any such appliances, you must contact the technical regulator immediately and make the installation safe.
- Other than room-sealed appliances, instantaneous water heaters must not be installed in a:
    - bedroom
    - bathroom
    - toilet
    - combined living/sleeping room.

The room-sealed continuous flow water heaters in Figure 13.15 have been manifolded together to provide hot water for a large hotel complex. These thermostatically controlled appliances are connected via a flow-and-return system to three storage tanks that act as a buffer supply during periods of peak demand.

Storage water heaters

Gas storage water heaters have been used for heating water for decades. These appliances can be split into two basic categories based upon the way they are internally flued.

- *Central flue* – the traditional centrally flued water heater features a central primary flue that runs up through the centre of the tank itself (see Figure 13.16). As the combustion products rise up this flue, heat is transferred through the walls of the tank into the water. One disadvantage of this design is that heat continues to be carried out from the primary flue after the burner shuts down.

FIGURE 13.14 Two appliances both classed as continuous flow water heaters despite considerable differences in mechanical operation

FIGURE 13.15 Room-sealed continuous flow water heaters manifolded together and connected to storage tanks to provide sufficient hot water for high demand

- *Condensing flue* – a condensing flue water heater is designed to direct products of combustion through a heat exchange jacket that envelops the outside of the water cylinder. Exposed to such a large surface area, considerable heat is extracted from the flue gases into the water. For the flue gases to exit the heater they must make their way up and then down through the heat exchanger, and in doing so must lose most of their original heat. These heaters are so efficient that the flue gases will fall below dew point and water will condense on the inside of the flue jacket and run out of the appliance via a drain point. A key feature of these heaters is that when the burner shuts down, the flue gases will cool inside the flue jacket and lock out, preventing further thermal losses.

The cross-section of a Rheem 'Stellar' condensing flue water heater in Figure 13.17 shows how the flue gases are exposed to a large heat exchange surface area. Notice how, after exiting the central primary flue, the gases are directed downwards, exchanging more heat until they exit from the front balanced flue terminal. Having given off most of its heat, the water vapour in the flue gases then condenses into liquid water and is drained away at the base of the appliance.

Installation requirements

Most installation needs for storage water heaters relate to flue terminal locations and general appliance requirements (see Figure 13.18). The Standard lists two specific appliance needs:

- The supporting base must be level and capable of supporting the appliance. You must remember that 1 litre of water weighs approximately 1 kg. Therefore, a 315 L storage water heater could weigh in excess of 400 kg when the cylinder and water are combined.
- When installed in a bathroom or toilet, a storage water heater must not exceed 40 MJ/h. Any more than this and such an appliance would become an asphyxiation hazard.

COMPLETE WORKSHEET 4

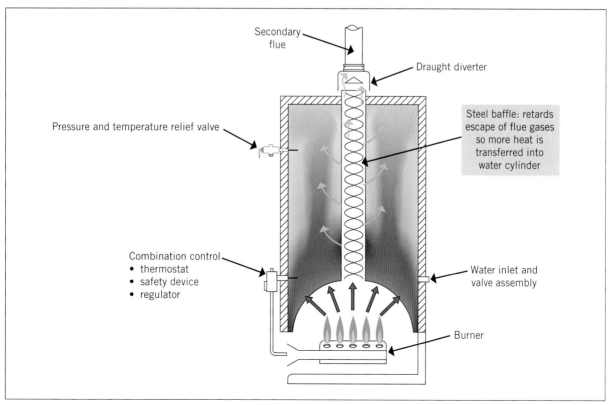

FIGURE 13.16 A centrally flued storage water heater

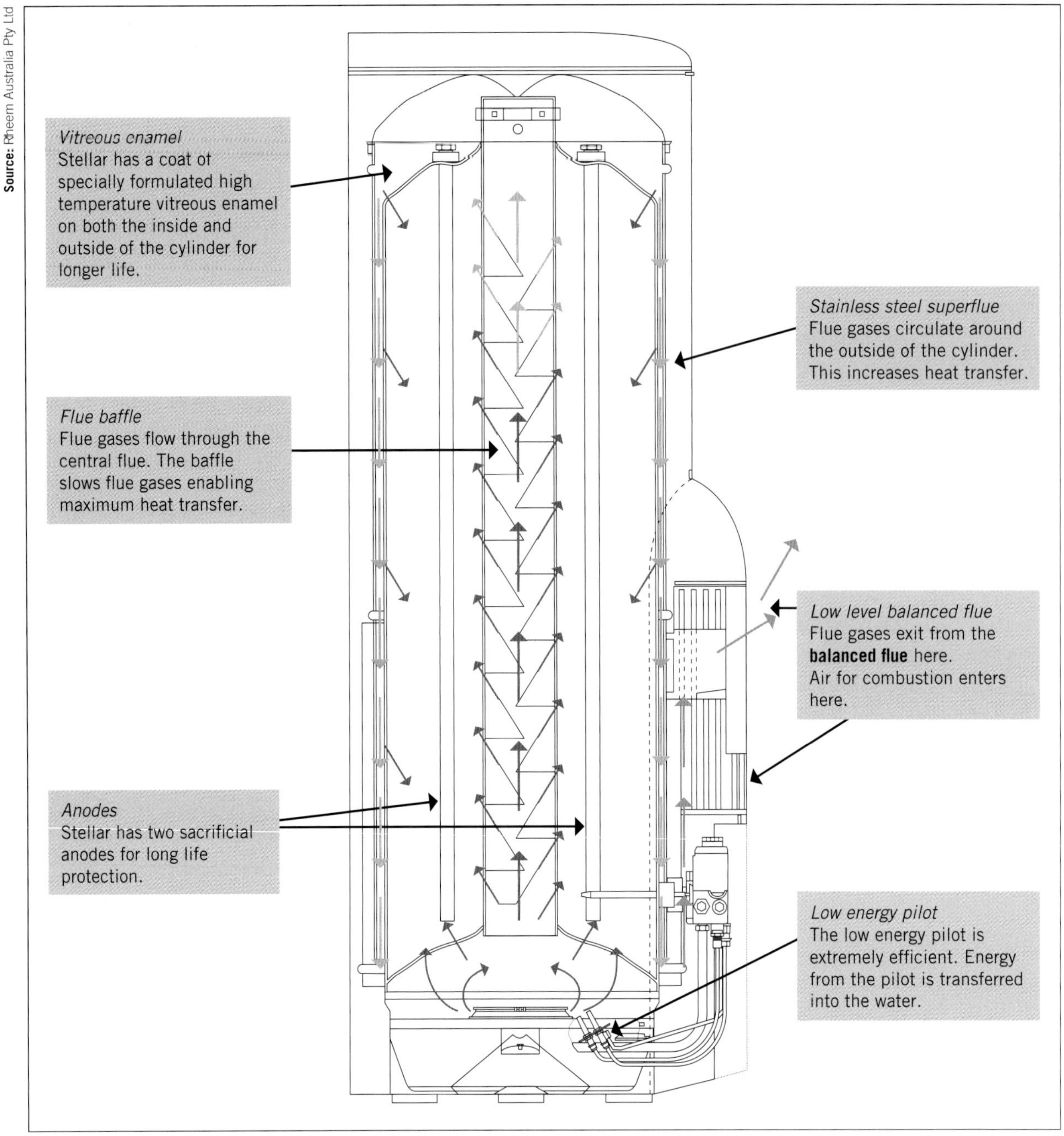

Vitreous enamel
Stellar has a coat of specially formulated high temperature vitreous enamel on both the inside and outside of the cylinder for longer life.

Stainless steel superflue
Flue gases circulate around the outside of the cylinder. This increases heat transfer.

Flue baffle
Flue gases flow through the central flue. The baffle slows flue gases enabling maximum heat transfer.

Low level balanced flue
Flue gases exit from the **balanced flue** here.
Air for combustion enters here.

Anodes
Stellar has two sacrificial anodes for long life protection.

Low energy pilot
The low energy pilot is extremely efficient. Energy from the pilot is transferred into the water.

FIGURE 13.17 Cross-section of a Rheem Stellar condensing flue water heater

Space heaters

Space heaters are designed and sized to heat a certain area or 'space' to a required temperature. Numerous models and applications allow the gas fitter to choose a heating solution for almost every situation. However, all gas-heating appliances can be grouped together in one of three categories:
- radiant heating
- convection heating
- combined radiant and convection heating.

Radiant heating

Radiant heat is the type of heat you can feel on your skin from a gas heater flame directly across an otherwise cold room. Radiant heat is a form of long-wave radiation that travels in straight lines from its source until it strikes a solid surface, which is able to absorb heat (see **Figure 13.19**). The solid surfaces then radiate this heat back outwards. Therefore, although radiant heat does not directly heat the air in between the source and the object, the air temperature will gradually rise through direct contact with the heated objects within the room.

FIGURE 13.18 A storage water heater designed for indoor installation.

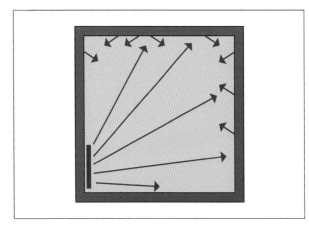

FIGURE 13.19 The operation of radiant heating

FIGURE 13.20 A portable flueless gas heater featuring ceramic tile elements that when heated provide a source of radiant heat

Gas appliances designed to create radiant heat use the gas fuel source to heat a ceramic or metal element that, in turn, radiates heat outwards (see Figure 13.20).

Factors to take into account when considering a radiant heat solution are that:

- it is ideal for areas subject to draughts (workshops)
- it provides almost instant warmth on surfaces
- customers derive satisfaction from the warm glow of the radiant ceramic elements
- heat distribution is uneven, especially after lighting
- the air temperature is slow to rise.

Convection heating

Convection heating relies upon a simple principle: hot air rises. Gas appliances designed to provide convective heat use the fuel gas to heat some form of heat exchanger. The room air is passed over this heat exchanger and becomes heated as it does so. Air heated by the gas appliance rises back into the room. As this air rises away from the appliance, it is replaced by cooler air. As the heated air cools again, it falls towards the floor and is drawn once again into the heater (see Figure 13.21). This process creates a convective circulation within the room.

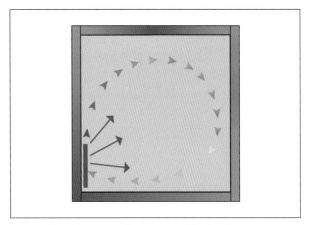

FIGURE 13.21 The operation of convection heating

Few gas appliances rely entirely on natural convection itself. Most modern convective heaters use some form of fan to increase heating efficiency; these are known as forced convection heaters. These

(c)

FIGURE 13.22 (Continued)

appliances draw air in, pass it over or around the heat exchanger and back out into the room. The room air does not mix with the products of combustion. Forced convection heaters may be natural draught flued or fan-assisted room-sealed appliances (see Figure 13.22).

Ducted forced convection heating

Some heating appliances are designed for installation with a system of ducts that carry the heated air around the building. This is generally referred to as gas-ducted central heating. Heat can be provided through the floor (see Figure 13.23) or ceiling.

Room air is drawn into the heater, where it passes around the heat exchanger. The air is then forced through a series of ducts that terminate at floor registers or ceiling downjets in each room. A centrally located 'return-air register' and duct carry the room air back to the furnace where it is heated once again.

FIGURE 13.22 Three forced convection heaters: (a) a room-sealed fan-assisted space heater, (b) an installed older natural draft flue wall furnace (c) a modern replacement power flue wall furnace

FIGURE 13.23 Cut-away image showing how a ducted central heating system can provide heated air to each room in the house through the floor

The factors to take into account when considering a convection heater are that:
- it will heat the room quickly
- many models are available to suit different requirements
- some customers may find continual air movement a nuisance, particularly during the heat-up period.

Combined radiant and convection heating

Some gas space heaters combine both radiant and convection characteristics into the one appliance. Typically, the appliance will feature a number of ceramic elements to directly radiate heat into the room as well as to provide a heat source for the appliance heat exchanger. Air is drawn in from the base of the unit by a fan assembly, passed around the heat exchanger and out through the front of the appliance.

Factors to take into account when considering a combined radiant–convection heater are that:
- it provides almost instant warmth on surfaces
- customers derive satisfaction from the warm glow of the radiant ceramic elements
- it will heat the room quickly
- many models are available to suit different requirements
- it is generally not as efficient as a dedicated convection appliance.

Flueless heaters

Flueless portable heaters are available in both radiant, convection and combination models. When there is no flue system at all, combustion products are released directly into the room space. This makes such appliances very efficient. However, correct ventilation of the room and commissioning is critical for the safe use of the appliance, to ensure an asphyxiation hazard does not arise.

Flueless space heaters fall into two classes:
- *Thermostatically controlled* – the appliance will either modulate the gas rate or turn on and off in response to changes in room temperature. In a cold room, it will run at full capacity but when the set temperature has been reached, gas supply to the burner will be reduced or cut off. Once the temperature drops, the gas supply will increase or re-ignite.
- *Non-thermostatically controlled* – once lit, these appliances will keep working at the set rate regardless of room temperature, and require manual adjustment.

You will use this form of classification to work out the maximum heater gas input for a certain room volume. Be aware that some authorities do not permit the use of flueless gas heaters, or require extra ventilation requirements in addition to those specified in the Gas Installation Standards. You must ensure you know what local rules apply.

Installation requirements

Specific requirements for domestic space heaters relate to location and the particular needs of flueless heaters. Where a space heater is intended for installation in a bedroom or bathroom it must be one of two types:

- *Room sealed appliance* – natural or fan-assisted room-sealed convection heaters can be installed virtually anywhere, as combustion air and flue gases are sourced from and disposed of outside the building.
- *Flued appliance with a flame-failure system* – combustion products are flued to outside, ensuring that room occupants are not exposed to dangerous contaminants.

Where any space heater is to be installed in an institution such as a school, aged care home or kindergarten, the heater must be secured in location and consideration given to the provision of guards around the appliance.

Flueless heaters must be selected to ensure they do not exceed the maximum gas consumption permitted for any given room volume.

> **REFER TO AS/NZS 5601.1 'RESTRICTIONS ON INSTALLING A FLUELESS SPACE HEATER'**

EXAMPLE 13.1

A customer has asked you to recommend a flueless space heater for use in the living room. The room is isolated by a door and measures 6 m × 4.5 m × 2.4 m. You want to provide a range of options to your customer including thermostatically controlled and non-thermostatically controlled appliances. Referring to your Gas Installation Standards, you apply the following formula:

Max gas consumption = Room volume m^3 × 0.4 (thermostatically controlled)
or × 0.2 (non-thermostatically controlled)

Max gas consumption (thermostatically controlled) = 64.8 m^3 × 0.4 = 25.92 MJ/h

Max gas consumption (non-thermostatically controlled) = 64.8 m^3 × 0.2 = 12.96 MJ/h

You will notice from this example that the maximum gas consumption of a non-thermostatically controlled appliance is only half that of the thermostatically controlled appliance. A greater safety margin has been applied to the appliance that can adjust burner input based upon room temperature.

When sizing flueless space heaters you also need to ensure that the room has fixed ventilation in accordance with the appliance manufacturer's minimum requirements and any relevant state-specific requirements specified in AS 5601.

> **REFER TO AS 5601 APPENDIX L 'SPECIAL LOCAL REQUIREMENTS'**

Overhead radiant tube heaters

An overhead radiant tube heater is a forced draught gas appliance used for heating commercial and industrial workspaces.

The burner of a radiant tube heater is connected directly to a long tube covered by a deflector plate that radiates heat downwards (see Figure 13.24). Combustion products are either flued to atmosphere or in large buildings they may be vented into the room.

FIGURE 13.24 A double pipe radiant tube heater designed to radiate heat downwards via an overhead deflector plate

Installation requirements

The primary installation requirements for radiant tube heaters relate to the following three considerations:

- *Clearances to combustible surfaces*. These industrial-grade heaters have a high heat output and therefore combustible surfaces above and below the heater must be protected.
- *Air supply*. All gas appliances require a quality air supply. If these heaters are used in an industrial environment, they must be protected from contaminants such as chemicals and dust.
- *Flueing*. Flueing requirements must be based upon the manufacturer's instructions and the standard calculations found in AS/NZS 5601.1. Care must be taken where radiant tube heaters are used in an area contaminated with industry-generated dusts, as high dust environments can be explosive.

Overhead radiant heaters

Overhead radiant heaters are a specific form of flueless space heater. They are radiant-style heaters designed principally for commercial and industrial workshop situations (see Figure 13.25).

FIGURE 13.25 An overhead radiant heater used in a workshop situation

Installation requirements

The key installation requirements for an overhead radiant heater relate to their height above floor level and the clearance to combustible surfaces above the heater. As well as observing the manufacturer's instructions, it is important to note that you must never install these heaters less than 2.5 m above floor level when installed indoors. Depending upon their energy (MJ) input, specific clearance must also be maintained above the appliance. These distances from combustible materials may only be reduced where some form of protection is used.

Decorative flame effect fires

Decorative flame effect fires are heating appliances designed with the primary goal of providing space heating with an aesthetically pleasing flame. Originally known as decorative gas log fires, these appliances now include products that incorporate imitation coal, river stones and other abstract flame uses, and as such the term 'decorative flame effect fire' is technically more correct.

REFER TO AS/NZS 5601.1 'DECORATIVE FLAME EFFECT FIRES, OTHER THAN FLAME EFFECT GAS SPACE HEATERS'

Installation requirements

Before determining the specific installation requirements for decorative flame effect fires, you must be able to understand the difference between the following two appliance classifications:

- *Type 1* – these are manufactured without an enclosure and are designed for installation in an existing fireplace with a chimney. The Type 1 decorative flame effect fire burner assembly shown in Figure 13.26 is inserted into an existing open fireplace where it is connected to the gas supply. The burner has no enclosure at all. Any existing fire damper must be removed or permanently fixed in an open position.

FIGURE 13.26 A Type 1 decorative flame effect fire burner assembly installed in an existing open fireplace

- *Type 2* – these are manufactured with an enclosure and are designed to be freestanding with a flue that vents flue gases to atmosphere. The burner assembly of this type of decorative flame effect fire is fully encased within the steel and glass of the appliance body. These fully enclosed appliances may be designed for insertion into existing fireplaces, or stand-alone on their own base, simulating a freestanding wood heater.

The burner assembly of the Type 2 decorative flame effect fire shown in Figure 13.27 is fully enclosed behind glass. The appliance itself is fitted with a flue system behind the wall or up through an existing chimney.

FIGURE 13.27 A Type 2 decorative flame effect fire

Ventilation of decorative flame effect fires

In addition to the primary ventilation requirements that apply to all appliances, decorative flame effect fires require supplementary fixed ventilation direct to outside, regardless of room size. One or more openings into the room not less than the equivalent

cross-sectional area of the flue cowl must be installed. The flue cowl–vent size relationship is as follows:
- For a Type 1 appliance, the flue cowl must be at least 225 mm in diameter. Therefore, the required ventilation must be equivalent to at least 39 760 mm^2.
- For a Type 2 appliance, the flue cowl must be at least the same nominal diameter as the flue. Most Type 2 appliances will have a 100 mm flue for normal installation requirements. This would require a vent size of approximately 7853 mm^2.

 COMPLETE WORKSHEET 5

Prepare for installation

All materials used for appliance installation and the appliances themselves must be checked for serviceability to ensure that manufacturing faults or pre-purchase damage has not occurred. Once an appliance has been delivered, check for any external damage to the packaging. Photograph and contact the supplier immediately if any damage is detected. As the packing is removed confirm that all components are present and once again do not delay in contacting the supplier if there is a problem.

Always make an extra effort to protect your materials during storage and handling.
- Keep dust and water out of pipes with caps or plugs.
- Keep fittings in an orderly and clean manner, ensuring that they are not contaminated with dirt and other foreign substances.
- Ensure that plastic and multilayer pipe is not stored in direct sunlight as this will degrade the pipe.
- Avoid rough handling of your pipe, fittings and materials.
- Ensure that brass flared fittings are always stored with the nut screwed on, to prevent damage to the mating faces of the flare.
- Ensure press-fit copper connectors are kept dirt free.

Appropriate tools and equipment

Apart from general hand tools used on almost all jobs, tools and equipment used for appliance installation will to a large extent be dictated by any appliance-specific needs. The following list shows just a selection of items that may be required.

Other equipment related to handling of the appliance and the proposed installation location may include:
- elevated work platforms
- ladders
- hand trolleys (stair climber), forklifts and hydraulic wheeled hoist.

Tools and equipment
• Oxy/acetylene brazing equipment
• Press-fit tool and matching jaws
• Flaring tools
• Copper pipe cutters and reamer
• Benders
• Internal and external bending springs
• Expanders
• Shifters
• Digital manometer
• Multi-meter
• Volt stick
• Bonding straps
• Drills
• Hole saw sets
• Jigsaw
• Thread paste
• Spirit level
• Soap and water solution
• Gas detector
• Incense sticks

Install and commission appliance

Many specific appliance characteristics have been identified so far in this chapter. In this section you will be introduced to the basic requirements related to appliance installation and subsequent commissioning.

An important aspect of installation that must be satisfied before any appliance is installed and commissioned is ensuring that any surrounding combustible surfaces are protected from heat generated by the appliance.

Protecting a combustible surface

The primary purpose of all gas appliances is to produce heat in some form. While most heat generated by combustion is absorbed in the way intended by the design of the appliance (water or air is heated, food is cooked, and so on), certain appliances may lose heat through the flue system or appliance body itself. In many cases the vertical and horizontal surfaces surrounding a gas appliance may therefore need to be protected from radiated or conducted heat.

As a gas fitter, you not only need to ensure that the appliance is correctly set up, but you are also responsible for ensuring that any combustible materials near the appliance are either removed or protected from exposure to temperatures in excess of the maximum limits. Follow the manufacturer's instructions as well as the associated requirements stipulated in AS/NZS 5601.1 to ensure that the installation remains safe.

What is a combustible surface?

'Combustible surface' is a defined term in AS/NZS 5601.1 and is regularly referred to throughout the standard.

REFER TO AS/NZS 5601.1 'COMBUSTIBLE SURFACE'

You must always be on the lookout for any materials around your gas installation that may ignite and burn. Examples of combustible surfaces that you may encounter near gas appliances and components include, but are not limited to:
- timber in any form
- sheet plasterboard
- kitchen bench top decorative and protective laminates
- some thermal insulation products
- some exterior cladding and roofing products
- most synthetic surfaces
- some decorative surface coatings, even if applied over a non-combustible surface.

Some of these surfaces may actually be clad or concealed by a non-combustible surface such as sheet metal. However, heat may still be conducted through the non-combustible material and cause the underlying combustible material to ignite. AS/NZS 5601.1 stipulates that any combustible surface adjacent to a gas appliance must not be subjected to temperatures in excess of 65°C above ambient.

REFER TO AS/NZS 5601.1 SECTION 6 'TEMPERATURE LIMITATION ON NEARBY COMBUSTIBLE SURFACE'

Unfortunately, many fires, injuries and fatalities have been caused by the incorrect protection of surrounding combustible materials from the heat of gas appliances and flueing systems. Of particular concern is the fact that even relatively low heat applied over a long period of time can cause a fire to start many months, or even years, after installation of the appliance. The condition leading up to such fires is known as pyrolysis.

What is pyrolysis?

Put simply, pyrolysis is the chemical decomposition of an organic substance caused by the application of heat. Where there is not sufficient clearance or protection between the heat source and the surface, ignition and combustion of the substance can occur without the direct application of a flame.

To examine this in more detail, consider a timber frame stud wall that is subjected to heat from an incorrectly installed appliance. Depending upon the species, timber will normally ignite at approximately 260°C. However, exposure to a heat source will cause significant changes to the molecular structure of the timber, with the effect of lowering the ignition temperature:
- At temperatures between 65°C and 100°C, the timber begins to lose structural strength.
- At temperatures above 100°C, the timber begins to dehydrate as chemical bonds begin to break apart, releasing flammable volatiles and tars into the surrounding area.
- As pyrolysis of the timber continues, the ignition temperature of the timber can fall to as low as 100°C! With sufficient air supply and application of the right ignition temperature, the timber is then likely to combust.

Where can pyrolysis occur?

Substances can suffer from pyrolysis wherever an appliance or its flue system is not installed with the mandatory clearances and/or correct physical protection of the surface. Some instances to be aware of include:
- cabinets and cupboards made from unprotected timber-based products surrounding gas wall ovens
- flue penetrations too close to timber rafters, cellulose-based insulation or sheet plaster
- cooktop burners too close to combustible vertical and/or overhead horizontal surfaces
- commercial catering appliances installed without adequate clearances to combustible wall materials
- combustible wall materials covered only in sheet metal or non-approved decorative glass splashbacks
- inappropriate use of sheet metal single-wall flue materials.

For example, the elevated cooker in Figure 13.28 has been installed incorrectly with a vertical combustible surface located too close to the periphery of the burner. As a result, the surface of the kitchen bench top has been charred and represents a significant fire hazard. How would you solve this problem?

You will not only need to ensure that you avoid these mistakes by following the Standards and the manufacturer's instructions, but you must also learn to identify poor installations when you encounter them on the job. All of these problems have a solution based on one of the following two categories: required clearances or the use of fire-resistant material.

Clearances

Many appliance installations require certain minimum clearances to be maintained between appliance bodies, flue systems and combustible materials. You must adhere to these requirements. AS/NZS 5601.1 lists required minimum distances from combustible materials for all appliances and flue installations. Examples include:
- domestic cookers
- commercial catering equipment

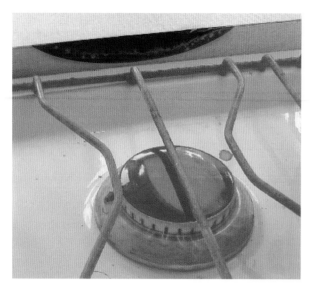

FIGURE 13.28 A charred vertical surface located too close to the periphery of a burner

- space heaters
- overhead radiant heaters
- radiant tube heaters
- refrigerators
- single wall flue systems
- twin wall flue systems.

For example, the overhead radiant heater in Figure 13.29 represents an extreme fire hazard. The heater has been installed in a public building with insufficient clearance to the combustible material situated vertically above it. You can see that the ceiling lining has shrunk and is beginning to discolour. The metal deflector plate is clearly inadequate. This appliance would have an approximate gas input in excess of 40 MJ/h. Use AS/NZS 5601.1 to determine the minimum clearance permitted between such an appliance and any overhead combustible material. What would you do if you came across such an installation?

 Always follow the manufacturer's specifications and the AS/NZS 5601.1 clearance requirements to prevent ignition of combustible materials!

What is fire-resistant material?

Fire-resistant material has a defined meaning in AS/NZS 5601.1 and refers to material that has the ability to retard thermal conduction through to the combustible surface.

 REFER TO AS/NZS 5601.1, 'FIRE-RESISTANT MATERIAL' AND APPENDIX C

The thermal insulation performance of any sheet product must comply with the requirements of AS/NZS 5601.1 and must never be less than 6 mm in thickness. The Standard provides a formula that enables you to determine the suitability of any particular product according to the relationship between its thermal conductivity and material thickness. However, as finding the correct material thermal conductivity data can be difficult, it is recommended that a gas fitter always consult the technical regulator for advice before selecting a product. The technical regulator may suggest examples of compliant fire-resistant materials suitable for gas installations, or provide guidance on what to look for from a manufacturer when selecting a product. See Figure 13.30 for the statement from Energy Safe Victoria. Check your local technical regulator for more information.

As a defined term, you will find fire-resistant material referred to frequently throughout AS/NZS 5601.1, and you need to use these products as and when required by the standard.

Do not use any product that does not meet the requirements of AS/NZS 5601.1 or that is not on a list recommended by the technical regulator. While some products do not support combustion themselves, they conduct too much heat and therefore should not be used. This includes most forms of standard cement fibre sheets (often known as 'fibro'). Such sheets can only be used in combination with other materials as per Appendix C.

> **ASBESTOS WARNING**
> Asbestos is a known human carcinogen (cancer-causing substance), and its removal and disposal is controlled by legislative requirements. Keep yourself and others safe by contacting a qualified asbestos removal contractor before any work proceeds.

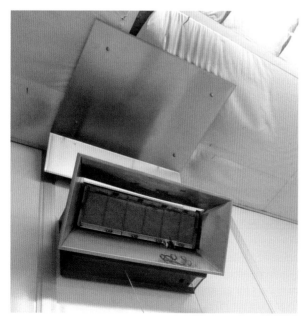

FIGURE 13.29 An overhead radiant heater located too close to overhead combustible material

FIGURE 13.30 An information sheet from Energy Safe Victoria

Protection around gas cookers

Combustible surfaces around gas cooking appliances require protection whenever minimum clearance requirements cannot be satisfied. AS/NZS 5601.1 Table C1 shows a number of facing and backing material combinations that you can choose from to suit individual installation needs. Some examples are listed:

- Sheet metal of at least 0.4 mm thickness can be used as the facing material as long as it has a backing of either a 12 mm fibre cement board, or a 6 mm fibre cement board overlaying a 10 mm gypsum-based wall board.
- Toughened safety glass of at least 5 mm thickness compliant with AS/NZS 2208 may be used on top of either a 10 mm gypsum-based wall board or a 6 mm fibre cement board.
- Ceramic tiles of at least 5 mm thickness may be used over either 10 mm gypsum-based wall board or a sheet of 6 mm fibre cement board.

REFER TO AS/NZS 5601.1 'PROTECTION OF A COMBUSTIBLE SURFACE NEAR A GAS COOKING APPLIANCE' AND APPENDIX C

Where you are asked to install an appliance before protective surfaces have been fixed by the builder and

FROM EXPERIENCE

Caution! Gas cooking appliances are a very common and popular choice for homeowners, professional cooks and builders alike. However, you must ensure that you work closely with all parties as early as possible in a project to ensure that the design of the kitchen will accommodate the requirements for gas appliance installation. You must be satisfied that everyone understands that there are restrictions on the use of certain materials and locations that must be adhered to.

you are not due to return to the job until it is ready for fit-out, you must ensure in writing that all parties understand what needs to be done in relation to the use of correct protection.

What about existing installations?

Often you will be asked to install appliances in existing installations, such as commercial or domestic kitchens. In many cases, the walls may already be sheeted in stainless steel, or appliances may be already in place.

If you arrive on a job to find that the proposed location of the appliance is not compliant with AS/NZS 5601.1 or you cannot satisfy yourself that approved fire-resistant material has been used where required, do not proceed until this can be confirmed and altered where necessary.

As a licensed gas fitter, you are responsible for everything related to the gas appliance installation, and this includes knowing that all combustible surfaces have been protected according to AS/NZS 5601.1.

LEARNING TASK 13.2

PROTECTING A COMBUSTIBLE SURFACE

1. What is pyrolysis?
2. What must be done when a gas appliance is to be installed prior to the installation of protective surfaces.

Commissioning

'Commissioning' can be defined as 'bringing an appliance into full working condition in accordance with manufacturer's instructions'. The gas fitter is required to correctly commission every gas appliance. Commissioning procedures must be followed during new installations and after an appliance has been re-installed or repaired.

Each manufacturer has different commissioning requirements for each product. However, where such instructions are inadequate or missing, a simple guide to basic commissioning is included in Appendix O of the Standards.

> **REFER TO AS/NZS 5601.1 'COMMISSIONING REQUIREMENTS' AND APPENDIX O**

Basic commissioning steps

The variety and number of appliances available now make it almost impossible to provide generic commissioning guides. You must always be guided by the requirements set out within the manufacturer's instructions. These manufacturer's instructions should be followed within the context of the following commissioning steps.

HOW TO

Step 1: Check for stray current
Use a suitable test instrument to ensure both the pipe and the appliance are safe to touch. Check and recheck throughout the procedure.

Step 2: Check for compliance
Locate the appliance data plate and certification sticker.

Step 3: Visually inspect the installation and the appliance
Check the installation for any damage and/or missing parts. Ensure that the appliance location, flueing, ventilation and clearances to combustible surfaces are correct.

Step 4: Test the consumer piping as per Appendix E
All installations are required to be pressure tested prior to the commissioning procedure.

Step 5: Purge the consumer piping of air and/or inert gas
Prior to lighting the appliance, you must ensure that all air is removed from the consumer piping. Purge until only pure fuel gas is left in the system.

Step 6: Light the appliance as per the manufacturer's instructions
Follow the manufacturer's instructions to ensure that the appliance is operating correctly.

Step 7: Instruct the customer and hand over the operating instructions
Where the customer is not available, you must leave the operating instructions in a suitable location where the customer can readily access them.

Purging a small installation through an appliance

Purging is a central part of the commissioning process. If you attempted to light an appliance without having purged the line, any air–gas mix within the pipe may ignite. This may result in damage to the appliance or an explosion at the gas meter.

Never try to light an appliance unless all air has been purged from the line!

Piping less than 30 L (0.03 m^3) is classed as a small installation. AS/NZS 5601.1 sets out three detailed appliance purge procedures:
- purging through an appliance fitted with an open burner (an open burner is a burner that has no flame failure device fitted)
- purging through an appliance fitted with an electronic flame safeguard
- purging through an appliance fitted with a thermoelectric flame safeguard.

> **REFER TO AS/NZS 5601.1 APPENDIX D**

Before commencing an appliance purge, consider the following:
- Plan the purge in a logical sequence so that the most pipe is purged in one go and no pockets of air are left in the pipe.
- Try to purge through an open burner first, as this is the quickest procedure.
- Avoid purging into a confined space, particularly on larger installations. Use a plastic hose to vent purged gas to atmosphere.
- Isolate any ignition device that may inadvertently activate when you turn on a gas control during the purge.

Purging through an open burner

Purging through an open burner is the easiest procedure of these three and will be covered here in detail. Traditionally this was almost always done at a cooking appliance. However, all cooking appliances must now be manufactured with either thermoelectric or electronic flame safeguard systems, therefore open burners will quickly become less prevalent as time goes on. There are of course many existing installations where this purging process can be employed and as such the process is described below:

HOW TO

1. Turn on one burner gas control until the presence of gas is detected. Put your ear to the burner and listen to the escaping air. As the gas comes through you will hear a change in pitch due to the difference in density between gas and air. Use your sense of smell too, although you will find this is not as reliable as you may think.
2. Let the gas flow for a few seconds longer then turn off and allow sufficient time for any accumulated gas to disperse before proceeding. Be particularly careful when purging LPG, as it is heavier than air and will tend to sit in appliance recesses. Fan with a piece of cardboard if necessary.

3 Turn on one gas control valve and keep a continuously burning flame at the burner until the gas is alight and the flame is stable.

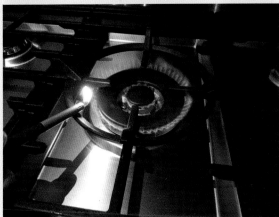

4 Continue to purge until gas is available at each appliance. Do this at each appliance before you adjust the pressure or aeration. You must ensure that all air is out of the system before continuing with the commissioning process.

Purging through an appliance with a flame safeguard device

Because flame safeguard devices prevent the release of any gas from a burner without the presence of a flame, you cannot efficiently purge the appliance from a burner. The only option is to purge at the appliance inlet connection. The amount of air left in the appliance internals is negligible and does not pose any hazard. The procedure is described as follows:

HOW TO

1 Isolate any electrical supply to the appliance.
2 Fit a bridging device (bonding strap) across the appliance inlet union connection.

3 Slacken the union to allow gas to flow out, but do not fully disconnect it. This procedure maintains a level of controlled gas release.
4 Turn on the appliance manual shut-off valve until the presence of gas is detected. Put your ear to the fitting and listen to the escaping air. As the gas comes through you will hear a change in pitch due to the difference in density between gas and air. Use your sense of smell too, although you will find this is not as reliable as you may think.
5 As soon as the presence of gas is detected, tighten the union and test it with soap and water solution.
6 Allow sufficient time for any gas to disperse.
7 Remove the bonding straps.
8 Turn on the power supply and activate the ignition source.
9 Remember ignition may not be successful immediately and lock-out may occur a number of times before combustion is satisfactory.
10 Allow sufficient time for any unburnt gas to disperse before resetting the system.

The procedure for purging an appliance fitted with a thermoelectric safety device is similar to that outlined above, with additional steps needed in order to light the pilot.

Testing for flue spillage

Under certain circumstances, mechanical ventilation can cause the products of combustion to be drawn out of the appliance or its flue back into the room itself. This is potentially very dangerous and a vital part of the commissioning procedure is to carry out a spillage test.

Where any flued appliance is to be installed in proximity to any form of mechanical extraction fan, you must carry out a spillage test in accordance with Appendix R of the AS/NZS 5601.1.

Following the method described in Appendix R, you must use some form of smoke candle (or incense stick) to observe if any combustion products are being drawn out of the appliance itself or its flue system by the nearby mechanical ventilation system. If spillage is detected, then you must following the procedure in Appendix R to negate the depressurisation effect of the mechanical extraction.

As you must have some knowledge of flue systems to understand this problem, more information on the spillage test is included in Chapter 14.

REFER TO AS/NZS 5601.1 APPENDIX R

Clean up the job

The job is never over until your work area is cleaned up and tools and equipment accounted for and returned to their required location in a serviceable condition.

Clean up your area

The following points are important considerations when appliance installation is completed:
- If not already covered in your site induction, consult with supervisors in relation to the site refuse requirements.
- Cleaning up as you go not only keeps your work area safe it also lessens the amount of work required at the end of the job.
- Dispose of rubbish, off-cuts and old materials in accordance with site-disposal requirements or your own company policy.
- Ensure all materials that can be recycled are separated from general refuse and placed in the appropriate location for collection. On some sites specific bins may be allocated for metals, timber etc. or if not, you may need to take materials back to your business premises.

Check your tools and equipment

Tools and equipment need to be checked for serviceability to ensure that they are ready for use on the job. On a daily basis and at the end of the job check that:
- all items are working as they should with no obvious damage
- high-wear and consumable items such as cutting discs, drill bits and sealants are checked for current condition and replaced as required
- items such as digital manometers, crimping tools, gas detection equipment etc. are working within calibration requirements
- where any item requires repair or replacement, your supervisor is informed and if necessary mark any items before returning to the company store or vehicles.

Complete all necessary job documentation

Most jobs require some level of documentation to be completed. Avoid putting it off till later as you may forget important points or times.
- Ensure all SWMS and/or JSAs are accounted for and filed as per company policy.
- Complete your timesheet accurately.
- Reconcile any paper-based material receipts and ensure electronic receipts have been accurately recorded against the job site and/or number.
- As per State or Territory requirements complete all necessary gas notices that must be submitted to the technical regulator, gas supplier and owner.

 COMPLETE WORKSHEET 6

SUMMARY

In this unit you have examined the requirements related to the installation of Type A gas appliances including:
- review of general appliance compliance requirements
- an overview of appliance types and specific installation needs
- protection of combustible surfaces
- commissioning procedures
- clean-up of job and return of equipment.

GET IT RIGHT

The photo below shows an incorrect practice when installing a cooktop near a vertical plasterboard surface. Identify the problem and provide reasoning for your answer.

_____

_____

_____

_____

_____

WORKSHEET 1

To be completed by teachers
Student competent ☐
Student not yet competent ☐

Student name: _____

Enrolment year: _____

Class code: _____

Competency name/Number: _____

Task

Review the section *General installation requirements* and answer the following questions.

1. Where should a general-purpose outlet (GPO) electrical connection be located for a gas appliance? Provide a reference from AS/NZS 5601 to support your answer.

   _____
   _____
   _____

2. You have been asked to convert an appliance from LPG to SNG. The data plate does not include any information relating to SNG. What would you do?

   _____
   _____
   _____

3. A six-burner range is fitted with castors to enable easier movement for cleaning. You plan to connect it with a 1200 mm long Class D hose assembly. What is the maximum length of the required restraint chain? Provide a reference from AS/NZS 5601 to support your answer.

   _____
   _____
   _____

4. What is the maximum gas pressure permitted where a push-on connector connecting a laboratory Bunsen burner is being used? Provide a reference from AS/NZS 5601 to support your answer.

   _____

5. Could a continuous flow water heater certified for outdoors installation only be installed in a multi-level car park with open sides? You will need to discuss this problem with your teacher.

   _____
   _____
   _____

6 Explain in detail why a flueless space heater is not permitted for installation in a bathroom.

_____
_____
_____

7 A builder has asked you to install a storage water heater in a cupboard. The Standard requires that you fix a warning notice adjacent to the appliance. What does it say?

_____
_____
_____
_____

8 You plan to install a central heater under the floor of a domestic dwelling. The subfloor space measures 800 mm between ground level and the floor joists. You would like to site the heater 1.5 m from the subfloor access opening. Would you be permitted to install the appliance under these conditions? Provide a reference from AS/NZS 5601 to support your answer.

_____
_____
_____
_____

9 Where an appliance is installed in a roof space or under a floor, the Standard requires that permanent fixed lighting be provided at the appliance. Where should you ask the electrician to install the light switch? Provide a reference from AS/NZS 5601 to support your answer.

_____
_____
_____
_____

10 Where an appliance is installed in a roof space, you are required to install a walkway to the appliance from the access opening. How wide is this walkway and, where an appliance sits on this platform, what are the dimensions around the appliance? Provide a reference from AS/NZS 5601 to support your answer.

_____
_____
_____
_____

11 A customer asks you to install a hydronic boiler in a residential garage. What are two key considerations that you must comply with in respect of appliance location? Provide a reference from AS/NZS 5601 to support your answer.

_____

_____

_____

_____

12 The owner of a commercial kitchen has purchased a new salamander and wants you to install it. He indicates his preferred position on the wall 700 mm above an existing char grill. Is this OK? Provide a reference from AS/NZS 5601 to support your answer.

_____

_____

_____

WORKSHEET 2

To be completed by teachers
Student competent ☐
Student not yet competent ☐

Student name: _____

Enrolment year: _____

Class code: _____

Competency name/Number: _____

Task

Review the section *Specific installation requirements* and answer the following questions.

1. What is the minimum distance permitted between the periphery, or edge, of the cooktop burner and a vertical combustible surface? Provide a reference from AS/NZS 5601 to support your answer.

   _____

2. To what height would you protect a vertical combustible surface around a cooktop? Provide a reference from AS/NZS 5601 to support your answer.

   _____

3. What is the minimum distance permitted between the top of the burner and any downward-facing combustible surface? Provide a reference from AS/NZS 5601 to support your answer.

   _____

   _____

   _____

4. A timber-framed plaster-sheeted wall is sited 150 mm away from the periphery of the cooktop burner. The builder plans to cover this with stainless steel. Is this sufficient protection for a combustible surface? Provide a reference from AS/NZS 5601 to support your answer.

   _____

   _____

   _____

5. You need to purchase a flexible hose assembly for an upright cooker with a high-level connection. The supplier provides you with a hose measuring 1500 mm long. Would such a hose assembly be suitable for this job? Provide a reference from AS/NZS 5601 to support your answer.

   _____

   _____

   _____

6. What Australian Standard and class do hose assemblies for cooker installations have to comply with? Provide a reference from AS/NZS 5601 to support your answer.

_____

_____

_____

7. Would you need to protect a combustible surface within 200 mm of the rear burners of an upright cooker that is fitted with a fixed splashback? Provide a reference from AS/NZS 5601 to support your answer. What if the splashback were detachable for cleaning?

_____

_____

_____

_____

_____

8. The flue outlet on an elevated cooker is situated at the top-rear of the appliance above the oven itself. What clearance must you maintain between this outlet and the nearest overhead surface? Provide a reference from AS/NZS 5601 to support your answer.

_____

9. Look carefully at the photo in Figure 13.8. You will notice that the side of the cooktop hob is below the level of the kitchen bench top. This combustible bench top surface is also within 200 mm of the periphery of the nearest burner. State two methods you would use to make this installation comply with the standard. Provide a reference from AS/NZS 5601 to support your answer.

_____

_____

_____

_____

10. Where should you ensure that the inbuilt oven electrical connection point is located? Provide a reference from AS/NZS 5601 to support your answer.

_____

_____

_____

11. An inbuilt oven may be certified for connection only with a hose assembly supplied by the manufacturer. Where this is not the case, what is the only other approved method of gas connection? Provide a reference from AS/NZS 5601 to support your answer.

_____

_____

_____

WORKSHEET 3

Student name: _____

Enrolment year: _____

Class code: _____

Competency name/Number: _____

To be completed by teachers
Student competent ☐
Student not yet competent ☐

Task

Review the section *Commercial catering equipment* and answer the following questions.

1 Where commercial catering equipment is installed, which Australian Standard would the overhead exhaust system need to comply with? Provide a reference from AS/NZS 5601 to support your answer.

_____

_____

2 What is the minimum clearance between the top of a char grill and an overhead grease filter? Provide a reference from AS/NZS 5601 to support your answer.

_____

3 Clarify the definition of a combination cooking range in your area with your teacher. Would any six appliances sharing a supply line be classed in this way, or do they have to be physically connected together behind a fascia?

_____

_____

_____

4 You have been asked to install a range next to a table with a timber preparation top. The top of the table is 50 mm below the top of the range. How close to this combustible surface is the range permitted to be? Provide a reference from AS/NZS 5601 to support your answer.

_____

WORKSHEET 4

To be completed by teachers
Student competent ☐
Student not yet competent ☐

Student name: _____

Enrolment year: _____

Class code: _____

Competency name/Number: _____

Task

Review the sections *Continuous flow water heaters* and *Storage water heaters* and answer the following questions.

1 You go to a job and find an old internal, unflued continuous flow water heater installed above a shower-bath. Explain what could happen that could lead to someone being asphyxiated.

_____

_____

_____

2 Would you be permitted to replace an unflued water heater with a new room-sealed appliance in a bathroom?

_____

_____

3 What are the two main primary flue systems used in the design of gas storage water heaters?

_____

_____

4 What is the maximum input (MJ/h) for a storage water heater installed in a bathroom? Provide a reference from AS/NZS 5601 to support your answer.

_____

5 You have been asked to install a storage water heater on a timber veranda, but you are concerned that the veranda might not be able to support the load. The cylinder weighs approximately 95 kg and holds 400 L of water. What is the approximate total mass of the filled cylinder?

_____

WORKSHEET 5

To be completed by teachers
Student competent ☐
Student not yet competent ☐

Student name: _____

Enrolment year: _____

Class code: _____

Competency name/Number: _____

Task
Review the sections *Space heaters* through to *Decorative flame effect fires* and answer the following questions.

1. What is the minimum vertical clearance permitted between the top of a flueless space heater and any overhead projection? Provide a reference from AS/NZS 5601 to support your answer.

   _____

2. When calculating flueless space heater requirements, what is the maximum combined gas consumption permitted for all heaters in a room? Provide a reference from AS/NZS 5601 to support your answer.

   _____

3. A customer has pre-purchased a flueless space heater and wants you to install a bayonet fitting so that it can be used in the kitchen. The heater is a non-thermostatically controlled appliance with a gas input of 13 MJ/h. The room measures 3 m × 4.5 m × 2.7 m. Are you permitted to install the bayonet fitting knowing that this appliance is intended for use? What are your options? Provide a reference from AS/NZS 5601 to support your answer.

   _____

   _____

   _____

   _____

   _____

   _____

4. Subject to technical regulatory requirements, what is the minimum distance between an overhead radiant tube heater and the floor? Provide a reference from AS/NZS 5601 to support your answer.

   _____

5. Where air contaminants exist, what must be done with the air supply and flue outlet of the overhead radiant tube heater? Provide a reference from AS/NZS 5601 to support your answer.

_____

_____

_____

6. You are asked to install an overhead radiant heater with a maximum gas consumption of 42 MJ/h. The position indicated by the owner would place the appliance 2.6 m above floor level with a clearance above the heater of 950 mm to a combustible ceiling surface. Would this location be permissible? If not, what are your options? Provide a reference from AS/NZS 5601 to support your answer.

_____

_____

_____

7. A customer wants to install two 24 MJ/h overhead radiant heaters in a backyard shed measuring 9 m × 6 m × 3.5 m. The appliances have no thermostatic control. Would this be permissible? Provide a reference from AS/NZS 5601 to support your answer.

_____

_____

_____

8. What is the maximum permitted gas consumption of a decorative flame effect fire? Provide a reference from AS/NZS 5601 to support your answer.

_____

9. Can a decorative flame effect fire be installed in a bathroom? Provide a reference from AS/NZS 5601 to support your answer.

_____

_____

10. What appendix in AS/NZS 5601.1 describes the procedure for carrying out a spillage test on a flued appliance?

_____

11. Where a Type 1 decorative flame effect fire is to be installed in an existing brick chimney, what is the minimum size of the square section chimney outlet? Provide a reference from AS/NZS 5601 to support your answer.

_____

_____

WORKSHEET 6

To be completed by teachers
Student competent ☐
Student not yet competent ☐

Student name: _____

Enrolment year: _____

Class code: _____

Competency name/Number: _____

Task
Review the sections *Protecting a combustible surface* and *Commissioning* and answer the following questions.

1 What is the definition of an *open burner*?

_____
_____

2 Why must bonding straps be used when purging through the appliance inlet union?

_____
_____
_____

3 Name two methods of detecting the presence of gas during a purge.

_____
_____

4 Why is appliance purging a required part of the commissioning procedure?

_____
_____
_____

5 List at least four types of combustible materials.

_____
_____
_____

6 What action would you take if you suspected that fire-resistant material surrounding an old boiler contained asbestos?

_____
_____
_____

INSTALL AND COMMISSION TYPE A GAS APPLIANCES

7 What is the minimum thickness permitted for any fire-resistant material?

8 Find out the following information using internet resources, information from a technical regulator and, if necessary, assistance from your teacher.

   i   Name at least two fire-resistant sheet products available in your local area.

   ii Where can you purchase them?

   iii What is the thickness of each product?

INSTALL TYPE A GAS APPLIANCE FLUES

Learning objectives

In this chapter, you will learn about the need for the correct flueing of gas appliances and how to interpret AS/NZS 5601 so that your flue installations are compliant for material selection, specific flue pipe installation and locations of flue terminals. Requirements related to the installation of a natural draught flue and fan-assisted flue are also covered in detail.

Areas addressed in this chapter include:
- why flues are necessary
- the various types of flues
- flue components and materials
- heat loss
- terminal locations
- flue pipe roof penetration and flashings
- fan-assisted flue installation
- flue spillage testing.

For students undertaking training at the Certificate IV level necessary for licensing, please refer to the Appendix, which covers the following:

Advanced skills – flue sizing:
- sizing individual appliance flues
- sizing horizontal common flue manifolds
- sizing vertical common flue manifolds.

Before working through this section, it is absolutely important that you complete or revise each of the chapters in Part A 'Gas fundamentals' with particular emphasis on Chapter 6 'Combustion principles'.

Overview

The purpose of a flue is to safely convey the products of combustion away from the gas appliance to the flue terminal. To understand the importance of the flue system, you need to remember some basic combustion principles.

'Complete combustion' describes a chemical reaction between a *fuel* (natural gas or LPG) and an *oxidiser* (oxygen), which becomes self-sustaining after the application of the correct amount of *heat* (a match or spark).

For complete combustion to take place, the proportion of fuel gas and air must be as follows:
- Natural gas – 1 m^3 of natural gas requires 9.5 m^3 of air
- LPG – 1 m^3 of LPG requires 24 m^3 of air

This mix of air and fuel gas is lit in the combustion chamber and in doing so will give off certain products of combustion.

Products of complete combustion

When a fuel gas is burnt in air under normal atmospheric conditions, the following products of complete combustion are the result:
- heat
- water vapour
- carbon dioxide.

In an area of sufficient ventilation, these products of combustion are harmless. However, if they were permitted to build up in a room or enclosure, a hazardous situation would arise:
- Room heat may rise excessively, and with it condensation and dampness from the trapped water vapour within the flue products.
- As carbon dioxide levels increase in the room, the potential availability of oxygen would be reduced and the burner flame would begin to be smothered as secondary air is reduced. Without sufficient oxygen, the fuel gas would burn incompletely.
- Incomplete combustion would produce the dangerous gas carbon monoxide (CO), which not only exacerbates poor combustion but also may endanger the lives of any people in the room.

Therefore, even complete combustion is compromised without an adequate flue system.

The effect of inadequate flueing on combustion

- A blocked or incorrectly sized flue will cause a build-up of combustion products, reducing the amount of fresh secondary air available for complete combustion to take place.
- When the supply of air is restricted, not enough oxygen will be present for the hydrocarbon fuel to be completely oxidised fully into CO_2 and water vapour. This lack of secondary air for incomplete combustion is called 'vitiation' (pronounced *vish-ee-ay-shon*).

Products of incomplete combustion

Incomplete combustion of a fuel gas will result in the creation of the following products:
- *carbon monoxide*
- soot (carbon)
- aldehydes

Poorly installed flues regularly cause injury, death and property damage. You must ensure that you follow the Gas Installation Standards and manufacturer's instructions to the letter.

Your job as a gas fitter is to use this basic knowledge of combustion principles and, adhering to your Standard, size, install and test flues that safely remove combustion products from the appliance to atmosphere.

LEARNING TASK 14.1

1. Circle which of the following combustion products are the result of incomplete combustion.
   a carbon dioxide
   b soot
   c aldehydes
   d nitrogen
   e carbon monoxide
   f water vapour
2. What term is used to describe incomplete combustion where an undersized flue causes a reduction in secondary air in the combustion chamber?
3. What is the primary purpose of an appliance flue?

Flue types

There are number of different flue types used with modern gas appliances. You need to recognise all flue types so that you can relate to their particular requirements as detailed in AS/NZS 5601.1.

Natural draught open flue

A natural draught open flue is designed to operate under normal atmospheric conditions. As hot flue gases are less dense than the surrounding air, they will rise of their own accord without any mechanical assistance. This natural rise of hot flue gases is known as aeromotive force (see Figure 14.1). A natural draught flue must be carefully sized and installed to ensure that the flue gases remain hot and maintain this aeromotive force.

Many gas appliances with a natural draught flue will not come with a flue kit or prescribed flueing solution, so you will need to use your knowledge and skills to select materials and design a flue to suit the installation.

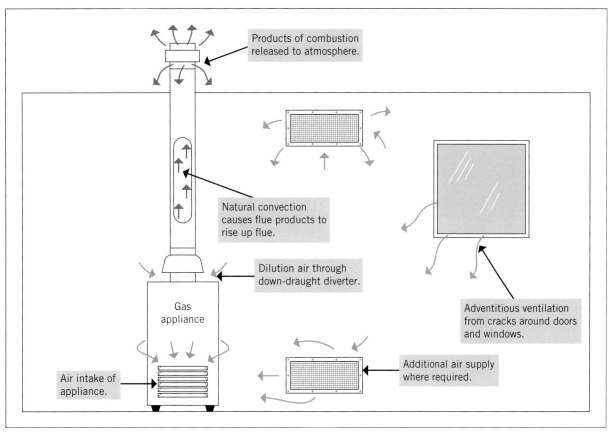

FIGURE 14.1 Operation of the natural draught open flue system

Balanced flue

A balanced flue is one in which the products of combustion and the incoming air for combustion enter into and are exhausted from the appliance at the same height and atmospheric pressure. Such a flue may be a feature of a room-sealed appliance or an external appliance with a balanced flue.

Room-sealed appliance

This type of appliance is designed so that all air required for combustion is drawn from outside the building and all products of combustion are exhausted to outside the building (see Figure 14.2). Internal air quality is unchanged. Some space heaters and water heaters are designed on this principle.

External appliance balanced flue

Some external appliances feature a balanced flue terminal. These flues are designed on the basis that, at the same atmospheric pressure, the hot and less dense combustion products will rise quickly once released from the flue, while colder fresh air is drawn in via a lower or separate inlet within the same terminal.

Co-axial flue

Co-axial flues are specially designed double-tube flue systems made by manufacturers for specific appliances (see Figure 14.3). Such flues feature an internal tube through which the products of combustion are forced by

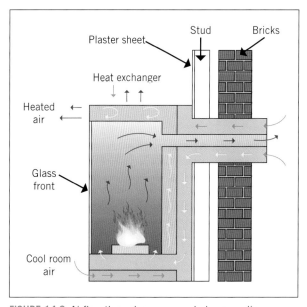

FIGURE 14.2 Airflow through a room-sealed gas appliance

the combustion fan. At the same time, the combustion fan draws in fresh air via the outer casing (see Figure 14.4).

Power flue

A power flue is defined as a flue in which the products of combustion are removed from the appliance via a fan situated in the flue itself. This is not to be confused with a fan-assisted flue in which an electric fan in the

FIGURE 14.3 The external and internal components of a fan-assisted, co-axial flue from a room-sealed heating appliance

FIGURE 14.4 A component from a co-axial flue system designed for internal continuous flow water heaters, showing an internal exhaust tube surrounded by an outer plastic casing through which the incoming combustion air flows

appliance provides either forced or induced draught combustion air into the combustion chamber.

Power flues must be carefully designed and sized to ensure that all products of combustion are effectively removed, and they must incorporate special gas supply interlocks to ensure the system fails safe in the event of a power flue breakdown.

> **LEARNING TASK 14.2**
>
> 1 What is the term used to describe the natural rise of hot products of combustion within a natural draught flue system?
> 2 What type of natural draught flue draws in its combustion air and exhausts its flue products from the same terminal?
> 3 What poisonous gas is produced where an incorrectly installed flue causes incomplete combustion?

Prepare for flue installation

Detailed preparation before job commencement can save many lost hours fixing problems that could have been easily identified through a methodical process.

Materials

The job specifications or material list provided by your employer should detail the primary requirements for the job. You should pay particular attention to the following:

- Ensure flue materials on order and those that have arrived match the requirements of the plans and specifications.
- Confirm that the supplied materials are suitable for the flue application and that the proposed flue location is compliant with the Standard.
- Flues that are pre-packaged as a kit should be checked to ensure the kit matches the appliance.
- Flue kits are designed for specific installation requirements of roof height, wall thickness and cladding type. Ensure that the kit supplied will be suitable for your particular installation requirements.
- Check that all flue components are certified for use. Current certification numbers can be checked against the online data provided by the relevant entities responsible, such as the Australian Gas Association and SAI Global. The Gas Technical Regulators Committee (GTRC) website provides a portal to assist in accessing confirmation of material and product certification – http://equipment.gtrc.gov.au.
- Check that all PPE is serviceable and where relevant check replacement dates on items such as height safety harnesses.

Required tools and test equipment

As with other gas fitting tasks, your choice of tools and equipment will be dependent upon the particular installation requirements and the type of flue system you are working with. Most flue installations will require use of some form of height access and protection. As a general guide this might include:
- ladders and/or elevated work platform
- assorted battery power tools relevant to the task
- spirit level and plumb bob
- laser levels
- measuring equipment
- smoke match/candles or incense sticks.

Wherever 240 V power tools might be required for a certain task, remember to check that the tool has been tested and take particular note of the retest date on the tag. If out of date, let your supervisor know and mark it in some manner that it is not to be used.

LEARNING TASK 14.3

Based upon a job you have recently done or a potential flue installation task indicated by your teacher, create a list of tools and equipment specific to that location and material requirements.

Job characteristics	Tools and equipment required
• Roof or wall penetration	

• Flue type:
  • Natural draught?
  • Fan-assisted?	

• Other considerations:	

Identify flue requirements

In the following sections you will learn about:
- the components that make up a flue system
- available and approved materials for flues and the definition of 'high heat loss' and 'low heat loss' materials
- flue locations in accordance with the Standard.

Flue components

Before you can move on to learn about more specific flue requirements, you need to know the correct terminology for all flue components and their purpose. Refer to Figure 14.5 to identify each component part.

Downdraught diverter

One of the most important components within a natural draught flue system is the downdraught diverter. The diverter has three key functions:
- to reduce aeromotive force
- to dilute the products of combustion
- to divert any downdraught away from the combustion area.

Review Figure 14.6 to find out more about the downdraught diverter.

What would happen if there were no downdraught diverter?

In almost all cases, natural draught flued appliances would be unable to operate effectively without the use of a downdraught diverter in the flue system. Figure 14.7 shows what would happen if a flue were fitted directly to the gas appliance.

Flue materials

A wide range of flue materials exists for different circumstances. All materials must be approved for use and selected in accordance with AS/NZS 5601.1.

REFER TO AS/NZS 5601.1 'FLUE DESIGN – MATERIAL'

Flue material selection will depend on some of the following factors:
- *Manufacturer's instructions*. All installations must comply with the manufacturer's instructions. Certain appliances have been granted appliance approval on condition that a particular flue system is installed. Where flue materials are prescribed in such a way, you are not permitted to use alternatives.
- *Durability*. Where sections of a flue installation may be concealed, you will need to ensure that the material you select is corrosion resistant.
- *Flue gas temperature*. Some materials can only be used up to a maximum temperature. Above this limit, another type of material would be required.
- *Flue diameter*. When using materials such as mild steel, you will find that the maximum permissible diameter of the flue depends upon the wall thickness of the material.

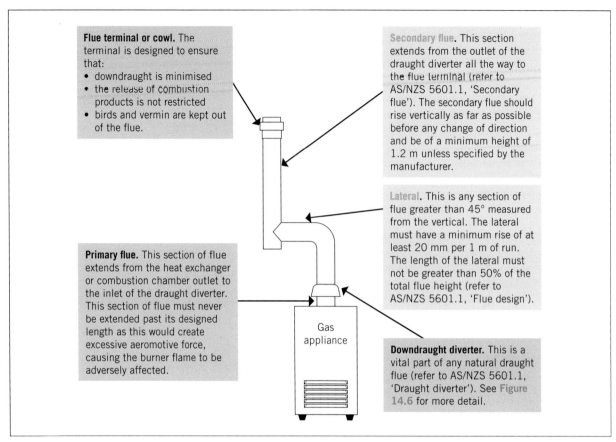

FIGURE 14.5 Natural draught open flue component identification

- *Heat loss.* Flue materials should be selected to ensure that the temperature of the flue gases is maintained and not lost prematurely. Some materials cannot be used where the flue exceeds a prescribed height.

Heat loss

Flue installations are classed as either 'high heat loss' or 'low heat loss'. This describes the amount of heat lost through the walls of the flue to the surrounding environment. The sizing of flues is to a large degree dependent upon this classification.

High heat loss installations and materials

A high heat loss installation or material is one in which the flue loses a lot of heat through the walls of the flue to the surrounding environment. Depending upon the efficiency and appliance type, the flue gas temperature of many natural draught appliances will range from approximately 100°C to 300°C. Some high-efficiency appliances may have flue temperatures that are much lower.

Stainless steel, galvanised steel and aluminium are materials commonly used for single wall metal flues. In the right circumstances they offer a cost-effective and adequately performing flue solution. However, you must be aware that high heat loss materials have a number of important disadvantages that need to be taken into account:

- If flue gases lose too much heat, the water vapour, which is normally carried away as one of the products of combustion, will start to condense on the inside of the flue. The temperature at which this happens is called the dew point of the flue gas. For most natural draught flues, the dew point is approximately 40°C. Flue gas condensate is quite acidic, as it contains both sulfuric and nitric acids, and has a highly corrosive effect upon flue materials such as galvanised or mild steel. This could lead to costly early replacement, or a hazardous situation in which a perforated flue allows products of combustion to accumulate inside a building.
- Flue gases rise up and out of the flue because they are hotter and less dense than the surrounding air. If the flue gases lose too much heat through the walls of the flue, this aeromotive force will slow and could stall, almost causing a 'plug' of cold air to form in the flue. When this happens, flue gases may spill from around the draught diverter into the room. As the condition continues, a build-up of flue gases in the combustion chamber will restrict the availability of secondary air to the burner, causing vitiation and producing carbon monoxide.

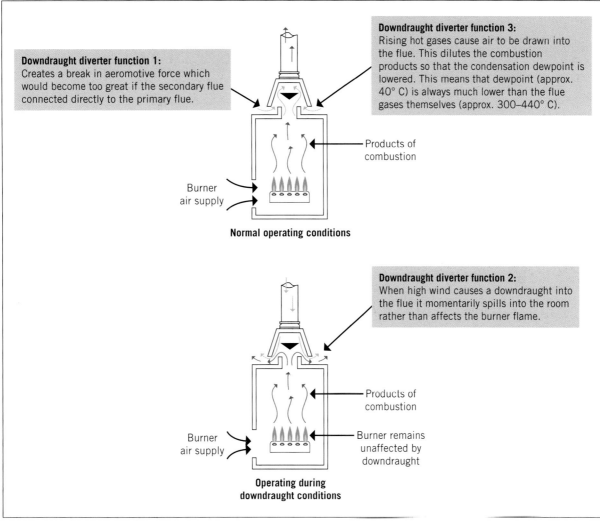

FIGURE 14.6 Functions of a downdraught diverter

- Single wall metal flues (see Figure 14.8) conduct a considerable amount of heat and must be installed in strict compliance with AS/NZS 5601.1 so that clearances from combustible surfaces are maintained. This may add to the installation cost. Note that:
- single wall metal flues installed outdoors are always classed as high heat loss installations
- single wall metal flues installed indoors and not subject to draughts can be classed as low heat loss installations, as long as not more than 1 m of single skin flue is exposed to the outside air. This 1 m rule, although not part of the Standard, has been based upon industry and manufacturer experience for many years. If in doubt, always consult with the manufacturer and technical regulator.

Low heat loss installations and materials

Low heat loss installations are those in which the flue is insulated against excessive heat loss. This may be due to the nature of the flue material itself or the location of the appliance and flue. Approved low heat loss materials include:
- fire bricks
- autoclaved fibre cement (asbestos-free)
- proprietary twin wall flue (see Figure 14.8).

Of these materials, the twin wall flue is most commonly used by gas fitters. A twin wall flue is made from an outer galvanised steel shell surrounding an internal aluminium core separated by an approximate 5 mm air gap. It is an approved proprietary flue system and must be installed according to the manufacturer's instructions. It offers the gas fitter the following advantages:
- simple twist lock jointing
- corrosion resistance
- able to be installed as close as 10 mm to a combustible surface
- can be used in concealed installations
- does not require heat-resistant silicone roof flashings

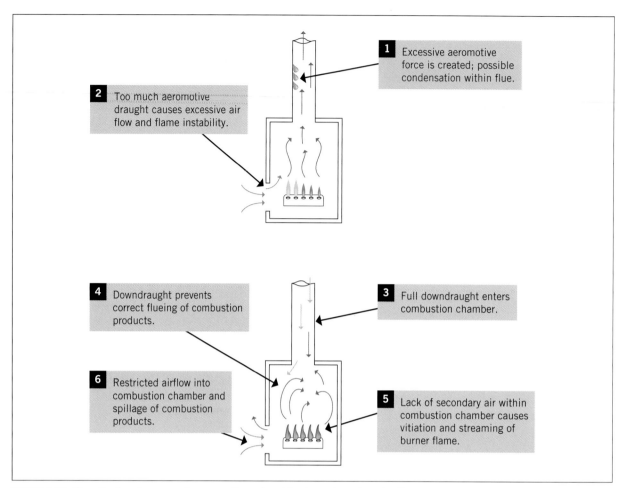

FIGURE 14.7 The effect on burner performance through not having a downdraught diverter

FIGURE 14.8 A high heat loss single wall flue (left) and a low heat loss twin wall flue (right)

- available in a wide range of lengths and components
- when installed in accordance with the manufacturer's specifications, it has a long warranty period.

Proprietary-brand twin wall flue systems are available in a range of diameters, lengths and special components, including round-to-oval transition fittings as seen in Figure 14.9.

FIGURE 14.9 A range of pipe lengths and transition fittings are available in twin wall flue

Note: Twin wall or insulated flues are classed as low heat loss irrespective of whether they are installed outdoors or indoors.

LEARNING TASK 14.4

1 Is a single wall flue installed indoors and not subject to draughts classed as a high heat loss or low heat loss situation?
2 What flue component of a natural draught flue is located between the primary and secondary flues?
3 For a flue system that has a lateral section 2 m long, what is minimum required rise in this section?

Flue installation and terminal locations

During the planning stages and at the time of installation you must always consider how the appliance is going to be flued. Getting it wrong can be catastrophic, as many fires have been started because of poorly installed flue systems. Adherence to AS/NZS 5601.1 and the relevant manufacturer's instructions will prevent this from happening. When planning for a flue installation you need to consider the following points:

- Is the flue clear of structural members?
- Is any combustible material present and what clearances are required?
- Is any part of the flue going to be concealed and/or inaccessible?
- How will the flue be supported?
- Can the appliance be removed without disturbing the flue itself?
- Where is the flue required to terminate?

Having evaluated your installation, you will find that AS/NZS 5601.1 has a solution to these variables.

Flue termination above a roof line

Flue terminals must be located to ensure that regardless of wind direction the flue will not be subject to downdraughts. While AS/NZS 5601.1 provides basic guidelines, you will also need to account for more extreme installations where building design may cause additional problems. High-pitched roofs are one example where significant pressure imbalances can lead to downdraught problems.

REFER TO AS/NZS 5601.1 'FLUE TERMINALS'

Required clearances are measured from the building structure to the end of the flue pipe *before* the cowl is fitted. Figures 14.10 to 14.12 illustrate some of the basic clearance requirements for flues that are terminated above a roof line. Figure 14.13 shows a chimney plate and a cowl.

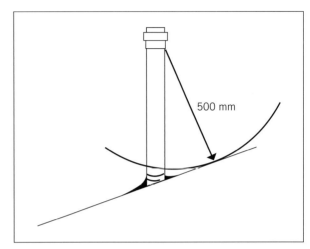

FIGURE 14.10 Clearance for a flue terminal at least 500 mm from the nearest part of a roof

Balanced flue terminal locations

Balanced flues rely on equal atmospheric conditions and differences in air density between hot gases and colder air to work effectively. There are clear guidelines in AS/NZS 5601.1 for balanced flue terminals to ensure that the building structure has no adverse effect upon flue performance and that terminals are positioned so that flue gases are prevented from entering building openings.

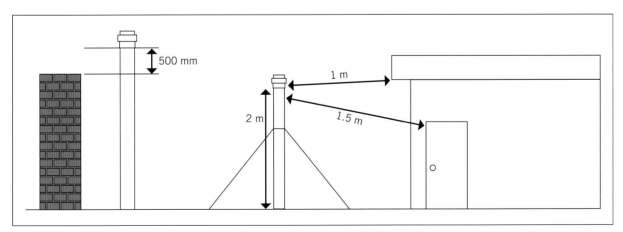

FIGURE 14.11 Clearance for a flue terminal 2 m above a trafficable roof and at least 500 mm above any parapet

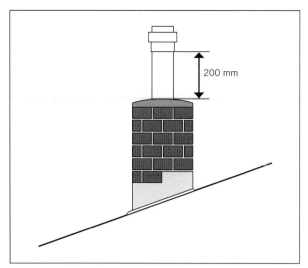

FIGURE 14.12 Clearance for a flue terminal at least 200 mm from the nearest part of a chimney

FIGURE 14.13 A chimney plate and cowl

AS/NZS 5601.1 prescribes clearances from both natural draught and fan-assisted flue terminals. These clearances are always measured from the nearest part of the flue terminal itself to the neighbouring structure, not from the side of the appliance.

Figure 14.14 shows how you would apply the balanced flue clearance requirements within AS/NZS 5601.1 to the installation of two common gas appliances, a continuous flow water heater and a fan-assisted space heater. Knowing the MJ/h input and proposed location, simply choose the relevant clearance from the standard. Each clearance distance is measured directly from the flue terminal itself.

You will often be asked by clients to install appliances under covered areas such as verandas, decks and pergola roofs, or in recesses. Whenever this is proposed, you need to carefully evaluate the area in reference to AS/NZS 5601.1 so that the appliance has enough cross-ventilation to prevent flue gases from contaminating the combustion air supply.

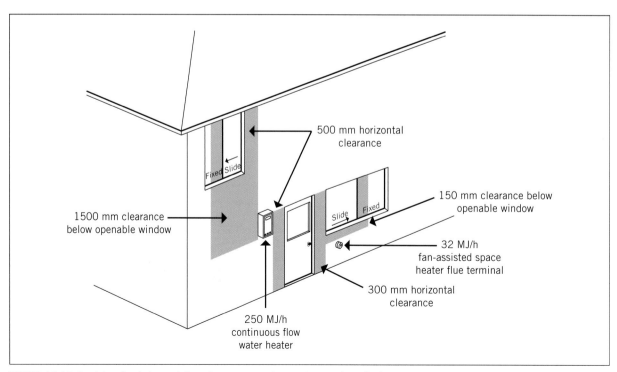

FIGURE 14.14 Applying the balanced flue clearance requirements to an installation

Where an appliance with a natural draught flue is to be installed in such a location, you need to ensure that the area is open on at least two sides and that free air flow is always available.

LEARNING TASK 14.5

1. The minimum distance between the end of a flue above a roof and the roof covering shall be no less than:
   a 1 m
   b 500 mm
   c 600 mm
   d 2 m
   e 1.5 m
   f 300 mm
2. List at least four items you need to consider when planning a flue installation?

COMPLETE WORKSHEET 1

Install and test flue

So far you have learnt where flue terminals may terminate, what materials can be selected and how to size the flue, all in accordance with your Standard. In this section, you will now look at some of the actual installation requirements associated with flueing and, importantly, learn about the mandatory testing process that must be carried out on a flue during the commissioning process.

Clearances to combustible surfaces

A vital part of any flue installation is to ensure that minimum clearances to combustible surfaces and materials are in accordance with AS/NZS 5601.1. Unfortunately, many buildings have been destroyed and lives lost because of flues that have been installed incorrectly.

'Combustible surface' is a defined term in AS/NZS 5601.1 and is regularly referred to throughout the Standard.

REFER TO AS/NZS 5601.1 'COMBUSTIBLE SURFACE'

Of particular concern is the fact that even relatively low heat applied over a long period of time can cause fires to start many months, or even years, after installation of the appliance and flue. The condition leading up to such fires is known as pyrolysis.

What is pyrolysis?

Put simply, pyrolysis is the chemical decomposition of an organic substance caused by the application of heat. Where sufficient clearance or protection is not provided between the heat source and the surface, ignition and combustion of the substance can occur without the direct application of flames.

To examine this in more detail, consider a timber roof rafter or ceiling joist being subjected to heat from an incorrectly installed single skin flue. Depending upon the species, timber will normally ignite at approximately 260°C. However, exposure to a heat source will cause significant changes to the molecular structure of the timber, with the effect of lowering the ignition temperature:

- At temperatures between 65°C and 100°C, the timber begins to lose structural strength.
- At temperatures above 100°C, the timber begins to dehydrate as chemical bonds begin to break apart, releasing flammable volatiles and tars into the surrounding area.
- As pyrolysis of the timber continues, the ignition temperature of the timber can fall to as low as 100°C! With sufficient air supply and application of the right ignition temperature, the timber is then likely to combust.

Where can pyrolysis occur?

Substances can suffer from pyrolysis wherever an appliance or its flue system is not installed with the mandatory clearances and/or correct physical protection of the surface.

You must always be on the lookout for materials in the vicinity of your flue installation that might ignite and burn. Examples of combustible surfaces that you may encounter near flues and components include, but are not limited to:

- timber in any form including studs, noggins, rafters, joists and battens
- the paper lining of sheet plasterboard
- some thermal insulation products, particularly older sarking and vapour membranes
- some non-metallic exterior cladding and roofing products
- most synthetic surfaces.

REFER TO AS/NZS 5601.1 'FLUE INSTALLATION'

Twin wall flues can be installed as close as 10 mm to any combustible surface, but single wall flues require very specific clearances to both protected and unprotected combustible surfaces. Common problem areas include:

- *flues adjacent to rafters and roof members.* Do not directly clip flues to combustible structures! If you need to support a flue pipe in a roof space, use proprietary hanging clips and where required fabricate lengths of sheet metal folded in an 'L' shape. You can cut these to length and support a flue between roof members without making contact.
- *wall and ceiling penetrations.* Ensure your penetration hole is large enough to provide the required clearance between the flue wall and any combustible material. Note that sheet plasterboard must be considered

combustible because it is covered with paper on each side. The purpose of the ceiling register is not only to hide the unsightly hole but also to help hold your flue pipe in the correct position to maintain minimum clearances (see Figure 14.15).

- *ceiling and wall cavity thermal barriers and insulation.* Ensure that where you penetrate any wall or ceiling, the flue pipe is kept well clear of any thermal or vapour wrap as well as paper-based insulation products.

When protecting any combustible surface use only the methods described in AS/NZS 5601.1. In particular, when you intend to use some form of fire-resistant material to provide protection, ensure that it is an approved product. Note that standard cement sheeting (commonly known as 'fibro') used by itself is not an approved product for the protection of combustible material, as it conducts too much heat.

Installing a space heater into an existing chimney

Some space heaters are designed and approved for installation into an existing chimney to replace a previously used open fire or wood heater (see Figure 14.16).

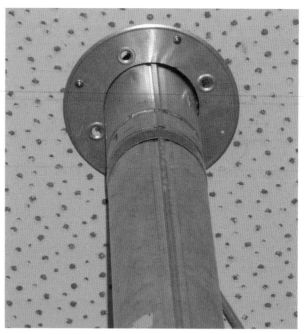

FIGURE 14.15 A flue ceiling register

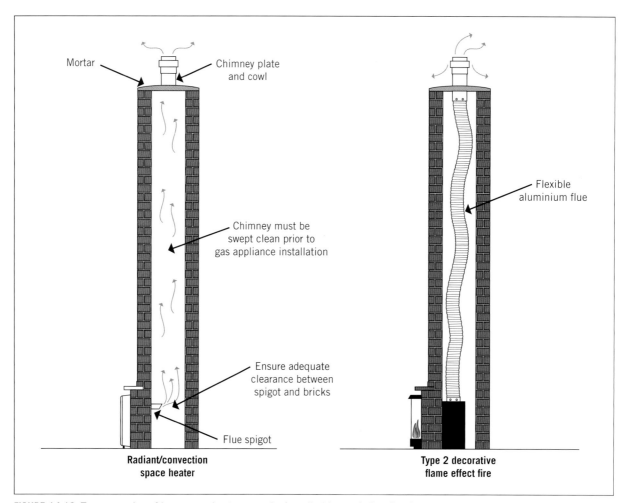

FIGURE 14.16 Two examples of how space heaters may be installed into existing fireplaces

> **REFER TO AS/NZS 5601.1 'USE OF EXISTING FLUE OR CHIMNEY'**

A chimney can often be used as the flue for the gas appliance. However, there are certain aspects of a chimney installation you need to consider:

- You need to ensure that all soot and creosote tars are swept from the chimney prior to gas appliance installation. These substances are deposited on the inside of chimneys through the slow combustion of wood, and under the right conditions can ignite, causing what is known as a 'flue fire'.
- Check that the chimney or flue is of an adequate size for the gas appliance being installed.
- Inspect the chimney for cracks or damage that may allow flue gases to escape. In particular, try to access the ceiling space.
- Damper plates must be permanently secured in the open position or removed altogether.

You should also be aware that very large or high brick chimneys represent a large thermal mass that can absorb a great deal of heat. Since the products of combustion from a gas appliance are much cooler than a wood fire, there is a danger that too much heat will be absorbed from the flue gases into the cold brickwork, causing a reduction in aeromotive force. This may lead to spillage around the appliance fascia. Appliances must be well sealed against the front of the fireplace and tested for spillage in accordance with the steps detailed in the AS/NZS 5601.1 (see the next section).

Where the manufacturer's specifications allow, it is advisable to insert a flexible aluminium chimney liner through the chimney to connect the appliance directly to the flue cowl (see Figure 14.16). This solves problems associated with flue gas temperature and potential spillage.

Flue pipe roof penetration flashings

Polymer penetration flashings are a very effective means of flashing flue pipes through both steel and tiled roofs. However, you need to select a flashing rated to resist the expected temperature that the flue pipe is likely to conduct. Standard type EPDM polymer flashings are suitable when using a twin wall flue (see Figure 14.17), as very little heat is conducted to the outer sheath of the flue pipe (EPDM is an abbreviation for a synthetic rubber compound called ethylene propylene diene monomer).

Whenever you need to flash around a single skin metal flue, you need to be aware that the flue may become very hot. Standard EPDM polymer flashings are rated to withstand approximately 115°C. Where you suspect that the flue may become hotter than this, it is

FIGURE 14.17 A twin wall gas flue penetration, soaker flashing and standard EPDM synthetic collar

safer to use an approved high-temperature flashing; these normally have an upper temperature rating of approximately 200°C.

Good plumbing practice should apply to all your penetration flashings:

- Ensure that your polymer flashing does not impede water flow and if necessary install a soaker flashing that enables water to flow freely on either side of the penetration.
- Be aware that the grooved seam on single wall sheet metal flues can draw in water through capillary action. Run a bead of silicone up the length of the grooved seam to prevent water ingress.

Sheet lap anti-capillary drain

Importantly, if your flue penetrates the lap of the roof sheet, a drain notch must be cut into the overlapping sheet immediately upstream of the flashing so that any water within the anti-capillary profile of the sheet is able to drain into the pan without flowing under the flashing and into the building (see Figure 14.18). If this is not clear, ensure you consult your teacher or supervisor to clarify how this is done.

Fan-assisted flue installations

The following steps relate to the installation of a flue system for a room-sealed space heater. The appliance has

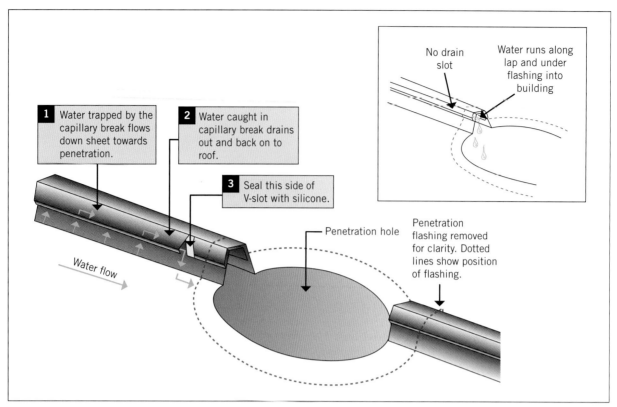

FIGURE 14.18 Cut a small V-slot just upstream of the flue flashing to drain the capillary break into the sheet pan

a fan-assisted co-axial flue that must be installed through a brick veneer wall. These steps serve as a guide only to the processes involved. Actual flue configuration will vary between manufacturers, as will installation instructions. Each installation process will also depend upon wall thickness and external cladding material.

Testing for spillage from flued appliances

Flued appliances must be designed to safely convey potentially hazardous products of combustion away from the appliance through a flue and terminal. In order for these combustion products to be displaced outside to atmosphere, the same amount of air must enter the building. Under favourable conditions, sufficient combustion and flue dilution air is made available through various forms of adventitious ventilation. However, problems can occur when some form of air extraction is used in the same zone as the flued appliance. This mechanical extraction can actually be more powerful than the aeromotive force or fan system in a flue system.

Flue spillage

Driven by modern construction materials and energy efficiency requirements, many buildings are now quite airtight and do not permit a great deal of adventitious air to flow inside. In such conditions, the use of kitchen range hoods, exhaust fans and air transfer systems operating in the proximity of a flued appliance may cause combustion gases to be sucked out of the flue system into the room. Any subsequent rise in carbon monoxide levels may be lethal. Be aware that these same hazards apply equally to caravan and marine craft installations.

To mitigate flue spillage hazards, the Gas Installation Standards details a very specific procedure to test for spillage and enables you to apply a prescribed solution.

REFER TO AS/NZS 5601.1 APPENDIX R 'SPILLAGE TESTS FOR FLUED APPLIANCES'

This test procedure can be broken up into three main sections:
1 Baseline test. Determine if spillage is occurring with neither appliances *nor air extraction fans* operating.
2 Smoke test. Appliances are still not operating. Determine if spillage is occurring with air extraction fans operating at their highest setting and determine whether additional ventilation area is required to prevent flue gases being drawn away from the appliance/s.
3 Spillage test. This test procedure requires the installation must be tested with appliances operating, having first used combustion-detection equipment to confirm background air quality readings.

It is important to note that for your own safety and that of others in the vicinity, you must only undertake flue spillage testing in the strict order detailed in your Standard. To assist you the Standard provides simple flow charts for each stage of this procedure.

HOW TO

INSTALL A FAN-ASSISTED SPACE HEATER FLUE

Step 1	a	Plot location of studs and noggins behind plaster with a 'stud finder'.
	b	The red lights on the stud finder show the location and approximate width of the hidden stud.
	c	Slide the stud finder over the entire area and discretely mark the location of all hidden framework.
Step 2	a	Establish the installation centre line and determine gas pipe penetration point.
Step 3	a	Refer to the manufacturer's instructions and determine the flue penetration position.
	b	This appliance flue outlet can be rotated from its axis point and gives sufficient flexibility to avoid studs.
	c	This zone has been highlighted in the diagram and shows its relationship to the adjacent stud.

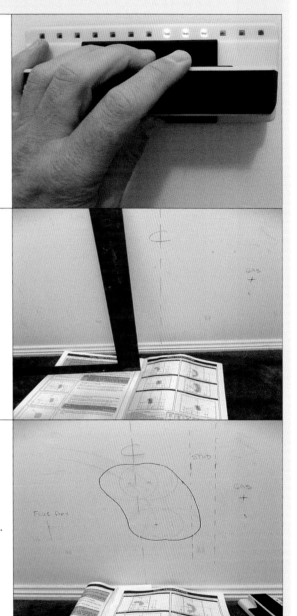

INSTALL TYPE A GAS APPLIANCE FLUES **325 GS**

Step 4	a	Having determined the range of flue movement and its proximity to the wall stud, determine the penetration diameter from the instructions and choose the location.	b	As a hint, any wall plugs to secure the internal flue fitting should be inserted BEFORE cutting out the hole; otherwise you might have instances where the edge of the hole breaks away when the plug is inserted.
	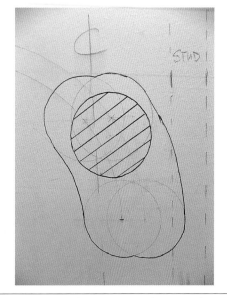		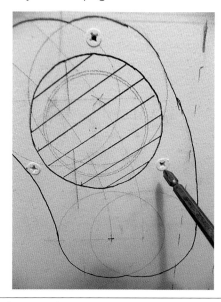	
Step 5	a	Use a plaster jab saw to neatly create the penetration hole.	b	Once you have access, ensure there are no undetected services or obstructing materials in proximity to the penetration.
			c	Carefully move insulation aside and make a minimal cut in building vapour wrap.
	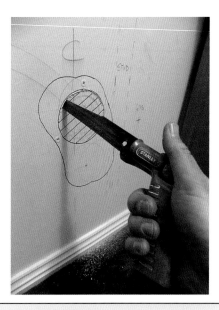		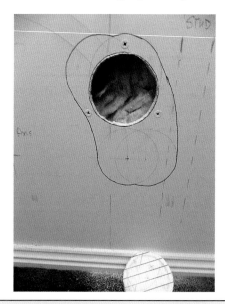	

Step 6	a	While not shown here, a diamond core drill of matching diameter is ideal for drilling a penetration through masonry.	b	Where a correctly sized core drill is not available, use a masonry bit to drill a pilot hole through the bricks to establish centre. Then from the outside, complete a series of holes around the circumference to enable the section to be knocked out with a hammer and chisel.

 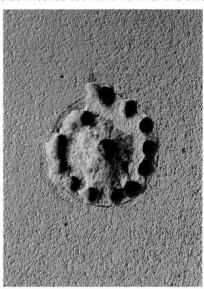

Step 7	a	After cutting the co-axial flue to suit the wall thickness in accordance with manufacturer's instructions, the internal flue assembly is inserted and secured.	b c	The outer flue terminal is then installed. High efficiency appliances will often create condensate in the flue. You will need to ensure the flue co-ax falls the specified amount to ensure any condensate drains to the outside.

Be aware that where required, flue spillage relief venting is in addition to any existing air supply vents normally installed to meet gas appliance air supply requirements. Work through the Standard in partnership with your teacher and/or supervisor to attain a firm understanding of this vital test procedure. Also access this link to test for spillage using CO-detection equipment: www.esv.vic.gov.au >Gas professionals >Technical information sheets >Sheet 38.

Refer to your Standard to gain an understanding of the process.

Safety considerations during test
You must use some form of CO testing equipment during the operational test to establish background readings and ensure that the presence of CO is detected well before it reaches hazardous levels.

Smoke generation options
The spillage test requires an efficient and simple form of smoke generation that creates enough smoke for the test without creating a nuisance within the building. Three main options exist for the gas fitter in this regard.

FIGURE 14.19 A hand-held CO gas detector used to monitor gas levels near the downdraught diverter during a spillage test

- Smoke match – smoke matches are simple to use and generate enough smoke for quick tests of draught diverters and appliance surrounds. However, depending upon the brand, their burn time is only 20-30 seconds and they require that you be present to strike and hold the match in position
- Smoke candles – smoke candles have the advantage that they can be lit and left in a tray adjacent to a draught diverter while you carry out the test. Some can produce a considerable amount of smoke for up to 2–5 minutes so you must be careful that smoke alarms are not set off unnecessarily or that building occupants are not affected
- Incense sticks – simple incense sticks available from chemists and homeware stores are perfectly suitable for most tests. They can be lit and left in position, do not create too much smoke and have a long burn time. These are recommended by the author for most testing requirements.

Regardless of what type of smoke generator you use, always ensure that any combustible surfaces are protected while placing them in position. Also check if any building occupants have any respiratory conditions that may be affected by the presence of smoke and discuss how the test may be carried out without risk to their health.

Clean up the job

Your flue installation task is not over until your work area is cleaned up and tools and equipment accounted for and returned to their required location in a serviceable condition.

Clean up your area

The following points are important considerations when finishing the job:
- If not already covered in your site induction, consult with supervisors in relation to the site refuse requirements.
- Cleaning up as you go not only keeps your work area safe it also lessens the amount of work required at the end of the job.
- Ensure any swarf left around your flue roof penetration is swept up to prevent rusting of the roof surface.
- Dispose of rubbish, flue offcuts and old materials in accordance with site-disposal requirements or your own company policy.
- Ensure all materials that can be recycled are separated from general refuse and placed in the appropriate location for collection. On some sites specific bins may be allocated for metals, timber etc. or if not, you may need to take materials back to your business premises.

Check your tools and equipment

Tools and equipment need to be checked for serviceability to ensure that they are ready for use on the job. On a daily basis and at the end of the job check that:
- all items are working as they should with no obvious damage
- high-wear and consumable items such as cutting discs, drill bits and sealants are checked for current condition and replaced as required
- items such as laser levels and measuring devices are working within calibration requirements
- where any item requires repair or replacement, your supervisor is informed and if necessary mark any items before returning to the company store or vehicles.

Complete all necessary job documentation

Most jobs require some level of documentation to be completed. Avoid putting it off till later as you may forget important points or times.
- Ensure all SWMS and/or JSAs are accounted for and filed as per company policy.
- Complete your timesheet accurately.
- Reconcile any paper-based material receipts and ensure electronic receipts have been accurately recorded against the job site and/or number.

LEARNING TASK 14.6

When cleaning up after a flue installation, you have a number of manufacturer steel flue offcuts left over. How should these be disposed of?

 COMPLETE WORKSHEET 2

SUMMARY

In this unit you have examined the requirements related to the sizing, installation and testing of gas appliance flues including:
- basic combustion principles
- a knowledge of flue types and applications
- job preparation
- specific installation requirements related to flues including terminal locations
- flue installation practice and protection of combustible materials
- flue spillage testing requirements
- clean-up of job and return of equipment.

GET IT RIGHT

The photo below shows an incorrect practice when installing a single skin appliance flue.
Identify the problem and provide reasoning for your answer.

_____
_____
_____
_____

INSTALL TYPE A GAS APPLIANCE FLUES 331 GS

WORKSHEET 1

Student name: _____

Enrolment year: _____

Class code: _____

Competency name/Number: _____

To be completed by teachers
Student competent ☐
Student not yet competent ☐

Task

Review the sections *Overview* through to *Flue installation and terminal locations* and answer the following questions.

1. What three functions does a draught diverter perform?

   i _____

   ii _____

   iii _____

2. If a flue has a vertical height of 6 m measured from the draught diverter to the terminal, what is the maximum permitted length of any lateral section? Provide a reference from AS/NZS 5601 to support your answer.

   _____

   _____

   _____

3. A boiler you are installing inside a plant room requires a natural draught flue. You intend to use single wall galvanised steel of 0.8 mm thickness.

   i What is the maximum flue diameter you can fabricate using this material? Provide a reference from AS/NZS 5601 to support your answer.

   _____

   ii What is the maximum height this flue must not exceed? Provide a reference from AS/NZS 5601 to support your answer.

   _____

   iii Where the flue terminates above the roofline, what is the required clearance between any part of the roof and the end of the flue? Provide a reference from AS/NZS 5601 to support your answer.

   _____

4 What is the minimum clearance between the terminal of a natural draught water heater of 127 MJ/h input and:

   i   an internal return corner of a building? _____

   ii an openable window? _____

   iii beneath an openable window? _____

   iv an external corner? _____

   v   beneath the building eaves? _____

5 What would be the minimum rise required in a lateral measuring 1800 mm long? Provide a reference from AS/NZS 5601 to support your answer.

   _____

6 Where a fan-assisted outdoor appliance is to be installed on a wall, what is the minimum clearance between the flue terminal and any structure in the direction of discharge? Provide a reference from AS/NZS 5601 to support your answer.

   _____

7 A flue with an internal diameter exceeding 150 mm must be at least 50 mm from an unprotected surface: true or false? Provide a reference from AS/NZS 5601 to support your answer.

   _____

8 Part of your flue installation is to pass through a concealed section of a roof space; therefore, you are considering having it fabricated out of stainless steel. Determine the following requirements:

   i   What grade of stainless steel should you specify to your sheet metal workshop?

       _____

   ii What is the maximum diameter permitted in 0.5 mm thick material?

       _____

9 You need to install a natural draught flue from a central heating furnace in the ceiling space beneath a low pitch roof. Flue height should be minimal. What does the AS/NZS 5601.1 state in relation to minimum flue height?

   _____

   _____

10 What is the minimum dimension of any free ventilation opening? Provide a reference from AS/NZS 5601 to support your answer.

   _____

WORKSHEET 2

To be completed by teachers
Student competent ☐
Student not yet competent ☐

Student name: _____

Enrolment year: _____

Class code: _____

Competency name/Number: _____

Task

Review the section *Install and test flue* and answer the following questions.

1. Why does soot need to be cleaned from a former solid fuel chimney prior to a gas appliance installation?

   _____
   _____
   _____

2. Where joints in a flue are subjected to the weather, their sockets should face downwards to prevent rainwater ingress. Why is the internal part of the flue required to have upwards facing sockets?

   _____
   _____
   _____

3. List at least four types of combustible material or products you might find around flue installations.

   _____
   _____
   _____

4. Where the spillage test indicates that relief venting is required, how do you determine the size of the new vent? Provide a reference from AS/NZS 5601 to support your answer.

   _____
   _____
   _____

5. Detail one approved method of protecting a combustible surface from the heat of a flue. Provide a reference from AS/NZS 5601 to support your answer.

   _____
   _____
   _____

6 A space heater is approved for installation in a chimney previously using a fuel other than gas. What are three things you must do before installing the heater? Provide a reference from AS/NZS 5601 to support your answer.

   i _____

   ii _____

   iii _____

7 You have cut a penetration through a roof to fit a new gas appliance flue. The penetration intersects the lap of the high profile roof sheeting. Do you cut in an anti-capillary drain V-slot immediately above, or below, the penetration flashing?

   _____

   _____

   _____

8 Does the flue spillage testing described in Appendix R of your Standard apply to a balanced flue room sealed appliance? Why?

   _____

   _____

   _____

9 In what condition are gas appliances during the baseline test procedure described in Appendix R? Circle the correct answer.

   Appliance burner is ON   Appliance burner is OFF

   _____

10 For how long must a Type 1 decorative gas flame effect appliance be operated before conducting the spillage test?

   _____

   _____

DISCONNECT AND RECONNECT TYPE A GAS APPLIANCES

15

Learning objectives

This chapter covers the disconnection and reconnection of Type A gas appliances operating on natural gas (NG), liquefied petroleum gas (LPG) or tempered liquefied petroleum gas (TLPG) up to a pressure of 200 kPa.

Areas addressed in this chapter include:
- tools and equipment required
- testing of installations
- appliance isolation
- appliance certification and gas type checks
- confirming safe operation
- operational tests.

The work described in this chapter is confined entirely to the disconnection and connection of like-for-like appliances.

Note: This chapter is aimed specifically at plumbing/gas fitting apprentices and requires considerable pre-requisite completion of other gas units and sections within this text coupled with appropriate work-based experience.

This chapter has *not* been designed or written as a stand-alone solution to meet any State or Territory licensing categories that permit other non-gas services trades to undertake such work.

To connect and disconnect gas appliances, you must have a firm grasp of the basic principles of gas and appliance installation. Before working through this section it is very important that you understand the basic fundamentals covered in Part A of this text and the following chapters from Part B:
- Chapter 8 Size consumer piping systems
- Chapter 10 Install gas pressure control equipment
- Chapter 12 Calculate and install natural ventilation for Type A gas appliances
- Chapter 13 Install and commission Type A gas appliances.

Overview

A common job for a gas fitter is the disconnection and reconnection of a gas appliance within an existing installation. Common reasons for this task include:

- replacement of an appliance that is beyond economical repair
  - Example: replacement of a domestic gas cooktop.
- upgrade of an appliance to the latest and often more-efficient model
  - Example: replacing an existing space heater with a new high efficiency model that consumes less gas while providing better heat output.
- temporary removal of an appliance so that renovation or associated work can be carried out on the room or building
  - Removal of a gas range so that new kitchen cupboards and benches can be installed.
- cleaning behind commercial appliances
  - Dangerous amounts of flammable fats and greases can sometimes build up in poorly-designed commercial kitchens and this may require the temporary removal of appliances to efficiently clean the area.
- alteration to associated services including water pipes, electrical supply, flues and mechanical services connections
  - Access and upgrade requirements of associated service connections may require temporary appliance removal.

In each of these instances, you must apply a rigorous and methodical approach to the work to ensure that each stage is completed safely and in accordance with the Standards and the manufacturer's requirements.

Prepare for work

Prior to undertaking any appliance work, some basic planning must be done so that the job may proceed in a safe and efficient manner.

Accessing appliance specifications and manuals

Once you have notification that a job is planned, wherever possible try to ascertain what type and brand of appliances are involved. This will enable you to access relevant appliance specifications and manuals to ensure that the appliance matches the requirements of the job, is approved for the intended use and that appliance certification is in place.

Work, health and safety considerations

As soon as you walk on to a job site or evaluate a set of plans, you should be mindful of potential hazards of all kinds. As a bare minimum you must have on hand basic personal protective equipment (PPE) necessary for most jobs. This includes:

- safety boots
- safety glasses
- hearing protection
- dust masks and/or respirators
- gloves
- overalls or relevant protective clothing
- hard hats.

In addition to the normal WHS considerations that cover all jobs, work on existing installations exposes you to significant electrical hazards that you must be aware of.

Electrical hazards

Electrical faults are an ever-present hazard for gas fitters and it is vital that you apply appropriate risk assessment processes in all your work. SWMS and JSAs should be prepared and evaluated before the commencement of any work. Ensure that you revise the electrical safety section within Chapter 5 'Gas industry workplace safety' before proceeding further.

Appropriate tools and equipment

Apart from general hand tools used on almost all jobs, tools and equipment used for the disconnection and connection of appliances relate principally to appliance handling and testing. Having identified the appliance type and location from a site visit or within plans and specifications, you must make a simple list of what you require prior to commencing the job. Depending upon the type of job, the table below shows just a selection of items that may be required.

	Tools and equipment
Handling and access	• Trolley • Hydraulic lift trolley • Scaffolding • Elevated work platform • Step or extension ladder
Appliance	• Multi-meter • Volt stick • Bonding straps • Power point tester • Soapy water or other leak detection product • Manometer • CO detector • Hydrocarbon gas detector • Measuring equipment • Thermometer (normally K Type plug-in for multi-meter) • Airflow testing instruments • Spirit levels • Portable lighting • Water heater drain hose • Protective floor mats • Power drill with masonry and timber bits • Assorted plugs or caps to isolate gas valves
Alterations to services pipework	• Oxy/Acetylene brazing equipment • Press-fit tool and matching jaws • Flaring tools

Tools and equipment (continued)	
Alterations to services pipework	• Copper pipe cutters and reamer • Benders • Internal and external bending springs • Expanders • Branch formers • Shifters

Test equipment

Regardless of what type of appliance you are working on, or the location in which it is being installed, your test equipment should be on every job and used at all times. The disconnection and reconnection of appliances exposes you to very real electrical and gas hazards; you must also use certain

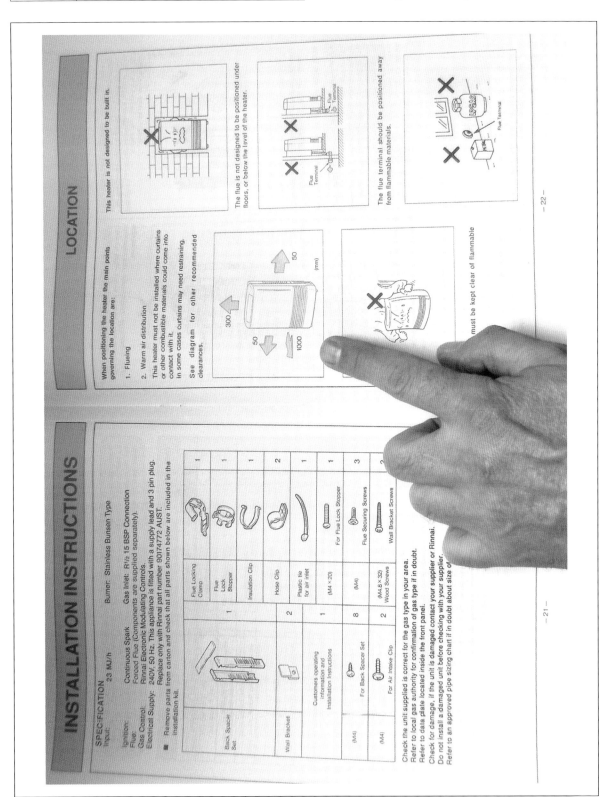

FIGURE 15.1 Confirm appliance installation requirements as soon as possible

instruments to commission the appliance once connected (see Figure 15.2).

FIGURE 15.2 A selection of basic test equipment that should be at hand on every job

Planning with other people associated with the work

Once you have evaluated the basic job requirements, you need to consider who you need to liaise with so that your part in the work can proceed with minimum disruption. Some considerations are outlined in the following table.

Planning stage or activity	Who is affected?
Site access	• Builder • Owner • Manager and other staff
Shut-down time and duration	• Builder • Owner and/or manager
Appliance supply and installation	• Appliance retailer • Transport company • Electrician • Refrigeration mechanic • Other plumbers

Although standard residential work is almost exclusively done during normal business hours, some jobs may restrict shutdown periods to specific times. Busy commercial restaurants commonly operate up to 20 hours per day, 7 days a week and any shutdown must be negotiated with managers and rostered chefs well ahead of time. In many instances you may need to conduct such work between midnight and 4 a.m. just to ensure the restaurant can continue to operate at all other times.

Industrial sites and busy commercial centres also require comprehensive negotiation and you need to list all the people who may be affected by temporary shutdown of gas, water and power supply.

> **LEARNING TASK 15.1 PREPARE FOR WORK**
>
> Give three examples of instance when a gas appliance would be required to be disconnected and reconnected.

Identify appliance requirements

Before any work should commence, you need to confirm that the appliance you are about to work on is compliant in all respects.

Confirm the appliance is safe to touch

Where you are required to work on an existing installation you must always check that the appliance and any associated pipework is electrically safe to touch. Most appliance data plates and stickers are in a location that will require you to move around the appliance or open it up; therefore checking for stray current is vital for your safety.

For basic inspections of installations you must always carry a non-contact voltage detector (volt stick) and apply it to all pipes, metal casings or appliance frames before physically touching the installation.

1. Check your electrical test instrument at a 'known earth'.
2. Ensure your body attitude will reduce potential contact points.
3. Carry out a test on all sections of piping and components, ensuring that it is safe to touch.
4. Test all appliances before touching them.
5. Test and re-test as you move around the job or conduct work.

Confirm appliance gas type

You are required to always check and ensure that an appliance gas type is correct for its intended or current installation.

FIGURE 15.3 Never touch any appliance without checking for stray current first!

FIGURE 15.4 Check that the appliance gas type matches the gas supply

New appliances – Do not assume that a new appliance is always set up to run on the right type of gas.
- Many owners do not understand the difference between gases and may have simply ordered the wrong type.
- Dispatch errors from suppliers may result in the wrong appliance being sent out.
- Some appliances will arrive by default as natural gas with the intent being that the installer convert to LPG with an attached kit as required.

Existing appliances – simply because an appliance may already have been installed does not mean that it is rated for that type of gas.
- Some existing appliances may have been installed illegally by an unlicensed person and can be running on the wrong type of gas.

A Type A gas appliance will have a sticker applied clearly noting the type of gas (this sticker may not be present on older appliances).The gas type is also identified on the appliance data plate.

Confirm appliance certification

Having confirmed that the gas type is correct for the installation, you must also check that the appliance is 'certified'. In years past, a simple sticker was applied to show that an appliance was certified or approved. Although certifying stickers are still used in many instances, it is advisable that you check appliance certification using the relevant identifying number.

Only certified Type A gas appliances can be installed. Review the steps in the following 'How to' section.

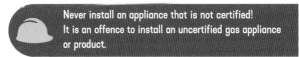
Never install an appliance that is not certified! It is an offence to install an uncertified gas appliance or product.

Confirm general installation compliance

Prior to commencing work you must conduct general compliance inspection of the installation to ensure that you will not be undertaking work on an installation that is fundamentally non-compliant.

Before beginning any quotation or work, walk around the site and work through a checklist of possible problem areas. The following is an example:
- Consumer-billing meter or LPG cylinders are in correct location with required clearances.
- Materials and components are approved.
- Piping and components are sized and installed correctly.
- Appliance data plates are identified.

HOW TO

CONFIRM APPLIANCE CERTIFICATION

Find the appliance data plate.

INSTRUCTIONS. IF OWNER'S MANUAL IS MISSING, CONTACT THE RETAILER OR MANUFACTURER.

CERTIFIED FOR INDOOR INSTALLATION ONLY.

AIRXCEL

DO NOT REMOVE

AUSTRALIA AND NEW ZEALAND GAS SAFETY CERTIFICATION

MODEL NO.	IW60A
STOCK NO.	5187
TYPE OF GAS	UNIVERSAL LPG
ORIFICE SIZE	1.05 & 1.1 MM DMS
MAX. INLET GAS PRESSURE	3.24 kPa
MIN. INLET GAS PRESSURE	2.74 kPa
TEST POINT PRESSURE	1.89 kPa
GAS INPUT	64.0 MJ/h
MIN. GAS INPUT	13.4 MJ/h
MAX. WORKING PRESSURE	1034 kPa
TEST PRESSURE	1200 kPa
POWER INPUT	60W
MAX. WATER TEMPERATURE	50°C

HEAT OUTPUT 17.5 kW
HEATING CAPACITY 7.5 L/min at 28°C RISE

CERTIFICATE NUMBER: AGA8353G
SERIAL NUMBER

IW60A   5187A
182700003Z
MADE IN CHINA   193645

Confirm the appliance certification number.

MAX. INLET GAS PRESSURE	3.24 kPa
MIN. INLET GAS PRESSURE	2.74 kPa
TEST POINT PRESSURE	1.89 kPa
GAS INPUT	64.0 MJ/h
MIN. GAS INPUT	13.4 MJ/h
MAX. WORKING PRESSURE	1034 kPa
TEST PRESSURE	1200 kPa
POWER INPUT	60W
MAX. WATER TEMPERATURE	50°C

HEAT OUTPUT 17.5 kW
HEATING CAPACITY 7.5 L/min at 28°C RISE

CERTIFICATE NUMBER: AGA8353G
SERIAL NUMBER

IW60A   5187A
182700003Z
MADE IN CHINA   193645

Access the website of the Gas Technical Regulators Committee (GTRC): www.gtrc.gov.au.

GTRC Gas Technical Regulators Committee

Home | Gas Industry | About | Gas Regulations | GTRC Links | Publications | Contact

▼ Member Login

AUSTRALIA AND NEW ZEALAND GAS SAFETY CERTIFICATION ™

Do you have an approved appliance?

To ensure safety to the public, all gas appliances sold in Australia and New Zealand are required to be tested and certified.

For information on certification requirements, contact your local Technical Regulator listed under Contacts, or click on the map below.

Learn More

National Database of Certified Gas Appliances and Components

The GTRC National Certification Database provides a listing of gas appliances and components that are or have been previously certified by the certification bodies that are recognised individually by the GTRC members. The GTRC National Certification Database provides all stakeholders with information in relation to certified product as well as the certification status of the product. This information is updated typically on a monthly basis. If in doubt in relation to the validity of the information provided then the recognised certification bodies also have their own certification listings that can be referred to via their respective websites for confirmation or otherwise of the information provided.

- AGA
- SAI Global
- IAPMO
- Global-Mark
- Vipac
- BSI

Source: Courtesy of the Gas Technical Regulators Committee (GTRC)

>> Go to the website link of the certifying entity relevant to the appliance.
In this example we are accessing the Australian Gas Association (AGA) website.

Source: Source: Australian Gas Association (AGA).

PART B

15

Access the database.
Here we have downloaded the Directory of AGA Certified Products

Source: Australian Gas Association (AGA). The Directory of AGA Certified Products can be found at www.aga.asn.au/product_directory

Find the appliance.
Use the contents pages to find the appliance and confirm that the number on the data plate matches the one listed in the directory.

DISCONNECT AND RECONNECT TYPE A GAS APPLIANCES **343 GS**

- Appliances are installed in correct location and with required clearances.
- Power supply points are adjacent to the appliance and accessible
- Flue system is sized correctly.
- Appliance ventilation is sufficient.

LEARNING TASK 15.2 COMPLIANCE CHECK

A compliance check of this existing installation reveals the incorrect use of multilayer pipe (PE/AL/PE). Refer to your Australian Standard and answer the following questions.
1 What is at fault?
2 How could this be rectified?
3 Australian Standard reference/s to support your answer

Where existing and proposed installation locations do not meet job set-out requirements or the Australian Standards, report full details to your supervisor immediately.

Carry out pre-work system pressure test

Prior to disconnection or connection work being carried out on any installation, you are required to conduct a leakage test in accordance with the Standard.

REFER TO AS/NZS 5601.1 APPENDIX E 'E6 LEAKAGE TEST FOR EXISTING INSTALLATIONS'

Leakage test

The requirement for a leakage test is not discretionary. It is a mandatory procedure designed to ensure that the installation is gas-tight prior to any work being carried out. It requires that gas be disconnected and the installation pressurised to operating pressure or 2 kPa (whichever is the greater). Following stabilisation, if there is no loss of pressure over a test period of 5 minutes then you are able to proceed with the job.

Any failure of the leakage test should be reported immediately to your supervisor if the cause is not readily identifiable and repaired.

COMPLETE WORKSHEET 1

LEARNING TASK 15.3

IDENTIFY APPLIANCE REQUIREMENTS

1 Before you start work, how can you confirm an appliance is safe to touch?
2 What is the purpose of a general compliance inspection?
3 At what pressure should a leakage test be carried out and for how long?

Disconnect and reconnect equipment

Plan the disconnection work to occur at a time of least inconvenience to building occupants and other workers. As part of this process you must identify who will be affected during the period the gas is turned off and notify all parties as appropriate.

Where an appliance is to be drained of water, you will need to identify a drain point and a means of removing the appliance without causing any water damage.

Electrical safety check

Once you are ready to start work, your first requirement is to ensure that electrical safety procedures are determined and applied as appropriate.
- Do not touch any part of the installation without carrying out a test for stray current.
- Ensure power is isolated.
- If necessary, have an electrician remove any 'hard-wired' connections and check that earthing and equipotential bonding is adequate.

Isolate gas supply

Where an appliance isolation valve is fitted immediately upstream of the appliance, isolate the gas and have on hand an appropriately sized plug or cap to seal off the valve once the disconnection is carried out. Remember, an isolation valve can be turned on accidentally at any time, so is regarded as an 'open end' unless capped or plugged.

If an appliance isolation valve is not in place (some space heaters and cookers), then you must isolate the supply at the next upstream consumer piping isolation valve, consumer billing meter or LPG supply point.

Disconnect the appliance

Some electrical faults will only become apparent once a break in the pipework is made and an alternative path to earth created. If you are holding on to this pipe when this happens you can be electrocuted.

Before disconnecting the appliance you must provide a path of electrical continuity around the intended break. This is done with a set of 'bonding straps' with each end firmly attached to a point either side of the intended disconnection point. This is not a choice – it is a clear direction in the Standard.

> **REFER TO AS/NZS 5601.1 'ELECTRICAL SAFETY BONDING OR BRIDGING'**

The bonding straps must remain in place until it has been proven safe to remove them or the connection re-made. When removing an appliance and having made the disconnection, you must use a multi-meter to ensure the installation remains safe.

Once the appliance is disconnected, immediately plug or cap the outlet of the isolation valve.

Hose assemblies

For some installations you may be removing an older appliance that was originally installed with a 'hard' connection, such as a fixed compression fitting or barrel union, but the replacement appliance may be suitable for use with a hose assembly. Where approved for use, a  connection is advantageous for both the gas fitter and the owner/operator. These advantages include:

- a flexible form of appliance connection free from vibration
- allowing movement of the appliance by the owner/operator
- when permitted for use in conjunction with the use of a quick-connect device, allows the owner/operator to disconnect and connect the appliance safely.

> **REFER TO AS/NZS 5601.1 'USE OF HOSE ASSEMBLIES'**

Hose assemblies must be considered an extension of the pipe installation itself and, as such, should be sized accordingly. Additionally, you need to ensure that hose assemblies are approved for use in gas installations. All hose assemblies must be manufactured according to AS/NZS 1869 and be of a class matching the supply pressure and temperature requirements (see Table 15.1). You must choose the right hose for the job.

TABLE 15.1 Gas hose assemblies compliant with AS/NZS 1869

Class	Max working pressure at 23°C ± 2°C	Max working temperature
A	7 kPa	−20° to 65°C
B	7 kPa	−20° to 125°C
C	2600 kPa	−20° to 65°C
D	2600 kPa	−20° to 125°C

Additional requirements apply to the use of hose assemblies, and depend upon the appliance type and manufacturer's instructions. If an appliance is not approved for use with a hose assembly then you cannot use one. In addition, you are only permitted to install a hose assembly connection point in certain locations.

LEARNING TASK 15.4
DISCONNECT AND RECONNECT EQUIPMENT

1. Why should an isolation valve be plugged or capped when an appliance is disconnected?
2. Can hose assemblies be used on all gas appliances?

> **COMPLETE WORKSHEET 2**

Test operation of equipment

Installation of an appliance requires a number of tests to be carried out to ensure safe operation.

Gas pressure testing

All gas consumer piping installations must be pressure tested to ensure that they are gas-tight. No leaks. There are no exceptions. It is one of your most important duties as a gas fitter.

Note: Appendix E, 'Testing for gas-tightness' of the Standards, is a *normative* appendix. This means you *must* follow its requirements! Whenever a new standard is introduced, you are obliged to follow the latest procedures. Refer to your teacher if new standards are introduced during the life of this edition.

REFER TO AS/NZS 5601.1 APPENDIX E 'TESTING FOR GAS-TIGHTNESS'

There are four main types of pressure test:
- *pipework test* – for testing consumer piping without appliances connected
- *installation test* – for testing consumer piping with appliances connected
- *leakage test* – for testing the existing consumer piping before starting work
- *final connection test* – for testing any connection after a pressure test has been completed.

Having replaced an existing appliance you are required to carry out an installation test prior to connection to the gas supply to ensure that the appliance connection and the internals of the appliance are gas-tight.

REFER TO AS/NZS 5601.1 APPENDIX E 'E5 INSTALLATION TEST PROCEDURE'

Installation test

- Ensure gas supply remains isolated.
- Check that all appliance isolation valves are open.
- Ensure that the internal burner valves of the appliance are off.
- Attach a manometer to any suitable connection or test point.
- Pressurise the system to operating pressure or 2 kPa, whichever is the greater.
- Isolate the pressure source and ensure no loss of pressure over a period of 5 minutes.

REFER TO AS/NZS 5601.1 'OPERATING PRESSURE'

EXAMPLE 15.1

If your supply pressure is 1.5 kPa out of the meter with a lock-up of 1.6 kPa, then this is deemed to be the *operating pressure*. As this is lower than 2 kPa, you would need to use a hand pump to raise the system pressure to the prescribed amount. If the normal system lock-up pressure is greater than 2 kPa then this is sufficient.

If the pressure drops during this period then you must find the source of the pressure loss. Use your soapy water to find the leak. You will need to look at every joint in the pipework and at each connection point for appliances. Remember that older appliances may also be leaking internally. Having found the leak, repeat the full test procedure from the beginning.

Once you are sure the system has passed the test, remove your equipment and seal the open end.

Apply a final connection test as outlined in AS/NZS 5601.1.

Final connection test

The final connection test is required for any connection made after the removal of test equipment. This normally relates to the connection point that the test gear was connected to. It must be tested at the *operating pressure* (lock-up pressure) of the system. Use soapy water to ensure that the joint is gas-tight.

Once testing is complete, fully seal all open ends and record any test data as required by company or gas regulator protocols.

REFER TO AS/NZS 5601.1 APPENDIX E E7 'TESTING A CONNECTION MADE AFTER A TEST PROCEDURE'

Check all associated appliance connections

Type A gas appliances are often connected to other service connections. These may include:
- cold and heated water connections, e.g. water heaters, hydronic boilers
- water drain connections. e.g. wok tables
- air connections. e.g. furnace flexible or rigid ducting.

Ensure all ancillaries are correctly connected in accordance with the manufacturer's instructions and ensure there are no leaks in the water connections. Where fitted, pressure-limiting valves and balancing valves should be checked for correct operation.

Appliance commissioning

Appliance commissioning is all about bringing an appliance into operational condition. This generally requires a number of procedures and settings to be made to ensure safe and efficient operation.

Purging

As part of the appliance commissioning process, the appliance must not be lit until all air has been purged from the system leaving only pure gas. There are two primary forms of appliance purge procedure as detailed in the Standard:
- purging through an open burner
- purging an appliance fitted with a flame safeguard device:
  - Electronic flame safeguard Thermoelectric.

Purging through an open burner

Purging through an open burner is the easiest procedure of these three and will be covered here in detail. Be aware that few new appliances will now have an open burner but the process is still relevant for appliance purging of existing installations.

HOW TO

1. Turn on one burner gas control until the presence of gas is detected. Put your ear to the burner and listen to the escaping air. As the gas comes through you will hear a change in pitch due to the difference in density between gas and air. Use your sense of smell too, although you will find this is not as reliable as you may think.
2. Let the gas flow for a few seconds longer then turn off and allow sufficient time for any accumulated gas to disperse before proceeding. Be particularly careful when purging LPG, as it is heavier than air and will tend to sit in appliance recesses. Fan with a piece of cardboard if necessary.
3. Turn on one gas control valve and keep a continuously burning flame at the burner until the gas is alight and the flame is stable.
4. Continue to purge until gas is available at each appliance. Do this at each appliance before you adjust the pressure or aeration. You must ensure that all air is out of the system before continuing with the commissioning process.

Purging through an appliance with a flame safeguard device

Because flame safeguard devices prevent the release of any gas from a burner without the presence of a flame, you cannot efficiently purge the appliance from a burner. The only option is to purge at the appliance inlet connection. The amount of air left in the appliance internals is negligible and does not pose any hazard. The procedure is described below.

HOW TO

1. Isolate any electrical supply to the appliance.
2. Fit a bridging device (bonding strap) across the appliance inlet union connection.

3. Slacken the union to allow gas to flow out, but do not fully disconnect it. This procedure maintains a level of controlled gas release.
4. Turn on the manual shut-off valve of the appliance until the presence of gas is detected. Put your ear to the fitting and listen to the escaping air. As the gas comes through you will hear a change in pitch due to the difference in density between gas and air. Use your sense of smell too, although you will find this is not as reliable as you may think.
5. As soon as the presence of gas is detected, tighten the union and test it with soap and water solution.
6. Allow sufficient time for any gas to disperse.
7. Remove the bonding straps.
8. Turn on the power supply and activate the ignition source.
9. Remember ignition may not be successful immediately and lock-out may occur a number of times before combustion is satisfactory.
10. Allow sufficient time for any unburnt gas to disperse before resetting the system.

The procedure for purging an appliance fitted with a thermoelectric safety device is similar to that outlined above, with additional steps needed in order to light the pilot if required.

Test and adjust burner pressure

Some appliances are pre-set in the factory and only require minimum inlet pressures to be checked; however, most Type A appliances still require some form of burner pressure adjustment to be carried out by the gas fitter.

Instrument choice

A water gauge manometer is still perfectly suitable for some appliances, but, increasingly, the sensitivity of pressures required by some manufacturers for the high and low modulating burner settings means that a digital manometer is generally the most suitable instrument for testing.

Procedure

The variety of different appliances means there can be no one simple procedure; however the basic requirements are as follows:

- Confirm burner pressure from appliance data plate.
- Use manufacturer's instructions to identify location of burner pressure test point. This could be:
  - On the gas supply rail to the burner controls
  - Integral with the combination control
  - At the outlet of the appliance regulator.
- Connect test instrument to the test point.
- Identify where the pressure is to be adjusted, e.g. at appliance regulator or combination control.
- Operate the appliance at maximum gas consumption (or in accordance with manufacturer's requirements).
- Observe manometer reading and adjust regulator:
  - Screw in to increase pressure (clockwise)
  - Screw out to decrease pressure (anti-clockwise).
- Once correct pressure is set, turn off burner and remove test instrument.
- Relight burner and soap-test the test point (see Figure 15.5).

Appliance operational settings

Once the burner pressure is confirmed, additional appliance checks will vary considerably according to the type of appliance being installed. Some basic requirements include:

FIGURE 15.5 Remember to check for leaks at test point

FIGURE 15.6 Low flame adjustment is an important step in commissioning a cooktop

- *aeration*. Most modern appliances have little or no adjustment to primary aeration settings but, where applicable, you must follow the instructions and adjust the aeration to achieve the best flame characteristics.
  - Flame must be entirely blue
  - There are no yellow tips
  - There is no flame lift-off
  - There is no streaming flame
  - Flame should readily ignite.
- *low flame adjustments*. Most cooking appliances will require some form of low flame adjustment. This requires adjustment of rail-cocks to ensure that the full range of flame height can be achieved between the high and low settings. This procedure will vary somewhat between manufacturers, so you must follow the specifications for each appliance as required (see Figure 15.6).
- *pilot lights*. Where a pilot light is used, ensure the pilot flame is impinging correctly on any thermocouple and remains stable when the main burner flame cuts in and out.
- *fan operation*. Check that convection fans operate on all power settings.
- *flame failure operation*. Confirm that thermoelectric flame failure systems shut down in the expected time period after turning off the flame and that rectification systems operate according to appliance operational requirements.
- *set thermostats*. Depending upon the type of appliance, set the thermostats according to the operational needs of the appliance or the customer requirements. Where necessary, confirm satisfactory modulation of burner flames in response to thermostat settings.

Instruct customer on operation

Once the appliance is running satisfactorily, you are required by the Standard to instruct the owner on the appliance operation. Where the owner or user is absent, the appliance instructions must be left in an obvious location. It is advisable to make a follow-up call and, if necessary, call in to ensure the owner is fully confident with the appliance.

LEARNING TASK 15.5
TEST OPERATION OF EQUIPMENT

1. What test is required before starting work on existing consumer piping?
2. What is the purpose of testing and adjusting burner pressure on an appliance and what instrument would be used to achieve this?
3. What must be observed when checking the operational appliance settings for an appliance with a pilot light?
4. What would you do if you were unable to instruct the customer on the safe use of the gas appliance after installation?

Clean up the job

Appliance disconnection and connection can result in considerable mess, materials management and disposal needs.

Clean up your area

The following points are important considerations when finishing the job:

- Cleaning up as you go not only keeps your work area safe it also lessens the amount of work required at the end of the job.
- Dispose of rubbish, off-cuts and old materials in accordance with site-disposal requirements or your own company policy.
- Ensure all materials that can be recycled are separated from general refuse and placed in the appropriate location for collection. On some sites specific bins may be allocated for metals, timber etc. or if not, you may need to take materials back to your business premises.
- Older appliances will generally require disposal at an approved waste transfer station or metal recycling centre.

Check your tools and equipment

Tools and equipment need to be checked for serviceability to ensure that they are ready for use on the job. On a daily basis and at the end of the job check that:

- all items are working as they should with no obvious damage
- high-wear and consumable items such as cutting discs, drill bits and sealants are checked for current condition and replaced as required
- items such as laser levels, manometers and gas analysers are working within calibration requirements
- where any item requires repair or replacement, your supervisor is informed and, if necessary, mark any items before returning to the company store or vehicles.

Complete all necessary job documentation

Most jobs require some level of documentation to be completed. Avoid putting it off till later as you may forget important points or times.

- Ensure all SWMS and/or JSAs are accounted for and filed as per company policy.
- Complete your timesheet accurately.
- Complete any gas notice paperwork and send copies to the technical regulator and/or gas supplier as required.

 COMPLETE WORKSHEET 2

SUMMARY

In this unit you have examined the requirements related to the disconnection and connection of Type A gas appliances including:
- electrical safety requirements
- appliance certification checks
- general and specific compliance requirements
- disconnection procedures
- connection procedures
- pressure testing
- commissioning.

GET IT RIGHT

The photo below shows an incorrect practice when connecting an appliance using a compression joint. Identify the incorrect method and provide reasoning for your answer.

_____
_____
_____
_____

 **WORKSHEET 1**

To be completed by teachers
Student competent ☐
Student not yet competent ☐

Student name: _____

Enrolment year: _____

Class code: _____

Competency name/Number: _____

Task

Review the sections *Prepare for work* and *Carry out pre-work system pressure test* and answer the following questions.

1. What type of gas pressure test are you required to perform prior to any work being carried out? Provide a reference from AS/NZS 5601 to support your answer.

   _____

   _____

   _____

2. Identify at least four items of PPE that may be required for work involving the disconnection of appliances.

   _____

3. What must you test for on any gas installation before touching pipework or appliances?

   _____

   _____

   _____

4. Where would you find the appliance gas type displayed on an appliance?

   _____

5. Which website would you access in order to confirm appliance certification?

   _____

   _____

   _____

6. List at least four things you would look for when conducting a general installation compliance check prior to work commencing.

   _____

   _____

   _____

WORKSHEET 2

Student name: _____

Enrolment year: _____

Class code: _____

Competency name/Number: _____

To be completed by teachers

Student competent ☐

Student not yet competent ☐

Task

Review the sections *Disconnect and reconnect equipment* and *Test operation of equipment* and answer the following questions.

1. During an installation test with all appliances connected, should manual shut-off valves be left open or closed? Why?

   _____

   _____

   _____

2. Is the temperature stabilisation period part of the five-minute test or in addition to it?

   _____

3. Having successfully completed an installation test, you remove your test instrument from the connection point. What do you need to do now? Provide a reference from AS/NZS 5601 to support your answer.

   _____

   _____

   _____

4. In relation to test procedures, what is the definition of the term 'operating pressure'?

   _____

   _____

   _____

5. Why must bonding straps be used when purging through the appliance inlet union?

   _____

   _____

   _____

6   Name two methods of detecting the presence of gas during a purge.

   _____

   _____

7   Why is appliance purging a required part of the commissioning procedure?

   _____

   _____

   _____

8   Where an appliance is fitted with a thermoelectric flame failure device, at what location do you carry out the pipework purge?

   _____

   _____

9   What is the maximum working pressure of a Type D hose and what Australian Standard must it comply with?

   _____

   _____

10 Having fully commissioned the appliance, what is the final thing you must do before leaving the completed installation?

   _____

   _____

INSTALL LPG STORAGE OF AGGREGATE STORAGE CAPACITY UP TO 500 LITRES

Learning objectives

This chapter details some of the basic skills and knowledge required to plan for and install LPG systems with a storage capacity up to 500 L. The primary focus is on the cylinder site itself and the associated connections.

Areas addressed in this chapter include:
- calculating vaporisation requirements using the 'DLK' formula
- cylinder site installation requirements
- how to install LPG cylinders
- pressure testing of LPG cylinder connection systems.

Before starting this chapter, you must ensure that you have worked through all of Part A of this text, with particular review of Chapter 7 'LPG basics' so that you have a good grasp of LPG fundamentals before you start.

It is also assumed that you have completed the following chapters from Part B:
- Chapter 8 Size consumer piping systems
- Chapter 9 Install gas piping systems
- Chapter 10 Install gas pressure control equipment.

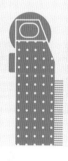

Overview

This section will provide you with specific guidelines relating to the site and installation requirements of LPG systems up to 500 L capacity. To work through this section successfully, you will need access to the latest version of AS/NZS 1596 'The Storage and Handling of LP Gas'. This is the normative standard that should be referenced for all LPG cylinder installations up to the outlet of the first regulator. All aspects of an LPG installation downstream of the first regulator should reference AS/NZS 5601 'Gas Installations'.

REFER TO AS/NZS 5601.1 'GAS INSTALLATIONS' APPENDIX J AND AS/NZS 1596 'THE STORAGE AND HANDLING OF LP GAS'

As you work through this section, you will also need to undertake some basic LPG calculations as required. Refer to Table 16.1 for the key facts.

TABLE 16.1 Propane key facts (some figures are subject to test conditions, temperatures and rounding)

Boiling point	−42°C
MJ per kg	49.6
MJ per litre	25
Litres per kg	1.96
Relative density	1.55
Heat value MJ (commercial)	96
Flammability range	2–10 per cent gas in air

Cylinder basics

Cylinders chosen for each LPG installation will depend upon a number of factors, including consumption, vaporisation requirements, portability, tanker-truck access and the site's physical characteristics.

Cylinders are generally identified according to both litres of water capacity and their capacity in kilograms of propane.

Water capacity (WC) is defined in the standard as the total volume of space enclosed within the cylinder expressed in either litres (L) or kilolitres (kL).

REFER TO AS/NZS 1596 'DEFINITIONS': 'CAPACITY'

Water capacity is also still commonly sized by the weight of water needed to totally fill the vessel. Therefore, a cylinder with a WC of 108 L may also be referred to as having a WC of 108 kg because a litre of water weighs approximately 1 kg.

Although most gas fitters in the field will refer to cylinders in terms of their capacity in kg of propane, your Standards will refer to cylinder sizes according to their WC in litres or kilolitres. The WC of any cylinder will always be stamped into the valve guard on top of the cylinder itself. Cylinders can be divided into three main categories (see Figure 16.1):

- *Consumer-owned cylinders.* These include any size up to 9 kg. For small installations, a customer may prefer to use their own cylinder rather than pay the

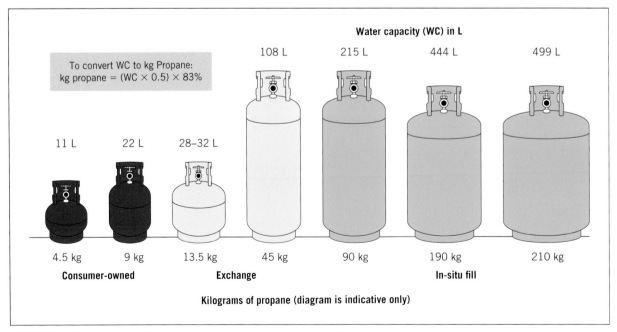

FIGURE 16.1 The standard cylinder sizing in both water capacity and kilograms of propane

rental fee for a gas supplier's cylinders. Although the gas cost is considerably higher, the low consumption rate may mean that overall costs are reduced. Consumer-owned cylinders are also most commonly found in caravans and marine craft.
- *Exchange cylinders.* When a consumer empties an exchange cylinder, the gas supplier will deliver a replacement full cylinder to the site and exchange it for the empty cylinder. Exchange cylinders are 13.5 kg and 45 kg in propane capacity.
- *In-situ fill cylinders.* These cylinders always remain on the site and are filled on a regular basis by a tanker-truck. The driver pulls out a long hose fitted with a special fill-gun and tops up the cylinder either on demand or as part of a contract schedule. In-situ fill cylinders are normally 90 kg, 190 kg or 210 kg. Where the customer base is large enough, some gas suppliers also install 45-kg in-situ fill cylinders. In-situ fill cylinders are normally required where gas consumption is consistently high, such as for central heating and in commercial catering installations.

Cylinder testing
All standard cylinders have a service life of 10 years. At that point, the gas supplier must inspect the vessel; if it is in good order, it will be recommissioned and stamped with a new expiry date. Exchange cylinders are regularly checked by the supplier as they are taken back to the depot for filling. In-situ fill cylinders are removed from the site after 10 years for testing. It is an offence for any person to fill any LPG cylinder after the stamped expiry date.

Vapour and liquid withdrawal
A gas fitter's normal range of work is entirely based upon the connection of gas systems to a vapour-withdrawal valve of the cylinder. However, you must be aware that 190-kg cylinders, 210-kg cylinders and most tanks are fitted with liquid-withdrawal valves as well. This is normally to provide a decanting facility so that smaller, consumer-owned cylinders can be refilled.

These liquid valves should be clearly marked and fitted with a security plug, but you must always double check that you are not connecting your pigtail to the wrong valve. Some 45-kg cylinders are also fitted with liquid-withdrawal valves for special decanting purposes.

Volume of a cylinder
You cannot determine the volume of gas within a standard LPG cylinder by reading a pressure gauge in the same way you are able to do with oxygen or acetylene cylinders. A full and an almost empty cylinder exposed to the same temperature would have the same pressure reading. Therefore, forklift cylinders and standard cylinders of 90 kg and above are fitted with special volume gauges that feature a mechanical float system that directly indicates the liquid level, read as a percentage of total water capacity. Figure 16.2 shows a gauge assembly from a forklift cylinder and Figure 16.3, a cross-section of the mechanism.

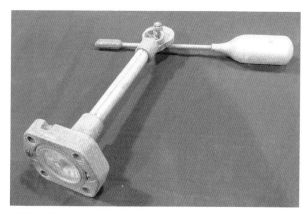

FIGURE 16.2 A gauge assembly from a forklift cylinder

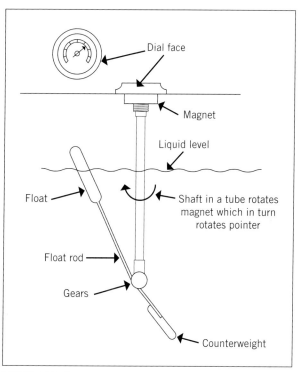

FIGURE 16.3 Cross-section of a mechanical float valve

Identify LPG storage requirements
The amount of gas vapour available from LPG cylinders depends upon two key factors:
- the wetted surface area of the cylinder (i.e. the steel in contact with the liquid propane)
- the ambient temperature.

Variations in ambient temperature and wetted surface area can have a dramatic effect on the ability of an installation to operate effectively.

In most climates encountered in Australia, the average domestic LPG installation of a space heater, cooker and hot water system can be run from a standard 2 × 45-kg cylinder site. Any decision to increase the number of gas cylinders in this case is more to do with delivery frequency than the vaporisation capacity of the cylinders.

However, in colder areas or where commercial or industrial installations have a high vapour draw-off, correct cylinder and tank sizing becomes very important. If the liquid is unable to vaporise into gas fast enough, the working pressure of the appliances may drop, resulting in incomplete combustion. In these cases, you will need to be able to calculate the amount of LPG vaporisation.

For an extreme example see the photograph in Figure 16.4, which was taken at an Australian Antarctic base. Winter temperatures in this area regularly fall to –20°C or colder. Such low temperatures act to severely limit the capacity of each cylinder to vaporise sufficient gas. To operate a standard commercial kitchen, the multiple cylinder manifold shown in the picture provides the necessary aggregate wetted surface area needed to ensure enough gas can be vaporised to run the cooking appliances. In mainland Australia, only two to four cylinders would be required to do the same job.

FIGURE 16.4 A multiple cylinder manifold in Antarctica

The next section will show you how to apply a simple formula to determine the vaporisation capacity of any proposed cylinder or tank installation at specific temperatures.

Vaporisation calculation – the DLK formula

The Gas Installation Standards provides basic vaporisation tables using specific assumptions for temperature, amount of LPG and load.

REFER TO AS/NZS 5601.1 APPENDIX J8

Where your requirements fall outside the scope of the tables provided in the Standard, you are still able to calculate the vaporisation capacity by applying what is known as the DLK formula. See Table 16.2:

Vaporisation capacity (MJ/h) = $D \times L \times K$

where:

D = diameter of the storage vessel (m)
L = length of the storage vessel (m)
K = a pre-calculated factor representing the relationship between temperature, volume and wetted surface area.

Determining DLK calculation percentage

You will note from Table 16.2 that you have a range of options related to the amount of LPG in the vessel expressed as a percentage of full. These options relate to the range of winter temperatures normally found in your area. The warmer the area the greater the vaporisation capacity of a vessel at lower percentages. While in some areas of Australia a 10% capacity is used, in other colder areas a higher figure such as 30% is often applied, so that the sizing solution will accommodate the reduced vaporisation characteristic of colder ambient temperatures. You will need to consult the local LPG supplier and find out what percentage should be applied in the area.

TABLE 16.2 Vaporisation formulas and temperature multipliers

Per cent of full	$D \times L \times K$ factor	Temperature multiplier								
		–18°	–15°	–12°	–9°	–7°	–1°	4°	10°	16°
60	$D \times L \times 163$	× 1.00	× 1.25	× 1.50	× 1.75	× 2.00	× 2.50	× 3.00	× 3.50	× 4.00
50	$D \times L \times 147$	× 1.00	× 1.25	× 1.50	× 1.75	× 2.00	× 2.50	× 3.00	× 3.50	× 4.00
40	$D \times L \times 130$	× 1.00	× 1.25	× 1.50	× 1.75	× 2.00	× 2.50	× 3.00	× 3.50	× 4.00
30	$D \times L \times 114$	× 1.00	× 1.25	× 1.50	× 1.75	× 2.00	× 2.50	× 3.00	× 3.50	× 4.00
20	$D \times L \times 98$	× 1.00	× 1.25	× 1.50	× 1.75	× 2.00	× 2.50	× 3.00	× 3.50	× 4.00
10	$D \times L \times 73$	× 1.00	× 1.25	× 1.50	× 1.75	× 2.00	× 2.50	× 3.00	× 3.50	× 4.00

EXAMPLE 16.1

A tank measures 3.92 m long × 1.22 m wide. To find the worst-case vaporisation capacity of the tank in MJ/h, you follow the left-hand column down to 10 per cent of full, then apply the formula on that row from the 'D × L × K factor' column:

$$\text{Vaporisation} = D \times L \times 73$$
$$= 1.22 \times 3.92 \times 73$$
$$= 349.1192$$

You then need to apply the temperature multiplier relevant to normal winter temperatures in your area. In this example, we will use 4°C, which has a multiplier of 3.00.

$$\text{Vaporisation} = 349.1192 \times 3.00$$
$$= 1047.3456 \text{ MJ/h}$$

In summary then, this installation can be expected to be able to vaporise approximately 1047 MJ/h at an ambient temperature of 4°C and 10 per cent full.

Although cylinders and tanks can be designed and ordered to suit different applications, they are normally supplied in standard sizes. As a general guide, Table 16.3 provides some indicative sizes that you can apply to your DLK formula.

TABLE 16.3 Standard cylinder and tank sizes

45 kg	1.115 high × 0.374 wide
190 kg	1.207 high × 0.776 wide
2.75 kL	3.330 long × 1.065 wide
4.2 kL	3.920 long × 1.220 wide
7.5 kL	6.620 long × 1.220 wide

You should be aware that such a calculation is a guide only and you need to consider other factors that may affect your choice of storage volume. These include:
- possible use of multiple smaller cylinders/tanks to achieve the same vaporisation rate
- gas company purchase/rental/delivery costs
- probable simultaneous demand.

In most cases, where installations call for anything more than 500 L of stored LPG, a site licence may be required, and you will need to work closely with the supplier on a solution. Regardless, your ability to calculate accurate vaporisation rates will give you confidence in your installation and demonstrate your professional skill to supplier and client alike.

FROM EXPERIENCE

A key attitude that you need to develop as a gas fitter is the preparedness to consult regularly and in detail with gas suppliers and technical regulators. Time spent consulting is better than time spent rectifying!

Calculating cylinder storage requirements

When determining the storage volume of cylinders for a particular site, you must be aware of a number of basic considerations to ensure you deliver the best solution for your customer:
- Too little storage and the customer will run low on gas too quickly.
- Too much storage and the customer may be paying unnecessary cylinder rental or may exceed site volume licence limits.
- Storage should be sufficient to ensure that enough gas is available to supply the customer between the scheduled visits of exchange cylinder deliveries or tanker refill cycles.
- Storage should be sized to ensure that vaporisation is sufficient for the probable demand of the system at the lowest expected temperatures.

The average small domestic installation of a cooker and a storage water heater is generally well served with a simple 2 × 45-kg cylinder installation. However, larger domestic, commercial and industrial installations will require more detailed planning.

Being mindful of these basic considerations, you can move on to the actual calculation of storage requirements.

How long will my gas last?

Customers will regularly ask how long their gas will last before an exchange cylinder delivery or refill is required. Unfortunately, there is no easy answer because customer usage patterns differ, and many appliances do not always run at the full gas rate. This variability makes it difficult to provide a definitive answer. However, by using your knowledge of basic propane characteristics you will be able to give an approximate answer. These calculations will determine how long the gas will last and whether the stored volume can vaporise enough gas on demand.

To answer both of these questions you need to determine three things:
- gross potential draw-off in MJ
- probable daily consumption
- probable simultaneous demand in MJ/h.

The volume of LPG vessels is often quoted in both litres and kilograms. However, most gas fitters are more familiar with the use of kilograms in relation to LPG volumes, so we will use this base unit for each calculation.

FROM EXPERIENCE

Gas consumption questions are very common. It is good business practice to have some generic consumption solutions already worked out, so you can provide clear and concise advice when required. Customers who are considering the switch to gas need to have confidence in your professional knowledge.

LEARNING TASK 16.1

1. What is the gross potential draw-off for the following cylinders?
   - 9 kg
   - 45 kg
   - 210 kg
2. How many litres are contained within 1 kg of LPG?
3. What are the three main categories of gas cylinders?

Gross potential draw-off

The gross potential draw-off represents the maximum amount of gas that could be used (drawn-off) from the supply cylinders/tanks. This calculation simply converts the total volume of propane in kg to MJ. You will see from Table 16.1 that 1 kg of propane is equivalent to approximately 49.6 MJ. For these calculations, we will round this up to 50 MJ/kg. Therefore:

Gross potential draw-off in MJ = Capacity of vessel in kg × 50 MJ

The next question you need to answer is how long a given supply will last based on probable daily consumption.

Probable daily consumption

Gas consumption depends upon how long the customer uses each appliance and the operational characteristics of the appliance itself. While some commercial and industrial installations may have predictable consumption patterns, most general appliances do not run at the full rate all the time and will either modulate automatically or be controlled directly by the operator.

For instance, a four-burner cooktop may have a total input of 30 MJ/h, but it is unlikely that all burners will be used when cooking. Most space heaters will also modulate the burner rate and turn themselves off and on in response to room temperature changes.

You may also be asked to determine consumption for a boiler that operates at set times each day at full burner rate. In this instance, your calculations are quite simple.

To work out probable daily consumption, you first need to establish the input of each appliance and then make a judgement on how long each appliance will be used each day and at what rate. You then divide this total daily consumption figure into the gross probable draw-off capacity to establish how many days a gas supply will last. Using Examples 16.2–16.4, we now work out how many days each supply will last.

Probable simultaneous demand and vaporisation capacity

Wherever you need to calculate the vaporisation capability of an installation, you must do so according to the probable simultaneous demand. Using total rated input is fine for fixed consumption appliances and processes, but if you base your calculations on the total appliance input where demand is variable, then you are likely to end up with an oversized storage capacity.

EXAMPLE 16.2

A 4 × 45-kg cylinder manifold installation supplies gas to a domestic central heater. The system is fitted with an automatic change-over regulator and draws off only two cylinders at once.

The heater has an input of 90 MJ/h and is connected to a 4 × 45-kg cylinder manifold (a two-cylinder draw-off). The customer tells you that they expect to use the appliance between 6–8 am and 6–10 pm: a total of six hours per day. As it is a well-insulated house, you judge that the appliance will operate at full rate for half that time and will be off for the remainder.

a Determine the gross potential draw-off.

Gross potential draw-off in MJ = (2 × 45 kg) × 50 MJ
= 90 kg × 50 MJ
= 4500 MJ

b Give an estimate of how long the gas supply will last.

Probable daily consumption = hours × MJ/h
= 3 hours × 90 MJ/h
= 270 MJ/day

$$\text{Probable supply duration} = \frac{\text{Gross potential draw-off in MJ}}{\text{Probable daily consumption (MJ/day)}}$$

$$= \frac{4500 \text{ MJ}}{270 \text{ MJ/day}}$$

= 16.6 days

At this consistent consumption rate, the automatic change-over regulator will swap to the other two cylinders after about 16–17 days, and the owners will need to arrange a cylinder exchange visit from their gas supplier.

EXAMPLE 16.3

A motorhome is fitted with two 4-kg cylinders. The regulator is fitted with a manual change-over valve and therefore only one cylinder can be used at once.

a Determine the gross potential draw-off.

$$\text{Gross potential draw-off in MJ} = 4 \text{ kg} \times 50 \text{ MJ}$$
$$= 200 \text{ MJ}$$

The motorhome is fitted with a 28 MJ/h cooktop and a 12 MJ/h storage water heater. The customer informs you that they would only use one 7 MJ/h cooktop burner for 30 minutes per day (equivalent 3.5 MJ/h) and that they turn on the small storage water heater for only one hour per day. You assume that the water heater is working at full rate for the whole hour. The installation draws off only one 4-kg cylinder at a time.

b Give an estimate of how long the gas supply will last.

$$\text{Probable daily consumption} = \text{hours} \times \text{MJ/h}$$
$$= 3.5 \text{ MJ/h} + 12 \text{ MJ/h}$$
$$= 15.5 \text{ MJ/h}$$
$$= 15.5 \text{ MJ/day}$$

$$\text{Probable supply duration} = \frac{\text{Gross potential draw-off in MJ}}{\text{Probable daily consumption (MJ/day)}}$$
$$= \frac{200 \text{ MJ}}{15.5 \text{ MJ/day}}$$
$$= 12.9 \text{ days}$$

Therefore, the owners of the motorhome would need to change over the cylinder after approximately 12–13 days.

EXAMPLE 16.4

Determine the gross potential draw-off for a two-tonne tank.

$$\text{Gross potential draw-off in MJ} = 2000 \text{ kg} \times 50 \text{ MJ}$$
$$= 100\,000 \text{ MJ}$$

A Type B industrial appliance with a 360 MJ/h input generally runs at full rate for eight hours per day and at half rate for four hours per day. How often must the two-tonne tank be topped up?

$$\text{Probable daily consumption 1} = \text{hours} \times \text{MJ/h}$$
$$= 8 \text{ hours} \times 360 \text{ MJ/h}$$
$$= 2880 \text{ MJ/day}$$
$$\text{Probable daily consumption 2} = \text{hours} \times \text{MJ/h}$$
$$= 4 \text{ hours} \times 180 \text{ MJ/h}$$
$$= 720 \text{ MJ/day}$$
$$\text{Total probably daily consumption} = 2880 + 720$$
$$= 3600 \text{ MJ/day}$$

$$\text{Probable supply duration} = \frac{\text{Gross potential draw-off in MJ}}{\text{Probable daily consumption (MJ/day)}}$$
$$= \frac{100\,000 \text{ MJ}}{3600 \text{ MJ/day}}$$
$$= 27.7 \text{ days}$$

In fact, a gas supplier would always ensure that a customer's tank is topped up well before the maximum duration is reached, but this calculation will ensure that the existing tank is sized adequately for stored volume.

Obviously, many variables may affect these calculations and you must ensure that the customer understands they are only a guide to their expected gas consumption.

EXAMPLE 16.5

You intend to install a 160 MJ/h continuous flow water heater and a 52 MJ/h upright cooker in an area that normally experiences 10°C average winter temperatures. In warmer parts of Australia, this would easily be serviced by a simple 2 × 45-kg cylinder installation. If you consider the probable simultaneous demand, it is quite likely that for some time in the evening and in the morning the water heater would be working at full rate and the cooker at about 30 MJ/h: a total of 190 MJ/h. Is the draw-off from one 45-kg cylinder enough?

If you revise the section on LPG vaporisation and apply the DLK formula, you can see that a 45-kg cylinder will vaporise approximately 106.54 MJ/h at 10 per cent full and 10°C. This would be insufficient for such an installation. Drawing off two cylinders at once would vaporise approximately 213 MJ/h. This means that you would need to install a 4 × 45-kg cylinder manifold or perhaps two 90-kg in-situ fill cylinders.

COMPLETE WORKSHEET 2

LEARNING TASK 16.2
IDENTIFY LPG STORAGE REQUIREMENTS

1. What are the two key factors for gas vapour availability for an LPG cylinder?
2. Why would a higher cylinder percentage figure be applied when using the DLK formula?
3. What must be considered when calculating cylinder storage requirements?
4. What is the difference between 'gross potential draw-off' and 'probable simultaneous demand'?

Cylinder locations and installation requirements

The requirements for cylinder locations are set out in both AS/NZS 5601.1 and AS/NZS 1596, and provide the gas fitter with all required clearances and limitations. However, the cylinder site recommendations within the AS/NZS 5601.1 are classed only as 'informative' and therefore the AS/NZS 1596 will be used here as the primary reference document.

You should familiarise yourself with the scope of this Standard and in particular its relationship demarcation with AS/NZS 5601.1

REFER TO AS/NZS 1596 'SCOPE AND GENERAL' AND FIGURE 1.1(E) REGULATOR DETAIL

The three types of cylinder installation we will look at in this section are:
1. twin cylinder exchange installations
2. in-situ fill installations
3. manifold installations.

REFER TO AS/NZS 1596 'CYLINDERS AND CYLINDER SYSTEMS'

General cylinder site requirements

Cylinder installations have many requirements, and it takes considerable practice and experience to confidently walk onto a job and choose a site quickly. In the meantime, you must consult the Standard at every opportunity.

Before you attend to required clearances and exclusion zones, here is a list of some prohibited locations for any cylinder site:

- Within a building – you cannot install any sized cylinder inside a building unless in an enclosure that is approved in AS/NZS 1596.
- Under a stairway – this often looks like a handy spot to locate some cylinders but it is explicitly not permitted. Stairways provide means of escape during house fires and cannot be compromised by flammable gas storage that may vent and prevent escape.
- Under a building – a cylinder cannot be installed under any building apart from the maximum permitted allowance past the building line of a house built up on stumps or piers.

There are a number of other prohibited zones that you will need to refer to within the Standard.

Cylinders are permitted for installation on a veranda but with very specific restrictions relating to the location of the veranda, cross-ventilation and maximum capacity of the cylinders.

REFER TO AS/NZS 1596 'CYLINDERS ON A VERANDA'

Where any cylinder site is exposed to possible damage from moving vehicles, you must provide protection in the form of steel bollards or guard railings in accordance with AS/NZS 1596. For domestic installations, the Standard advises using your discretion; however, it is strongly recommended you provide protection wherever vehicle movement is likely to result in cylinder impact.

Twin-cylinder installations – exchange cylinders

One of the most common LPG installations is the standard twin-cylinder installation (see Figure 16.5).

- The site must be free from any flammable material within 1 m of the cylinder.
- The base for an exchange cylinder installation must be made from a non-combustible material, and must be firm and level. It is permitted to sit the cylinders directly on a non-combustible surface (concrete) as long as it is well drained. Where such a surface is not available, you will need to provide a pre-cast concrete pad/s to form a base at least 900 mm × 450 mm × 50 mm thick.
- Exchange cylinders should be secured in such a way that they cannot be knocked out of position. A chain looped around the valve guard and secured to the wall is the most common method.
- The regulator must be fixed to an independent and rigid support.
- The regulator must be located so that any liquid that may form in the copper or flexible pigtail is able to drain freely back into the cylinder. This means that the pigtail connection to the regulator and the pigtails themselves must be higher than the cylinder valve. At the time of writing, AS/NZS 1596 does not require the actual regulator outlet to be higher than the cylinder valve.

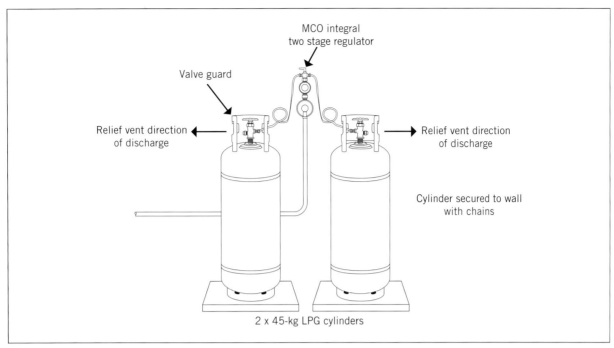

FIGURE 16.5 A standard twin-cylinder installation. Locate the regulator to allow free drainage of any liquid back to the cylinder valves

In-situ fill cylinder sites

When planning for an in-situ fill cylinder installation, there are some additional requirements you must be aware of:
- In-situ fill cylinders must be supported on a base at least 50 mm higher than the surrounding ground level. This base is normally a pre-fabricated concrete slab of a size suitable for the cylinder diameter.
- Cylinders of less than 200 L water capacity must be restrained and prevented from falling.

REFER TO AS/NZS 1596 'ADDITIONAL REQUIREMENTS FOR CYLINDERS FOR IN SITU FILLING'

One additional consideration is the proximity of the site for the filling process. You will need to confirm these details with your local supplier. Generally, the following will apply:
- The site must be no greater than 30 m from where the tanker truck is parked. This relates to the common length of fill hose fitted to most tankers.
- Tankers must have street access acceptable to the LPG supplier including road width, vehicle turn-around dimensions and road incline.
- The person filling the cylinder may have to have a direct line of sight between the cylinder location and the truck itself. This is particularly the case for single person operation.

Do not assume an in-situ fill cylinder site is appropriate without having checked with the supplier first.

Hinged protective hoods

In some localities, the technical regulator will require you to install protective hoods over the regulator assembly and valves of both cylinders (see Figure 16.6). Where this is the case, ensure the relief vents can freely discharge away from the cylinders and the building if required.

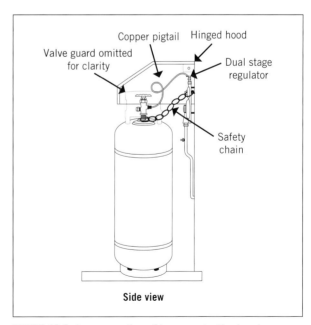

FIGURE 16.6 A cross-section of how a protective hood assembly should be installed

Some technical regulators do not require any form of protective hood; however, you must still ensure that the regulator vent is directed downwards and that the installation is generally protected.

Cylinder clearances and exclusion zones
Clearances from openings into buildings and caravans

All cylinders must be situated so that a minimum distance of 1 m is maintained between the side of the cylinder and any opening into a building.

Openings into buildings include:
- doors
- openable windows
- vents of any description

Although AS/NZS 1596 states that weep holes in brickwork are not deemed to be an opening into a building, you should still refer to your local LPG supplier and check if the company has any additional requirements on this point.

The exclusion distances are measured using a 'tight-wire' method from the closest part of the cylinder to the edge of the opening. Where the minimum clearance cannot be achieved directly, a vapour-proof wall may be built adjacent to the cylinders to effectively increase the clearance to the opening.

In Figure 16.7, for example, you can see that the distance between the cylinder and the opening is too short. As there is no other place for the cylinders to be situated, one option here would be to build a vapour-proof wall to increase the distance to the required amount.

REFER TO AS/NZS 1596 'FIREWALLS AND VAPOUR BARRIERS'

Where exchange cylinders are situated under windows, a clearance of at least 150 mm must be maintained. This is measured from the top of the cylinder valve to the opening.

In-situ fill cylinders must have at least 500 mm between the top of the valve and the bottom edge of the window opening. When these cylinders are being topped up, a considerable amount of gas vapour is released, and so the required clearance beneath windows is therefore much greater.

Clearances from drain openings

Propane is heavier than air and if released it will sink to the ground and accumulate in low points. For this reason, any open drains must be at least 1 m from the side of the nearest cylinder.

Some technical regulators permit gully trap risers within the zone on the assumption that the trap water seal will keep any gas out of the drain. However, you would be advised to avoid this wherever possible as many gully traps incorporate drain connections above the water seal. You will need to confirm your own local requirements with the technical regulator in relation to gully traps.

Ignition source exclusion zones

No ignition sources of any kind are permitted within prescribed zones around any cylinder site. Ignition sources include:
- power sockets
- appliance flue terminals
- appliance openings where escaping gas may be lit from electrical circuits or burners
- air conditioner compressors
- electric lights.

Ignition source measurements are taken from the centreline of the cylinder. The exclusion zone around in-situ fill cylinders is much larger because of the gas vapour released during the filling process (see Figure 16.8).

Points to note about ignition sources

When checking clearances from ignition sources, be very careful to consider the following two points:
- An opening such as a door or window may be more than 1 m away from a cylinder but still within the ignition source exclusion zone. This is fine as long as you realise that if a door or window is opened, the exclusion zone may extend *into the building*. Look out for ignition sources inside the building as well (cooktops, power sockets, etc.).
- Where cylinders are situated close to property boundaries, remember to look over the fence into the adjacent property. The ignition source exclusion zone does not stop at fence lines. You may find an

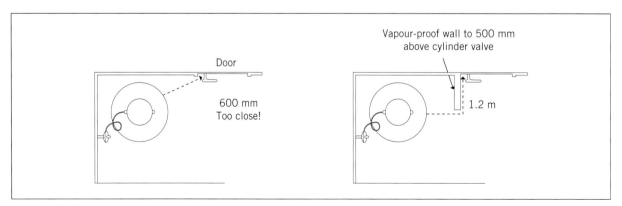

FIGURE 16.7 Plan view of a situation in which a vapour-proof wall might be required

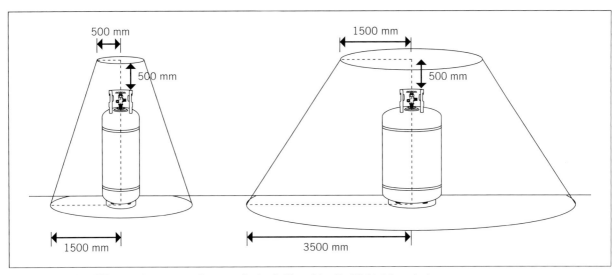

FIGURE 16.8 The difference between exchange cylinder (left) and in-situ fill (right) exclusion zones

FIGURE 16.9 A domestic water heater located at a safe distance from the exchange cylinders

external GPO or flue terminal only a couple of metres away from your cylinder site.

Note in the domestic installation shown in Figure 16.9 that the terminal of the water heater and appliance power connection are well outside the exchange cylinder/ignition source exclusion zone.

Cylinder manifolds

Where required due to vaporisation needs or if the size of the installation dictates a larger volume of gas to be stored on-site, cylinders can be manifolded together. This must be done using some form of proprietary-brand manifold kit. Essentially, each cylinder is connected with a pigtail to a wall-mounted manifold. This in turn is connected with two other pigtails to the regulator inlet (see Figure 16.10).

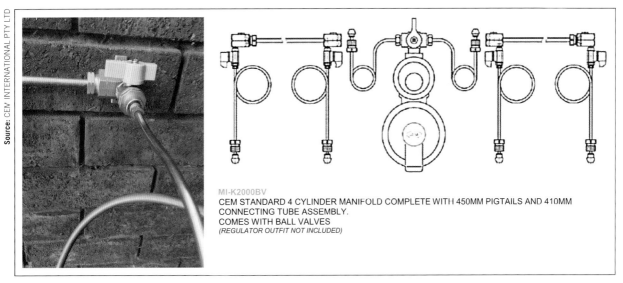

MI-K2000BV
CEM STANDARD 4 CYLINDER MANIFOLD COMPLETE WITH 450MM PIGTAILS AND 410MM CONNECTING TUBE ASSEMBLY.
COMES WITH BALL VALVES
(REGULATOR OUTFIT NOT INCLUDED)

FIGURE 16.10 The configuration of a four-cylinder manifold kit

Maximum aggregate capacity of cylinders

Cylinders are generally installed adjacent to a building of some sort. You need to be aware that there are limits on the maximum amount of gas permitted to be stored in each cylinder group. The total, or aggregate, capacity of a cylinder group must not exceed 2500 L.

REFER TO AS/NZS 1596, 'CYLINDERS IN USE'

For example, this may equate to a maximum of five 190-kg (2220 L) cylinders in one group. Cylinders in each group may not necessarily service the same installation. Where installation requirements mean that the aggregate capacity of a group may be exceeded, you will need to provide 3 m of separation between each cylinder group.

LEARNING TASK 16.3

CYLINDER LOCATIONS AND INSTALLATION REQUIREMENTS

1. What two standards should be referred to when determining the location of an LPG cylinder?
2. Can an LPG cylinder be installed underneath a building?
3. What is the minimum distance that an LPG cylinder must be installed away from an open drain?

How to choose a suitable LPG regulator

Like all gas components, LPG regulators must be currently listed and certified products. You cannot use or re-use any regulator that does not have a current certification listing.

When choosing a new or checking an existing LPG regulator you need to consider the following information:

- *Capacity.* This relates to the maximum amount of gas vapour the regulator is designed to pass. If you have installed a 60 MJ/h cooker, a 35 MJ/h space heater and a 42 MJ/h storage water heater, you would most likely choose a regulator with a capacity of around 150 MJ/h. All Australian 'certified' regulators will now be rated in MJ/h; however, you may find existing regulators to be marked in kg/h, L/h or BTU/h depending upon where they were made. You will often need to apply conversion calculations to be sure.
- *Inlet pressure.* While standard integral two-stage cylinder regulators are designed to handle full cylinder pressure, a second stage regulator may have a maximum inlet pressure of 70 kPa and must be matched with an upstream first stage regulator.
- *Adjustment range and outlet pressure.* Do you need a working pressure of 2.75 kPa, 70 kPa or 100 kPa? Dual stage, first stage or second stage? Choose the regulator to suit the application.
- *Connection type.* There is a wide range of inlet connections. Some regulators are fitted with manual change-over (MCO) valves with inverted-flare connections, inverted-flare tee blocks, POL tee blocks, female POL inlets, single inverted flare, ¼″, ⅜″, ½″ female NPT (national pipe thread taper) and so on. A POL tee (see Figure 16.11) is an ideal way to connect in-situ fill cylinders to the regulator. POL connections are made gas-tight very easily. You would use two POL × POL pigtails to connect from the cylinder valves to the tee inlets.

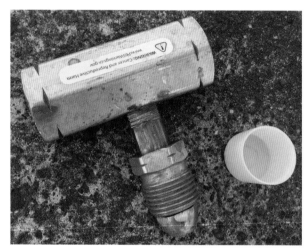

FIGURE 16.11 A POL tee

- *Manual or automatic change-over?* MCO regulators are fine for installations where there is little inconvenience for the owner in changing from empty to full cylinders. However where non-supervised appliances are installed on an exchange cylinder site, it is recommended that an automatic change-over regulator be specified so that the appliance operation is not interrupted when the regulator changes from the empty bank to the full bank of cylinders.

In-situ fill cylinders are normally connected to an open regulator tee with no change-over function, as the tanker refill cycle is timed to prevent the owner from running low on gas.

Once you have decided what type of regulator you require, you need to access the manufacturer's information to choose the product and have it ordered through your supplier. For example, Figure 16.12 shows an extract from a distributor's comprehensive product catalogue. From this, you can choose the type of regulator you need, check its certification and order from your supplier or direct from the company itself.

FIGURE 16.12 An extract from a distributor's product catalogue

Prepare for installation

All materials and components used for cylinder installations must be checked for serviceability to ensure that manufacturing faults or pre-purchase damage has not occurred. If you find a fault in a component, you are not permitted to repair it. It must be replaced. Only the best quality materials should be used in gas installations.

> **REFER TO AS/NZS 5601.1 'MEANS OF COMPLIANCE – MATERIALS, FITTINGS AND COMPONENTS'**

Always make an extra effort to protect your materials during storage and handling.

- Do not open regulator packets until you are ready to install.
- Never allow pigtails to be contaminated with dirt, and ensure POL fitting faces are capped until ready for use to avoid scratches.
- Keep fittings in an orderly and clean manner, ensuring that they are not contaminated with dirt and other foreign substances.
- Avoid rough handling of your components, fittings and materials.

Appropriate tools and equipment

The table below shows a selection of items that may be required for LPG cylinder installations.

Tools and equipment
• Measuring equipment
• Shifters
• Power drill and various sized masonry/timber bits
• Impact driver for general screwing tasks
• Bolt cutters to cut restraining chain
• Trolley to manoeuvre slabs and cylinders
• Spirit level
• Shovel and crowbar where cylinder based requires minor excavation
• Soapy water solution and spray bottle

Install and test LPG storage system

When installing either exchange or in-situ fill cylinders, one very important requirement is to ensure that the relief valve is pointing in the right direction. This is a vital safety requirement to ensure that in the event of fire the cylinder can safely vent without creating further problems. This is explained further in the section below.

Cylinder pressure relief

If a cylinder is overfilled or exposed to excessively high temperatures, the internal gas pressure will rise to unacceptable levels. To avoid a cylinder rupture, relief valves will release gas as required to keep internal pressures within safe design limits.

Cylinder pressure relief valves incorporate a spring-loaded valve as part of the main control valve assembly.

Figure 16.13 shows the control valve of a 45-kg cylinder. The integral pressure relief valve spring can be easily seen at the right of the photograph. Where cylinders are exposed to fires or very high ambient temperatures, as often found in the northern and inland areas of Australia, the relief valve will open and close repeatedly to keep internal pressures at a safe level.

When choosing a site for your cylinder installation, you must ensure that no structure or other cylinder is situated in the direction of discharge from the cylinder relief valve. If the jet of escaping high-pressure gas vapour were to ignite, poor valve alignment could cause the flames to impinge upon the side of the cylinder and possibly result in a BLEVE.

> **LEARNING TASK 16.4**
>
> **INSTALL AND TEST LPG STORAGE SYSTEM**
>
> 1. What is the purpose of an LPG cylinder pressure relief valve?
> 2. At what pressure will the relief valve open?
> 3. What is the purpose of LPG cylinder manifold kits?

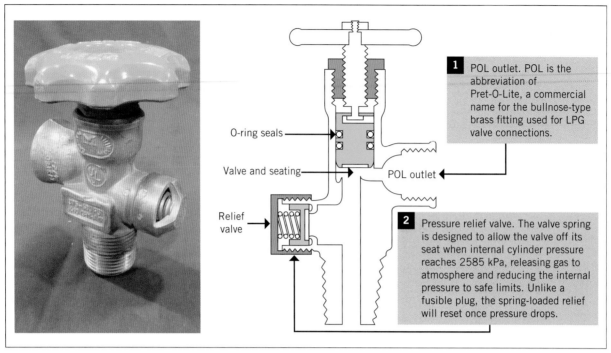

FIGURE 16.13 The control valve of a 45-kg cylinder

Clean up the job

The job is never over until your work area is cleaned up and tools and equipment accounted for and returned to their required location in a serviceable condition.

Clean up your area

The following points are important considerations when finishing the job:
- If not already covered in your site induction, consult with supervisors in relation to the site refuse requirements.
- Cleaning up as you go not only keeps your work area safe it also lessens the amount of work required at the end of the job.
- Dispose of rubbish, off-cuts and old materials in accordance with site-disposal requirements or your own company policy.
- Ensure all materials that can be recycled are separated from general refuse and placed in the appropriate location for collection. On some sites, specific bins may be allocated for metals, timber etc. or, if not, you may need to take materials back to your business premises.

Check your tools and equipment

Tools and equipment need to be checked for serviceability to ensure that they are ready for use on the job. On a daily basis and at the end of the job check that:

- all items are working as they should with no obvious damage
- high-wear and consumable items such as cutting discs, drill bits and sealants are checked for current condition and replaced as required
- items such as laser levels, pressure testing equipment etc. are working within calibration requirements
- where any item requires repair or replacement, your supervisor is informed and if necessary mark any items before returning to the company store or vehicles.

Complete all necessary job documentation

Most jobs require some level of documentation to be completed. Avoid putting it off till later as you may forget important points or times.
- Ensure all SWMS and/or JSAs are accounted for and filed as per company policy.
- Complete your timesheet accurately.
- Reconcile any paper-based material receipts and ensure electronic receipts have been accurately recorded against the job site and/or number.
- Complete all necessary gas supplier paperwork associated with cylinder site installations and forward according to local requirements.

SUMMARY

In this unit you have examined the requirements related to the installation of LPG cylinder sites up to 500 L in capacity. These requirements include:
- cylinder storage capacity calculations based upon:
  - vaporisation capacity
  - peak load and draw-off needs
- cylinder location requirements
- regulator selection
- cylinder installation including
  - twin cylinder sites
  - in-situ fill
  - manifold systems
- cylinder site testing requirements.

GET IT RIGHT

The photo below shows an incorrect practice when installing an LPG regulator on a twin cylinder installation. Identify the incorrect method and provide reasoning for your answer.

_____
_____
_____
_____
_____

WORKSHEET 1

To be completed by teachers
Student competent ☐
Student not yet competent ☐

Student name: _____

Enrolment year: _____

Class code: _____

Competency name/Number: _____

Task

Review the section *Vaporisation calculation – the DLK formula* and answer the following questions.

1 Calculate the vaporisation rate of a tank measuring 6.62 m long × 1.22 m wide at an ambient temperature of 10°C and 10 per cent full.

_____

_____

_____

2 A 4 × 45-kg cylinder manifold is required to supply gas to a central heater at 130 MJ/h and an upright cooker at 50 MJ/h. In this region, it is not uncommon for winter night ambient temperatures to average 4°C. As the installation draws off two cylinders simultaneously, will sufficient vaporisation be achieved with both cylinders at 10 per cent full?

_____

_____

_____

WORKSHEET 2

To be completed by teachers
Student competent ☐
Student not yet competent ☐

Student name: _____

Enrolment year: _____

Class code: _____

Competency name/Number: _____

Task

Review the section *Calculating cylinder storage requirement* and answer the following questions.

1. What is the gross potential draw-off for a 190-kg cylinder in MJ?

   _____
   _____
   _____

2. i. You intend to replace existing storage water heaters with a manifold of four 199-MJ/h continuous flow water heaters for a football club shower block. The units would work at full rate for approximately three hours per day on Saturday and Sunday and two hours on Wednesday. What is the probable demand in kg for propane per week?

   _____
   _____
   _____

   ii. The football club is situated in an area where winter temperatures are often about 4–6°C. The existing LPG installation draws off four 190-kg cylinders manifolded together. Will this manifold provide adequate vaporisation for the expected simultaneous demand?

   _____
   _____
   _____

   iii. At the expected rate of use, how many weeks' worth of gas supply would the football club have before it ran out of gas?

   _____
   _____
   _____

3 How many hours of continuous use would a 90-kg cylinder provide to a continuous flow water heater rated at 220 MJ/h?

WORKSHEET 3

To be completed by teachers
Student competent ☐
Student not yet competent ☐

Student name: _____

Enrolment year: _____

Class code: _____

Competency name/Number: _____

Task
Review the section *Cylinder locations and installation requirements* and answer the following questions.

1. List at least six prohibited locations for a cylinder site. Provide a reference from AS/NZS 1596 to support your answer.

   i _____

   ii _____

   iii _____

   iv _____

   v _____

   vi _____

   _____

   _____

2. What is the maximum permissible distance a cylinder can be installed underneath a house built on piers? Provide a reference from AS/NZS 1596 to support your answer.

   _____

3. What sort of support base is required for an in-situ fill cylinder site? Provide a reference from AS/NZS 1596 to support your answer.

   _____

   _____

   _____

4. If you had no alternative to installing a twin-cylinder installation in a driveway down the side of a commercial building, what approved method would you use to protect the cylinders from vehicle impact? Provide a reference from AS/NZS 1596 to support your answer.

   _____

   _____

   _____

5. What is the radius of the ignition source exclusion zone at the base of an in-situ fill cylinder? Provide a reference from AS/NZS 1596 to support your answer.

_____

6. Are weep holes between bricks considered to be openings into buildings?

_____

_____

_____

7. How many 210-kg cylinders could you locate in one group area? Provide a reference from AS/NZS 1596 to support your answer.

_____

_____

_____

8. Detail two methods you could use to ensure the security of a cylinder/s installed in a public place. Provide a reference from AS/NZS 1596 to support your answer.

_____

_____

_____

_____

9. The maximum distance permitted between the lower edge of a window opening and the top of an in-situ fill cylinder valve is 150 mm: true or false? _____

_____

_____

10. What is the maximum kg of LPG permitted to be kept on a veranda? Provide a reference from AS/NZS 1596 to support your answer.

_____

11. When fixing a regulator to a wall for a twin-exchange cylinder installation, what key consideration must you comply with in relation to regulator height?

_____

_____

_____

17

INSTALL LPG SYSTEMS IN CARAVANS, MOBILE HOMES AND MOBILE WORKPLACES

Learning objectives

Areas addressed in this chapter include:
- installing caravan gas systems
- gas appliances for caravans.

Before working through this section it is very important that you understand the basic fundamentals covered in Part A of this text and the following chapters from Part B:
- Chapter 8 Size consumer gas piping systems
- Chapter 9 Install gas piping systems
- Chapter 10 Install gas pressure control equipment
- Chapter 11 Purge consumer piping
- Chapter 13 Install and commission Type A gas appliances
- Chapter 16 Install LPG storage of aggregate storage capacity up to 500 litres.

The requirements for both caravan and marine craft gas installations are to be found within AS/NZS 5601.2 (Gas Installations). In this chapter we are going to focus on the installation requirements related specifically to caravans as defined in the Standard.

Overview

As with all terms used in the Gas Installation Standards, you need to be clear about how a caravan is defined in AS/NZS 5601.2 before proceeding.

REFER TO AS/NZS 5601.2 'SECTION 1 SCOPE AND GENERAL'

Because of the quite broad use of the term 'caravan', it is very important that you have a clear understanding of the scope and application of this Standard before proceeding further.

A caravan is more than just a towed structure intended for human habitation. The same installation requirements will apply to all of the following:
- self-propelled motor homes – new recreational vehicles, converted buses and 4WDs
- camper trailers
- pop-top caravans
- mobile catering vehicles – towed or self-propelled
- any attached permanent or temporary annex to such vehicles.

Be aware that wherever commercial catering equipment is to be installed in any of these vehicles, then these appliances should be installed in accordance with Part 1 of the AS/NZS 5601 and not Part 2.

REFER TO AS/NZS 5601.2 SECTION 1 'EXCLUSIONS'

In addition to the exclusion of commercial catering equipment, a large transportable home or office is not generally regarded as a caravan, although this may not be entirely clear for some smaller structures. If in doubt, contact your technical regulator for clarification.

REFER TO AS/NZS 5601.2 SECTION 1.3 'CARAVAN' AND 'CARAVAN (DOMESTIC)'

Caravans have some very specific gas installation requirements that are quite different from those you may be familiar with for general construction-type gas work. The definitions of caravan detail specific differences between domestic caravans and those used commercially. Prior to starting your first caravan job, you will need to review the Standards in detail to remind yourself of these differences.

Identify LPG system requirements

Prior to commencing any work you must fully plan the job to ensure all requirements of the Standard are met. This section reviews each of these main considerations.

Work, health and safety considerations

Caravan gas installations share many of the installation hazards and considerations found in other forms of gas fitting work and it is very important that you have reviewed Chapter 5 'Gas industry workplace safety' in Part A of this text.

In addition to these hazards, additional emphasis should be applied to the issue of electrical safety.

Dual power supply

The majority of caravans run a dual power supply for both lighting and appliances.
- 240 V – When staying in a caravan park and using a powered site, the caravan can be plugged into the site 240 V power supply for the duration of the stay. Lighting, heaters and refrigerators can then be selected to or switched to run on this supply.
- 12 V – When camping in remote locations where a 240 V supply is not available, some caravan electrical items can be switched to operate on the tow vehicle primary 12 V battery. Many vehicles and vans will also have an ancillary battery and solar charging system to prevent discharging the tow vehicle main battery.
- Appliances such as refrigerators can be specified to run on gas as well as 12 V and 240 V (see Figure 17.1).

FIGURE 17.1 Appliances such as this refrigerator have both 12 V and 240 V power supplies

This feature of caravans means that you will encounter wiring layouts around the caravan and within appliances that will be dedicated to the 12 V and 240 V power supplies. The identification of these systems is not always clear, particularly with older vans and those that have been altered many times by owners.

- Before working on gas appliances with dual power supply connections and anywhere in the vicinity of caravan wiring always ensure the 240 V connection is fully disconnected.
- Test all wiring and appliances for stray current with a non-contact voltage detector or multi-meter before *and* during work.
- Never assume a 12 V supply or component is safe to touch, as you have no guarantee that it has not been accidentally crossed over with the 240 V system.

Always move and work around caravans with the same caution and respect for electrical hazards as you are required to do on any building site. Complacency can get you killed.

System layout and gas load

When preparing to undertake caravan gas work, you must first confirm the proposed layout of the LPG cylinder supply and the location of all gas appliances.

Confirm the following details from a site inspection or floorplan and elevation drawings:
- Cylinder location. The cylinder/s may be located externally on the drawbar or internally within a compartment located at the front or side of the vehicle.
- Appliances. Determine what gas appliances are to be installed and confirm the:
  - location within the vehicle
  - appliance certification number
  - appliance MJ/h input rating from the data plate and/or manufacturer's specifications

Note: Second-hand appliances must only be installed if they meet the safety requirements of your State or Territory technical regulator – do not install such appliances without checking first!

Every aspect of cylinder, piping and appliance locations is subject to requirements detailed within the Standard. These are discussed in more detail in the following sections.

LEARNING TASK 17.1 ELECTRICAL SAFETY

1. What are the two power systems that may be used in a caravan?
2. Why should a 12 V appliance always be tested for stray current?

Regulator selection and capacity

Regulators for gas appliances in caravans must be certified products in accordance with AS 4621 or UL 144. You must check that any product you purchase complies with these Standards. Furthermore, regulators must have built-in over-pressure protection and be of the dual-stage (integral two-stage) configuration.

 Single stage barbecue-style regulators that screw directly into the cylinder are not permitted for use in caravan installations.

If you are not sure the regulators supplied by your plumbing hardware are adequate, it is your responsibility to source the information from the manufacturer or supplier.

MANUAL CHANGE OVER REGULATOR

Marshall Excelsior Integral Two Stage High Capacity manual change-over regulator outfit complete with copper pigtails, mounting bracket & screws at capacities shown.

(Also available with Rubber Pigtails)

PART NO.	CAPACITY MJ	INLET CONNECTION	OUTLET CONNECTION	ADJUSTMENT RANGE	OUTLET PRESSURE SETTING	CARTON LOTS
ME-290-00HCMC	290 MJ	1/4" INV. FLARE	3/8" FNPT	9 – 13" WC	11" WC	25

Source: CEM International Pty Ltd

FIGURE 17.2 Use supplier product websites to match compliant regulator MJ/h capacity to your installation requirements

A range of both manual change-over and automatic change-over regulators are available to suit various installation needs. Automatic change-over regulators are particularly useful where the customer intends to use gas-heated water for a shower.

Cylinder compartments often have restricted space and it can be difficult to mount the regulator at the correct height so that any liquid propane caught in the pigtails can drain back to the cylinder. Offset low-profile regulators are available that are suited to this purpose (see Figure 17.3).

FIGURE 17.3 An offset, low-profile regulator especially useful for the restricted space within compartments

> **REFER TO AS/NZS 5601.2 'MEANS OF CONFORMANCE – GAS PRESSURE REGULATORS'**

LEARNING TASK 17.2

REGULATOR SELECTION AND CAPACITY

1. Are single stage LPG barbecue regulators permitted to be used in a caravan installation?
2. What are the two types of change-over regulators that can be installed in a caravan installation?

Pipe sizing

Pipe sizing for caravans follows the same basic process as applied to other gas installations. When pipe sizing for caravans, you are to use the propane pipe sizing tables for copper at maximum pressure drop of 0.25 kPa and supply pressure of around 3 kPa as per AS/NZS 5601.1 Appendix F or AS/NZS 5601.2 Appendix C. Other permitted materials are to be sized using recognised formulas and methods.

Caravan LPG cylinder selection and location

Caravan LPG cylinders must be located and mounted in accordance with AS/NZS 5601.2 and meet the cylinder design requirements of AS 2030.1. Importantly, this Standard details the level of corrosion resistance for cylinders in various environments.

The majority of LPG cylinders used for caravans are still made from traditional steel; however, options also include stainless steel and lightweight, translucent cylinders made from polymer composites (see Figure 17.4). Although these are more commonly found in marine gas applications, it can be expected that composite cylinders, in particular, with their much lighter drawbar weight will gradually become more common for caravan use.

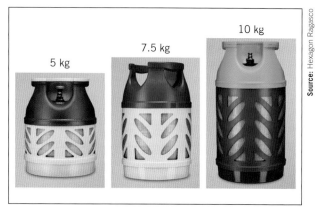

FIGURE 17.4 Composite cylinders offer a translucent, corrosion-free and lightweight alternative to traditional steel

Coatings on steel cylinders include:
- painted coatings
- zinc-rich anti-corrosive coatings
- zinc galvanised.

Cylinders used for caravan applications and particularly those mounted externally on the drawbar are exposed to a very harsh environment. As well as by normal weather conditions, corrosion of a steel cylinder

LEARNING TASK 17.3 — PIPE SIZING

The exercise below can be completed as revision of the pipe sizing processes you have learned previously.

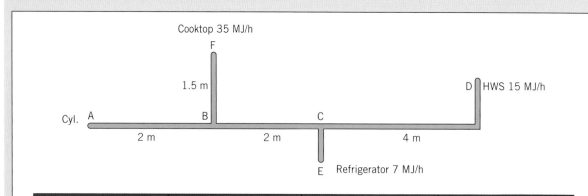

Section	Main run	Load MJ/h	DN size
AB		35 + 7 + 15 = 57	
BC	AD	7 + 15 = 22	
CD	8 m	15	
CE		7	
BF		35	

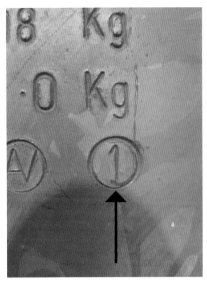

FIGURE 17.5 You must check the corrosion coating ID number on the cylinder guard

may be accelerated by stone chips and a range of aggressive road contaminants.

Coatings on cylinders that may be fine for backyard barbecues are not resilient enough for caravan applications. Each coating and the cylinder resistance to corrosion is afforded a rating based upon a number from 1 to 4, with number 1 being the most resistant. This number is stamped on the cylinder guard (Figure 17.5). Cylinders marked with a '4' identification number must not be used for any caravan or mobile application.

Cylinder location

The same location and mounting requirements apply to both connected and spare cylinders. When checking plans or preparing to work on an existing caravan, the following requirements apply:

- Cylinders must always be stored in the upright position with the valve uppermost, ensuring that the valve is always in contact with the vapour space of the cylinder.
- Cylinders must be located:
  - externally and mounted on the chassis of the caravan in a protected location (see Figure 17.6)
  - within a vapour-proof cylinder compartment mounted within the profile of the van or on the drawbar (see Figure 17.7)
  - in a recess on the chassis and under the skirt of the van that is vapour-proof to the inside of the van (see Figure 17.8)
  - in a location that has no sources of ignition within the exclusion zone above and to the side of the cylinder as detailed within the Standard.

Note: The relief valves of cylinders mounted on the drawbar must be directed forwards in the direction of travel at 45°, as detailed within the Standard.

FIGURE 17.6 A twin 9-kg cylinder drawbar installation

Cylinder mounting and restraint

Cylinders must be rigidly secured in a carrier capable of withstanding a load of at least four times the weight of the filled cylinder from any direction. The Standard provides a guide to the construction requirements of this structure. The restraint bands must also incorporate insulating strips to prevent abrasion of the cylinder protective coating.

REFER TO AS/NZS 5601.2 'MEANS OF CONFORMANCE – CYLINDERS'

FIGURE 17.7 Cylinders installed within a vapour-proof compartment behind the drawbar of the caravan

FIGURE 17.8 Two 4-kg cylinders mounted within a vapour-proof side recess of a mobile home

Cylinders must be sited so that:
- the valve is at least 150 mm beneath any opening into the van
- no air vent or opening is within 1 m horizontally from the centre line of the cylinder.

REFER TO AS/NZS 5601.2 'REQUIRED CLEARANCES AROUND A CYLINDER'

Cylinder compartments

Compartments allow cylinders to be installed so that they are protected from theft, damage and exposure to the elements. However, you must realise that special requirements apply where cylinders are installed within compartments.
- The compartment must be sealed to prevent any gas from entering the caravan.
- The compartment must be waterproof and corrosion resistant.
- The cylinder position within the compartment must not obstruct the drain hole.
- The compartment drain hole must be located in or within 25 mm of the compartment base and have a minimum opening of 500 mm² clear area. It must also terminate at a point facing away from the caravan direction of movement so that air is not forced into the compartment during travel. Compartments with high and low vents are an alternative as detailed in the Standard.
- The drain outlet must terminate at least 1 m away from any opening into the caravan unless the opening is at least 150 mm above the outlet. The outlet is to be at least 1.5 m away from any ignition source unless the ignition source is at least 500 mm above the outlet.
- Compartments must only hold cylinders and associated equipment (regulator assembly, spanner, soapy water, isolation valve and a non-sparking shut-off device if fitted).

In some circumstances, the external mounting of cylinders may not be possible. In this instance a compartment that is accessible from the inside of the van may be used provided that total cylinder capacity does not exceed 2 × 15 kg (36 L) cylinders and the bottom edge of the door opening is not less than 50 mm above floor level.

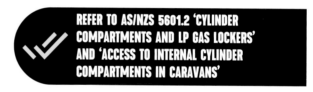

REFER TO AS/NZS 5601.2 'CYLINDER COMPARTMENTS AND LP GAS LOCKERS' AND 'ACCESS TO INTERNAL CYLINDER COMPARTMENTS IN CARAVANS'

A warning sign (compliant with AS/NZS 5601.2) must be affixed in a position adjacent to the cylinder compartment.

LEARNING TASK 17.4

CARAVAN LPG CYLINDER SELECTION AND LOCATION

1. Which cylinder coating, based on its corrosion resistance is not permitted to be installed in a caravan installation?
2. Why must a cylinder compartment be sealed?
3. What can be stored in a cylinder compartment?

Caravan appliance types and locations

Caravan appliances generally fall into the following categories:
- cooktops
- upright cookers
- ovens
- refrigerators
- storage water heaters
- room-sealed continuous flow water heaters
- gaslights (now rarely used but still referred to in the Standard)
- room-sealed space heaters.

> **REFER TO AS/NZS 5601.2 'MEANS OF CONFORMANCE – GAS APPLIANCES'**

All appliances used in caravans must be fitted with flame safeguard systems to all burners! When working on older vans fitted with original open burner cooktops, it would be a good policy to advise customers on the safety benefit of upgrading to modern flame safeguard protection.

As with general Type A installations, all appliances installed in caravans must be certified for use and installed in accordance with all general and specific manufacturer's requirements and the Gas Installation Standards. For example, the combined cooktop and griller shown in Figure 17.9 should comply with all requirements for clearances and the protection of combustible surfaces.

FIGURE 17.9 A combined cooktop and griller installed in a caravan

Clearances to a non-diesel fuel filler cap or vent

Minimum clearance requirements exist for vehicles fuelled by other than diesel. Fuel filler caps and vents can release a considerable volume of vapour. If fuel vapour were to be exposed to a hot flue outlet or drawn into an appliance air intake, a fire or explosion could occur. To prevent this, the minimum distance between a petrol filler cap/vent and a combustion air intake or flue outlet of an appliance with a continuous source of ignition is 1 m.

Caravan piping

In the Standard, the pipework between the cylinder valve and the regulator is classed as high-pressure piping while the pipework installed from the outlet of the regulator is referred to as low-pressure piping.

> **REFER TO AS/NZS 5601.2 SECTION 5 'MEANS OF CONFORMANCE – PIPING AND FITTINGS'**

High-pressure pipe installation requirements

Three methods and materials are available for the connection between the cylinder and the regulator. These are:
1. POL × flare copper pigtail formed into a loop
2. a hose assembly with a maximum length of 600 mm featuring an integral excess flow valve (see Figure 17.10); such hoses must be Class F in accordance with AS/NZS 1869
3. stainless steel – 316 or 304 Grade.

FIGURE 17.10 Hose assembly showing excess flow valve

Normal hose assemblies are not acceptable. As stated, the hose must incorporate an excess flow valve in the POL fitting that will act to restrict gas flow in the event of damage to the gas system. Double check the manufacturer's information to confirm that you are purchasing the correct hose.

> **REFER TO AS/NZS 5601.2 'PIPING SPECIFICATION'**

Low-pressure pipe installation requirements

Caravan pipe installations have particular needs deriving from the mobile nature of the vehicle and the environment in which it operates. You must carry out every installation with a view to ensuring that the pipe is protected from damage and vibration.

Materials

Pipe materials for caravans from the outlet of the regulator are restricted to the following:
- a certified hose assembly compliant with AS/NZS 1869 Class A, B, C or D

- stainless steel – 316 Grade
- *fully annealed* copper pipe – Type A or B.

It is important to note that the only acceptable fittings and joint types to be used with copper tube are flared compression, silver brazed capillary or silver brazed expanded tube sockets. Press-fit crimped end connectors are not permitted. Other prohibited fittings are also listed in the Standard.

REFER TO AS/NZS 5601.2 'PROHIBITED TYPES OF PIPING, JOINTS AND FITTINGS'

Press-fit crimped fitting systems are not permitted in any manner for use on caravan installations. All joints in copper pipe must be copper alloy flared compression or silver-brazed capillary.

Of these three approved pipe types, the most commonly installed material is fully annealed copper tube. Although not required under AS/NZS 5601.2, it is a common industry practice to install plastic-sleeved annealed copper pipe (see **Figure 17.11**). This will prevent any chance of electrolysis or abrasion damage between the pipe and vehicle structure.

FIGURE 17.11 Plastic-coated DN10 fully annealed copper

Pipe fixings should be of a non-ferrous material compatible with the pipe material being used.

LEARNING TASK 17.5 PIPING MATERIALS

1. What three materials are approved for use in a caravan?
2. Are press fit crimped fitting systems permitted to be installed in caravan systems?

Prepare for installation

All materials used for caravan installations must be checked for serviceability to ensure that manufacturing faults or pre-purchase damage has not occurred. Faults in components cannot be repaired and the component must be replaced. Only the best quality materials should be used in gas installations.

Always make an extra effort to protect your materials during storage and handling.
- Keep dust and water out of pipes with caps or plugs.
- Keep fittings in an orderly and clean manner, ensuring that they are not contaminated with dirt and other foreign substances.
- Avoid rough handling of your pipe, fittings and materials.
- Ensure that brass flared fittings are always stored with the nut screwed on, to prevent damage to the mating faces of the flare.

Appropriate tools and equipment

Apart from general hand tools used on almost all jobs, tools and equipment used for caravan pipe installation will, to a large extent, be dictated by the type of pipe material and jointing method being used. In most cases copper pipe, silver-brazed and compression flare fittings are used. The table below shows just a selection of items that may be required for the installation of copper piping.

Tools and equipment
• Oxy/acetylene brazing equipment (see **Figure 17.12**)
• Flaring tools
• Copper-pipe cutters and reamer
• Benders
• Internal and external bending springs
• Expanders
• Shifters

Other considerations are related to handling of the pipe and the proposed installation location. Working under caravans installing pipe can be very tight and restrictive. Items may include:
- mechanics' wheeled creeper platform
- mirrors
- LED lighting.

FIGURE 17.12 Oxy/acetylene brazing equipment will be required for almost all copper pipe caravan installations

FIGURE 17.13 A dual stage manual change-over regulator (note the use of a test point fitting at the regulator outlet, which simplifies pressure-testing procedures)

Install LPG system, including flue and ventilation

Installing the regulator

Specific requirements relate to the choice, location and mounting of LPG regulators.

Regulator choice

As stated earlier in this chapter you are required to install certified dual stage regulators incorporating the required over-pressure protection design. These regulators must be adjustable for pressure setting and be of a manual change-over or automatic configuration (see **Figure 17.13**).

REFER TO AS/NZS 5601.2 'MEANS OF CONFORMANCE – GAS PRESSURE REGULATORS'

Regulator mounting

The mounting and installation of the regulator must comply with the following points:
- The regulator must be independently and securely fixed to the caravan.
- The regulator outlet pressure must be set to a maximum of 3 kPa with all appliances operating.
- Regulators must be located so that any liquid that forms in the pigtails is able to drain back into the cylinder. This means that you must fit the regulator so that the regulator inlet is higher than the cylinder valve and there is an overall fall in the pigtails from the regulator back to the cylinder valve (The regulator itself *does not* need to be entirely above the valve).

REFER TO AS/NZS 5601.2 APPENDIX G

Pipework installation

Some key differences from standard gas fitting work apply to caravan pipe installations and these are covered in the following sections.

Design and location of pipe

A principal requirement of caravan installations is that in all cases the main run of piping must be located outside the caravan with only individual branches permitted to enter the vehicle adjacent to the appliance each branch services. As such, this externally located pipe must be situated so that it is not exposed to probable damage below the lowest point of the main chassis or frame.

For example, the end-elevation cross-section of a caravan chassis in **Figure 17.14** shows how the pipe can

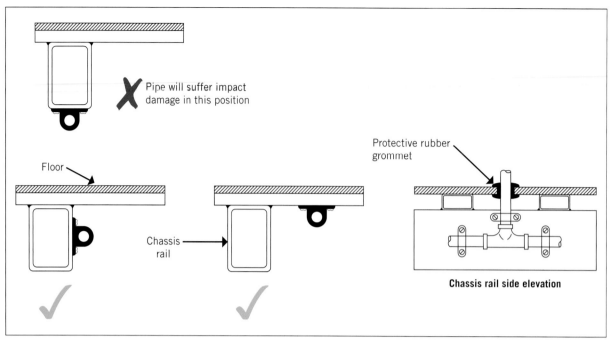

FIGURE 17.14 End- and side-elevation cross-section of a caravan chassis

be clipped to the floor or side of the chassis. It cannot run along or cross under the chassis as it is liable to impact damage. The side elevation shows that the pipe is protected with a grommet where it penetrates the floor.

Pipe fixing intervals are much closer than specified for pipe installations in buildings because of the vibrations to which caravans are exposed. Pipe must also be supported with clips at a maximum of 150 mm from each tee or change of direction. Due to the corrosive environment that pipe and clips will be subjected to underneath a caravan, you should also specify stainless steel screws when installing clips.

 **REFER TO AS/NZS 5601.2 'INSTALLATION OF PIPING'**

Use of hose assemblies for appliance connections

Hose assemblies can be used between rigid piping and the appliance, where approved by the manufacturer. Where only one appliance is installed, you are permitted to use a hose assembly between the outlet of the regulator and the appliance (again subject to the appliance manufacturer's specifications). In all cases the hose assembly must be:

- compliant with AS/NZS 1869 to the class specified in AS/NZS 5601.2
- of a continuous and minimum practicable length
- where applicable, fixed with compatible materials in accordance with the maximum distance specified in AS/NZS 5601.2.

Multiple appliance pipe installations

Where more than one appliance is installed, you need to install your pipe in accordance with the following requirements:

- The main run of the pipe must be installed outside the caravan.
- Each appliance must be connected to the main run with a separate branch. The tee must remain outside the caravan.
- All joints and unions must remain accessible.
- Pipe penetrations of floors must be protected from any edge abrasion using a grommet.
- All appliances must be fitted with a readily accessible quarter-turn isolation valve prior to the appliance inlet.

Quick-connect devices

Quick-connect devices or 'bayonet fittings' are not permitted for use inside a caravan as they would then allow consumers to connect flueless space heaters. As these appliances are prohibited in the small confines of a caravan, so too are quick-connect devices. This prohibition also extends to that wall area of the caravan where an annexe is installed.

LEARNING TASK 17.6 INSTALL LPG SYSTEM

1. Is the main run of the gas piping system permitted to be installed in the walls of the caravan?
2. Can the gas piping system be installed on the underside of a chassis rail?
3. How are pipe penetrations through the flooring protected from abrasion?

Ventilation

In addition to any specific appliance ventilation requirements, all caravans require at least two fixed, high- and low-level ventilation openings. This also applies to pop-top style caravans, even when the roof is in the down position. The vents are required to be installed at either end and/or at opposite sides of the area to be ventilated to provide sufficient cross-flow of air.

Total free ventilation area must be at least 4000 mm^2 or in accordance with the formula shown in Figure 17.15, whichever is the greater:

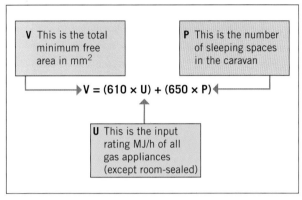

FIGURE 17.15 Total free ventilation

When calculating the total MJ/h load, only include non-room-sealed appliances such as cookers and ovens. Refrigerators are required to be installed in a room-sealed recess while all approved water heaters now operate with room-sealed, balanced flue systems.

The upper natural ventilation opening must provide 50 per cent of total free ventilated air supply and be situated within 150 mm of the ceiling. The lower opening must also provide 50 per cent of total free ventilated air supply and be situated within 150 mm of the floor.

REFER TO AS/NZS 5601.2 'MEANS OF CONFORMANCE – VENTILATION'

Flueing/venting of caravan gas appliances

Flueing systems for caravans are generally appliance specific and determined by the manufacturer. The mobility of caravans makes normal flue solutions inadequate. Instead, caravan appliances are often designed to flue from the side of the vehicle or through specially designed outlets (see Figure 17.16). The flue may be incorporated in the appliance ventilation system.

Where you need to fabricate or specify the use of particular flue materials, these must be in accordance with those listed in the Standard. All flue/venting systems must be tested for spillage.

EXAMPLE 17.1

A customer has asked you to install a cooker and fridge in a caravan. Upon inspection, you record the following information:

- Number of sleeping spaces = 4
- Cooker = 45 MJ/h
- Fridge = room sealed – not included in calculation

Calculate the total ventilation.

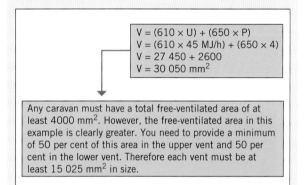

$V = (610 \times U) + (650 \times P)$
$V = (610 \times 45 \text{ MJ/h}) + (650 \times 4)$
$V = 27\,450 + 2600$
$V = 30\,050$ mm^2

Any caravan must have a total free-ventilated area of at least 4000 mm^2. However, the free-ventilated area in this example is clearly greater. You need to provide a minimum of 50 per cent of this area in the upper vent and 50 per cent in the lower vent. Therefore each vent must be at least 15 025 mm^2 in size.

REFER TO AS/NZS 5601.2 SECTION 8 'MEANS OF CONFORMANCE – FLUEING'

FIGURE 17.16 A pop-top van featuring a storage water heater accessed via a sealed recess in the side of the vehicle (note the burner, PTR valve and side outlet flue terminal)

The most commonly installed appliance with a required flue system is the gas refrigerator. These appliances must be installed in a sealed recess, ensuring that combustion products are kept from the inside of the caravan. The products of combustion are generally flued in a method similar to one of the options seen in **Figure 17.17**. Subject to installation instructions, the flueing system may be integral with the required ventilation system for the appliance.

 **REFER TO AS/NZS 5601.2, 'REFRIGERATORS'**

Figure 17.18 shows the bottom vent of a gas refrigerator. Air for combustion is drawn in via the bottom vent and the products of combustion are flued out through the top. Vents such as these can be purchased from specialist gas and caravan retailers.

Prohibited appliances

Caravans are generally small structures that often incorporate combined sleeping and living areas. With restricted air supply and relatively confined spaces, restrictions apply to the use and installation of many appliances.

Appliances prohibited for use in a caravan include:
- water heaters other than room-sealed types
- space heaters other than room-sealed types
- gas appliances designed to operate on an unregulated supply or in excess of 2.75 kPa (camping-style stoves and heaters).

LEARNING TASK 17.7 VENTILATION

1. When installing upper and lower ventilation in a caravan, at what distance should the vent be from the ceiling and the floor?
2. Do flued appliances within a caravan require testing for flue spillage?

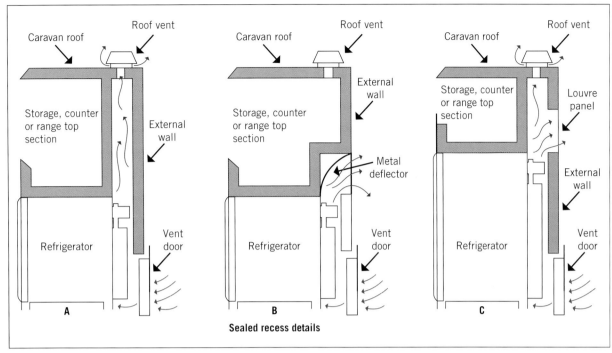

FIGURE 17.17 Some options for flueing a gas refrigerator

FIGURE 17.18 Refrigerator vents are sized according to the appliance storage capacity. Refer to AS/NZS 5601.2 Table 6.12.4

Test and commission LPG system

Like all gas fitting work, caravan LPG installations also require comprehensive testing.

Conducting a pipework test

Pressure testing a caravan pipe installation involves using the test equipment and testing processes described in Chapter 9 'Install gas piping systems'. AS/NZS 5601.2 provides detailed steps for a 14 kPa pipework test. However, some technical regulators may require a higher test pressure, so it is advisable to check your local requirements before proceeding.

You are required to conduct a 14 kPa pressure test for a period of 5 minutes with no detectable loss of pressure. A standard water gauge manometer would be unsuitable for this test and, therefore, you will require a calibrated digital manometer.

Conducting a gas tightness test

Following installation of the appliances, you must carry out a gas tightness test to prove that the appliance connections right through to the appliance control valves themselves are gas-tight at nominal operating pressure.

The Standard requires that with all appliances operating, the LPG regulator must be set to supply no greater than 3 kPa gas pressure and have no loss of pressure over a period of 5 minutes.

REFER TO AS/NZS 5601.2 APPENDIX E AND TO YOUR LOCAL TECHNICAL REGULATOR

If system testing is successful, proceed with appliance commissioning procedures as per the manufacturer's instructions and the Standard.

REFER TO AS/NZS 5601.2 APPENDIX H

Consumer instruction labels

Caravans are required to have explicit instructions and warning signs applied adjacent to cylinder compartments/mountings and appliances. These signs are available through specialist gas fitting suppliers and caravan manufacturers.

When working on older vans, you will usually need to replace all existing signs, as they are often missing or illegible. Specialist gas fitting suppliers will be able to supply you with the appropriate range of compliant signs. Some caravan spare parts retailers will also carry a range of signs. If you have customised signs printed, ensure that the exact wording, font size, dimensions, colour and layout comply with the specific requirements detailed in the Standard. All labels and signs must be permanent, durable and legible. Some regulators now require them to be etched onto metal and will no longer accept simple stickers – check for local requirements.

 REFER TO AS/NZS 5601.2 SECTION 10 AND TO YOUR LOCAL TECHNICAL REGULATOR

Clean up

The job is never over until your work area is cleaned up and tools and equipment accounted for and returned to their required location in a serviceable condition.

Clean up your area

The following points are important considerations when finishing the job:
- Cleaning up as you go not only keeps your work area safe it also lessens the amount of work required at the end of the job.
- Dispose of rubbish, off-cuts and old materials in accordance with site-disposal requirements or your own company policy.
- Ensure all materials that can be recycled, such as copper pipe offcuts, are separated from general refuse and placed in the appropriate location for collection.

Check your tools and equipment

Tools and equipment need to be checked for serviceability to ensure that they are ready for use on the job. On a daily basis and at the end of the job check that:
- all items are working as they should with no obvious damage
- high-wear and consumable items such as cutting discs, drill bits and sealants are checked for current condition and replaced as required
- where any item requires repair or replacement, your supervisor is informed and if necessary mark any items before returning to the company store or vehicles.

Complete caravan compliance plate and all necessary job documentation

Most jobs require some level of documentation to be completed. Avoid putting it off till later as you may forget important points or times.
- Ensure that any relevant State or Territory gas installation notice is completed fully and submitted as per local protocols.
- Engrave and fix a metallic gas compliance plate to the caravan as per the Standard.

 **REFER TO AS/NZS 5601.2 SECTION 10**

- Ensure all SWMS and/or JSAs are accounted for and filed as per company policy
- Complete your timesheet accurately
- Reconcile any paper-based material receipts and ensure electronic receipts have been accurately recorded against the job site and/or number.

 COMPLETE WORKSHEET 1

SUMMARY

While sharing many of the same characteristics of standard gas fitting work, you have now learned a number of key differences that apply to the installation of LPG in caravans. Of note these include:
- restricted range of piping materials – no press fit fittings permitted
- specific cylinder mounting and location requirements
- special regulator fitting and pressure requirements
- external separation of the main pipe run from all pipework branches
- a 14 kPa pipework test
- specific ventilation calculation processes.

GET IT RIGHT

The photo below shows an incorrect practice in a cylinder compartment in a caravan.

Identify the incorrect method and provide reasoning for your answer.

WORKSHEET 1

To be completed by teachers

Student competent ☐

Student not yet competent ☐

Student name: _____

Enrolment year: _____

Class code: _____

Competency name/Number: _____

Task

Review this chapter and answer the following questions.

1. For installation within a caravan, are any pipe joints permitted between the branch point and a manual shut-off valve? Provide a reference from AS/NZS 5601 to support your answer.

   _____

2. Where must the main run of caravan gas pipe be located? Provide a reference from AS/NZS 5601 to support your answer.

   _____

   _____

   _____

3. With all appliances running at maximum gas consumption, what is the maximum outlet pressure of a caravan cylinder regulator? Circle the correct answer.

   3.5 kPa   3 kPa   7 kPa

4. List three locations approved for the installation of a caravan gas cylinder. Provide a reference from AS/NZS 5601 to support your answer.

   i _____

   ii _____

   iii _____

5. Is an unflued space heater fitted with an oxygen depletion pilot permitted for use in a caravan? Provide a reference from AS/NZS 5601 to support your answer.

   _____

   _____

   _____

6. What would be the minimum required area in mm² for a lower opening of a caravan designed to accommodate six people and having appliances with a total input of 90 MJ/h? Provide a reference from AS/NZS 5601 to support your answer.

___

7. What is the maximum spacing between clips securing horizontally installed rigid pipe? Provide a reference from AS/NZS 5601 to support your answer.

___

8. Which caravan appliances require a manual shut-off valve at the inlet connection of the appliance? Provide a reference from AS/NZS 5601 to support your answer.

___

9. Where a hose assembly is used to connect a single appliance from the cylinder regulator, what class of hose must it be? Provide a reference from AS/NZS 5601 to support your answer.

___

10. How would you protect a pipe from abrasion where it penetrates the floor?

___

11. Where should the upper and lower vents be located? Provide a reference from AS/NZS 5601 to support your answer.

___

12 Find a ventilation solution for a caravan designed to sleep four people and having appliances with a total input of 52 MJ/h. Provide a reference from AS/NZS 5601 to support your answer.

_____

_____

_____

_____

13 Your boss has asked you to undertake a caravan installation for a local manufacturer. Upon arrival, the manager asks you to use a batch of press-fit crimped fittings because he has to save money on the pipe installation. Would this be acceptable? Provide a Standards reference to support your answer.

_____

_____

_____

14 At what height must an LPG regulator be installed in relation to the cylinders? Provide a Standards reference to support your answer.

_____

_____

_____

15 What is the maximum length of any hose assembly connecting a regulator to a cylinder? Provide a Standards reference to support your answer.

_____

_____

_____

INSTALL LPG SYSTEMS IN MARINE CRAFT

Learning objectives

Areas addressed in this chapter include:
- installing marine craft gas systems
- gas appliances for marine craft
- marine craft gas detection system requirements.

Before working through this section it is very important that you understand the basic fundamentals covered in Part A of this text and the following chapters from Part B:
- Chapter 8 Size consumer gas piping systems
- Chapter 9 Install gas piping systems
- Chapter 10 Install gas pressure control equipment
- Chapter 11 Purge consumer piping
- Chapter 13 Install and commission Type A gas appliances
- Chapter 16 Install LPG storage of aggregate storage capacity up to 500 litres.

The requirements for marine craft gas installations are to be found within the AS/NZS 5601.2 (Gas Installations). While Part 2 covers both caravan and marine craft, this chapter will cover those installation requirements specifically related to marine applications.

Overview

LPG is used extensively in marine craft applications. When installed and used correctly it is a safe and reliable fuel source for on-board gas appliances. Depending upon local State and Territory licensing requirements, you will have an opportunity to provide gas fitting services in a wide range of marine applications.

Marine craft installations have some very specific requirements that are quite different from those you may be familiar with from general construction-type installations, and where you are not undertaking this work regularly you will need to review the Standards in detail to remind yourself of these differences.

While the term 'marine craft' as a category includes all types of waterborne vessels, the Standard also adopts the term 'boat' for most general references. Due to the quite broad understanding and use of the term 'boat', it is very important that you have a clear understanding of the scope and application of this term within the Standard before proceeding further.

REFER TO AS/NZS 5601.2 'SECTION 1 SCOPE AND GENERAL'

Essential terminology

As defined in the Standard, the term 'boat' includes a range of private and commercial vessels including:
- yachts, both inshore and ocean-going
- motor boats, trailer sailers, pleasure cruisers and houseboats (check with local requirements)
- commercial vessels such as fishing boats, trawlers, floating restaurants, ferries and tour boats.

Exclusions

If you are undertaking any gas fitting work on certain commercial vessels, you may need to refer to the Uniform Shipping Laws Code of Australia. This code has been adopted by the Australian Transport Council as the foundation for uniform legislation between the Commonwealth, States and Territories. If in doubt, contact your technical regulator before starting any work on commercial marine craft.

Be aware that wherever commercial catering equipment is to be installed in any of these vessels, then these appliances should be installed in accordance with Part 1 of the AS/NZS 5601 and not Part 2.

REFER TO AS/NZS 5601.2 SECTION 1 'EXCLUSIONS'

Marine craft terms

Before you begin gas fitting work on marine craft, you need to be familiar with some basic nautical terms. Some of these terms may not apply to all vessels and may depend upon the size, structure and vessel application:
- bilge – this area is the lowest internal area of a boat or ship
- lower deck – generally regarded as the last trafficable deck above the bilge
- upper deck – normally the outer trafficable deck exposed to the elements (does not include the cabin top)
- main deck – usually the most used deck and often incorporates an external section where the helm or tiller is located
- cabin top – a protective structure above or on the upper deck
- cockpit – on yachts, the recessed area from where the vessel is steered with the use of a helm or tiller
- galley – the nautical term for a kitchen
- head – the nautical term for a toilet
- bow – the front of the vessel (the pointy end!)
- stern – the back of the vessel
- starboard – as you face forwards, the right-hand side of the vessel
- port – as you face forwards, the left-hand side of the vessel.

Identify LPG system requirements

Prior to commencing any LPG work on marine craft you must fully plan the job to ensure all requirements of the Standard are met. This section reviews each of these main considerations.

Work, health and safety considerations

Gas installations on marine craft share many of the installation hazards and considerations found in other forms of gas fitting work, and it is very important that you have reviewed Chapter 5 'Gas industry workplace safety'.

In addition to these hazards, additional emphasis should be applied to the issue of electrical safety.

Dual power supply

The majority of marine craft run a 12 V power system, but can also incorporate a dual power supply to run lighting and appliances when moored in a marina.
- 240 V – When docked in a marina, the vessel can be plugged into the wharf/pontoon 240 V power supply for the duration of the stay. Lighting, heaters and refrigerators can be then selected to or switched to run on this supply.

- 12 V – When underway or anchored in remote locations where a 240 V supply is not available, some marine craft electrical items can be switched to operate on a 12 V battery often charged by the vessel's motor.

This dual voltage characteristic means that you will encounter wiring layouts around the boat and within appliances that will be dedicated to the 12 V and 240 V power supply. The identification of these systems is not always clear, particularly with older vessels and those that have been altered many times by owners.

- Before working on gas appliances with dual power supply connections and anywhere in the vicinity of boat wiring, always ensure the 240 V connection is fully disconnected.
- Test all wiring and appliances for stray current with a non-contact voltage detector or multi-meter before *and* during work.
- Never assume a 12 V supply or component is safe to touch, as you have no guarantee that it has not been accidentally crossed over with the 240 V system

Always move and work around marine craft with the same caution and respect for electrical hazards as you are required to do on any building site.

Working below deck on a commercial vessel can be classed as a confined space workplace. This may require that you have undertaken training to deal with the hazards associated with working in such an environment. Check with your technical regulator for local requirements.

Dangerous atmospheres

You should treat any vessel with enclosed spaces as intrinsically dangerous. This mindset is necessary particularly in relation to work to be carried out on older vessels. You will have no idea what dangers you might have to deal with once you go below. Hazards include:

- fuel gas leaks. LPG is heavier than air and will sink to the lowest part of the boat. In most cases this is the bilge and will not disperse unless removed. The presence of this gas will become a permanent hazard and you should always do a precautionary check with a gas detector before commencing work.
- carbon monoxide. CO gas is potentially deadly. It is invisible and odourless and on many vessels can be leaking from engine compartments. It is highly recommended that you adopt the use of a personal CO detector that remains on at all times you are below decks.
- oxygen depleted atmosphere. This is a particular hazard of steel boats. Larger commercial vessels of steel construction often incorporate steel-lined holds and possible poorly ventilated enclosures. Any unpainted areas will corrode and in doing so consume oxygen in the process. NEVER enter such spaces without being trained in and carrying out atmospheric gas testing.

System layout and gas load

When preparing to undertake gas work on marine craft, you must first confirm the proposed layout of the LPG cylinder supply and the location of all gas appliances.

Confirm the following details from a site inspection or floorplan and elevation drawings.

- Cylinder location. The cylinder/s may be located externally on the deck or internally within a compartment.
- Appliances. Determine what gas appliances are to be installed and confirm the following:
  - location within the vessel
  - appliance certification number
  - appliance MJ/h input rating from the data plate and/or manufacturer's specifications
  - whether the appliance will have a continuously burning flame (pilot light) – in this instance a gas detection and shut-off system will be required.

Note: Second-hand appliances must only be installed if they meet the safety requirements of your State or Territory technical regulator – do not install such appliances without checking first!

Confirm whether the appliance will have a continuously burning flame (pilot light) – in this instance a gas detection and shut-off system will be required.

Every aspect of cylinder, piping and appliance locations is subject to requirements detailed within the Standard. These are discussed in more detail in the following sections.

Regulator selection and capacity

Regulators used on marine craft must be certified products in accordance with AS 4621 or UL 144. You must check that any product you purchase complies with these Standards. Furthermore, regulators must have built-in over-pressure protection and be of the dual-stage (integral two-stage) configuration.

Single stage barbecue-style regulators that screw directly into the cylinder are not permitted for use in marine craft installations.

If you are not sure the regulators supplied by your plumbing hardware supplier are adequate, it is your responsibility to source the information from the manufacturer or supplier.

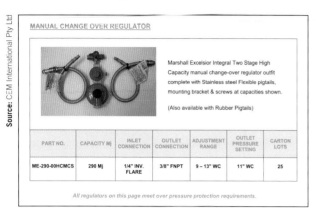

FIGURE 18.1 Use supplier product websites to match compliant regulator MJ/h capacity to your installation requirements

A range of both manual change-over and automatic change-over regulators are available to suit differing installation needs. Automatic change-over regulators are particularly useful where interruption of the gas supply will be inconvenient.

Pipe sizing

Pipe sizing for marine craft follows the same basic process as applied to other gas installations. When pipe sizing for boats, you are to use the propane pipe sizing tables for copper at maximum pressure drop of 0.25 kPa and supply pressure of around 3 kPa as per AS/NZS 5601.1 Appendix F or AS/NZS 5601.2 Appendix C. Other permitted materials are to be sized using recognised formulas and methods.

Marine craft LPG cylinder selection and location

Marine craft LPG cylinders must be located and mounted in accordance with AS/NZS 5601.2 and meet the design requirements of AS 2030.1. Importantly, this Standard details the level of corrosion resistance for cylinders in various environments identified with a number from 1 to 4 on the cylinder guard (Figure 18.2). Cylinders marked with a '4' identification number must not be used for any marine craft application.

Marine craft operate in particularly harsh and corrosive environments. Therefore, it is highly recommended that all LPG cylinders have a galvanised protective coating, or be fabricated from stainless steel or composite polymer. Painted steel cylinders will rust extremely quickly, particularly in proximity to salt water.

Cylinder capacity

Cylinders used in non-commercial applications must not exceed a nominal capacity of 25 L (10 kg). Larger cylinders may be used in commercial applications subject to all technical regulator requirements and relevant Standards.

Cylinder location

Very strict restrictions are placed on the location of LPG cylinders on board marine craft. As you will recall, LPG is heavier than air, so the cylinder must be installed in a location where there is no chance of escaping gas accumulating in the lower decks or bilge of the vessel.

- Cylinders must always be stored in the upright position with the valve uppermost, ensuring that it is always in contact with the vapour space of the cylinder.
- Cylinders must not be installed inside the vessel unless within a vapour-proof compartment.

LEARNING TASK 18.1 PIPE SIZING

The exercise below can be completed as revision of the pipe size processes you have learned previously.

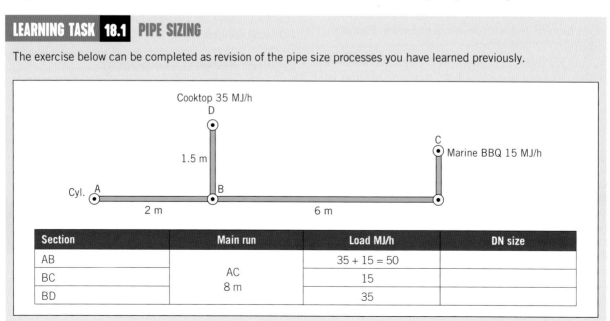

Section	Main run	Load MJ/h	DN size
AB	AC 8 m	35 + 15 = 50	
BC		15	
BD		35	

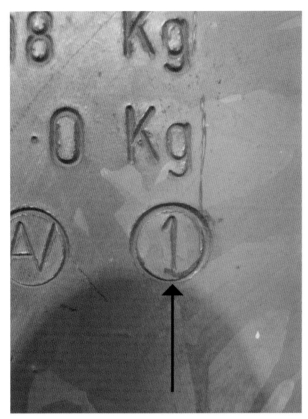

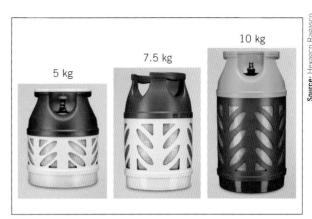

FIGURE 18.2 (Continued)

- Cylinders must be located:
  - externally on the cabin top, on the upper deck or in a self-draining cockpit
  - in a cylinder compartment
  - in a location where they will not impede movement
  - at least 1 m away from any horizontal opening into the vessel
  - at least 1 m away from any opening below the cylinder site
  - at least 150 mm below any opening into the vessel, as measured from the top of the cylinder valve and the lowest edge of the opening.

The ignition source exclusion zone around a marine craft cylinder installation is the same as for all LPG exchange and in-situ fill cylinder locations.

REFER TO AS/NZS 5601.2 'MEANS OF CONFORMANCE – CYLINDERS'

Cylinder compartments

Compartments allow cylinders to be installed so that they are protected from theft, damage and exposure to the elements. However, you must realise that special requirements apply where cylinders are installed within compartments.

- The compartment must be sealed to prevent any gas from entering the vessel.
- The compartment must be waterproof and corrosion resistant.
- The compartment drain hole must be a minimum of 19 mm in diameter and be drained to a point below the locker bottom but where it cannot be submerged.
- The compartment must not contain any electrical equipment. This does not include a non-sparking (encapsulated) shut-off device – e.g. a gas-detection shut-off solenoid valve.
- The compartment must be clearly identified with a durable label as stipulated in AS/NZS 5601.2.

FIGURE 18.2 Fully galvanised, stainless steel or polymer composite cylinders should be selected for marine applications

> **LEARNING TASK 18.2**
>
> **IDENTIFY LPG SYSTEM REQUIREMENTS**
>
> 1 What are three WHS considerations regarding dangerous atmospheres on a marine craft?
> 2 When LPG cylinders are to be installed internally in a marine craft, how should they be located?
> 3 When LPG cylinders are installed externally, what is the minimum distance the cylinder must be away from any horizontal opening into the vessel?

Marine craft appliance types

Marine craft appliances generally fall into the following categories:
- cooktops
- upright cookers
- ovens
- refrigerators
- water heaters
- room-sealed continuous flow water heaters
- room-sealed space heaters.

REFER TO AS/NZS 5601.2 'MEANS OF CONFORMANCE – GAS APPLIANCES'

Prohibited appliances

The relatively small, enclosed spaces of marine craft dictate that certain appliances are not permitted. These include:
- water heaters other than room-sealed types
- appliances designed to operate on an unregulated supply or in excess of 2.75 kPa
- space heaters other than room-sealed types.

Depending on local requirements, allowances may be made for the use of flueless space heaters on houseboats. You will need to confirm this with your technical regulator.

Appliance connections

Appliances can be connected directly to the consumer piping with a manual shut-off valve at the appliance inlet. Where a cooker is to be connected with a hose assembly, the hose must be AS/NZS 1869 certified and be sized correctly.

Some cooking appliances are fitted with gimbals and surrounds that act to compensate a little for the movement of the vessel. This lessens the chance of hot liquids being splashed out of pots and prevents pots and pans dislodging from their position while cooking. Where an appliance is fitted with gimbals, it must be connected to the consumer piping with a hose assembly as in AS/NZS 1869.

REFER TO AS/NZS 5601.2 'COOKING APPLIANCES'

Flame safeguard requirements

All new gas appliances used in marine craft must be fitted with a flame safeguard system to all burners. In the past this was not the case and although the Standards are not retrospective, if you plan to work on an existing marine installation you may need to upgrade appliances or, at the very least, recommend an upgrade for safety reasons.

Marine craft piping

In the Standard, the pipework between the cylinder valve and the regulator is classed as high-pressure piping, while the pipework installed from the outlet of the regulator is referred to as low-pressure piping.

REFER TO AS/NZS 5601.2 SECTION 5 'PIPING AND FITTINGS'

High-pressure pipe installation requirements

Three methods and materials are available for the connection between the cylinder and the regulator:
1 POL × flare copper pigtail formed into a loop
2 a hose assembly certified to AS/NZS 1869 Class F with a maximum length of 600 mm
3 stainless steel pipe – 316 or 304 grade.

Normal hose assemblies are not acceptable. The hose must incorporate an excess flow valve in the POL fitting that will restrict gas flow in the event of damage to the gas system. Double check manufacturer's information to confirm that you are purchasing the correct hose.

REFER TO AS/NZS 5601.2 'MEANS OF CONFORMANCE – PIPING AND FITTINGS'

Low-pressure pipe installation requirements

Marine craft pipe installations have very particular needs that derive from the mobile nature of the craft and the harsh environment in which they operate. You must carry out every installation with a view to ensuring the pipe is protected from damage and vibration.

Materials

For boats, pipe materials from the outlet of the regulator are restricted to the following:
- a certified hose assembly compliant with AS/NZS 1869 Class A, B, C or D
- stainless steel – 316 grade
- *fully annealed* copper pipe, fully sleeved with plastic applied by the manufacturer – type A or B.

It is important to note that the only acceptable fittings and joint types to be used with copper tube are flared compression, silver-brazed capillary or expanded tube sockets. Press-fit crimped end connectors are not permitted. Other prohibited fittings are also listed in the Standard.

> **REFER TO AS/NZS 5601.2 'PROHIBITED TYPES OF PIPING, JOINTS AND FITTINGS'**

> Press-fit crimped fitting systems are not permitted in any manner for use in marine craft installations. All joints in copper pipe must be copper alloy flared compression or silver-brazed capillary.

Of the three approved pipe types, the most commonly installed material is fully annealed, plastic coated copper tube (see Figure 18.3). Pipe fixings should be of a non-ferrous material compatible with the pipe material being used.

FIGURE 18.3 Fully annealed copper *must* have a plastic coating applied by the manufacturer

LEARNING TASK 18.3
MARINE CRAFT APPLIANCE AND REQUIREMENTS

1. Name three types of appliances that are not permitted to be installed on a marine craft.
2. Name three approved materials for the connection between the cylinder and the regulator.
3. Are Class A gas hose assemblies permitted to be used between the cylinder and regulator?

Prepare for installation

All materials used for marine craft installations must be checked for serviceability to ensure that manufacturing faults or pre-purchase damage has not occurred. Faults in components cannot be repaired and the component must be replaced. Only the best quality materials should be used in gas installations.

Always make an extra effort to protect your materials during storage and handling.

- Keep dust and water out of pipes with caps or plugs.
- Keep fittings in an orderly and clean manner, ensuring that they are not contaminated with dirt or other foreign substances.
- Avoid rough handling of your pipe, fittings and materials.
- Ensure that brass flared fittings are always stored with the nut screwed on, to prevent damage to the mating faces of the flare.

Appropriate tools and equipment

Apart from general hand tools used on almost all jobs, tools and equipment used for marine craft installations will to a large extent be dictated by the type of pipe material and jointing method being used. In most cases copper pipe, silver-brazed and compression flare fittings are used. The table below shows just a selection of items that may be required for the installation of copper piping.

Tools and equipment
• Oxy/acetylene or MAPP gas brazing equipment
• Flaring tools
• Copper pipe cutters and reamer
• Benders
• Internal and external bending springs
• Expanders
• Shifters

Other considerations are related to handling of the pipe and the proposed installation location. Working within marine craft installing pipe can be very tight and restrictive, and in some locations you must also monitor oxygen levels, carbon monoxide levels and the presence of fuel gas. Items may include:

- inspection cameras
- mirrors
- LED lighting
- personal gas detection sensors.

Determine and install LPG gas systems

Gas appliances on marine craft have very specific installation requirements that reflect the challenging environment in which they operate.

FIGURE 18.4 Inspection cameras enable you to visually access an otherwise enclosed space

Regulator choice

You are required to install certified integral dual-stage regulators incorporating the required overpressure protection design. These regulators must be adjustable for pressure setting.

Regulator mounting

The mounting and installation of the regulator must comply with the following points:

- The regulator must be securely fixed to the marine vessel with the vent pointing downwards. Regulators that connect directly to the cylinder with a POL fitting are not permitted for use, as they are prone to damage.
- The regulator must be located so as to allow any liquid that forms in the pigtails to drain back to the cylinder.
- Regulator outlet pressure must be set to a maximum of 3 kPa with all appliances operating.

REFER TO AS/NZS 5601.2 'MEANS OF CONFORMANCE – GAS PRESSURE REGULATORS'

FIGURE 18.6 A dual stage manual change-over regulator

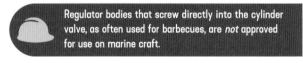

Regulator bodies that screw directly into the cylinder valve, as often used for barbecues, are *not* approved for use on marine craft.

FIGURE 18.5 Oxy/acetylene or MAPP gas brazing equipment will be required for almost all copper pipe marine craft installations

Installing the regulator

Regulators installed in marine craft must be installed according to specific requirements.

Design and location of pipe

As marine craft are exposed to considerable and sometimes extreme movement, the pipe must be located and installed with flexibility to account for structural movement and vibration. The pipe should also be adequately protected so that it is not exposed to possible damage, and be outside of any structure or false bottom area.

Penetrations of bulkheads, partitions and decks should be vapour-proof and use some form of rubber grommet to protect the pipe against abrasion.

The line should be constructed in continuous lengths with no joints between a branch and a manual shut-off valve unless it can be shown to be impractical. Pipe lines must always be continuous where passing through an engine room or sleeping area.

Pipe should also be supported with compatible clips at the maximum distances stipulated in AS/NZS 5601.2 and within 150 mm from each tee or change of direction.

REFER TO AS/NZS 5601.2 'INSTALLATION OF PIPING'

Using hose assemblies for appliance connections

Hose assemblies can be used between rigid piping and the appliance where approved by the manufacturer. Where only one appliance is installed, you are permitted to use a hose assembly between the outlet of the regulator and the appliance (again subject to the appliance manufacturer's specifications). In all cases the hose assembly must be:
- compliant with AS/NZS 1869 to the class specified in AS/NZS 5601.2
- of a continuous and minimum practicable length
- where applicable, fixed with compatible materials in accordance with the maximum distance specified in AS/NZS 5601.2.

Quick-connect devices

Quick-connect devices or 'bayonet fittings' are not permitted for use inside a boat, as they would then allow consumers to connect flueless space heaters. As these are prohibited appliances in the small confines of a boat, so too are quick-connect devices. Contact the technical regulator in those jurisdictions where such appliances may be permitted for installation on a houseboat.

Marine craft ventilation

Marine craft ventilation solutions are provided in AS/NZS 5601.2. The relatively small enclosure spaces in most marine craft mean that these installations have some particularly specific requirements. Importantly you must note that there is no allowance made in AS/NZS 5601.2 for adventitious ventilation. Therefore, where non-room sealed appliances are used you *must* provide an additional source of air supply.

Vent location and type

Marine craft on which gas appliances are installed must have at least two additional ventilation openings installed direct to the outside (see Figure 18.7). To provide a cross flow of air, the vents are required to be installed at either end and/or on opposite sides of the area to be ventilated. This may

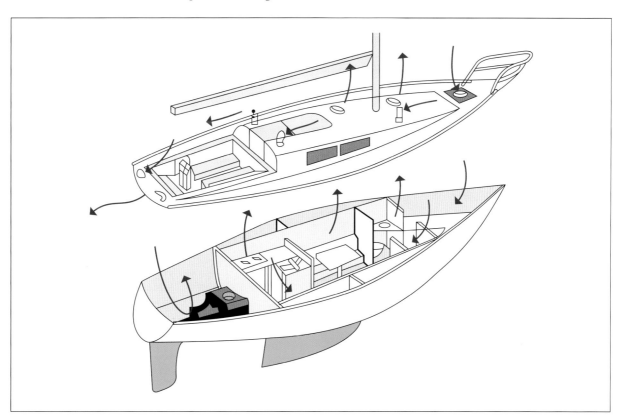

FIGURE 18.7 Marine craft require good cross-ventilation where gas appliances are installed

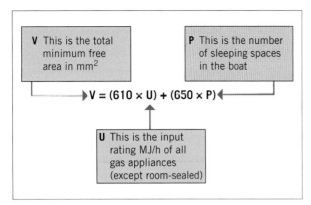

FIGURE 18.8 Use this formula to determine ventilation requirements for marine craft

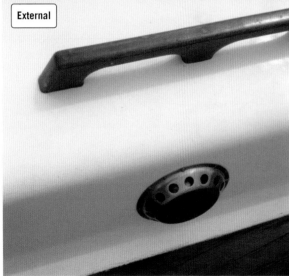

FIGURE 18.9 Marine 'mushroom' style ventilation

be the cabin or, preferably, the deckhead to the underside of the deck.

Total free ventilation area must be at least 4000 mm² or in accordance with the formula in Figure 18.8, whichever is the greater:

Many vents now used in marine craft will incorporate mesh screens that reduce the actual free ventilated opening area. Where such screens are used, it is recommended in the Standard that the ventilation solution be doubled.

Marine craft operate in a particularly harsh environment and the vents you choose for the task must not only comply with the minimum free-ventilated area, but also be watertight and resistant to corrosion.

- Mushroom style vents are particularly common on smaller vessels and yachts. Externally they feature a low profile that will not foul rigging and sheets (ropes) (see Figure 18.9).
- Dorade-style vents are more commonly found on larger vessels. These vents normally feature a curved cowl mounted on an integral or additional drainage box. Any water that enters the cowl is drained away while still allowing air to enter the cabin (see Figure 18.10).

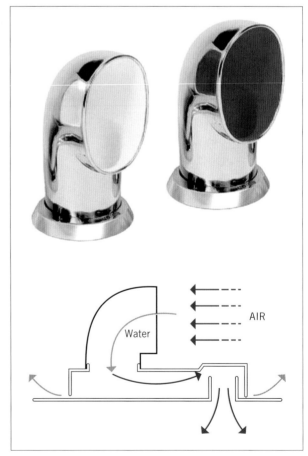

FIGURE 18.10 Dorade cowl and drainage box ventilation components

Such vents may be available at your local chandler (boat and yacht supplier). A wide range of vents is available, and you will need to work closely with the boat owner to ensure that a suitable model is chosen to suit the style of craft.

REFER TO AS/NZS 5601.2 'VENTILATION'

LEARNING TASK 18.4

DETERMINE AND INSTALL LPG GAS SYSTEMS

1. Are LPG regulators permitted to be connected directly to the cylinder?
2. Are quick-connect devices or 'bayonet fittings' permitted to be installed on a marine craft.
3. When providing ventilation for a marine craft, what is the minimum total free ventilation area required?
4. What are the two types of vents commonly used on marine craft?

Test and commission LPG and detection system

Like all gas fitting work, marine craft LPG installations also require comprehensive testing.

Conducting a pipework test

Pressure testing a marine craft pipe installation involves using the test equipment and testing processes described in Chapter 9 'Install gas piping systems'. AS/NZS 5601.2 provides detailed steps for a 14 kPa pipework strength test. However, some technical regulators may require a higher test pressure, so it is advisable to check your local requirements before proceeding.

You are required to conduct a 14 kPa pressure test for a period of 5 minutes with no detectable loss of pressure. A standard water gauge manometer would be unsuitable for this test and, therefore, you will require a calibrated digital manometer.

Conducting a gas tightness test

Following installation of the appliances, you must carry out a gas tightness test in order to prove that the appliance connections right through to the appliance control valves themselves are gas-tight at nominal operating pressure.

The Standard requires that with all appliances operating the LPG regulator must be set to supply no greater than 3 kPa gas pressure and have no loss of pressure over a period of 5 minutes.

REFER TO AS/NZS 5601.2 APPENDIX E AND YOUR LOCAL TECHNICAL REGULATOR

If system testing is successful, proceed with appliance commissioning procedures as per manufacturer's instructions and the Standard.

REFER TO AS/NZS 5601.2 APPENDIX H

Marine craft gas detection system requirements

Not all marine craft with LPG appliances will require a gas detection system. The Standard states that a gas detection system is required in any marine craft where an appliance with a continuously burning flame is installed below the upper deck and there is no low-level ventilation in the area in which it is situated. Of course these are minimum requirements and there is no reason why a detection system should not be installed even for appliances without a continuously burning flame. Local regulations may also require a broader use of detection systems and you should always consult your technical regulator.

A gas detection system must be a certified product designed for a marine craft application. Whether you purchase a gas detection system from a yacht chandler or a gas fitting supplier, always double check that the device is certified and complies with the requirements listed in AS/NZS 5601.2.

FIGURE 18.11 Marine craft gas detection system components

REFER TO AS/NZS 5601.2 'GAS DETECTION SYSTEM REQUIRED – BOATS ONLY'

Installation instructions for a gas detection system are detailed by the manufacturer and primary requirements are specified in AS/NZS 5601.2.

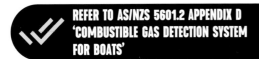
REFER TO AS/NZS 5601.2 APPENDIX D 'COMBUSTIBLE GAS DETECTION SYSTEM FOR BOATS'

A marine craft gas detection system is designed to shut off the supply of gas whenever the concentration of gas in air exceeds 25 per cent of the lower explosive limit (LEL).

The system is made up from a minimum of four primary components:
- a solenoid valve
- a control panel
- sensors
- an alarm.

Locating the solenoid valve
The solenoid valve must be located between the cylinder outlet and the regulator. It should be of the non-sparking type and rated for the pressure applied. Some manufacturers will supply the solenoid pre-fitted to the outlet of the regulator.

Locating the control panel
The control panel will incorporate a visual alarm as well as other system indicators and therefore should be located in a clearly observable and accessible location, usually adjacent to all other vessel controls. It should also be protected from the elements.

Locating the sensors
The Standard requires that a minimum of two sensors be installed:
- The first sensor is installed at the lowest point where leaking gas is likely to accumulate. This is normally the bilge or lower deck.
- The second sensor is to be located near the appliance, below the level of the lowest flame.

Sensors must be mounted in the manner prescribed by the manufacturer, to ensure that they are able to detect the presence of gas.

Locating the alarm
The key consideration in deciding the location of the alarm is to ensure that it would be heard in all locations on the craft, even when the craft is working under normal operating conditions.

Consumer instruction plates
All switch indicators, control panel markings and consumer instructions must be permanently fixed to the vessel and of a durable material (see Figure 18.12). The use of adhesive labels is not permitted. Some markings are integral with the control panel as per Figure 18.12.

FIGURE 18.12 This control panel incorporates permanent labelling indicating each of the system functions

REFER TO AS/NZS 5601.2 SECTION 10 AND YOUR LOCAL TECHNICAL REGULATOR

LEARNING TASK 18.5
TEST AND COMMISSION LPG AND DETECTION SYSTEM

1 What is the minimum test pressure and time period required when conducting a pipework test on a marine craft?
2 Why should the technical regulator be consulted in your local area prior to conducting a pipework test?
3 When is a gas detection system required to be installed on a marine craft?
4 What are the four primary components that make up a gas detection system?
5 What is the minimum number of sensors required on a marine craft with a gas detection system?
6 Where shall the audible alarm on a gas detection system be located? Provide a reference from AS/NZS 5601.2 to support your answer.

Clean up

The job is never over until your work area is cleaned up and tools and equipment accounted for and returned to their required location in a serviceable condition.

Clean up your area

The following points are important considerations when finishing the job:
- Cleaning up as you go not only keeps your work area safe it also lessens the amount of work required at the end of the job.
- Dispose of rubbish, off-cuts and old materials in accordance with site-disposal requirements or your own company policy.
- Ensure all materials that can be recycled, such as copper pipe offcuts, are separated from general refuse and placed in the appropriate location for collection.

Check your tools and equipment

Tools and equipment need to be checked for serviceability to ensure that they are ready for use on the job. On a daily basis and at the end of the job check that:
- all items are working as they should with no obvious damage
- high-wear and consumable items such as cutting discs, drill bits and sealants are checked for current condition and replaced as required
- where any item requires repair or replacement, your supervisor is informed and if necessary mark any items before returning to the company store or vehicles.

Complete marine craft compliance plate and all necessary job documentation

Most jobs require some level of documentation to be completed. Avoid putting it off till later as you may forget important points or times.

- Ensure that any relevant State or Territory gas installation notice is completed fully and submitted as per local protocols.
- Engrave and fix a metallic gas compliance plate adjacent to the cylinder location in the boat as per the Standard (see Figure 18.13).

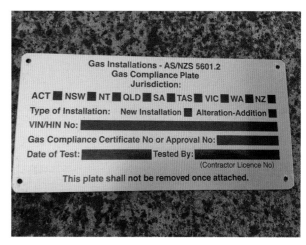

FIGURE 18.13 A compliance plate must be fitted to all installations

REFER TO AS/NZS 5601.2 SECTION 10

- Ensure all SWMS and/or JSAs are accounted for and filed as per company policy.
- Complete your timesheet accurately.
- Reconcile any paper-based material receipts and ensure electronic receipts have been accurately recorded against the job site and/or number.

COMPLETE WORKSHEET 1

SUMMARY

While sharing many of the same characteristics of standard gas fitting work, you have now learned of a number of key differences that apply to the installation of LPG in marine craft. Of note these include:
- restricted range of piping materials
- specific cylinder mounting and location requirements
- special regulator fitting and pressure requirements
- external separation of the main pipe run from all pipework branches
- a 14 kPa pipework test
- specific ventilation calculation processes
- requirement for a gas detection system for some installations.

GET IT RIGHT

The photo below shows an incorrect practice when installing a gas detector for a gas detection system. Identify the incorrect method and provide reasoning for your answer.

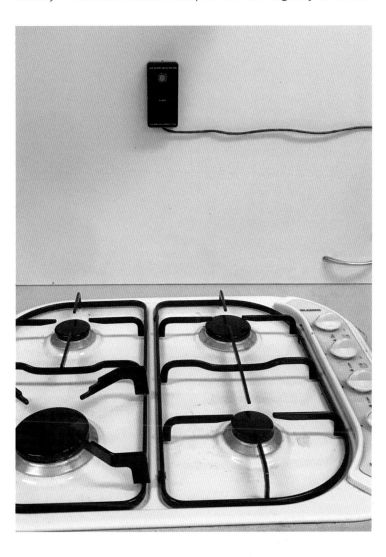

_____
_____
_____
_____
_____

WORKSHEET 1

To be completed by teachers	
Student competent	☐
Student not yet competent	☐

Student name: _____

Enrolment year: _____

Class code: _____

Competency name/Number: _____

Task

Review this chapter and answer the following questions.

1. What two documents should you refer to if planning to undertake an LPG installation on a fishing boat?

    i _____

    ii _____

2. List two locations in boats in which joints in pipes are not permitted. Provide a reference from AS/NZS 5601 to support your answer.

    i _____

    ii _____

3. What is the maximum vertical spacing between pipe supports used to restrain a hose assembly? Provide a reference from AS/NZS 5601 to support your answer.

    _____

4. What minimum test pressure do you apply for a pipework test in a marine craft installation?

    _____

5. Which of the following appliances are prohibited for installation on a marine craft? Circle the answers.
    - cooktops
    - upright cookers
    - flueless space heater
    - ovens
    - refrigerators
    - high-pressure camping cookers
    - water heaters
    - room-sealed continuous flow water heaters

- flueless water heaters
- room-sealed space heaters

6. What is the minimum clearance between a refuelling point or fuel tank vent outlet and an appliance flue terminal? Provide a reference from AS/NZS 5601 to support your answer.

7. What are the requirements for a hose assembly used to connect an appliance? Provide a reference from AS/NZS 5601 to support your answer.

8. Complete the following: The regulator must have an outlet pressure of no greater than _____ kPa.

9. What is the minimum distance separating a cylinder valve and an opening porthole below the cylinder location? Provide a reference from AS/NZS 5601 to support your answer.

10. Complete the following: The drain from a gas cylinder compartment on a boat must be at least _____ mm in diameter and terminate _____

11. To where must the combustion products of a refrigerator be directed? Provide a reference from AS/NZS 5601 to support your answer.

12. What is the exclusion zone required between an exchange cylinder and an ignition source?

13. What is the minimum number of gas detection sensors required?

14 Where is the upper gas detection sensor located?

_____

_____

_____

15 Is the solenoid fitted between the regulator and the appliances or between the regulator and the cylinder outlet?

_____

_____

_____

16 At what percentage of the LEL is a gas detector designed to sense the presence of gas?

_____

17 Where should the lower gas detection sensor be located?

_____

_____

_____

18 What formula do you use to determine vent sizes for a marine craft? Provide a reference from AS/NZS 5601 to support your answer.

_____

_____

19 Where the piping is exposed, are you permitted to use press-fit crimped fittings on marine craft? Provide a reference from AS/NZS 5601.2 to support your answer.

_____

_____

20 What are the two key requirements related to the selection of copper pipe for use in marine craft?

i _____

ii _____

INSTALL GAS SUB-METERS

Learning objectives

In this chapter you will cover the skills and knowledge necessary for the safe and compliant installation of gas sub-meters.

Areas addressed in this chapter include:
- the purpose and application of gas sub-meters
- identification of system requirements
- installation processes and requirements
- purging and testing procedures
- clean-up of site and equipment.

Before working through this section, it is absolutely important that you complete or revise each of the chapters in Part A 'Gas fundamentals', with particular emphasis on the electrical safety section within Chapter 5 'Gas industry workplace safety' and the characteristics of natural gas and LPG combustion as found within Chapter 6 'Combustion principles'.

Overview

Meters are used throughout the gas industry to measure the volume of gas being used by a consumer. Meters can be found in all sectors of a reticulated gas network, measuring the gas supply of domestic, commercial and industrial customers.

Meter applications

There are two main applications in the use of meters that you need to understand before proceeding further. These are:
- consumer billing meters
- sub-meters.

Consumer billing meters

The consumer billing meter measures the amount of gas passing from the distributor's reticulation network and used by the consumer. The gas distributor or retailer uses this measurement to calculate how much the consumer is going to be charged. Consumer billing meters are principally associated only with gases that are supplied through a reticulation network. In most cases LPG is generally supplied via exchange or in-situ fill cylinders and is priced according to the bulk kilograms or litres delivered at the time.

Sub-meters

A sub-meter is located downstream of the consumer billing meter within the actual consumer piping system. It is used to measure the use of gas for different applications or appliances within the same site so that a pro-rata quantity and cost can be applied to each point of use. Sub-meters are used with all gas types. There are a number of sub-meter applications:
- In shopping centres the centre management is often the primary account holder and pays for the total gas supply to the site based upon the amount flowing through the consumer billing meter. Each individual restaurant, food outlet, hairdresser etc. within the site may have a sub-meter installed and they each pay their proportion of the total gas used (see Figure 19.1).
- Industrial/commercial sites may use sub-meters for different tenancies. Individual businesses may also meter different appliances and applications to more efficiently monitor internal production costs and performance.
- Multi-residential sites may use an internal embedded reticulation network where each dwelling has its own sub-meter measuring individual consumption.

For most reticulation networks a gas fitter is not permitted to install or work on the consumer billing meter unless approved by the distributor. However, the installation and use of sub-meters on many commercial

FIGURE 19.1 Two sub-meters within a large food court complex (each meter serves a different restaurant)

and multi-residential sites sits entirely within your scope of work.

REFER TO AS/NZS 5601.1 'SCOPE AND GENERAL'

The remainder of this chapter will now focus specifically on sub-meter installation requirements.

Determination of system requirements

The first step in determining system requirements is to source the relevant information. This information may come from a range of sources including:
- job sheet details obtained from your employer
- site plans and specifications that detail sub-meter installation requirements
- gas distributor information relating to available pressures
- manufacturer product data and instructions
- company quality assurance requirements that may pertain to your gas installation and may include:
  - product purchasing protocols
  - job timesheet protocols
  - job safety documentation processes.

Basic work health and safety considerations

As with most gas work, the installation of gas sub-meters is likely to require the use of the following basic personal protective equipment (PPE):
- Safety glasses – These are used at any time you are using power, hand tools or equipment that may cause flying debris or contaminants.
- Hearing protection – Both long-term and sometimes immediate damage can happen to your hearing

through excessive noise caused by a range of construction tools and processes.
- Safety boots and protective clothing – These should be used every day on every site. Where hot work is to be undertaken, ensure use of non-combustible clothing. Cotton-drill shirts, trousers and overalls are one option. Safety boots must meet Australian Standard design requirements in order to protect your feet from crush damage, heat and puncture wounds.
- Safe work method statements (SWMS) – Familiarise yourself with both company and site SWMS prior to work commencement. Conduct job safety assessments (JSAs) in order to determine hazard and risk mitigation measures.

Safe working habits

The conduct of any gas installation requires consistent safe working habits to prevent hazards and potential injury to yourself and others. These include:
- good housekeeping. Keep tools, equipment and materials tidy and clean up as you go to prevent trip-hazards and unnecessary obstructions.
- regular electrical safety checks.
- body attitude. Avoid placing your body in a position that may result in injury or make you subject to electrical hazards.
- never leaving open ends in gas lines.

Basic gas and combustion principles

Installation and reading of gas sub-meters requires recall of basic gas and combustion principles. Meters are not gas-specific and only record the passage of gas in litres and cubic metres, allowing them to be used equally on LPG or natural gas. It is your knowledge of gas constituents that enables you to read the sub-meters effectively.

Gases produce a fixed amount of potential energy for each cubic metre. This is known as the heat value of the gas:
- natural gas – 38 MJ/m^3
- LPG – 96 MJ/m^3

This difference in heat value means that for an appliance of the same MJ rating, a natural gas meter will pass considerably more gas than an LPG meter for the appliance to do the same job.

When this gas is burned fully in a correctly adjusted appliance, it is known as complete combustion. The combustion process breaks down the air/gas mixture into three principal products of complete combustion:
1. water vapour
2. carbon dioxide (CO_2)
3. heat.

The products of complete combustion are essentially harmless. However, where an appliance is not correctly commissioned or becomes faulty, the combustion process is often compromised and, as a result, the air–gas mixture is not correctly burned. This process is potentially very dangerous and results in three products of incomplete combustion:
1. carbon monoxide (CO)
2. carbon (soot)
3. aldehydes.

The presence of soot and the ammonia-like smell of aldehydes indicates a poor combustion process. However, it is the presence of CO that presents the greatest hazard. CO is a highly flammable, poisonous, colourless and odourless gas. You must ensure that before, during and after your sub-meter installation work the installation is performing to requirements in order to maintain safe and compliant complete combustion.

Matching the meter to system specifications

In most instances the job plans and specifications will detail the meter requirements. As the installing gas fitter, you still have a responsibility to understand how this meter may have been selected and sized so that you can identify whether, for any unforeseen reason, the specified meter may be unsuitable.

There are two main types of meters available:
- diaphragm meters
- rotary meters.

Diaphragm meters (positive displacement meters)

The diaphragm meter is the most common style of meter found in domestic and commercial gas installations (see Figure 19.2). This form of meter still uses much the same principle of operation as it did when it was first introduced in 1844. The meter is separated into three compartments; these contain the mechanisms that enable the flow of gas to be accurately measured as a volume.

Diaphragm meters are also known as 'positive displacement meters' because of the characteristic cyclical oscillations of their diaphragms, by which gas is alternately drawn in and expelled to determine flow rate.

While some diaphragm meters are designed to handle gas flow rates of up to 140 m^3/h, it is more common to find such meters use up to approximately 85 m^3/h, after which rotary meters are normally specified. Diaphragm meters vary considerably in size, according to the volume of gas they are designed to measure.

Rotary meters

If an installation is likely to have very high gas consumption, the capacity of a diaphragm meter may be insufficient and a rotary meter may be specified instead (see Figure 9.3). Based originally on water pump design,

FIGURE 19.2 Diaphragm sub-meter used to measure an industrial appliance gas consumption

FIGURE 19.3 A rotary meter for a high-volume industrial application

rotary meters were first used for gas installations in the United States in 1920. These meters are designed to accurately measure large volumes of gas by counting the revolutions of a rotating mechanism.

Gas entering the meter flows into a chamber that houses two counter-rotating, lobed impellers in a 'figure 8' configuration (see Figure 19.4). The gas is trapped between the impellers and the side wall of the chamber. For each complete rotor cycle, four defined volumes of gas move through the meter and are recorded by the index counter.

Although rotary meters are generally associated with the high-volume gas consumption normally found in large commercial or industrial installations, low-volume units are also available. These small compact meters are often specified in commercial areas where bulky diaphragm meters are unable to be installed or where, for aesthetic reasons, meter installations need to be unobtrusive (such as in heritage areas).

Meter selection

The kind of sub-meter used for an installation will usually depend upon a number of factors:
- the total gas load of combined installation appliances
- the maximum flow capacity of the meter
- the probable simultaneous demand of gas consumption per hour
- physical size of the meter
- turn-down ratio (rangeability)
- price.

Of these, the turn-down ratio is the most important in most instances.

What is turn-down ratio?

Meter selection is normally based upon the manufacturer's stated turn-down ratio. Also known as 'rangeability', the turn-down ratio of a meter describes the ratio of the maximum to minimum rates of controlled flow that a meter is designed to measure accurately. If the expected demand of a particular installation is outside the designed turn-down ratio of the meter, then you must select a different model or type of meter for the job.

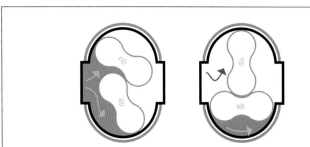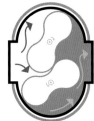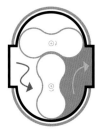

FIGURE 19.4 Internal mechanism of a rotary meter

EXAMPLE 19.1

If the consumption patterns of an industrial gas appliance were to vary from a 'high fire' rate of 100 m^3/h right down to a 'low fire' rate of 10 m^3/h, then you would need a meter able to deal with a maximum flow rate of at least 100 m^3/h and with a turn-down ratio of at least 1:10 (preferably 1:20 or 1:30).

EXAMPLE 19.2

A small rotary meter has a turn-down ratio of 1:30. For a maximum flow capacity of 120 m^3/h, it would accurately read flow rates as low as 4 m^3/h.

EXAMPLE 19.3

A standard diaphragm meter suitable for small commercial installations has a turn-down ratio of 1:600. For a maximum flow capacity of 10 m^3/h (NG: 380 MJ/h), it is able to accurately read flow rates as low as 0.0167 m^3/h (NG: 0.64 MJ/h).

As a generalisation, you should choose a rotary meter for a high-consumption installation but small turn-down ratio. You would choose a diaphragm meter for a lower consumption installation for which a greater turn-down ratio is required. However, manufacturers are offering a wider range of high-performance products all the time, and you must always check first before making a decision.

The examples above are provided to help you understand that there is more to meter selection than just purchasing whatever a supplier gives you. As with any component or system for gas installations, meters must be sized correctly. Meter manufacturers are able to provide you with comprehensive meter selection software that will enable you to determine your installation needs. Alternatively, you can ask their sales and technical representatives to provide a solution based upon your basic installation data.

Meter location

On the subject of sub-meter location, it is perhaps better to consider where they *cannot* be located.

REFER TO AS/NZS 5601.1 SECTION 5 'SUB-METERS'

AS/NZS 5601.1 is quite explicit on prohibited locations for sub-meters. While many of these prohibited locations are specific sites, the Standard also describes prohibited conditions that may be found in otherwise acceptable locations. You must try to think ahead to predict how a certain area may be used, and in what way this will affect your sub-meter.

EXAMPLE 19.4

You may want to site a sub-meter adjacent to an industrial burner within a large manufacturing plant. Initially this appears to be fine; however, further enquiries reveal that the proposed meter location would be subjected to very high temperatures during the production process. It is therefore unsuitable.

Recessed meter locations

Installing a sub-meter in a recess may prevent the meter assembly from being damaged or from being an obstruction. Where the recess is cut into any form of cavity wall, you must ensure that the recess is large enough to hold the meter and its associated parts, and that it is made from a non-combustible material. No part of the wall cavity must be open to the recess box and it must be ventilated directly to the outside.

Ventilation of sub-meter enclosures

Where a sub-meter is to be housed in an enclosure primarily designed to hold gas components, it must be ventilated. Vent sizing and location are determined by following the simple formulas in AS/NZS 5601.1.

REFER TO AS/NZS 5601.1 'VENTILATION OF GAS EQUIPMENT'

LEARNING TASK 19.1 COMBUSTION PRINCIPLES

1. Circle which of the following combustion products are the result of incomplete combustion.
   a. Carbon dioxide
   b. Soot
   c. Aldehydes
   d. Nitrogen
   e. Carbon monoxide
   f. Water vapour
2. If a natural gas appliance has an input of 325 MJ/h, what would be its gas consumption in cubic metres?
3. Is a sub-meter permitted for installation in the foundation area under a building so that it can be close to the gas appliance?

Prepare for sub-meter installation

Detailed preparation before job commencement can save many lost hours fixing problems that could have been easily identified through a methodical process.

Materials

The job specifications or material list provided by your employer should detail the primary requirements for the job. You should pay particular attention to the following:

- Ensure materials on order that have arrived match the requirements of the plans and specifications.
- Check that all gas components are certified for use. Current certification numbers can be checked against the online data provided by relevant entities responsible; examples include the Australian Gas Association and SAI Global. The Gas Technical Regulators Committee (GTRC) website at http://equipment.gtrc.gov.au provides a portal to assist in accessing confirmation of material and product certification.
- Check that all PPE is serviceable and, where relevant, check replacement dates on items such as height safety harnesses.

Required tools and test equipment

As with other gas fitting tasks, your choice of tools will be dependent upon the particular installation requirements and the type of pipe system you are working with. As a general guide this may include the following key items:

- Oxy/acetylene equipment
- Press-fit tool and selection of jaws
- Pipe cutters suitable for type of pipe to be used
- Assorted battery power tools relevant to the task
- Spirit level
- Measuring equipment
- Manometer and pipe test equipment
- Soapy water.

Wherever 240 V power tools might be required for a certain task, remember to check that the tool has been tested and take particular note of the re-test date on the tag. If out of date, let your supervisor know and mark it in some manner that it is not to be used.

Install and test gas sub-meter

The installation process includes a number of basic steps:
- application of appropriate electrical safety requirements
- conduction of a leakage test to confirm system is gas-tight
- isolation of gas pipework
- purging of gas from line to commence work
- installation of sub-meter in reference to the Australian Standard requirements
- testing and rectification of any leaks
- purging of pipework and sub-meter.

These procedures are detailed in the following sections.

LEARNING TASK 19.2 JOB REQUIREMENTS

Based upon a job you have recently done or a potential sub-meter position indicated by your teacher, create a list of tools and equipment specific to that location and material requirements.

Job characteristics	Tools and equipment required
Wall construction material where meter is to be mounted:	
Existing or intended pipe material:	
Other considerations:	

Electrical safety before and during work

Basic electrical safety requires both the testing of pipework for stray current and the specific requirements related to use of bonding straps where a break in pipework is being made. These procedures are required in order to protect you from very real and common hazards.

Test for stray electrical current

Pipework and appliances may be subject to stray electrical current at any time. Never assume an installation is safe to touch without undertaking basic and regular tests. As a bare minimum you should always have a simple volt stick for initial and ongoing tests (see Figure 19.5). Be aware that a volt stick has its limitations, and that with some form of training the use of a multi-meter is the preferred instrument not only for basic testing but also for a range of other applications.

Bonding straps

The removal of a sub-meter from existing pipe work is a hazardous task. A pipe installation may show no sign of stray current unless a break is made in the system and a pathway to earth results. This means that if you remove a valve, a section of pipe, a regulator or meter from a pipe installation, you may actually provide the required path to earth and suffer an electric shock.

The only way to protect yourself is to provide an alternative current path between the two separated halves of the system (see Figure 19.6). By using bonding straps, the gas fitter working on this pipe system has provided an alternative path for any stray current that may become a hazard after the cut is made. The straps must remain in place until the job is completed. Any potential current will then have no path to earth, allowing you the opportunity to test each side before continuing with the work.

FIGURE 19.5 Testing for stray electrical current using a volt stick

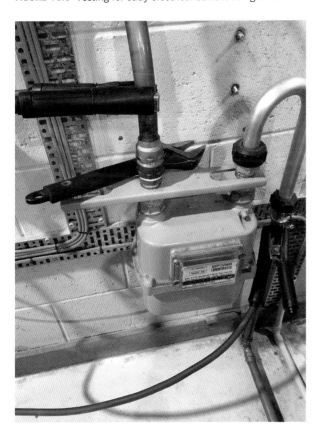

FIGURE 19.6 Bonding straps must be fitted prior to removal of a sub-meter

Many plumbers and gas fitters often neglect to use bonding straps, but it must be emphasised that the risk of electric shock is very real, and many tradespeople have suffered injury and death as a result of poor work practices. The Gas Installation Standards explicitly directs you to use bonding straps. As Australian Standards are specifically referred to ('called up') by legislation, your obligation to use bonding straps is a legal requirement and not a choice.

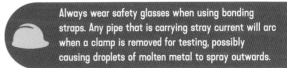

REFER TO AS/NZS 5601.1 'ELECTRICAL SAFETY BONDING OR BRIDGING'

In the following procedure, you can see how bonding straps are used to safely disconnect consumer piping from a sub-meter.

1. Connect one end of the bonding strap to the inlet side of the sub-meter.
2. Connect the other end to the outlet side of the sub-meter.
3. Ensure you scratch your way through to bare metal in both cases.
4. Make the break in the consumer piping at the outlet of the meter.
5. Without touching any part of the system, and ensuring you are wearing safety glasses, remove the end connected to the consumer piping and, using a suitable test instrument, test both the pipe and the bonding strap for any stray current.
6. Replace the bonding strap immediately.
7. If you detect current, ensure no one approaches the area and call an electrician or electrical authority.
8. Leave the bonding straps in position until all work is complete.
9. Check and re-check as required.
10. Only remove bonding straps once the sub-meter has been replaced.

> Always wear safety glasses when using bonding straps. Any pipe that is carrying stray current will arc when a clamp is removed for testing, possibly causing droplets of molten metal to spray outwards.

Commencing work

Having liaised with anyone affected by the work and ensuring that any existing pipework and components are electrically safe you are able to commence work. Initial cut-in of the meter will require some or possibly all of the following steps:

- Undertake a pre-work leakage test in accordance with Appendix E of the Standard to ensure that the pipework is gas-tight.
- Isolate the relevant section of pipe.

- If necessary, carry out a purge of gas from the affected section of pipe in accordance with Appendix D of the Standard.
- Cut into the line where the sub-meter is to be located using safe electrical bridging practices.
- Install support bracket.

Meter support

Sub-meters must be supported independently of the pipe installation, and must impart no strain on the pipe or components. You must normally use a manufacturer-supplied proprietary bracket, or fabricate a compatible bracket or support yourself.

> **EXAMPLE 19.5**
>
> You plan to install a sub-meter in a commercial catering kitchen that is part of a larger restaurant complex. The existing gas inlet is located low on the wall, close to floor level, and you realise that the sub-meter would be unsuitable in that position because the kitchen floor is being constantly wetted and washed out with chemical cleaners. This is likely to corrode the meter, and so you decide to locate it further up the wall.

Figure 19.7 demonstrates what can happen when a sub-meter is not located in accordance with AS/NZS 5601.1. In this instance, the meter was installed on the ground and was frequently wet. Electrolysis has caused the lower casing to be corroded away and, had it not been found by an alert gas fitter, could have resulted in a dangerous, even life-threatening, incident.

FIGURE 19.7 Corrosion to meter casing caused through incorrect location on wet floor area

Accessibility and connection

The meter should be installed in an accessible location, ensuring that the dial face can easily be seen for reading purposes. Always avoid locating a meter in a position that may be unsafe due to its height or cramped conditions. The meter must also be identified with an adjacent sign indicating the gas installation to which it is connected.

On close inspection of a meter, you will notice that the outlet has a non-standard external thread. This means that your standard BSP fittings will not be suitable for direct connection without some form of adaptor. Not all manufacturers will supply the outlet fittings with the meter and you will need to confirm the preferred methods to obtain and fit them. A meter bend (see **Figure 19.8**) allows a degree of movement from side to side so that perfect alignment and location of your consumer piping during installation is not necessary. If you choose to purchase a meter bend assembly, be aware that the connection nuts do not have BSP (British Standard Pipe) threads and you will require an adaptor to connect to your consumer pipe.

Standard BSP adaptor fittings (**Figure 19.9**) can be used to connect directly to your consumer piping. Be sure to always provide a means of disconnection so that the meter can be easily removed if necessary.

Transition from the meter outlet size to large-diameter consumer piping should be made as close as possible to the meter connection to minimise frictional pressure losses.

Sub-meter isolation requirements

You are required to install a gas-certified manual shut-off valve as close as practicable to the inlet of the sub-meter (see **Figure 9.10**; also note the use of a standard meter bracket to support the meter). This is to facilitate maintenance and replacement of the sub-meter whenever required without disturbing the rest of the installation.

Testing

Upon completion of component and pipework installation, you must undertake a pressure test to ensure that the entire installation and meter are gas-tight. Subject to the scope of work, the pipework may need to be subjected to a full pipework test and, as a minimum, an installation test of the section and meter itself.

A water or digital manometer is necessary for the testing procedure and you are to ensure that there is no loss of pressure over a minimum period of 5 minutes in addition to the stabilisation period. Where the instrument indicates pressure loss, you will need to track down the source of the leak. The most effective method is to use

FIGURE 19.9 Meter outlet connection using standard BSP thread fittings

FIGURE 19.8 Meter outlet connection using an aluminium meter bend and adaptors

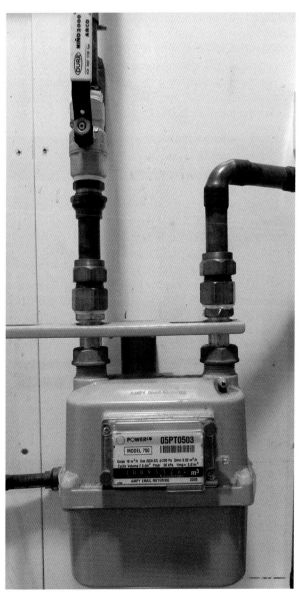

FIGURE 19.10 An isolation valve must be installed upstream of a sub-meter

INSTALL GAS SUB-METERS **431 GS**

an ammonia-free solution of soapy water on all joints. Bubbles will indicate the presence of a leak.

Once you have found the leak, you must pressurise the system once again, allow time for the pressure to stabilise and apply the full test procedure once again.

Sub-meter purging

When you have installed the sub-meter, it is vital to ensure that the pipe section and meter are correctly purged of any potentially dangerous air–gas mixture. If a mix of air and gas were to flow to the appliance and be lit, it is possible that a flashback could occur all the way back to the meter, with explosive results.

No appliance should ever be lit unless the meter and all piping have been fully purged of air, leaving only pure gas. Failure to do so may result in an explosion!

Purge requirements

The Standard requires that a volume of purge medium equal to five times the volume held by the sub-meter be passed through the meter to ensure that any and all dangerous air–gas mixture has been removed.

The purge medium can be the gas itself if commissioning an installation, or may be an inert gas such as nitrogen if making the installation safe for new work.

REFER TO AS/NZS 5601.1 APPENDIX D 'PURGING A SUB-METER'

Before you can purge a sub-meter you need to know how to read the dial face of the meter so that you can determine how much gas is flowing through the meter. Almost all meters found today are metric, and how to read such meters is covered in the next section. Some older gas installations may still be serviced by meters that use imperial measurements, but these are gradually being phased out.

How to read a gas meter

Every meter has important information that you need to find before you carry out any calculations. As a gas fitter, you need to identify the following information:
- maximum flow rate of natural gas in m^3
- cyclic volume of the meter in cubic decimetres (dm^3)
- test dial (if fitted) rotation volume in m^3.

Reading a metric meter requires you to understand and apply some simple mathematical principles. The first thing you need to know is the volume relationship between cubic metres and litres:

There are 1000 litres in one cubic metre (m^3).

This is important to know because a meter is measured in both litres and cubic metres. One thousand litres of gas must pass through the meter before it registers an increase of 1 m^3.

To convert cubic metres to litres, multiply by 1000.
To convert litres to cubic metres, divide by 1000.
Here are some examples:

$1\ m^3 = 1000\ L$
$0.05\ m^3 = 50\ L$
$0.036\ m^3 = 36\ L$
$0.005\ m^3 = 5\ L$

Meter manufacturers use the following volume measurement to describe how much gas has passed through the meter during one cycle of its internal mechanism.

One cubic decimetre (dm^3) equals one litre.

Figure 19.11 shows close-up detail of a domestic-sized meter index face with a simple odometer-style digital register. Important information is explained in each of the surrounding boxes.

Figure 19.12 shows close-up detail of a larger commercial meter index face with a characteristic rotating test dial and digital register. As is common with this style of meter, information is found on both the register face and a separate data plate. Important information is explained in each of the surrounding boxes.

EXAMPLE 19.6

If a sub-meter has a cyclic volume of 2 dm^3 (2 litres), then to carry out a purge you need to ensure that at least 10 litres of purge medium passes through the meter. A complete rotation of the first graduated indicator would be equivalent to the required 10 litres. Do this and the meter purge is complete.

EXAMPLE 19.7

A commercial-sized sub-meter with a cyclic volume of 5.6 L would require at least 28 litres to pass through the meter to effect the purge. This is easily read by observing the rotation of the test dial.

How to determine installation gas rate

The gas rate of an installation or appliance is usually quoted in MJ/h. The meter measures gas in cubic metres and litres per hour and you need to apply a simple formula in order to convert this to MJ/h.

However, this does not mean that you would need to sit down in front of the meter for a whole hour to work out the gas rate! There are two simple ways of working out the gas rate.

If you are on a job and cannot remember the formula in method 1, then try to bring everything to litres per hour first.

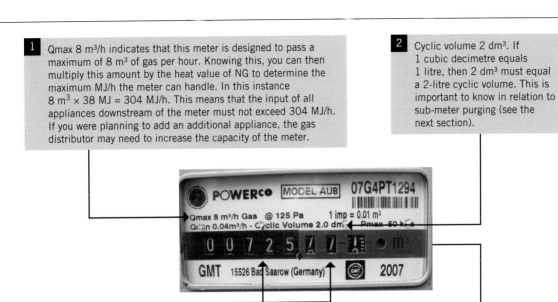

1 Qmax 8 m³/h indicates that this meter is designed to pass a maximum of 8 m³ of gas per hour. Knowing this, you can then multiply this amount by the heat value of NG to determine the maximum MJ/h the meter can handle. In this instance 8 m³ × 38 MJ = 304 MJ/h. This means that the input of all appliances downstream of the meter must not exceed 304 MJ/h. If you were planning to add an additional appliance, the gas distributor may need to increase the capacity of the meter.

2 Cyclic volume 2 dm³. If 1 cubic decimetre equals 1 litre, then 2 dm³ must equal a 2-litre cyclic volume. This is important to know in relation to sub-meter purging (see the next section).

3 The register on this meter is broken into two sections. As you face the meter, the left-hand figures in black and white are cubic metres. On the right-hand side you can see three figures surrounded by red shading. These digits show the volume in litres. This photograph indicates that this meter has so far passed 725 m³ and 771 L. Remembering that there are 1000 litres in 1 cubic metre, you can see that when the litre figures pass 999 L, the cubic metre figures will change to 726 m³.

4 Look carefully and you can see that the last figure on the far right-hand side of the meter is graduated. Each one of these graduations is equivalent to 1 litre. Therefore, one single rotation of this dial equals 10 litres of gas.

FIGURE 19.11 Close-up detail of a domestic-sized meter index face

HOW TO

METHOD 1

1. Using your watch or phone as a stopwatch, observe how many litres of gas pass through the meter over a period of 30 seconds (or 15 seconds if you wish). For this example, let's say 35 litres of gas pass in 30 seconds.
2. Apply the following formula:

$$\text{Litres per hour} = \frac{3600 \text{ s/h}}{30 \text{ s}} \times \frac{35 \text{ L}}{1}$$
$$= 4200 \text{ L/h}$$

$$\text{Cubic metres per hour} = \frac{4200 \text{ L/h}}{1000 \text{ L}}$$
$$= 4.2 \text{ m}^3/\text{h}$$

$$\text{Gas rate MJ/h} = 38 \text{ MJ(HV)} \times 4.2 \text{ m}^3$$
$$= 159.6 \text{ MJ/h}$$

HOW TO

METHOD 2

1. Observe how many litres of gas pass through the meter in 30 seconds (35 L).
2. Lay out your calculation in the following way:

   30 seconds = 35 L
   1 minute   = 70 L   (35 L × 2)
   1 hour   = 4200 L   (70 L × 60 minutes)

3. From this point, simply divide the litres per hour by 1000 to convert to cubic metres:

$$\text{Cubic metres per hour} = \frac{4200 \text{ L/h}}{1000 \text{ L}}$$
$$= 4.2 \text{ m}^3/\text{h}$$

$$\text{Gas rate MJ/h} = 38 \text{ MJ(HV)} \times 4.2 \text{ m}^3$$
$$= 159.6 \text{ MJ/h}$$

INSTALL GAS SUB-METERS

Smaller volumes of gas can be seen to be passing through the meter via the test dial. This small dial rotates counter-clockwise as indicated, and one complete revolution of the pointer is equivalent to $0.05 \, m^3$ or 50 L of gas. Each time this dial rotates twice, the hundreds digit on the main register will move to the next highest number. In this instance it would move from 00 019.5 to 00 019.6 cubic metres.

This meter is designed to pass a maximum of $12 \, m^3$ of gas per hour at its lower pressure supply range and $25 \, m^3$ at its upper pressure limit. Knowing this, you can then multiply this amount by the heat value of NG to determine the maximum MJ/h the meter can handle. Depending upon supply pressure, this meter could supply NG at a range of 456–950 MJ/h. Supplied with LPG, the range could be 1152–2400 MJ/h. Obviously, you will always need to confirm supply pressures in a given area with the gas distributor.

Designed to handle larger volumes of gas than a standard domestic meter, this register measures only in cubic metres on the left and hundreds of litres on the right. Measuring smaller quantities is not necessary. Currently the meter in the example has passed $19 \, m^3$ and approximately 500 L.

Unlike the domestic meter in the previous example, this data plate does not use cubic decimetres (dm^3) as a unit of measurement, and instead refers to $0.0056 \, m^3$/revolution. Converting this to litres by multiplying by 1000 L means that the cyclic volume of the meter is 5.6 L/revolution.

FIGURE 19.12 Close-up detail of a larger commercial meter index face and data plate

LEARNING TASK 19.3   METER READING

1. If an LPG meter passes 3100 L of gas in one hour, how many MJ is being supplied to the appliance?
2. A sub-meter has a 7.5 L cyclic volume. How many litres of purge medium must pass through the meter to comply with the requirements of the Standard?

Clean up the job

The job is never over until your work area is cleaned up and tools and equipment accounted for and returned to their required location in a serviceable condition.

Clean up your area

The following points are important considerations when finishing the job:
- If not already covered in your site induction, consult with supervisors in relation to the site refuse requirements.
- Cleaning up as you go not only keeps your work area safe it also lessens the amount of work required at the end of the job.
- Dispose of rubbish, off-cuts and old materials in accordance with site-disposal requirements or your own company policy.
- Ensure all materials that can be recycled are separated from general refuse and placed in the appropriate location for collection. On some sites specific bins may be allocated for metals, timber etc. or, if not, you may need to take materials back to your business premises.

Check your tools and equipment

Tools and equipment need to be checked for serviceability to ensure that they are ready for use on the job. On a daily basis and at the end of the job check that:
- all items are working as they should with no obvious damage

- high-wear and consumable items such as cutting discs, drill bits and sealants are checked for current condition and replaced as required
- items such as laser levels, crimping tools, measuring devices etc. are working within calibration requirements
- where any item requires repair or replacement, your supervisor is informed and if necessary mark any items before returning to the company store or vehicles.

Complete all necessary job documentation

Most jobs require some level of documentation to be completed. Avoid putting it off till later as you may forget important points or times.

- Ensure all SWMS and/or JSAs are accounted for and filed as per company policy.
- Complete your timesheet accurately.
- Reconcile any paper-based material receipts and ensure electronic receipts have been accurately recorded against the job site and/or number.

LEARNING TASK 19.4

During the installation of the sub-meter, you accidentally knock over your tripod and cross-line laser level. What action should you take?

SUMMARY

In this unit you have examined the requirements related to the installation of sub-meters including:
- a knowledge of meter types and applications
- job preparation
- specific installation requirements related to sub-meters including isolation and ventilation
- pressure testing requirements
- purging requirements including how to determine that 5 times the cyclic volume of the meter is passed
- clean up of job and return of equipment.

GET IT RIGHT

The photo below shows an incorrect practice when installing a sub-meter.

Identify the incorrect method and provide reasoning for your answer.

_____
_____
_____
_____

WORKSHEET 1

To be completed by teachers
Student competent ☐
Student not yet competent ☐

Student name: _____

Enrolment year: _____

Class code: _____

Competency name/Number: _____

Task
Review this chapter and answer the following questions.

1. How many litres are there in 1 m³ of gas?

   _____

   _____

   _____

2. List three factors that might determine the type of meter used in any particular installation.

   i _____

   ii _____

   iii _____

3. An LPG sub-meter passes 25 litres of gas in 20 seconds. What is the gas rate of the installation in MJ/h? Show all workings.

   _____

   _____

   _____

   _____

   _____

4. What needs to be installed as close as practicable upstream of a sub-meter inlet? Provide the relevant clause number from the Standard.

   _____

   _____

   _____

WORKSHEET 19

5. In Australia, how would a gas fitter determine the requirements for the location of consumer billing meters?

    _____

    _____

6. If a sub-meter had a cyclic volume of 6 litres, how many litres would need to be purged from the meter to comply with the Gas Installation Standards? What Gas Installation Standards clause did you refer to?

    _____

    _____

    _____

7. An NG meter passes 12 litres in 40 seconds. What is the gas rate of the installation in cubic metres? Show all workings.

    _____

    _____

    _____

    _____

    _____

8. A domestic NG gas meter has a maximum rate of 6 m^3/h. Determine if this meter would be suitable for a planned installation with the following appliances:
    - a continuous flow hot water system at 188 MJ/h
    - a space heater at 35 MJ/h
    - a cooktop at 27 MJ/h

    _____

    _____

    _____

    _____

    _____

9. What might be the consequence of an air–gas mixture being left in a meter?

    _____

    _____

    _____

10. Where a sub-meter is to be installed in an enclosure measuring 1.2 m (H) × 1.2 m × 0.6 m, where must you install the required ventilation?

INSTALL LPG STORAGE OF AGGREGATE STORAGE CAPACITY EXCEEDING 500 LITRES AND LESS THAN 8 kL

20

Learning objectives

This chapter details some of the basic skills and knowledge required to determine the installation requirements of LPG tank storage systems between 500 L and 8 kL. The primary focus of this chapter is the tank site itself and associated connections.

Areas addressed in this chapter include:
- calculating vaporisation using the 'DLK' formula
- the function of vaporisers
- LPG tank instalment requirements
- pressure testing of LPG tank connection systems.

Before starting this chapter, you must ensure that you have worked through all of Part A of this text, with particular review of Chapter 7 'LPG basics', so that you have a good grasp of LPG fundamentals before you start.

It is also assumed that you have completed the following chapters from Part B:
- Chapter 8 Size consumer gas piping systems
- Chapter 9 Install gas piping systems
- Chapter 10 Install gas pressure control equipment
- Chapter 16 Install LPG storage of aggregate storage capacity up to 500 litres.

Overview

This section will provide you with specific guidelines relating to the site and installation requirements of LPG cylinders and tanks. To work through this section successfully you will need access to AS/NZS 1596 'The Storage and Handling of LP Gas'.

REFER TO AS/NZS 5601.1 'GAS INSTALLATIONS' APPENDIX J AND AS/NZS 1596 'THE STORAGE AND HANDLING OF LP GAS'

As you work through this section, you will need to undertake some basic LPG calculations. Refer to Table 20.1 for the key facts.

TABLE 20.1 Propane key facts (some figures are subject to test conditions, temperatures and rounding)

Boiling point	−42°C
MJ per kg	49.6
MJ per litre	25
Litres per kg	1.96
Relative density	1.55
Heat value MJ (commercial)	96
Flammability range	2–10 per cent gas in air

Tank basics

A 210 kg cylinder is often regarded as a small tank, and manifolded 210 kg cylinders are commonly found for many smaller commercial and industrial sites. Generally, however, LPG required for high-consumption installations where the refilling of cylinders is too frequent is stored on the consumer's site in one or more tanks. The vessel chosen will depend upon a number of factors, including consumption patterns, peak-load vaporisation requirements, tanker-truck access and the site's physical characteristics.

Like cylinders, tanks are generally identified according to both litres of water capacity (WC) and to their capacity of propane in tonnes.

Water capacity is defined in the Standard as the total volume of space enclosed within the tank or cylinder expressed in either litres (L) or kilolitres (kL).

REFER TO AS/NZS 1596 DEFINITIONS: 'CAPACITY'

Tanks can range in size from less than 1 tonne right up to 90 tonnes and higher, often installed in multiples (see Figure 20.1). These tanks are sized according to storage needs, consumption requirements and vaporisation capability.

FIGURE 20.1 A tank that has been installed specifically to fill tankers at a gas supplier's main depot

Although most LPG tanks are installed horizontally, some can be designed for vertical installation, particularly where space is restricted by buildings or proximity to boundaries (see Figure 20.2).

FIGURE 20.2 A vertical LPG tank

Tank testing

Like cylinders, tanks have a standard service life of 10 years. At that point, the gas supplier must carry out a formal and detailed inspection of the vessel. This may entail an internal inspection, external painting and the replacement of all operation and safety relief valves.

Once this work has been completed, the tank will be recommissioned and stamped with a new expiry date. Suppliers will often take the tank away for service at their depot, leaving a new or reconditioned tank to ensure the customer is not inconvenienced any more than necessary during the process.

Vapour and liquid withdrawal

A gas fitter's normal range of work is entirely based upon the connection of gas systems to a vapour withdrawal valve of the tank. However, you must be aware that tanks may be fitted with liquid withdrawal valves as well.

These liquid valves should be clearly marked and sealed with a security plug, but you must always double check that you are not connecting your pigtail to the wrong valve. If you are unsure always contact the supplier for advice.

LEARNING TASK 20.1 TANK BASICS

1 What factors must be considered when choosing an LPG tank for the site?
2 How are some tanks installed where there may be space restrictions?
3 What is the standard service life of tanks?

Identify installation requirements

The amount of gas vapour available from LPG tanks depends upon the following two factors:
- the wetted surface area of the cylinder (i.e. the steel in contact with the liquid propane)
- the ambient temperature.

Variations in ambient temperature and wetted surface area can have a dramatic effect on the ability of an installation to operate effectively.

In colder areas, or where commercial or industrial installations have a high vapour draw-off need, correct tank sizing becomes very important. Inadequately sized tank systems can often be identified during draw-off periods by the formation of ice around the tank and its associated piping and connections. This is caused when so much heat is being absorbed from the atmosphere that water vapour is frozen on contact with the tank. If the liquid petroleum is unable to vaporise into gas fast enough, the working pressure of the appliances may drop, resulting in incomplete combustion. In these cases, you will need to be able to calculate the amount of LPG vaporisation required.

Vaporisation calculation – the DLK formula

The Gas Installation Standards provides basic vaporisation tables using specific assumptions for temperature, amount of LPG and load.

REFER AS/NZS 5601.1 APPENDIX J8

Where your requirements fall outside the scope of the tables provided in the Standard, you are still able to calculate the vaporisation capacity by applying what is known as the DLK formula (see Table 20.2).

Vapour (MJ/h) = $D \times L \times K$

where:
D = diameter of the storage vessel (m)
L = length of the storage vessel (m)
K = a pre-calculated factor representing the relationship between temperature, volume and wetted surface area.

Determining DLK calculation percentage

You will note from the DLK table that you have a range of options related to the amount of LP in the vessel expressed as a percentage of full. These options relate to the range of winter temperatures normally found in your area. The warmer the area the greater the vaporisation capacity of a vessel at lower percentages. While in some areas of Australia a 10% capacity is used, in colder areas a higher figure such as 30% is applied so that the sizing solution will accommodate the reduced vaporisation characteristic of colder ambient temperatures. You will need to consult your local LPG supplier and find out what percentage should be applied in your area.

TABLE 20.2 Vaporisation formulas and temperature multipliers

Per cent of full	$D \times L \times K$ factor	Temperature multiplier								
		−18°	−15°	−12°	−9°	−7°	−1°	4°	10°	16°
60	$D \times L \times 163$	× 1.00	× 1.25	× 1.50	× 1.75	× 2.00	× 2.50	× 3.00	× 3.50	× 4.00
50	$D \times L \times 147$	× 1.00	× 1.25	× 1.50	× 1.75	× 2.00	× 2.50	× 3.00	× 3.50	× 4.00
40	$D \times L \times 130$	× 1.00	× 1.25	× 1.50	× 1.75	× 2.00	× 2.50	× 3.00	× 3.50	× 4.00
30	$D \times L \times 114$	× 1.00	× 1.25	× 1.50	× 1.75	× 2.00	× 2.50	× 3.00	× 3.50	× 4.00
20	$D \times L \times 98$	× 1.00	× 1.25	× 1.50	× 1.75	× 2.00	× 2.50	× 3.00	× 3.50	× 4.00
10	$D \times L \times 73$	× 1.00	× 1.25	× 1.50	× 1.75	× 2.00	× 2.50	× 3.00	× 3.50	× 4.00

EXAMPLE 20.1

A tank measures 6.62 m long × 1.22 m in diameter. To find the worst-case vaporisation capacity of the tank in MJ/h, you follow the left-hand column down to 30 per cent of full, then apply the formula on that row from the '$D \times L \times K$ factor' column:

$$\text{Vaporisation} = D \times L \times 114$$
$$= 1.22 \times 6.62 \times 114$$
$$= 920.7096$$

You then need to apply the temperature multiplier relevant to normal winter temperatures in your area. In this example, we will use −1°C, which has a multiplier of 2.50.

$$\text{Vaporisation} = 920.7096 \times 2.50$$
$$= 2301 \text{ MJ/h}$$

In summary then, this installation can be expected to be able to vaporise approximately 2301 MJ/h at an ambient temperature of −1°C when 30 per cent full.

Although tanks can be designed and ordered to suit different applications, they are normally supplied in standard sizes. As a general guide, Table 20.3 provides some indicative sizes that you can apply to your DLK formula.

TABLE 20.3 Standard tank sizes

2.75 kL	3.330 m long × 1.065 m diameter
4.2 kL	3.920 m long × 1.220 m diameter
7.5 kL	6.620 m long × 1.220 m diameter

You should be aware that such a calculation is a guide only and you need to consider other factors that may affect your choice of storage volume. These include:
- possible use of multiple smaller tanks to achieve the same vaporisation rate
- probable simultaneous demand.

In most cases where installations call for anything more than 500 L of stored LPG, a site licence may be required, and you will need to work closely with the supplier on a solution. Regardless, your ability to calculate accurate vaporisation rates will give you confidence in your installation and demonstrate your professional skill to supplier and client alike.

Some installations will require a high vapour draw-off, but for a number of reasons, including site dimensions, site licensing restrictions or the prohibitive cost of large tanks, it will not be possible to install correctly sized vessels to achieve the required rate. In such cases you can consider the use of a 'vaporiser' to artificially provide the necessary heat to maintain the rate of vaporisation.

LEARNING TASK 20.2

VAPORISATION CALCULATION – THE DLK FORMULA

Calculate the vaporisation rate of a 4.2 kL tank at an ambient temperature of 4°C that is 30 per cent full.

What are vaporisers?

Vaporisers are mechanical devices that are designed to provide a heat source for vaporisation to occur when vessel size and/or ambient temperature are insufficient to do the job naturally. The gas supplier normally handles the sizing and selection of vaporisers, though you may be contracted to help install the system. The two most common vaporisers are described below.

Direct-fired vaporiser

This form of vaporiser is usually supplied as a self-contained, pre-fabricated system that incorporates an LPG cylinder heated directly with an atmospheric burner. The advantage of this system is its compact, pre-packaged size enabling it to fit easily into the larger plant installation.

Indirect-fired vaporiser

An indirect-fired vaporiser uses hot water (or other medium) to provide the necessary heat for vaporisation (see Figure 20.3). In this system, a gas-fired boiler heats water circulating through a flow-and-return pipe system. This water passes through the actual vaporiser via multiple tubes immersed in LPG. As it does so, it exchanges its heat with the liquid LPG before returning to the boiler for reheating. The circulating system will usually turn off and on in response to changes in the tank vapour pressure.

The indirect system offers the advantage that it can be added to a system where demand has become greater than the existing tank can supply. Although most indirect systems stand alone from the tank, some vaporiser heat exchangers are inserted into specially designed tanks, providing a more compact solution where space is limited. Hot water is still provided via a separate heating source.

Calculating tank storage requirements

When determining the storage volume of tanks and cylinders for a particular site, you must be aware of a number of basic considerations to ensure you deliver the best solution for your customer:
- Too little storage and the customer will run low on gas too quickly.
- Too much storage and the customer may be paying unnecessary tank costs.

FIGURE 20.3 A small indirect-fired vaporiser assembly at the end of an LPG tank

- Storage should be sufficient to ensure that enough gas is available to supply the customer between the scheduled tanker refill cycles.
- Storage should be sized to ensure that vaporisation is sufficient for the probable demand of the system at the lowest expected temperatures.

Being mindful of these basic considerations, you can move on to the actual calculation of storage requirements.

How long will my gas last?

Customers will regularly ask how long their gas will last before a refill is required. Unfortunately, there is no easy answer, because customer usage patterns differ and many appliances do not always run at the full gas rate. This variability makes it difficult to provide a definitive answer. However, by using your knowledge of basic propane characteristics you will be able to give an approximate answer. These calculations are used to determine how long the gas will last and whether the stored volume can vaporise enough gas on demand.

To answer both of these questions you need to determine three things:

- gross potential draw-off in MJ
- probable daily consumption
- probable simultaneous demand in MJ/h.

Recall that the volume of LPG vessels is often quoted in both litres and kilograms. However, most gas fitters are more familiar with the use of kilograms in relation to LPG volumes, so we will use this base unit for each calculation.

Gross potential draw-off

The gross potential draw-off represents the maximum amount of gas that could be used (drawn-off) from the supply cylinders/tanks. This calculation simply converts the total volume of propane in kg to MJ. You will see from Table 20.1 that 1 kg of propane is equivalent to approximately 49.6 MJ. For these calculations, we will round this up to 50 MJ/kg. Therefore:

Gross potential draw-off in MJ = Capacity of vessel in kg × 50 MJ

The next question you need to answer is how long will a given supply last based on probable daily consumption.

Probable daily consumption

Gas consumption depends upon for how long the customer uses each appliance and the operational characteristics of the appliance itself. While some commercial and industrial installations may have predictable consumption patterns, most general appliances do not run at the full rate all the time and will either modulate automatically or be controlled directly by the operator.

For instance, a four-burner cooktop may have a total input of 30 MJ/h, but it is unlikely that all burners will be used when cooking. Most space heaters will also modulate the burner rate and turn themselves off and on in response to room temperature changes.

However, you may be asked to determine consumption for a boiler that operates at set times each day at full burner rate. In this instance, your calculations are quite simple.

To work out probable daily consumption, you need first to establish the input of each appliance and then make a judgement on how long each appliance will be used each day and at what rate. You then divide this total daily consumption figure into the gross probable draw-off capacity to establish how many days a gas supply will last.

Example 20.2 looks at the consumption characteristics of a Type B industrial installation with a 360 MJ/h burner and a two-tonne tank supply.

EXAMPLE 20.2

a Determine the gross potential draw-off for a two-tonne tank.

   Gross potential draw-off in MJ = 2000 kg × 50 MJ
   = 100 000 MJ

b A Type B industrial appliance with a 360 MJ/h input generally runs at full rate 8 hours per day and half rate for 4 hours per day. How often must the two-tonne tank be topped up?

   Probable daily consumption 1 = hours × MJ/h
   = 8 hours × 360 MJ/h
   = 2880 MJ/day

   Probable daily consumption 2 = hours × MJ/h
   = 4 hours × 180 MJ/h
   = 720 MJ/day

   Total probably daily consumption = 2880 + 720
   = 3600 MJ/day

   $$\text{Probable supply duration} = \frac{\text{Gross potential draw-off in MJ}}{\text{Probable daily consumption MJ/day}}$$
   $$= \frac{100\,000 \text{ MJ}}{3600 \text{ MJ/day}}$$
   $$= 27.7 \text{ days}$$

In fact, a gas supplier would always ensure that a customer's tank is topped up well before the maximum duration is reached, but this calculation will ensure that the existing tank is sized adequately for stored volume.

Obviously, many variables affect these calculations and you must ensure that the customer understands they are only a guide to their expected gas consumption.

Probable simultaneous demand and vaporisation

Wherever you need to calculate the vaporisation capability of an installation, you must do so according to the probable simultaneous demand. Using total rated input is fine for fixed consumption appliances and processes but if you base your calculations on the total appliance input where demand is variable, then you are likely to end up with an oversized storage capacity.

  COMPLETE WORKSHEET 1

LEARNING TASK 20.3
CALCULATING TANK STORAGE REQUIREMENTS

Determine the probable supply duration for two 1000 MJ/h industrial appliances that operate at 10 hours each day using a 4.2 kL tank supply.

Tank locations and installation requirements

Tank installations are almost always under the control of the gas supplier. However, it is important for you to have a basic understanding of LPG tank location considerations.

FROM EXPERIENCE

A gas fitter is often consulted by a customer well before discussions are commenced with a supplier. You need to be very familiar with tank installation requirements as detailed under the Standard, so that you can provide accurate and important advice from the very beginning of the project.

You need to be aware of two definitions that relate to tank location clearance zones:
1 A *public place* is any place other than a private property. This includes roads and streets.
2 A *protected place* is almost any form of building and includes dwellings, schools, factories, workshops, warehouses and vessels at berth. For a full definition, refer to AS/NZS 1596.

 REFER TO AS/NZS 1596 'DEFINITIONS'

Tanks must be located so that minimum clearances to both public and protected places are maintained.

 **REFER TO AS/NZS 1596 'TANK SPACING AND SEPARATION DISTANCES'**

Tanks in groups

Specific requirements relate to locating tanks in groups. In particular, you should note that where a number of tanks are grouped together, they must be all aligned so that the long axes of the tanks are parallel to each other and directed away from adjacent storages of flammable liquids and gases.

> **REFER TO AS/NZS 1596 'TANKS IN GROUPS'**

In the rare event of a BLEVE, the domed ends of tanks have been known to be hurled like a projectile from the explosion and can cause considerable damage. Therefore, they must never be directed at another tank that is in close proximity.

How to choose a suitable LPG regulator for tank installations

Like all gas components, LPG regulators must be currently UL 144 listed and certified products. You cannot use or re-use any regulator that does not have a current certification listing.

When choosing an LPG regulator for tank applications, you need to consider the following information:

- *Capacity.* This relates to the maximum amount of gas vapour the regulator is designed to pass. If the installation provides gas for a 1500 MJ/h agricultural seed dryer, you would most likely choose a regulator with a capacity of about 2000 MJ/h. All modern regulators are now rated in MJ/h, kg/h and L/h, but you may find some older versions rated in BTU/h, depending upon where they are made. You will often need to apply conversion calculations to be sure.
- *Inlet pressure.* While standard integral two stage cylinder regulators are designed to handle full cylinder pressure, a second stage regulator may have a maximum inlet pressure of 70 kPa.
- *Adjustment range and outlet pressure.* Do you need a working pressure of 2.75 kPa, 70 kPa or 100 kPa? A dual stage, first stage or second stage regulator? Choose the regulator to suit the application.
- *Connection type.* There is a wide range of inlet connections. Most regulators will feature either an inverted-flare or a POL connection. POL connections are robust fittings that are made gas-tight very easily and are a commonly optioned regulator inlet. You would use a POL × POL pigtail to connect from the tank outlet valve to the regulator inlet (see **Figure 20.4**).

Once you have decided on what type of regulator you require, you need to access manufacturer's information to choose the product and have it ordered through your supplier. For example, **Figure 20.5** shows an extract from a distributor's comprehensive product catalogue. From this, you can choose the type of regulator you need, confirm inlet and outlet connection types and check its certification. You can then order from your supplier or direct from the company itself.

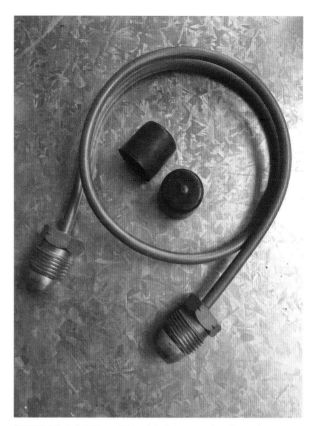

FIGURE 20.4 A POL × POL pigtail is a good option when connecting the tank valve to the regulator

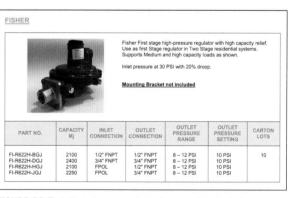

FIGURE 20.5 A first stage regulator from a distributor's product catalogue

LEARNING TASK 20.4

TANK LOCATIONS AND INSTALLATION REQUIREMENTS

1. When establishing clearance zones, what is the difference between a public place and a protected place?
2. What information needs to be considered when choosing first stage and second stage regulators?

INSTALL LPG STORAGE OF AGGREGATE STORAGE CAPACITY EXCEEDING 500 LITRES AND LESS THAN 8 kL

Prepare for installation

All materials and components used for cylinder installations must be checked for serviceability to ensure that manufacturing faults or pre-purchase damage has not occurred. If you find a fault in a component, you are not permitted to repair it; it must be replaced. Only the best quality materials should be used in gas installations.

REFER TO AS/NZS 5601.1 'MEANS OF COMPLIANCE – MATERIALS, FITTINGS AND COMPONENTS'

Always make an extra effort to protect your materials and components during storage and handling:
- Do not open regulator packets until you are ready to install.
- Never allow pigtails to be contaminated with dirt and ensure POL fitting faces are capped until ready for use to avoid scratches.
- Keep fittings in an orderly and clean manner, and ensure that they are not contaminated with dirt and other foreign substances.
- Avoid rough handling of your components, fittings and materials.

Appropriate tools and equipment

The following list shows a selection of items that may be required for LPG tank installations. Be aware that many tanks are quite large and you will require a ladder to access the valve turret at the top of the tank.
- Measuring equipment
- Shifters
- Power drill and various sized masonry bits
- Impact driver for general screwing tasks
- Spirit level
- Shovel and crowbar where regulator post hole requires minor excavation
- Soapy water solution and spray bottle
- Angle grinder with metal cut-off wheels to cut steel regulator support post
- Ladder to access tank valve turret on top of tank

Install and test LPG tank storage system

Although tank installations are large in size, their connection requirements are relatively simple as long as you adhere to supply company and Australian Standard guidelines at all times.

Tank pressure relief

If a tank is exposed to excessively high temperatures, the internal gas pressure will rise to unacceptable levels. To avoid a vessel rupture, tank relief valves will release gas as required to keep internal pressures within safe design limits.

While cylinder relief valves are built into the outlet valve, LPG tanks are generally designed with separate relief valves. These are located in the top vapour space of the tank as part of a cluster of other valves associated with filling and liquid/vapour draw-off.

Figure 20.6 shows LPG tank relief valves of various sizes. These are inserted directly into the tank. As shown in the accompanying diagram, spring pressure keeps the valve shut until internal gas pressure is high enough to cause the valve to open and relieve the pressure to atmosphere. Once the pressure is reduced, the spring forces the valve shut again.

Tank relief valves are installed in the vapour space at the top of the tank and in no circumstances must there be any obstruction placed or built above the valve itself.

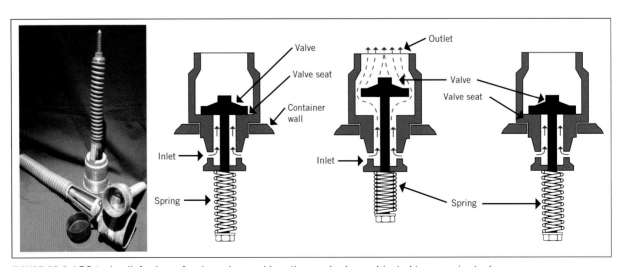

FIGURE 20.6 LPG tank relief valves of various sizes and how they work when subjected to excessive tank pressure

Tank connections

There are two main ways of connecting a tank to the regulator:

- *Copper pigtail* – Where a pigtail is used to connect the tank with the regulator, it is commonly a POL × POL configuration or a ¼″ NPT flare × POL. The pigtail is not to be greater than 1 m in length
- *Direct valve to regulator connection* – A male POL × ¼″ NPT adaptor is fitted to the regulator inlet and the complete assembly is then screwed into the tank outlet valve (see Figure 20.7). This can only be done where the full assembly can be contained within a lockable dome.

FIGURE 20.7 POL adaptor used to connect first stage regulator to tank valve

REFER TO AS/NZS 1596 'CONTROL OF LP GAS OUTFLOW'

When the regulator is connected to the tank via a pigtail, you must mount the regulator on an adjacent post, ensuring that the top of the regulator is higher than the tank valve so that any liquid that forms in the pigtail can drain back to the tank.

Pressure testing LPG tank connections

The test procedures as detailed under AS/NZS 1596 relate specifically to the connections and pipework associated directly with the scope of the Standard. Of relevance to your duties as a Type A gas fitter, this specifically means between the tank outlet and the connection to the regulator.

REFER TO AS/NZS 1596 APPENDIX I 'CONNECTIONS BETWEEN TANK OR CYLINDER AND FIRST STAGE REGULATOR'

All connections between the tank and the regulator must be checked by the gas fitter at tank pressure using an ammonia-free soap and water solution.

Pipe testing procedures downstream of the first LPG regulator fall under the requirements of the AS/NZS 5601 Gas Installation Standards.

LEARNING TASK 20.5

INSTALL LPG TANK STORAGE SYSTEM

What are the two main ways of connecting a tank to regulator?

Clean up the job

When you have completed work on a tank installation, the site must be cleared of all tools, equipment, off-cuts and general rubbish.

Clean up your area

The following points are important considerations when finishing the job:

- If not already covered in your site induction, consult with supervisors in relation to the site refuse requirements.
- Cleaning up as you go not only keeps your work area safe it also lessens the amount of work required at the end of the job.
- Dispose of rubbish, off-cuts and old materials in accordance with site-disposal requirements or your own company policy.
- Ensure all materials that can be recycled are separated from general refuse and placed in the appropriate location for collection. On some sites specific bins may be allocated for metals, timber etc. or if not, you may need to take materials back to your business premises.

Check your tools and equipment

Tools and equipment need to be checked for serviceability to ensure that they are ready for use on the job. On a daily basis and at the end of the job check that:

- all items are working as they should with no obvious damage
- high-wear and consumable items such as cutting discs, drill bits and sealants should be checked for current condition and replaced as required

- items such as laser levels, pressure-testing equipment etc. are working within calibration requirements
- where any item requires repair or replacement, your supervisor is informed and if necessary mark any items before returning to the company store or vehicles
- the tank regulator dome is shut and locked and the key returned to the owner or supplier.

Complete all necessary job documentation

Most jobs require some level of documentation to be completed. Avoid putting it off until later as you may forget important points or times.

- Ensure all SWMS and/or JSAs are accounted for and filed as per company policy.
- Complete your timesheet accurately.
- Reconcile any paper-based material receipts and ensure electronic receipts have been accurately recorded against the job site and/or number.
- Complete all necessary gas supplier paperwork associated with tank site installations and forward according to local requirements.

 COMPLETE WORKSHEET 2

SUMMARY

In this unit you have examined the requirements related to the installation of LPG storage sites with capacity from 500 L to 8 kL. These requirements include:
- tank storage capacity calculations based upon:
  - vaporisation capacity
  - peak load and draw-off needs
- tank location requirements
- regulator selection
- tank connection requirements
- cylinder site testing requirements.

GET IT RIGHT

A builder has provided you with a concept design for the location of a gas tank at a petrol/service station. The plan shows a 2.5 tonne (4900 litre) above-ground tank, located four metres away from the main building on the premises. The tank is used for the withdrawal of liquid LPG.

Identify the problem and provide reasoning for your answer.

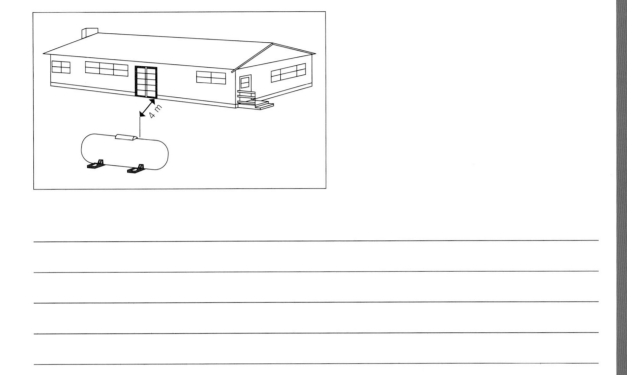

_____
_____
_____
_____
_____

WORKSHEET 1

To be completed by teachers
Student competent ☐
Student not yet competent ☐

Student name: _____

Enrolment year: _____

Class code: _____

Competency name/Number: _____

Task
Review the sections *Background* and *Vaporisation calculation – the DLK formula* and answer the following questions.

1. Calculate the vaporisation rate of a tank measuring 6.62 m long × 1.22 m in diameter at an ambient temperature of 10°C and 10 per cent full.

   _____

   _____

   _____

2. What is the simultaneous vaporisation rate of two tanks measuring 3.92 m long × 1.22 m in diameter at an ambient temperature of 4°C and at 30 per cent full?

   _____

   _____

   _____

3. What are the two main forms of vaporiser?

   _____

   _____

   _____

WORKSHEET 2

To be completed by teachers
Student competent ☐
Student not yet competent ☐

Student name: _____

Enrolment year: _____

Class code: _____

Competency name/Number: _____

Task
Review the section *Tank locations and simple installations* and answer the following questions.

1 What is the minimum distance permitted between a 2 kL vapour withdrawal tank and a protected place? Provide a reference from AS/NZS 1596 to support your answer.

_____

2 You meet a customer and a gas supplier representative on-site to discuss the location of a new LPG tank. You notice that the neighbouring property has a domestic heating oil tank installed adjacent to the house. This oil tank is 5 m from the intended LPG tank site. Is this permissible? Provide a reference from AS/NZS 1596 to support your answer.

_____

_____

_____

3 When connecting a regulator on a post to a tank, what key consideration must you comply with in relation to regulator height?

_____

_____

_____

4 What is the maximum permitted length of a copper pigtail for a tank installation? Provide a reference from AS/NZS 1596 to support your answer.

_____

_____

_____

MAINTAIN TYPE A GAS APPLIANCES

Learning objectives

This chapter covers the basic skills, knowledge and processes required to undertake the necessary routine maintenance tasks expected of you as a gas fitter. Depending upon the scope of the work, simple maintenance is something that a gas fitter may need to do on a daily basis but, in the context of this book, you should not confuse this with gas appliance servicing. Servicing gas appliances is a detailed and broad subject that is outside the scope of this text.

In order that you can undertake any form of gas appliance maintenance, you must investigate each of the following control systems.

Areas addressed in this chapter include:
- pressure control using regulators
- safety devices
- temperature control using thermostats
- ignition systems
- combination controls.

Completing the review questions at the end of the chapter will also prepare you for the practical application of this knowledge further into your training.

Before starting this chapter, you must ensure that you have worked through all of Part A of this text, with particular emphasis on Chapter 5 'Gas industry workplace safety' and Chapter 6 'Combustion principles'.

It is also assumed that you have completed the following chapters from Part B:
- Chapter 8 Size consumer gas piping systems
- Chapter 9 Install gas piping systems
- Chapter 10 Install gas pressure control equipment.

Overview

In the same way that people often go to the doctor for a health 'check-up', gas appliances also need regular preventative maintenance so that they continue to work safely and efficiently for as long as possible. These check-ups may be as simple as a yearly visit to blow out dust with compressed air from a domestic space heater, or large-scale scheduled maintenance contracts requiring regular weekly, monthly or annual visits for major commercial customers. Normally, simple maintenance procedures are carried out while the appliance remains connected to the gas supply.

Whatever your situation, there are some simple guidelines and knowledge that you can use to safely assess the maintenance requirements of each appliance, and to carry out subsequent maintenance tasks.

Pre-maintenance preparation

While every job is different, there are some basic planning steps that you need to follow to ensure that the task proceeds as smoothly as possible. Once you have been given the job, you need to gather as much information as possible, perhaps using a checklist to ensure that you do not omit anything.

Table 21.1 shows a simple list of items that you may need to think about, information that you may need to source, and equipment that you might need to put together before starting appliance maintenance work.

What equipment do I need for gas appliance maintenance?

There is always the odd job for which you will require something out of the ordinary. It may be an appliance-specific tool, for example, or perhaps heavy-lifting equipment. However, apart from your normal trade tools needed for almost every job, Table 21.2 gives a list of gear that would be a good start.

If you are unfamiliar with any of this equipment, ask your instructor to provide an example and demonstration where applicable.

With the basic planning done, it is time to get the job moving and to look at what needs to be done before starting work.

> **LEARNING TASK 21.1**
>
> **PRE-MAINTENANCE PREPARATION**
>
> 1 What information would you obtain from the job specifications?
> 2 What equipment would be required to measure gas pressure?

Identify maintenance requirements

Before undertaking any work on an existing gas installation, you should observe some basic procedures before commencing.

> It is counter-productive and often dangerous for yourself and others to start work on any individual appliance until you are satisfied that *the entire installation* is safe and in compliance with AS/NZS 5601.1 Gas Installation Standards.

Regardless of time pressures, approach each and every job with safety as your primary concern. Before you even consider the appliance itself, check the following job site conditions.

Gas type and supply pressure

Do not assume you know what type of gas is being used. Check it out!

TABLE 21.1 Appliance maintenance planning checklist

Item	Details	Tick
Job specifications from supervisor	• Location • Gas type • Appliance types • Maintenance needs	
Service manuals	• Maintenance guides • Parts list • Troubleshooting guides	
Tools required	• Any specialised tool and equipment requirements	
Company standard operating procedures (SOPs) and quality assurance regulatory obligations	• Required documentation • Work notices or completion certificates if required	
Third party involvement	• Owner/manager • Other trades • Possible shut-down times • Site access	

TABLE 21.2 Equipment needed for gas appliance maintenance

Tool or instrument	Considerations
Personal protective equipment (PPE)	You should always have access to and use PPE appropriate to every task. In particular, for gas maintenance you should ensure you have at least the following: • safety glasses • non-combustible clothing • safety boots.
Digital multi-meter	For the modern gas fitter, a good quality auto-ranging, digital multi-meter is almost indispensable. Using a multi-meter is the safest way to check for stray current and to carry out voltage, resistance and continuity tests.
Volt stick	Even if you use a multi-meter when testing for stray current, you should still have a volt stick available for use when needed. Most volt sticks enable you to trace wires behind plaster sheeting and their small size and portability enable them to be used in awkward spaces.
Power point tester	This simple plug-in device uses different coloured lights to determine the safety of a standard, single phase GPO. It normally detects poor earth and incorrect polarity.

MAINTAIN TYPE A GAS APPLIANCES

Tool or instrument	Considerations
Bonding straps	At times, it is necessary to disconnect appliances from the gas supply. Always use a set of 100 amp insulated bonding straps to provide electrical continuity across the break.
Water and digital manometers	For many years, a water manometer was sufficient for most jobs, and it still has many advantages today; in particular, it does not require calibration, unlike other instruments. However, a good digital manometer offers a greater flexibility for maintenance and servicing checks on modern gas appliances. Some appliances require differential pressure testing and this is best achieved with a digital manometer.
Digital thermometer	A digital thermometer has the advantage of giving an instant response to temperature changes. You will use these to check thermostat settings, duct temperatures, water temperatures and so on. Most quality multi-meters now have the capacity to plug in a thermometer lead.
Torch	You will need a good bright torch on nearly every job to enable you to look in dark appliance recesses. A rechargeable type is ideal.

Tool or instrument	Considerations
Jet drills and pilot cleaners	A comprehensive set of jet drills (often called number drills or thumb drills) and pilot cleaners are needed to work on the fine orifice sizes found on gas injectors and pilot assemblies.
Air compressor and/or vacuum cleaner	A small, portable air compressor or vacuum cleaner that you can take on the job is invaluable for cleaning dust out of appliances, particularly space heater fans and internal circuitry.
Video inspection borescope	A small handheld instrument with a camera mounted at the end of a snake-like probe transmitting an image back to the handpiece. These are invaluable for cavity inspections and looking into awkward locations within appliances.
Smoke candles (or incense sticks as shown)	These are required to enable a visual check of flue spillage, as detailed in Appendix R of AS/NZS 5601.1.

Source: Imagery courtesy Milwaukee Tool

Tool or instrument	Considerations	
Carbon monoxide detector	This device is vital for flue spillage tests or any time you suspect a CO hazard exists.	

In many areas, it is not uncommon to have LPG used in the same locality as reticulated natural gas (NG). Simulated natural gas (SNG) may also be used. It may not be immediately obvious which type of gas is being used, particularly in some commercial areas.

In areas where NG is being rolled out, it is common to find different types of gases in the same street, sometimes even in the same premises! In the example shown in Figure 21.1, a restaurant has been using a temporary LPG supply until such time as NG becomes available. A pipe has been connected to the NG meter, but you would need to confirm that the LPG has been disconnected and the appliances converted. Make no assumptions – always double check!

FIGURE 21.1 A temporary LPG supply to a restaurant

Not only do you need to know what type of gas is being used, you must also know at what pressure it is being supplied. This will vary from region to region and sometimes from street to street, particularly where the old low-pressure reticulation system is gradually being replaced with new high-pressure mains. If you are not sure, ask the customer and confirm details with your employer, gas supplier and regulator if necessary.

Electrical safety

Electrical safety is a vital part of every plumber and gas fitter's daily work. Undertaking appliance maintenance tasks can be particularly hazardous, as *you cannot assume any job is safe to touch*. Electrical hazards can be intermittent and unexpected; therefore, you must ensure that you carry out *regular* electrical tests, as and when required.

Revise the section on 'Basic electrical safety' in Chapter 5 if necessary.

Once you have established the installation gas type and supply pressure, you must conduct a basic electrical safety test prior to touching any pipework, piping components or appliances.

 Before carrying out electrical safety testing, locate the main power isolation switchboard so that you can quickly turn off the power supply if required.

1 Check your electrical test instrument at a 'known earth'.
2 Ensure your body attitude will reduce potential contact points.
3 Carry out a test on all sections of piping and components, ensuring that it is safe to touch.
4 Test all appliances before touching them.
5 Test and re-test as you move around the job or conduct work.

FIGURE 21.2 Protect yourself from unknown appliance faults with a portable power board fitted with a residual current device (RCD)

What do I do if I find a fault?

Unfortunately, it is not uncommon to find electrical faults affecting a gas or plumbing system. As most gas appliances now incorporate electrical functions and components in their design, it is almost inevitable that you will encounter an electrical hazard at some stage in the future, particularly on older or poorly maintained installations.

Even a well-installed and serviced installation can present a danger. A gas system can become 'live' through incorrectly earthed water pipes and faulty wiring somewhere else on the premises. Your gas pipe could even become live as a result of an electrical fault from a house down the street!

If you do find a fault:
1. Don't panic!
2. Avoid all contact with any part of the system.
3. Warn all onlookers, occupants or other workers of the hazard.
4. Where possible, isolate the power supply and apply an approved 'danger tag'.
5. Call an electrician or the electrical authority immediately.
6. Do not conduct any work on the system until it has been made safe by an authorised person.

General installation compliance

Before commencing any appliance maintenance, you need to check as far as possible that the installation is compliant with the Gas Installation Standards. As a licensed person, you have a 'duty of care' to identify faulty or non-compliant work and make it safe. If you are not sure what this means, check with your instructor or supervisor for a definition.

If you were to complete work on an appliance and leave the job knowing that it or another appliance was installed incorrectly, you could be held responsible, *even if you did not carry out the original installation.*

Refer to the AS/NZS 5601.1 Gas Installation Standards to check that:
- the consumer billing meter or LPG cylinders are in the correct location with the required clearances
- materials and components are approved
- piping and components are installed correctly
- appliance data plates are identified
- appliances are installed in the correct location and with the required clearances
- the flue system is sized correctly
- appliance ventilation is sufficient.

What do I do if the installation is not compliant?

On issues of installation compliance, you will need to contact your technical regulator to find out the procedures to follow. As these will vary, there is no single answer. However, at the very least you should do the following:
- Where you identify an installation that presents a clear danger to life and property, you need to inform the owner of the installation and make it safe. If you are unable to do so, contact the technical regulator immediately.
- You may be able to offer the owner your services to fix the problem. Where the owner agrees, you then have the opportunity to bring the installation up to standard. Where the owner does not agree, you should provide advice of the problem *in writing* to the owner and, depending upon local requirements, inform the technical regulator of the problem.

You will need to seek advice from your instructor and/or technical regulator about any specific requirements for your locality that you need to follow.

System leakage test

You must never start work on an existing installation until you have conducted a pressure test. You cannot assume that the piping and appliances are gas-tight, regardless of how new or old the installation is. Conducting maintenance checks on a gas system that is leaking may lead to asphyxiation, fire and explosions.

By carrying out a simple pressure test, you are also able to confirm the supply pressure and ensure that the lock-up pressure of the system is not excessive. Where a problem is found, you also have the chance then to inform the owner and make any adjustments in price to quoted work in order to rectify the situation. Few customers are happy about alterations in price after work has begun. In fact, you will find that by carrying out a safety pre-check, most owners will gain confidence in you as a tradesperson through this demonstration of professional competence.

What sort of pressure test do I carry out?
At this point in the job, you want to confirm only that the installation is gas tight. Therefore, you need to conduct a pre-work Leakage Test as per the mandatory requirements within the Standard.

REFER TO AS/NZS 5601.1, 'APPENDIX E6'

This test allows you to confirm the integrity of both the pipe system and the appliances themselves, right up to the control valves.

Check that the test pressure is within the required range for the section of gas system you are testing and ensure that the pressure does not fall at all during the prescribed time.

Pre-maintenance checks – summary
In this section, we have discussed the importance of ensuring the safety and compliance of the system through the following basic pre-work checks:
- gas type and supply pressure
- electrical safety
- general installation compliance
- leakage test (Appendix E).

With practice, these simple procedures will become almost second nature and provide you with a higher level of safety and confidence to move on to the maintenance or servicing tasks required.

COMPLETE WORKSHEET 1

LEARNING TASK 21.2
IDENTIFY MAINTENANCE REQUIREMENTS

Before commencing maintenance on an appliance, you should perform a general installation compliance check. What points should be considered before work starts?

Basic control systems

All gas appliances have control systems of some form. As a gas fitter, you must develop a basic understanding of all these control systems to allow you to undertake standard commissioning and maintenance tasks.

If you were to examine an upright domestic cooker, you would find that it has an appliance regulator to control the gas pressure, a thermostat to keep the oven temperature at desired levels, one or more safety devices to shut off the gas if the flame is blown out and an ignition system to light the burners. Not all appliances will have each of these control systems, but most will have one or more. For general maintenance purposes you must have a sound understanding of the following control systems:
- Regulators – covered in Chapter 10 of this text
- Safety devices
- Thermostats
- Ignition systems
- Combination controls

Safety devices

Safety devices in various forms are fitted to almost all Type A gas appliances so that if something does go wrong, the appliance will be rendered safe in some way. Appliances can be categorised into two groups:
1. Supervised gas appliances are appliances that are expected to be supervised by an operator during use. A gas cooktop is a good example; the appliance is in use only while someone is actually present. Barbecues and ring burners are other examples. Supervised appliances may not necessarily have safety devices fitted because of the presence of the operator.
2. Unsupervised gas appliances such as space heaters, water heaters and ovens are designed to operate automatically and without constant supervision. Because of this, all unsupervised appliances must have safety devices fitted in some way.

As a gas fitter, you need to be familiar with the main types of safety systems. These systems fall into four main categories:
1. flame-failure devices (FFD)
2. flame supervision devices (FSD)
3. energy cut-out systems (ECO)
4. oxygen depletion pilots (OD pilots).

Unsafe conditions
The safety devices that fall into each of these four categories are designed to deal with certain unsafe conditions:
1. *Flame loss:* Where the flame is extinguished for any reason, flow of the air–gas mix must be stopped before it builds up in the appliance or surroundings as an explosive hazard.
2. *Overheat conditions:* Appliances that are overheated may become damaged or cause fires and explosions.
3. *Oxygen depletion:* The depletion of oxygen within a room to below a certain limit may result in the injury and death of the room's occupants where flueless appliances are used.

A basic principle of all gas appliance safety devices is that they must cause the appliance to fail-safe 100 per cent of the time. In other words, they *must* work every time an unsafe condition occurs.

In the following sections, the basic system of operation will be covered for each category of safety device. While fault finding, maintenance and repair of such devices falls outside the scope of this text, simple hints relevant to normal Type A gas fitting practice will be provided.

Principles of bi-metallic operation

Before examining each safety device in detail, it is necessary to build a basic understanding of bi-metallic principles of operation. Many modern safety devices rely on the characteristics of bi-metallic operation to enable them to fail-safe.

At any given temperature, metals will expand at different rates. Compare the expansion differences between the three common metals shown in Table 21.3. You can see from this table how, under the same conditions, different metals have different expansion rates.

TABLE 21.3 Examples of the linear expansion of metals

Material	Initial length	Degree rise °C	Expansion
Steel	6 m	50	4.8 mm
Copper	6 m	50	5.1 mm
Aluminium	6 m	50	6.9 mm

A bi-metallic strip is made from two dissimilar metals fused together. When the strip is heated, the difference in the expansion rates causes the strip to bend in one direction. In this way, a bi-metallic strip can turn a change of temperature into mechanical displacement (see Figure 21.3).

Valves or switches connected to bi-metallic strips can be opened or closed by the application or removal of heat. Many energy cut-out devices use a bi-metallic strip to shut off the gas supply during overheat conditions.

Early continuous flow water heaters used bi-metallic strips to operate the main gas valve (see Figure 21.4).

FIGURE 21.4 An old continuous flow water heater featuring a bi-metallic flame-failure device

However, these are no longer used because of problems associated with strip-metal fatigue and the slow shut-down time.

The old 'Ascot' continuous flow water heater in Figure 21.4 was manufactured with a bi-metallic flame-failure device. Lighting the pilot would cause the strip to move downwards, causing the main control valve to open and allowing gas through to the main burner.

Thermoelectric flame-failure devices

Thermoelectric flame-failure devices have been the most common form of safety device found in Type A gas appliances for many years. In its simplest form, gas flow to the burner is controlled through the operation of a thermoelectric flame-failure valve. The operation of the valve itself depends on two main components:
1 a thermocouple
2 an electromagnet.

Refer to Figure 21.5 to see how each of these components work.

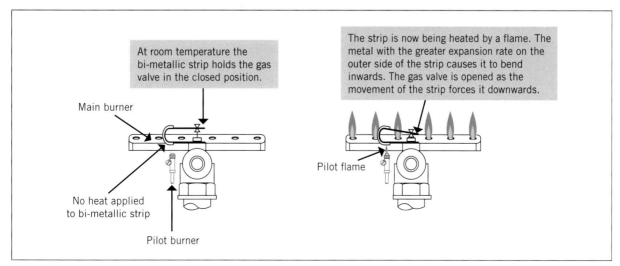

FIGURE 21.3 A bi-metallic strip converting a temperature differential into mechanical movement

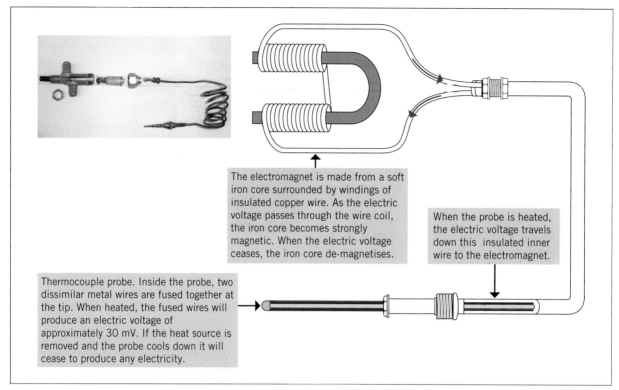

FIGURE 21.5 Cut-away diagram of a thermocouple and an electromagnet

Figure 21.6 shows how the thermocouple and electromagnet are used to operate the valve itself. Follow the sequence of operation by reading each information box in order.

You will encounter thermoelectric flame-failure assemblies in a wide range of appliances and often in many different configurations. However, they all have the same basic principles of operation. The images in Figure 21.7 show some examples of flame-failure devices. All of the appliances feature a thermoelectric flame-failure system to shut off the supply of gas in the event of flame loss.

Hints:
- Where the position of the thermocouple or flame can be adjusted, the flame should impinge only on approximately the last 3 mm of the thermocouple probe. Heating more than this may reduce the life of the outer sheath and does not produce extra electric voltage.
- Although a thermocouple will potentially produce approximately 30 mV, while installed its operational range is more likely to be between 12 and 18 mV. If a thermocouple produces less than this amount it should be replaced. (Ask your teacher how an in-service thermocouple can be tested using an interrupter block and multi-meter.)
- A thermocouple with a damaged outer sheath (see Figure 21.8) should be replaced even if it is still working, as its life will be short.
- A faulty thermocouple should be replaced with one having a probe of similar mass so that cool-down times are the same.
- Never over-tighten a thermocouple, as the delicate fibre washer separating the bulb and outer sheath may be broken, causing the assembly to malfunction. Hand-tight plus an eighth of a turn with a small shifter is normally sufficient.

Thermopiles

In Figure 21.9 you will see a probe that looks something like a thermocouple but much larger. This is actually a thermopile. (In the figure you can see the size difference between the two.) A thermopile is a thermoelectric component that is made up from a number of thermocouples joined together in series. By doing this, a thermopile is able to produce much higher electric voltage than a standard thermocouple. Depending on the application, thermopiles generate between 350 and 750 mV.

A thermopile can be used to energise the electromagnet of a flame-failure device, and to provide energy to operate other gas valve components. Thermopiles are commonly found on deep fryers and in some decorative flame effect fires.

Flame rectification systems

Flame rectification is an increasingly common form of flame supervision that is based upon the ability of a flame to conduct electricity. During the combustion process, the intermediate zone of the flame becomes electrically charged or ionised. This ionisation process allows the flame to conduct an electric current of between 2 and 4 microamps. The basic principle behind this is that unless an electric current is detected passing

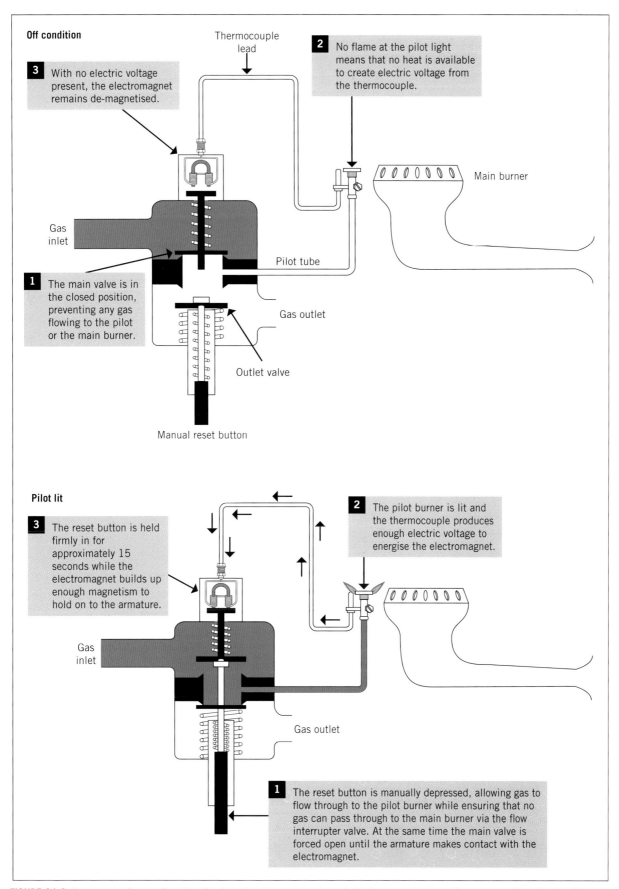

FIGURE 21.6 Sequence of operation showing how the thermocouple and electromagnet are used to operate the valve itself

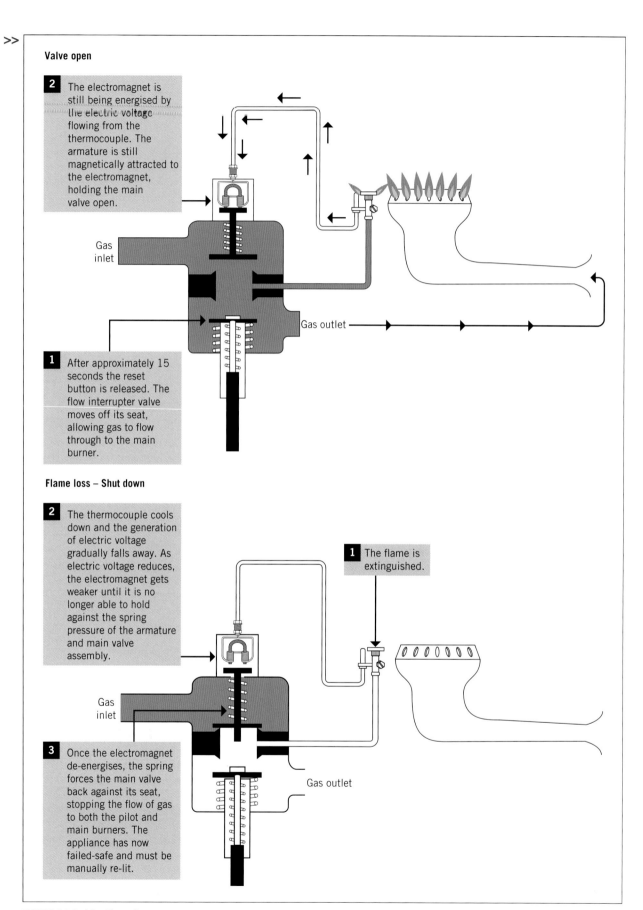

FIGURE 21.6 (*Continued*)

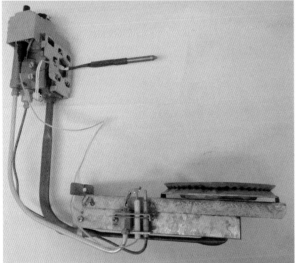

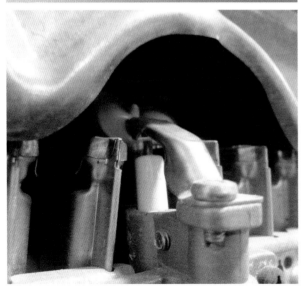

FIGURE 21.7 Thermoelectric flame failure devices on a cooker burner, storage water heater and continuous flow water heater

FIGURE 21.8 A thermocouple with a damaged outer casing at its tip

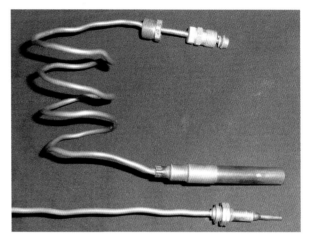

FIGURE 21.9 A thermopile (top) and a thermocouple

through the flame, the burner's supervision control will shut off the supply of gas at the solenoid valve.

However, simple conduction of an electric current is not enough to prove that there is still a flame on the burner. The flame could be extinguished and the electric current could still be detected through the earth system. To counter this, the electric current is partially changed from AC to DC as it passes through the flame, a process known as 'rectification'. (The reverse process of converting DC to AC is known as 'inversion'.)

To prove the presence of a flame, an AC electrical current is passed, via a flame rod, through the flame to the earthed burner head. As the electrical charge passes through the flame, more current flows in one direction than the other, partially rectifying the current from AC to DC. This can only happen if there is a flame present as the current passes through.

The burner supervision system is designed to accept and amplify only a DC electrical charge as proof of flame presence and to energise the gas inlet solenoids. If AC current were to leak directly to earth, this would indicate a loss of flame, and with no DC current present the solenoids would shut down, isolating the gas supply.

Figure 21.10 shows how the flame rectification system works when the flame is present and what happens when the flame is extinguished.

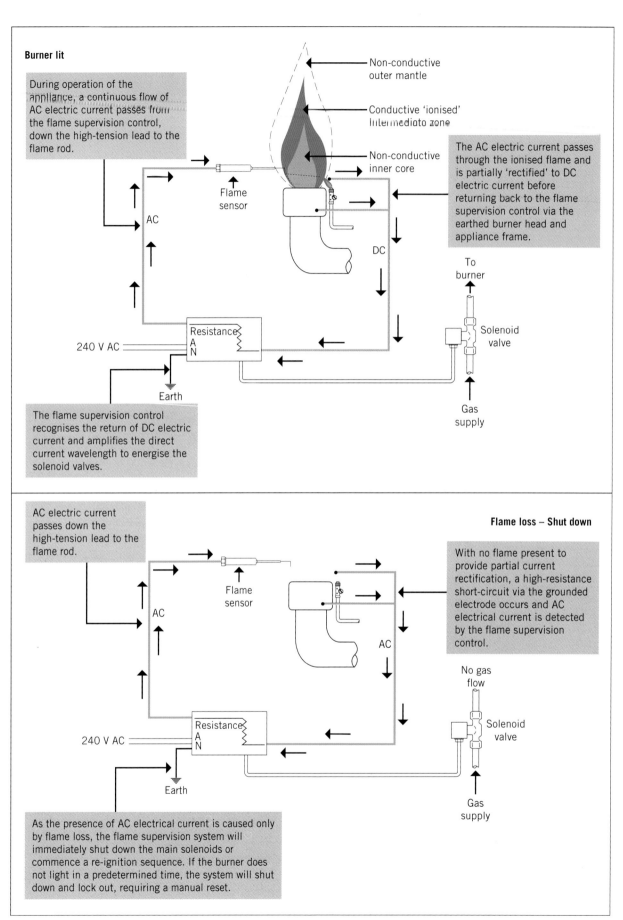

FIGURE 21.10 Flame rectification system when flame is present and when not

Hints:
1 Flame supervision systems are supplied with 240 V power. Do not work on such systems unless you have been trained to do so.
2 Simple checks of the rectification circuit include ensuring that flame rods are correctly positioned in the flame, high-tension leads are not earthing out on the appliance frame, and that the flame rod ceramic insulation is not cracked.

Energy cut-out devices

Energy cut-out (ECO) devices are supplementary safety shut-offs usually installed as part of the primary safety system. While the primary system deals with main burner supervision, these devices monitor other potentially hazardous appliance functions and conditions. They are normally found in unison with thermoelectric flame-failure systems and act to open the thermoelectric circuit in the event of a malfunction. This section includes a short description of the most common ECOs found in Type A gas appliances.

But first, what is a switch…?

We all know what a switch is, don't we? You use them all the time to turn lights off and on, and to operate power tools and appliances. However, when it comes to ECOs you need a deeper understanding of how switches work and the associated terminology.

ECOs all incorporate some form of switch. A switch is a mechanical device that enables an electrical circuit to be connected or disconnected. It consists of two metal 'contacts' that, when touching, create a circuit. When the switch is operated in the other direction, the circuit is broken. In this way, a switch can be abstractly thought of as a 'gate' that can be normally open or normally closed.

You need to have a clear understanding of the following terms:
- *Open circuit*. In this condition, electricity cannot pass through the switch because the 'gate' is open. An electrical diagram symbol drawn in this way indicates that such a switch is in the 'normally open' condition (see Figure 21.11).

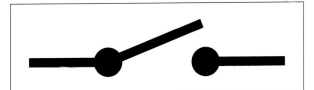

FIGURE 21.11 Symbol indicating an electrical switch in the 'normally open' condition

- *Closed circuit*. In this condition, electricity can pass through the switch because the 'gate' is closed. An electrical diagram symbol drawn this way indicates that such a switch is in the 'normally closed' condition (see Figure 21.12).

FIGURE 21.12 Symbol indicating an electrical switch in the 'normally open' condition

Knowing this, you will now be able to better grasp how ECOs use switches to ensure the safe operation of appliances.

High temperature cut-out

Also known as overheat switches, high temperature cut-outs act to shut down the appliance when some part or function of the appliance reaches too high a temperature. Two common examples of appliances that use a high temperature cut-out switch are central heaters and water heaters.

The central heater shown in Figure 21.13 uses a flame supervision system incorporating a high temperature cut-out in circuit with the control system. This ECO is mounted on the heat exchanger of the central heater and is designed to operate when the heat exchanger becomes too hot.

FIGURE 21.13 Energy cut-out device in a central heating furnace

The switch is in the 'normally closed' position, allowing voltage to flow under normal circumstances.

If the heat exchanger becomes too hot, the bi-metallic switch inside the device will heat up and move away from the contact, thereby opening the circuit. Without voltage returning to the control system, the solenoid valves will shut off the supply of gas to the burner.

Figure 21.14 shows the location of a high temperature cut-out as used on the heat exchanger of a continuous flow water heater. If water in the heat exchanger was to become too hot it could flash to steam causing a rupture or explosion. Therefore all water heaters incorporate some form of ECO to shut down the appliance in the event of an overheat condition.

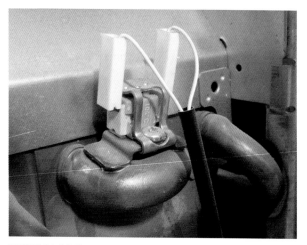

FIGURE 21.14 Energy cut-out device in the heat exchanger of a continuous flow hot water system

This ECO is in the 'normally closed' condition allowing normal thermoelectric operation. If the water temperature were to become too high, the bi-metallic switch would open the circuit, instantly cutting off the thermoelectric voltage from the electromagnet.

Note: Depending on the appliance, some high temperature cut-outs may be simple, heat-activated switches that reset themselves once the temperature drops, or they may be fusible links that must be replaced once the circuit is opened.

Tilt switch

A tilt switch maintains a closed circuit while the appliance is in the upright position and is designed to open the thermoelectric circuit when an appliance tilts past a predetermined angle.

Figure 21.15(a) shows a common flueless space heater and the location of its tilt switch ECO. If a flueless heater were to fall over and keep operating, a fire would immediately result. To avoid this, manufacturers install a tilt switch ECO to open the circuit of the thermoelectric system once it passes a certain point.

While the heater is in the upright position, the tilt switch is in the 'normally closed' condition. This circuit would 'open' if the heater were knocked over, cutting off the milli-voltage supply to the electromagnet.

Early versions of tilt switches used mercury as the conductor between two contacts. As soon as the appliance was tilted past a certain angle, the mercury would flow away from one of the contacts and the circuit would be opened. **Figure 21.15(b)** shows an older mercury tilt switch as found in an industrial portable radiant heater. This switch also acts as a shut-off mechanism when the pull cord is used.

Other forms of tilt switch use a metallic ball inside a sealed housing that rolls away from the contacts when tilted.

Pressure switch

Appliances such as ducted central heating furnaces and air curtains must maintain a minimum airflow to ensure the appliance does not overheat. In addition to a high temperature cut-out, such appliances often incorporate combustion fan/chamber pressure switches and duct air pressure switches. The pressure switch is connected to the combustion fan or other part of the appliance air duct via neoprene tubes. Air pressure acts upon an internal diaphragm attached to a switch. Unlike most ECOs, the pressure switch is designed to be in the 'normally open' condition. This means that combustion fan/chamber or duct air pressure must be high enough

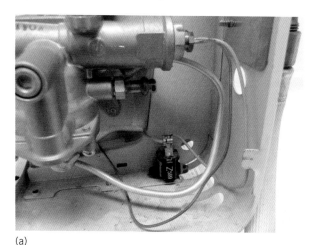

(a)

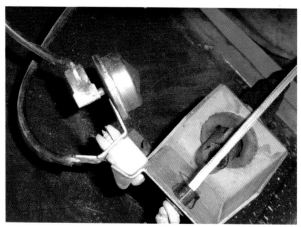

(b)

FIGURE 21.15 Tilt switches found in flueless space heaters

to cause the pressure switch to close the circuit. Some appliance burner control systems are designed to ensure that both combustion chamber and duct air switches maintain a constant 'proving' of the combustion fan airflow to allow continued operation. This form of interlock ensures that the appliance is always working in a safe condition as designed.

The pressure switch assembly in Figure 21.16(a) is from a central heating furnace. The micro switch as seen in the centre of the device is connected directly to the internal diaphragm. The images in Figure 21.16(b) show the pressure switch as it is installed in the appliance. Look carefully and you can see the pressure switch tube connected to the combustion fan outlet. Figure 21.17 shows a pressure switch acting as a proving mechanism for the combustion fan of an industrial burner.

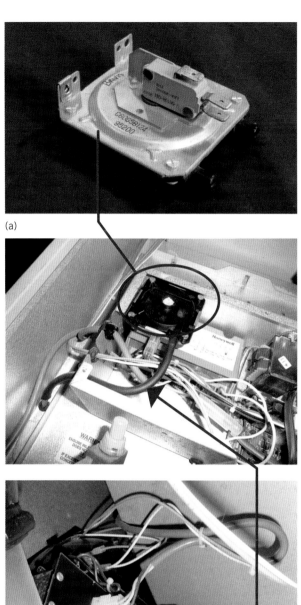

FIGURE 21.16 A pressure switch assembly (a) from a central heating furnace and (b) installed in the appliance

FIGURE 21.17 A pressure switch in an industrial burner

Oxygen depletion devices (OD pilots)

If a flueless space heater was to be used in a room with insufficient ventilation, the oxygen content of the room would start to fall as it is used up during the combustion process. Air contains approximately 21 per cent oxygen; if this percentage falls by more than 2 per cent, incomplete combustion will result, placing the room's inhabitants in danger of carbon monoxide poisoning and/or hypoxia.

To prevent this from happening, a number of manufacturers have developed specialised pilot burners that will ensure the appliance is shut down before oxygen depletion leads to incomplete combustion. Models include the Copreci pilot, the Sourdillon pilot and a device by OP Manufacturing. Of these, the Sourdillon device is used as an example here (see Figure 21.18). An oxygen depletion device is also known as an OD pilot.

Most Type A gas burners are designed to operate with high excess air content, so that safe and efficient

FIGURE 21.18 A Sourdillon pilot burner

combustion is maintained despite normal wear and tear. Such burners are supplied with more air then they can burn. In comparison, the Sourdillon pilot is a finely tuned burner that has been designed to only produce a specifically shaped flame with a precise amount of air. This flame will lift off the thermocouple and open the thermoelectric circuit if combustion conditions are not ideal.

You will also notice that the Sourdillon pilot uses a bi-metallic strip to control the size of the primary air port. This is often known as the 'cold-start choke'.

This bi-metallic strip will alter the primary airflow rate in relation to changes in appliance heat. Figure 21.19 shows the sequence of operation of the pilot.

Note that flueless appliances accumulate a lot of dust, and sensitive OD pilots will cause the appliance to shut down if dust blocks the pilot burner air supply. Use only compressed air to clean the pilot air port and injector, as both components are easily damaged.

LEARNING TASK 21.3 SAFETY DEVICES

1. What are the four main categories of safety devices?
2. What are the two categories of Type A gas appliances regarding safety devices?
3. What are the two main components of a thermoelectric flame-failure device?
4. How does a tilt switch operate?

Temperature control – thermostats

A thermostat can be defined as an automatic device that is designed to maintain the temperature of a system at a predetermined set point by controlling the flow of gas to

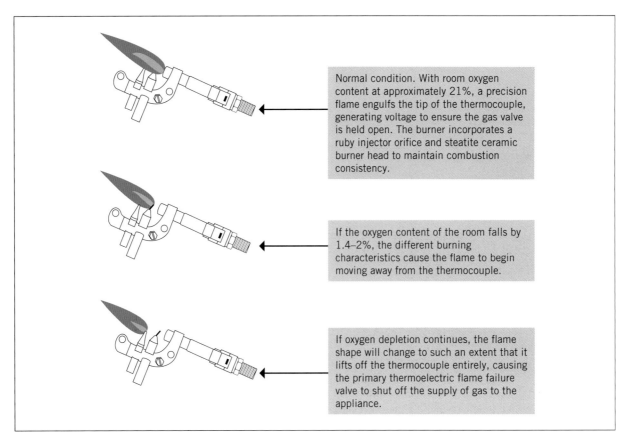

FIGURE 21.19 Oxygen depletion pilot sequence of operation

the burner. There are many forms of thermostat used to control heating and cooling appliances, some of which fall outside the scope of this text. Of primary concern to the novice gas fitter are the following basic thermostats:

1. *Modulating rod and tube thermostats.* These were once used in ovens but are now largely obsolete in modern Type A gas appliances. However, their principle of operation still applies to other devices.
2. *Snap-action rod and tube thermostats.* These are used in storage water heaters.
3. *Liquid expansion thermostats.* This is the most common form of thermostat found in many appliances, including space heaters, ovens, fryers and in some continuous flow and storage water heaters.

Modulating rod and tube thermostats

Early gas cookers used this type of thermostat to control the temperature of the oven compartment.

The cast iron upright gas cooker shown in Figure 21.20 was made in 1932. Its oven was controlled with a modulating rod and tube thermostat that was located directly in the top-front of the oven cabinet.

Thermostatic control of oven temperature was a major advance in early gas cookers sold during the 1800s. The basic rod and tube thermostat, as seen in Figure 21.20 was of a simple and robust design that was used in domestic cookers right up until the 1960s.

The basic principle of operation of these thermostats relies upon the differential of expansion between two dissimilar metals. The thermostat cross-sections seen in

FIGURE 21.20 An old cast iron upright gas cooker with modulating rod and tube thermostat

Figures 21.21 and 21.22 show how differential expansion can cause a valve to automatically modulate open and shut. Follow the information boxes in sequence.

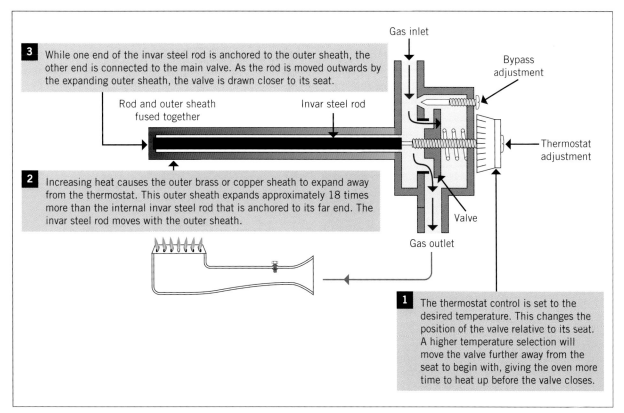

FIGURE 21.21 An oven heating up

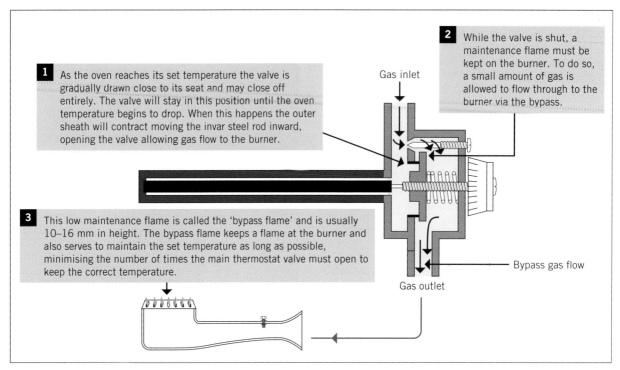

FIGURE 21.22 The oven reaches temperature (note the bypass flame on the burner)

Rod and tube thermostats needed to be a complete assembly and had to be inserted directly into the oven cabinet. This meant that the adjustment knob was often inconveniently located quite low down on the oven cabinet. As a result, manufacturers were quick to phase out oven rod and tube thermostats after the introduction of liquid-expansion thermostats with their characteristic remote-sensing bulb and capillary tube.

Snap-action rod and tube thermostats

Although modulating thermostats are ideal for cooking and heating appliances, they would not be very effective if used in a storage water heater. The thermostat response and water heater's recovery time would be too slow. A storage water heater requires that water be heated as quickly as possible with the gas burner operating at full capacity before shutting down entirely once the required temperature is reached.

A snap-action thermostat is designed to operate at full gas rate until the set temperature is reached, then shut off entirely. The storage water heater in Figure 21.23 is equipped with a snap-action thermostat as part of a combination control (in this case, a Robertshaw 'Unitrol'). This rod and tube is screwed into the water cylinder and is in direct contact with the water. Look carefully at the cut-away of the rod and tube bulb and you will see the internal Invar steel rod and the ECO module beneath it.

The thermostat cross-sections in Figures 21.24 and 21.25 show how the snap-action thermostat operates. Follow the information boxes in sequence.

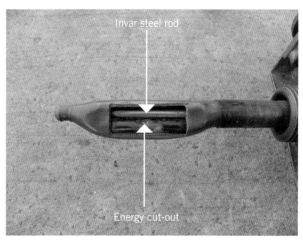

FIGURE 21.23 A storage water heater equipped with a snap-action thermostat, with close-up of the rod and tube bulb

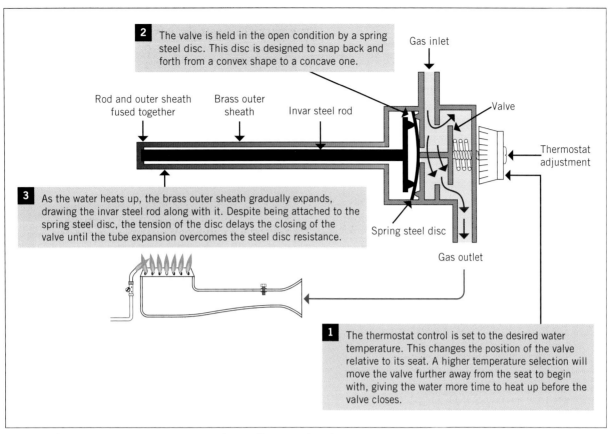

FIGURE 21.24 Water heating up (note that the burner flames are large)

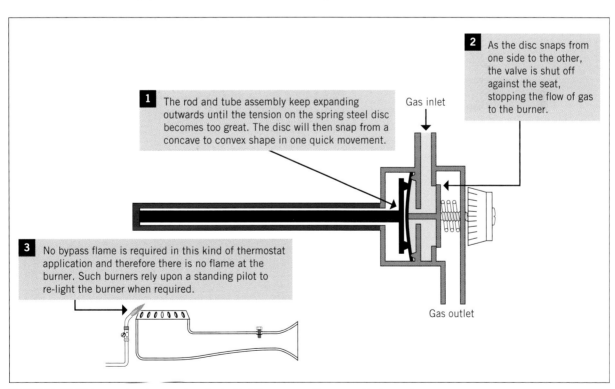

FIGURE 21.25 Water reaches temperature (note that there are no flames on the burner)

The temperature difference between the on condition and the off condition is called the 'temperature differential'. The range of the differential will vary according to the appliance type and its designed performance requirements. Water temperature may have to drop by 7–11°C before the thermostat opens and allows gas back through to the burner. Space heater differentials operate within a much smaller margin and may be no greater than 1°C.

Liquid expansion thermostats

The liquid-expansion thermostat (see Figure 21.26) is the most common form of thermostat found in modern Type A gas appliances. This type of thermostat relies on the expansion and contraction of a liquid held within a copper bulb and capillary tube to effect a change in valve position.

The cross-section diagrams in Figures 21.27 and 21.28 show how the liquid-expansion thermostat operates. Follow the information boxes in sequence.

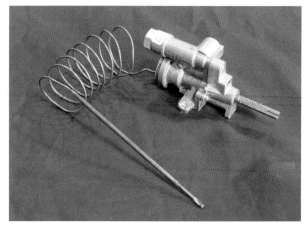

FIGURE 21.26 A domestic cooker liquid-expansion thermostat assembly

Liquid expansion thermostats are available in both modulating and snap-action designs. They also offer appliance designers the flexibility of being able to locate the sensing bulb well away from the actual thermostat

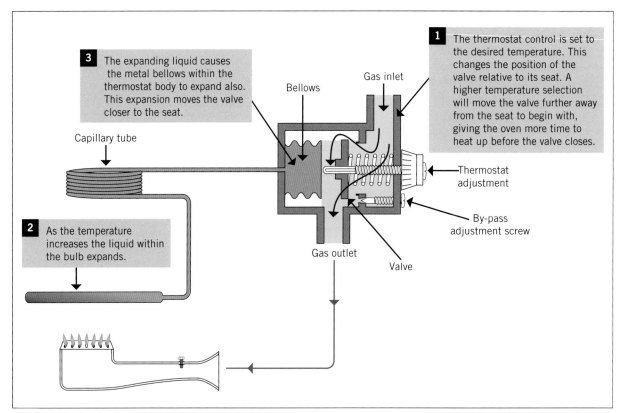

FIGURE 21.27 An appliance heating up (the flames are large)

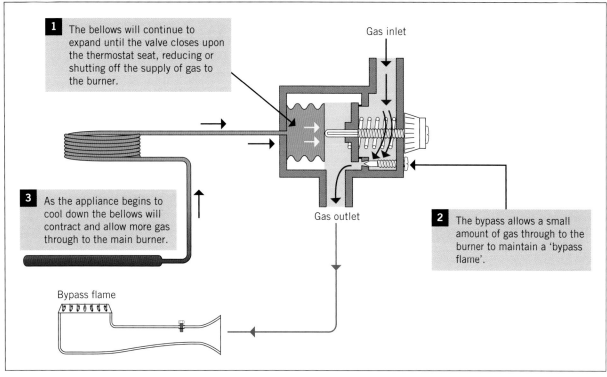

FIGURE 21.28 The appliance reaches temperature (note bypass flames on burner)

control body, thereby improving the ergonomic and safety features of gas appliances.

Rather than use a rigid snap-action rod and tube thermostat, the 'Eurosit' combination control in Figure 21.29 features a liquid-expansion snap-action system to control the temperature of a storage water heater. The sensing bulb is slid into a tube within the cylinder and is not in direct contact with the water.

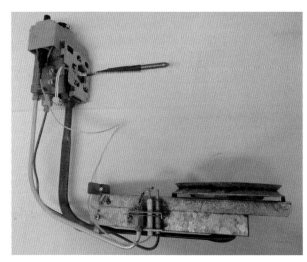

FIGURE 21.29 A 'Eurosit' combination control thermostat sensing bulb

LEARNING TASK 21.4 THERMOSTATS

1 What is the purpose of a thermostat?
2 What are the three thermostats that are used for Type A gas appliances?
3 What type of thermostat is used for gas storage hot water systems?

Ignition systems

Early gas appliances were lit manually whenever required; today almost all modern Type A gas appliances feature some form of ignition system that makes lighting the appliance easier and safer for the user.

In this section, you will be introduced to the following four basic systems:
- piezo ignition
- standing pilot ignition
- electronic ignition
- hot surface ignition.

Piezo ignition

Most people would be familiar with the push-button type piezoelectric ignition system found on many gas barbecues (see Figure 21.30). This form of ignition is also used on some Type A gas appliances, such as storage water heaters. Some crystals, such as quartz, and some synthesised piezo ceramics are known to possess piezoelectric qualities, meaning that when they are

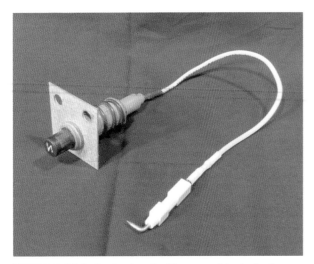

FIGURE 21.30 A simple piezo ignition assembly similar to those found in storage water heaters, older continuous flow water heaters, boilers and barbecues

suddenly compressed they discharge a single surge of electric current.

The igniter itself consists of a piezoelectric crystal held between an anvil and a spring-loaded hammer (see Figure 21.31). When you press the igniter button, the hammer is suddenly released, striking the crystal. This causes the crystal to compress slightly and generate a surge of electricity that may be up to 20 000 V (amperage is very low, ensuring that the device is safe to use).

The electric current travels down a high-tension lead to an igniter probe that is separated from the burner via a small air gap of approximately 3 mm. To complete the circuit and return to its source, the electric current is forced to cross this air gap. Jumping the air gap creates the spark that lights the gas from the burner.

Each compression of the piezo crystal produces only one spark.

Standing pilot ignition

Once a pilot light is lit, its role is to provide a permanent standing pilot flame adjacent to the burner so that when the thermostat or operator sends gas to the main burner it will immediately ignite.

Pilot lights can be found in storage water heaters, continuous flow water heaters (see Figure 21.32), commercial ranges, deep fryers, ovens and central heating furnaces. As such, they come in many varied designs and configurations to suit different appliance needs (see Figure 21.33).

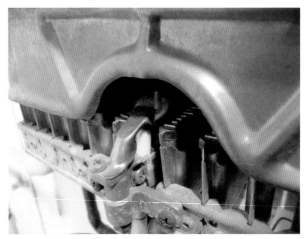

FIGURE 21.32 A standing pilot light in a continuous flow water heater

Electronic ignition

Many gas appliances are fitted with some form of electronic ignition system that is designed to provide continuous sparking to single or multiple burners until a flame is established (see Figures 21.34 and 21.35). Depending upon appliance type and design, the source of electricity may be a battery or a 240 V mains supply.

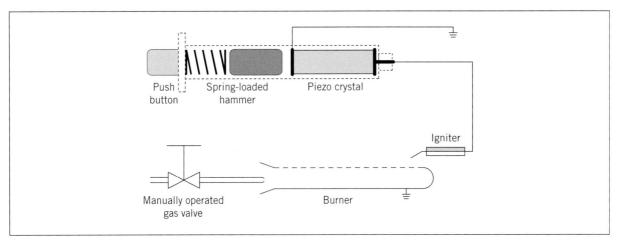

FIGURE 21.31 The piezoelectric circuit

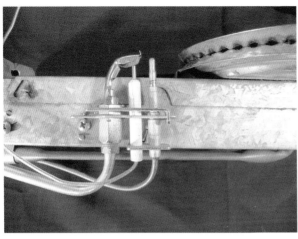 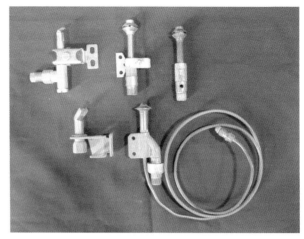

FIGURE 21.33 A small selection of different forms and applications of pilot lights

FIGURE 21.34 An electronic ignition system within a gas cooktop

The system consists of an ignition unit, a high-tension lead and an igniter probe. Activation of the system causes regular pulsating electrical surges of 10 000–15 000 V to be conducted via the high tension lead to the igniter probe. Jumping the air gap between the probe and the burner causes a spark to light the air–gas mix.

Some types of electronic ignition systems will also incorporate a re-ignition function (see Figure 21.36). Where the flame is extinguished due to air movement or a momentary loss of gas supply, these systems can detect the flame loss and initiate a re-ignition sequence. An electric current is continuously sent from the re-ignition unit to the flame sensor/igniter electrode, through the flame itself to the burner and then back to its source.

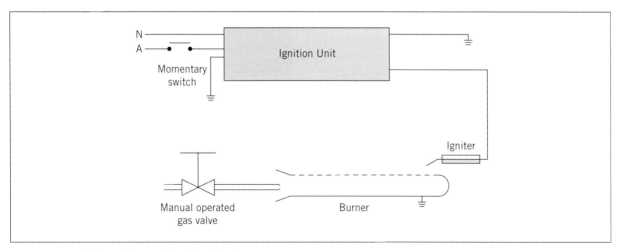

FIGURE 21.35 A simple electronic ignition circuit

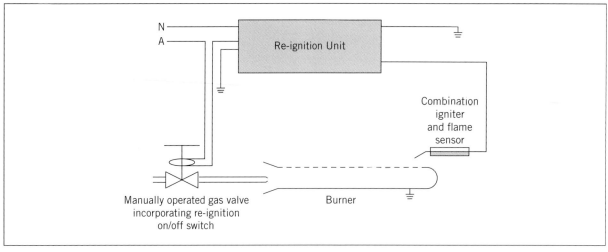

FIGURE 21.36 Re-ignition circuit

Although no rectification takes place, the presence of the flame is the only way that the current can be conducted and complete its circuit, and therefore a loss of flame will immediately initiate the re-ignition process.

Hot surface ignition

Many heating furnaces and cooking appliances rely upon a system called 'hot surface ignition' (HSI). Figure 21.37 shows an example of one of these.

This electronically controlled system consists of a ceramic and carbide ignition element located above the burner (see Figure 21.38). Upon a call for gas from the main controller/thermostat, an electric current is passed through the element, heating it to approximately 1500°C and making it glow almost white-hot. Gas is introduced to the burner after any purge cycle has been completed and is ignited on contact with the hot element. Once the presence of flame is 'proved' via a separate flame sensor, power to the HSI element is stopped and it cools down.

GREEN TIP

The introduction of HSI systems in many appliances has reduced running costs, as standing pilot lights are no longer required. This should be a consideration when recommending appliances to clients.

Combination controls

Basic gas appliance control systems and components include:
- regulators – to control pressure to the appliance

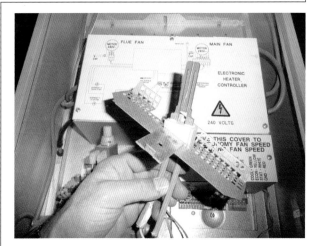

FIGURE 21.37 A hot surface ignition element

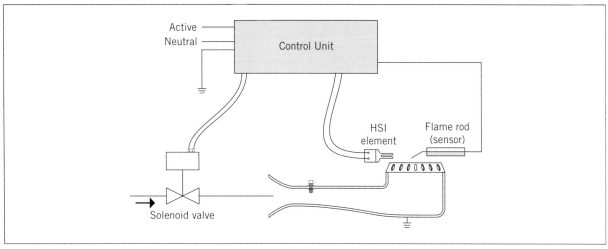

FIGURE 21.38 A hot surface ignition circuit

- safety devices – to ensure that the appliance operates safely, and to render it safe if it does not
- thermostats – to provide an automatic control of the appliance operating temperature.

As you have seen in this chapter, these systems are often made up of individual components and located in different parts of the appliance. Items such as constant pressure appliance regulators may be entirely separate from the appliance itself and require installation by the gas fitter. However, many appliance and component designers have combined these separate functions into one compact device. This is known as a combination control.

Figure 21.39 shows a combination control commonly found on deep fryers. The surrounding control components are all incorporated in the one design.

Combination controls offer the advantages of being compact in size, minimising installation time, and simplifying maintenance in some instances. Some of the more common combination controls found in many Type A gas appliances are shown in Figure 21.40, so that you can begin to recognise them during your commissioning practice.

LEARNING TASK 21.5 COMBINATION CONTROLS

What are the three basic components of a combination control?

Carry out maintenance

Having gained a fundamental understanding of basic Type A gas appliance control systems, it is now time to look at their actual maintenance, with a key

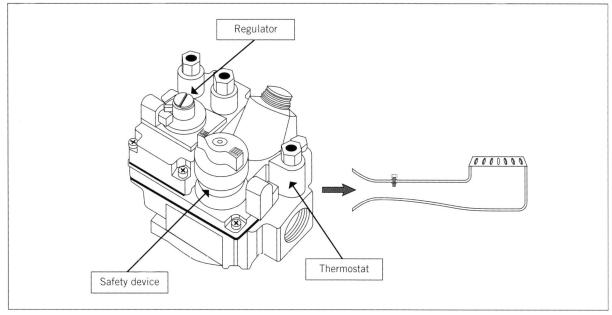

FIGURE 21.39 A combination control incorporating a number of control features into one component

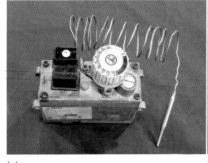  

(a) (b) (c)

FIGURE 21.40 A selection of common combination controls: (a) a SIT control (space heaters/commercial ovens), (b) a White Rogers control (fryer) and (c) a Robertshaw control (decorative flame effect fire)

emphasis on taking a methodical and safe step-by-step approach.

Maintenance – appliance evaluation

The number and types of gas appliances that have been produced over the years and that are on the market today are huge. Even an experienced gas fitter is unable to know everything about all appliances – there are simply too many of them. However, the experienced gas fitter knows how to 'look' at, or evaluate, an appliance in a way that a novice is yet to learn.

How to examine a gas appliance

You will probably feel somewhat nervous on your first maintenance job (especially if the customer is watching!). To gain some sense of control and confidence, pause for a moment and *identify what you do find familiar about the appliance*.

Remember, most Type A gas appliances share similar control systems. Although these control system components often differ considerably in appearance or configuration, by following some simple guidelines you will soon be able to mentally break up appliances into their main operational sections. For example, at first glance the components within the appliances shown in Figure 21.41 may appear quite daunting for the new gas fitter to work on. Identifying the primary control systems and components will help you 'simplify' the appliance.

As a guide, most, if not all, Type A gas appliances can be broken up into the following operational sections:
- ignition system
- pressure control system
- temperature-control system
- safety shut-off system
- combustion system.

When you first encounter a new or unfamiliar gas appliance, you may feel at a loss as to where to begin. A good strategy is to first identify the most obvious features of the appliance. This can be difficult on some appliances, as many parts are hidden within the casing. It also means that sometimes it will be necessary to familiarise yourself with the appliance *at the same time as you work through the maintenance procedure itself*. Whatever the situation, try to work your way through the appliance in the following manner:

1. Turn off the power supply and unplug the appliance.
2. Test the appliance metal casing and pipe connection again for stray current.
3. Locate where the gas comes in and follow this pipe right through to the burner itself. By doing this you will almost certainly find the pressure control components at some point. This may be as simple as an appliance regulator on a domestic cooker, or a more sophisticated combination control on a commercial catering appliance. Follow the gas line out of the pressure control device and you must eventually run into the main burner and pilot assembly (if fitted) forming the bulk of the combustion system.
4. Once you have found the main burner, you can't be too far from the ignition assembly. Tracking a high-tension lead back from an ignition probe will often reveal more of the ignition system.
5. Returning your attention to the burner once again, you should be able to identify some form of flame-failure device. This may be a thermocouple, or the flame rods from an electronic flame supervision

FIGURE 21.41 Looks complicated! Start by identifying key components and systems

system. Follow this back all the way to the shut-off mechanism itself and, before you know it, you have uncovered most of the safety shut-off system. Along the way you might also encounter various ECOs such as overheat and tilt switches.

6 Sometimes the temperature-control components may be a little more difficult to find, but in most cases, they will be incorporated into a combination control. Look for a capillary tube coming out of this control and follow it through to the sensing bulb. Some controls may incorporate an integral rod and tube thermostat, hidden from view. You should still be able to identify the location of a thermostat by the presence of the adjustment control.

Finding the primary operational components of the appliance is the important first step in proceeding to the actual maintenance process.

Maintenance procedure

Is there a standard appliance maintenance procedure?

Well ... almost. If every appliance were nearly the same, then we could approach maintenance procedures in the same way every time. However, this is not the case and you will need to be flexible in how you work on each appliance, which is why it is so important to read all the available literature and try to familiarise yourself with the particular features of each appliance. In doing so, you are likely to uncover variations and changes in sequence that will have an impact on the job itself.

While the sequence may change and some appliances may not require certain steps to be carried out, the following guide should provide a basis upon which to undertake appliance maintenance.

HOW TO

1 CHECK FOR STRAY CURRENT

Don't touch a thing until you check for stray current! Make sure you check both the gas pipe and the frame or body of the appliance. Avoid painted surfaces, making good contact with bare metal in a number of different locations.

General things to consider:
- Test both sides of any insulated components in the piping, valve train or appliance.
- Do not trust thermostatically operated devices that may start unexpectedly.
- Be wary of wiring insulation on older appliances.

2 EXAMINE OVERALL APPLIANCE CONDITION

Record any obvious faults and let the customer know before work begins. This avoids being held responsible for any damage or faults that were already there. At the same time, determine if the appliance needs to be disconnected from the gas line and removed for maintenance to be carried out.

3 CHECK THE MANUFACTURER'S SPECIFICATIONS

Having checked the general condition of the appliance, spend some time reviewing the manufacturer's specifications. In particular, look for information on parts identification, commissioning procedures and any steps related to maintenance. The quality of appliance information varies considerably. Don't hesitate to call the manufacturer for assistance if needed.

General things to consider:
- Check that your specifications are related directly to the actual model of appliance you are testing. It is not uncommon for component or specification changes to occur within a model run.

- While on-site, take the opportunity to record appliance model numbers and serial numbers for future reference. Use your phone to photograph each data plate for later download into a database of customer appliance details that will allow faster and more efficient service in the future.

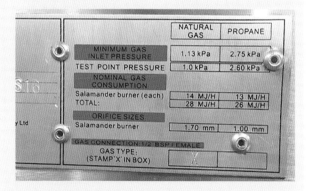

4 CLEAN OUT

It is important to clean the functional parts of an appliance so that a build-up of dust, dirt or grease does not eventually affect the operation and combustion performance of the appliance. Although cooking appliances are prone to excessive grease build-up, the number one enemy for space heaters is dust.

Space heaters are particularly prone to dust accumulation. If allowed to build up, this will severely impede appliance performance and reduce working life. Dust removal is best achieved with a small vacuum or compressed air blower.

>>

5 CHECK THE WORKING PRESSURE

Before any settings or adjustments can be made to an appliance, you must check the working pressure first. There is little point in trying to work out problems with aeration, safety devices or temperature-control systems if you have not ensured that the working pressure is what it should be.

General things to look for:
- Ensure that the working pressure is sufficient when other appliances are operating.
- Check that the appliance regulator holds consistent pressure and does not fluctuate unnecessarily.

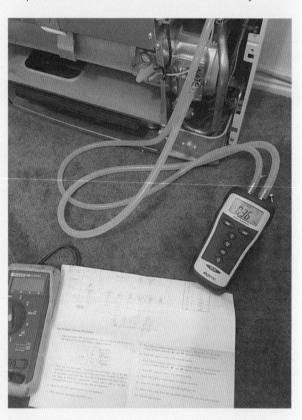

General things to consider:
- Remove all dust from appliance body and air intake grilles.
- Carefully clean dust from printed circuit boards, liquid-expansion thermostat bulbs and thermistor probes.
- Ensure fan vanes are cleaned of all dust and grime.
- Look for excessive grease build up around ignition and pilot assemblies on cooking appliances.
- Blow dust and/or soot out of burners and pilot assemblies.
- On external appliances, look for any build-up of leaves, grass and dirt near flue outlets and air intakes.

6 CHECK THE COMBUSTION SYSTEM AND FLUE OPERATION

With the working pressure confirmed, you are able to check that the flame characteristics demonstrate adequate primary and secondary aeration. Where flame shape and colour are unsatisfactory, make the necessary adjustments to the primary aeration shutters and interrupter screws where fitted. You will also need to check flue performance and ensure that no flue spillage is evident, as per Appendix R of the Gas Installation Standards.

General things to look for are:
- poor flame shape
- yellow flame tips
- flame lift-off

- flame streaming
- carbon deposits
- pilot light blocked by dust or grease
- flue spillage.

7 CHECK THE IGNITION SYSTEM

Even though you have already lit the appliance, the ignition system is usually checked after the working pressure, as incorrect pressure may have an effect on flame ignition. Knowing that you have correct pressure means that you can eliminate this cause while making any subsequent adjustments. Of course, if you were unable to light the appliance in the first place, something would need to be done before proceeding. Such a malfunction would indicate that an in-depth service was required as opposed to a simple maintenance check. Be very cautious of appliances that are difficult to light, as they may do so unexpectedly and any build-up of unburnt gas can be particularly hazardous.

General things to look for are:
- quick and efficient ignition of the flame
- explosive ignition
- effective burner cross-light
- cracks in ignition lead insulation or ceramic ignition probes
- corrosion at the ends of ignition leads
- broken piezo igniters
- slow or dysfunctional ignition packs
- an adequately sized pilot flame.

8 CHECK SAFETY DEVICE SYSTEMS

The majority of modern gas appliances incorporate some form of safety system. These will generally take the form of thermoelectric flame-failure devices, electronic flame-safeguard systems, oxygen depletion devices and various forms of energy cut-outs (ECOs). The gas fitter needs to check that whatever flame-failure device is used, it must shut off the gas supply in an appropriate time. For example, when checking an oven thermoelectric system you should ensure that you hear the electromagnet release the gas valve within one minute of extinguishing the flame.

General things to look for:
- Where suspect, ensure thermocouples are providing the correct mV readings.
- Ensure that the pilot flame impinges upon the thermocouple tip correctly.
- Ensure that the thermocouple tip is not damaged and the lead is not kinked.
- When checking thermocouple output, remember you will get a different reading depending on whether it is under load or disconnected from the electromagnet.
- Ensure that flame-failure devices shut off the gas supply in an appropriate time.
- Check that the oxygen depletion pilots are free of dust.

9 CHECK THE TEMPERATURE-CONTROL SYSTEM

For general maintenance purposes, your main objective is to ensure that the temperature-control system is providing a gas shut-off and/or modulation as intended.

General things to look for:
- The burner flame should cut out or modulate according to the consumer or design setting.
- When testing an oven thermostat, ensure that a correctly sized and stable bypass flame is achieved.
- Check that the thermostat capillary tube is not kinked.

LEARNING TASK 21.6   CARRY OUT MAINTENANCE

1. What are the five key components that make up a Type A gas appliance?
2. When maintaining Type A gas appliances, what are the nine maintenance procedure steps that should be considered?

Clean up the job

Following maintenance work, you must ensure that the area is cleaned.

Clean up your area

The following points are important considerations when finishing the job:
- Check that no small components or screws are left lying around.
- Clean around the appliance so that all grease, soot and dust is removed from floors, benches and the appliance itself.

Check your tools and equipment

Tools and equipment need to be checked for serviceability to ensure that they are ready for use on the job. On a daily basis and at the end of the job check that:
- batteries in your multi-meter, manometer and any other electronic device are charged
- if any item requires repair or replacement, your supervisor is informed and, if necessary, mark any items before returning to the company store or vehicles.

Complete all necessary job documentation

Most jobs require some level of documentation to be completed. Avoid putting it off till later as you may forget important points or times.

- Ensure all SWMS and/or JSAs are accounted for and filed as per company policy.
- Complete your timesheet accurately.
- Reconcile any paper-based material receipts and ensure electronic receipts have been accurately recorded against the job site and/or number.
- Complete all necessary gas supplier/technical regulator paperwork associated with gas appliance maintenance processes. This will vary between jurisdictions.

COMPLETE WORKSHEET 4

SUMMARY

The steps shown in the 'How to' box above are indicative of the approach taken in most maintenance jobs. Of course, you are going to occasionally encounter appliances that have different needs or require a change of sequence. However, regardless of any specific requirements, always approach maintenance tasks in a methodical manner, ensuring that you:

- carry out standard pre-maintenance checks
- evaluate individual appliance needs through physical examination and by sourcing all available literature and specifications
- work on each appliance in a logical sequence specific to its needs
- be on a constant lookout for safety hazards.

GET IT RIGHT

The photo below shows an incorrect practice when maintaining a Type A gas appliance. Identify the problem and provide reasoning for your answer.

WORKSHEET 1

Student name: _____

Enrolment year: _____

Class code: _____

Competency name/Number: _____

To be completed by teachers
Student competent ☐
Student not yet competent ☐

Task
Review the section on *Pre-maintenance checks* and answer the following questions.

1 List three items of equipment you can use to detect stray current.

_____
_____
_____

2 If you needed to disconnect an appliance from the gas supply, how would you protect yourself from any potential electrical hazard?

_____
_____
_____

3 If you were asked to carry out a maintenance check in a busy commercial kitchen, what would be two key pieces of information you would ask the owner or manager?

  i _____
     _____

  ii _____
     _____

4 List the three key pre-maintenance checks that you need to make before commencing work.

  i   _____
  ii _____
  iii _____

5 What would be your actions if you found that the installation you are planning to work on was not compliant with AS/NZS 5601? If unsure, check with your teacher or technical regulator.

_____
_____
_____

MAINTAIN TYPE A GAS APPLIANCES

6 What are two reasons for conducting a leakage test (as per Appendix E of the Gas Installation Standards) on the gas system?

   i _____

   ii _____

7 Regarding your own work area:

   i What are the available gas types?

   _____
   _____
   _____

   ii What NG supply pressure is provided from a domestic service regulator?

   _____
   _____
   _____

8 With the assistance of your teacher or supervisor, conduct a maintenance exercise on two different Type A gas appliances, recording activity details as follows:

	Appliance 1	Appliance 2
Type (HWS, space heater, etc.)		
Pre-maintenance checks		
Gas type		
Supply pressure		
Electrical safety – what instrument did you use?		
Installation compliance	Yes No	Yes No
System leakage test – what was the lock-up pressure?		
Maintenance procedure	Notes	Notes
Electrical safety		
Appliance general condition		
Appliance specifications – what type?		
Clean out		
Working pressure		
Combustion check – characteristics?		
Ignition system check – what type?		
Safety devices – what type? What is their shut-down time?		
Thermostat system – what type?		

WORKSHEET 2

Student name: _____

Enrolment year: _____

Class code: _____

Competency name/Number: _____

To be completed by teachers

Student competent ☐

Student not yet competent ☐

Task

Review the section on *Safety devices* and answer the following questions.

1. Name the three unsafe conditions that gas appliance safety devices are designed to prevent.

    i _____

    ii _____

    iii _____

2. Name the parts of this flame-failure device (FFD) where indicated by the arrows.

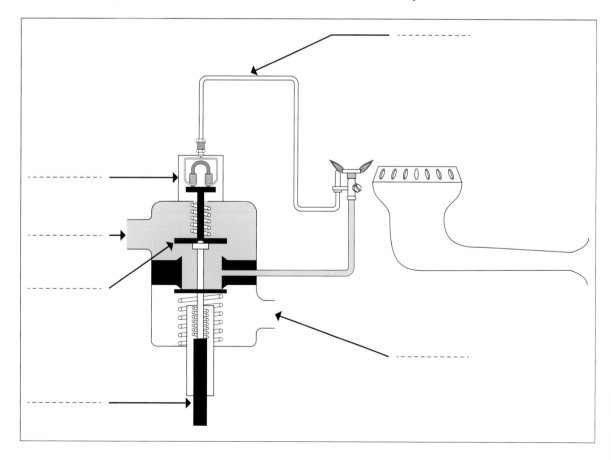

MAINTAIN TYPE A GAS APPLIANCES   **493 GS**

3 What is the nominal output of a thermocouple?

   _____

   _____

   _____

4 Indicate how the process of 'rectification' converts current. Circle the correct answer.

   Rectification changes: DC to AC   AC to DC

5 What kinds of gas appliances use OD pilots?

   _____

6 List three kinds of ECO.

   i   _____

   ii _____

   iii _____

7 Some storage water heaters incorporate a bi-metallic switch in circuit with the thermoelectric system. Under normal operating conditions, would this ECO maintain a closed circuit or an open circuit?

   _____

   _____

8 Would the high temperature cut-out on the heat exchanger of a decorative flame effect fire be in the 'normally open' position or the 'normally closed' position under normal operating conditions?

   _____

   _____

9 In your own words describe in three steps how a flame rectification system proves the presence of flame at the burner.

   i   Step 1

   _____

   _____

   ii Step 2

   _____

   _____

iii Step 3

10 What causes the FFD of a flueless space heater to shut down when oxygen content in a room falls by 2 per cent?

11 Faulty or damaged thermocouples can often be replaced with universal thermocouple kits available from plumbing and gas suppliers. State two things you need to consider when replacing a thermocouple.

   i

   ii

12 What is the normal 'operating' mV range for a standard thermocouple?

13 Name two appliances that may use a thermopile.

   i

   ii

14 Some thermocouples and/or pilot flames are adjustable, allowing you to change the position of the probe if necessary. What is the recommended amount of flame impingement on the tip of the thermocouple probe?

WORKSHEET 3

Student name: _____

Enrolment year: _____

Class code: _____

Competency name/Number: _____

To be completed by teachers
Student competent ☐
Student not yet competent ☐

Task

Review the section on *Temperature control – thermostats* and answer the following questions.

1. What does the term *temperature differential* mean?

   _____

   _____

   _____

2. As the temperature of an oven rises and gets closer to the set point, does the valve get closer to or further away from the thermostat seat?

   _____

3. Do storage water heaters use a snap-action or modulating thermostat?

   _____

4. What is the purpose of the bypass feature found in oven thermostats?

   _____

   _____

   _____

5. As an appliance cools down, does the valve move closer to or further away from the thermostat seat?

   _____

   _____

   _____

WORKSHEET 4

Student name: _____

Enrolment year: _____

Class code: _____

Competency name/Number: _____

To be completed by teachers

Student competent ☐

Student not yet competent ☐

Task

Review the sections on *Ignition systems* and *Combination controls* and answer the following questions.

1 Name at least two appliances that may feature hot surface ignition (HSI).

   i _____

   ii _____

2 What is the recommended air gap between an ignition electrode and the burner head?

   _____

   _____

   _____

3 Ask your teacher to nominate at least three different types of combination control and list how many separate functions are carried out by each one.

   i _____

   _____

   _____

   _____

   ii _____

   _____

   _____

   _____

   iii _____

   _____

   _____

   _____

INSTALL TYPE B GAS APPLIANCE FLUES

Learning objectives

In this chapter, you will learn about the requirements related to the correct flueing of Type B gas appliances and how to interpret AS/NZS 5601 so that your flue installations are compliant for material selection, specific flue pipe installation and locations of flue terminals. Once you have covered these basic requirements you can then move on to the actual sizing of natural draught flue systems suitable for atmospheric Type B appliances. This section does not cover the design of power flue systems or flues for induced or forced-draught burners.

Areas addressed in this chapter include:
- what are Type B appliances
- why flues are necessary
- types of flues used for Type B appliances
- flue components and materials
- heat loss
- terminal locations
- sizing individual appliance flues
- sizing horizontal and vertical common flue manifolds
- flue pipe roof penetration flashings
- flue installation.

Note to teachers: The relevance of Type B flueing is entirely subject to State or Territory licensing and 'scope of work' regulations. As such, the information contained in this chapter simply relates to generic requirements and the skills and knowledge needed to size and install simple, natural draught flue systems only.

Contact your local technical regulator for advice.

Before working through this section, it is absolutely important that you complete or revise each of the chapters in Part A 'Gas fundamentals' with particular emphasis on Chapter 6 'Combustion principles'.

Overview

Before examining the flue installation requirements of Type B gas appliances, it is important to have a clear understanding of what these types of appliances are and the basic combustion processes they use.

Type B gas appliances

Type A gas appliances are mass-produced in accordance with specific design requirements and if to be sold in Australia must be 'certified' by a certification body as meeting a strict level of minimum operational and safety performance. In comparison, Type B appliances are those with a gas consumption in excess of 10 MJ/h for which a certification scheme does not exist. In most cases, Type B appliances are used for very specific industrial and large commercial installations. Some appliances are mass-produced while others are custom designed as one-off solutions for individual applications.

Examples of some Type B appliances include:
- powder coating ovens
- pottery kilns
- process drying burners (seeds, bentonite, ore, sugar, etc.; see Figure 22.1)
- metallurgical process burners
- boilers (see Figure 22.2).

Who can install a Type B gas appliance and its flue system?

Regulatory requirements relating to Type B gas appliance installations vary around Australia. In most jurisdictions, a fully qualified and licensed Type B gas fitter is the only person able to install, commission and maintain a Type B appliance.

Similarly, the scope of work that defines Type B installation and whether this includes its flue system also varies. The Type B gas fitter is generally responsible for the appliance from the gas inlet isolation valve to the flue spigot. In some cases, it is entirely possible that the flue for an atmospheric boiler may well fall under the scope of a Type A gas fitter's licence, while in other instances the relative complexity of a forced-draught boiler flue and integral economiser could be fully under a Type B gas fitter's control, or be assembled as part of engineered works.

There is no single answer and as such you must consult with your teacher as to the specific requirements in your State or Territory.

(a)

(b)

FIGURE 22.1 A Type B appliance designed to dry bentonite clay used as a binder in foundry casting moulds

FIGURE 22.2 These Type B boilers feature large input atmospheric burners and natural draught flues

Source: Used with the permission of Rheem Australia

LEARNING TASK 22.1 — LICENSING REQUIREMENTS

Requirements in your State or Territory	Brief description
Who is permitted to install a Type B appliance?	
What is defined as Type B work?	
Who is permitted to install the flue of a Type B appliance?	

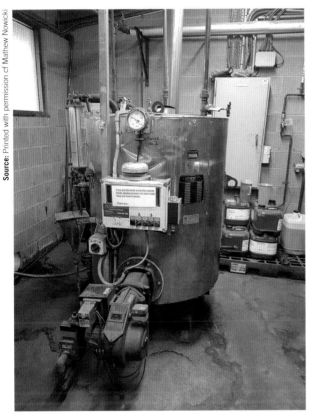

FIGURE 22.3 This vertical tube boiler is an example of a common forced-draught Type B gas appliance

Basic combustion principles

The basic combustion principles described in this section are general in nature and relate to the operation of atmospheric burners and natural draught flues. Type B gas appliances often employ more complex combustion and flueing processes, but it is beyond the scope of this unit to describe them here.

The purpose of a flue is to safely convey the products of combustion away from the gas appliance to the flue terminal. To understand the importance of the flue system, you need to understand some basic combustion principles.

Complete combustion describes a chemical reaction between a *fuel* (natural gas or LPG) and an *oxidiser* (oxygen), which becomes self-sustaining after the application of the correct amount of *heat* (a match or spark).

For complete combustion to take place the proportion of fuel gas and air must be as follows:
- Natural gas – 1 m^3 of natural gas requires 9.5 m^3 of air.
- LPG – 1 m^3 of LPG requires 24 m^3 of air.

This mix of air and fuel gas is lit in the combustion chamber and will give off certain products of combustion.

Products of complete combustion

When a fuel gas is burnt in air under normal atmospheric conditions, the products of complete combustion that result are:
- heat
- water vapour
- carbon dioxide.

Figure 22.4 shows a cut-away of a natural gas hot water cylinder. A correctly installed and commissioned appliance will ensure that complete combustion takes place and the products of combustion are able to exit the appliance via the flue system.

What about the nitrogen? You will notice that the 8 m^3 of nitrogen that enters the appliance still exists as 8 m^3 at the flue outlet. Nitrogen is not a product of combustion. It is an inert gas and will not burn; it simply goes in one end and comes out the other with no real change in quantity.

In an area of sufficient ventilation, these products of combustion are harmless. However, if they were permitted to build up in a room or enclosure, a hazardous situation would arise for the following reasons:
- Room heat may rise excessively, and with it condensation and dampness from the trapped water vapour within the flue products.
- As carbon dioxide (CO_2) levels in the room increase, the potential availability of oxygen would be reduced and the burner flame would begin to be smothered as secondary air is reduced. Without sufficient oxygen, the fuel gas would burn incompletely.
- Incomplete combustion would produce the dangerous gas carbon monoxide (CO), which not only exacerbates poor combustion but also may endanger the lives of any people in the room.

The effect of inadequate flueing on combustion

A blocked or incorrectly sized flue will cause a build-up of combustion products, reducing the amount of fresh secondary air available for complete combustion to take place.

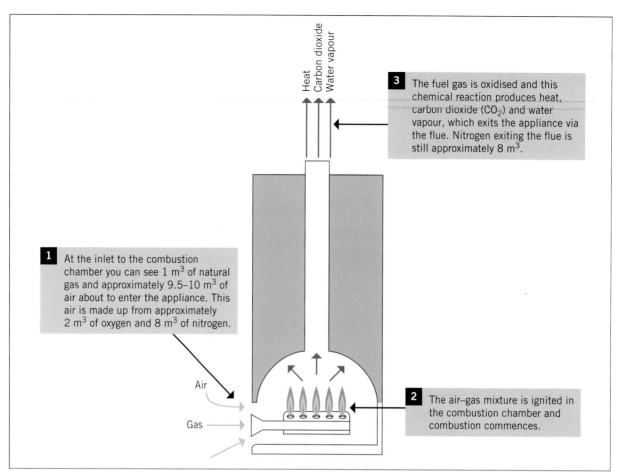

FIGURE 22.4 A cut-away of a NG storage hot water system, showing how air and fuel combust to create combustion products

When the supply of air is restricted, not enough oxygen will be present for the hydrocarbon fuel to be completely oxidised fully into CO_2 and water vapour. This lack of secondary air for incomplete combustion is called 'vitiation' (pronounced *vish-ee-ay-shon*).

Products of incomplete combustion

Incomplete combustion of a fuel gas will result in the creation of the following products:
- *Carbon monoxide*. The toxic result of incomplete oxidation of the fuel into CO_2 and water vapour, CO is colourless, odourless, tasteless, very toxic and highly flammable.
- *Soot (carbon)*. During combustion, fuel molecules will break down and change in a series of complex chemical reactions. In any flame, there is a 'soot-growth zone', in which particulate solids can form. During complete combustion, these particles normally oxidise completely into CO_2 and water vapour; but where combustion is incomplete, they will stabilise as solid particles in the flame. These particles of hot, glowing soot give any flame its yellow colour. The particles then cool as they escape the flame and will deposit as black carbon on any nearby surface. Many authorities regard soot as a carcinogenic (cancer-causing) substance. Soot can quickly block flues and quickly worsen the process of incomplete combustion.
- *Aldehydes*. These are a class of organic compounds formed during the partial or incomplete oxidation of hydrocarbon fuels. Aldehydes have a strong ammonia-like smell that although not toxic in small doses does provide a warning that incomplete combustion is taking place and that it is highly likely that CO is also present.

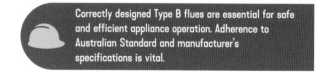

Correctly designed Type B flues are essential for safe and efficient appliance operation. Adherence to Australian Standard and manufacturer's specifications is vital.

As a Type B gas fitter, you will learn about the detailed calculations required in flue gas analysis and the specific performance requirements for appliance flues. However, at this stage identify the basic fundamental flue characteristics that follow.

LEARNING TASK 22.2 BASIC COMBUSTION

1. Circle the combustion products that are the result of complete combustion.
   a Carbon dioxide
   b Soot
   c Aldehydes
   d Nitrogen
   e Heat
   f Carbon monoxide
   g Water vapour
2. What term is used to describe incomplete combustion where an undersized natural draught flue causes a reduction in secondary air in the combustion chamber?
3. What is the primary purpose of an appliance flue?

Flue types

There are a number of different flue types used with Type B gas appliances. You need to recognise these flue types so that you can relate to their particular requirements as detailed in AS/NZS 5601.1.

Natural draught open flue

A natural draught open flue is designed to operate under normal atmospheric conditions. As hot flue gases are less dense than the surrounding air, they will rise of their own accord without any mechanical assistance. This natural rise of hot flue gases is known as aeromotive force (see Figure 22.5). A natural draught flue must be carefully sized and installed to ensure that the flue gases remain hot and maintain this aeromotive force.

Most Type B gas appliances with a natural draught flue will not come with a flue kit but are approved by the technical regulator with a strictly prescribed flueing solution. As the installer, you will need to use your knowledge and skills to confirm atmospheric flue size where required, select materials and install the flue according to specifications.

Power flue

A power flue is defined as a flue in which the products of combustion are removed from the appliance via a fan situated in the flue itself. This is not to be confused with a fan-assisted flue where an electric fan in a Type A appliance provides either forced or induced draught combustion air to the combustion chamber.

Power flues must be carefully designed and sized to ensure that all products of combustion are effectively removed. They must incorporate special gas supply interlocks to ensure the system fails safe in the event of a power flue breakdown.

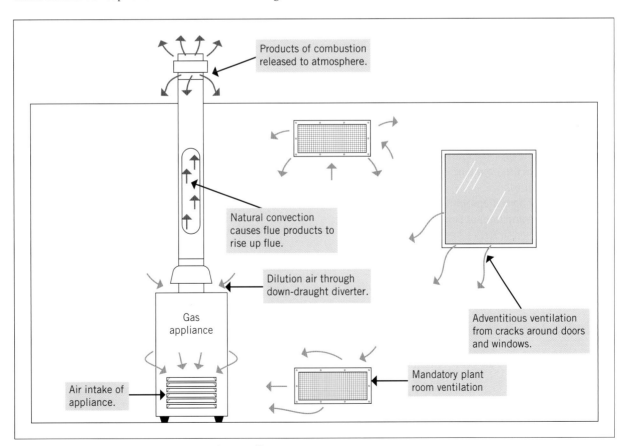

FIGURE 22.5 Operation of the natural draught open flue system

> **LEARNING TASK 22.3**
>
> 1 What is the term used to describe the natural rise of hot products of combustion within a natural draught flue system?
> 2 What poisonous gas is produced where an incorrectly installed flue causes incomplete combustion?

Prepare for flue installation

Detailed preparation before job commencement can save many lost hours fixing problems that could have been easily identified through a methodical process.

Access plans, specifications and work instructions

Type B appliances must be installed strictly in accordance with their approved plans and specifications (see Figure 22.6). These must be studied and regularly referred to throughout the entire installation process.

Regulatory and statutory requirements

Unless you are a Type B gas fitter yourself, you are unlikely to be required to seek flue system design approval in isolation from the overall Type B design approval process. In almost all instances, no work can commence until all aspects of a Type B installation meet the approval of the technical regulator. Therefore, at the stage when you could be required to install a Type B flue, all necessary appliance-specific approvals for the work will have already been carried out.

However, you would still need to check the following points:
- Confirm that the approval documentation has actually been signed off and has not expired. You must always work under a current approval.
- Ensure that the proposed flue installation also meets any additional Environment Protection Authority (EPA) guidelines.
- Confirm your obligations with the local authority that may require some form of Start of Works notification and/or Completion Notice for the project.
- Check with the local plumbing licensing regulator to confirm if licensed personnel are required to undertake large-diameter roof penetration and flashing work.

Pre-installation check of requirements

The job specifications or material list provided by your employer should detail the primary requirements for the job. You should pay particular attention to the following:
- Ensure flue materials on order and those that have arrived match the requirements of the plans and specifications.
- Confirm that the supplied materials are suitable for the flue application and that the proposed flue location is compliant with the Standard.
- Check that all flue components are certified for use. Current certification numbers can be checked against the online data provided by relevant entities responsible. Examples include the Australian Gas Association and SAI Global. The Gas Technical Regulators Committee (GTRC) website provides a portal to assist in accessing confirmation of material and product certification – http://equipment.gtrc.gov.au

Plan tasks with others involved or affected by the work

A Type B flue installation will invariably involve cooperation with a number of other people. Your work sequence and smooth job progression relies upon good communication with all parties. Always make a simple list of those you may need to work with during the project. Examples could include:
- owner and/or site supervisor
- site safety representative
- Type B gas fitter
- electricians
- project engineers and flue designers
- Type B appliance designer
- technical regulator
- flue component manufacturer
- transport company (for large flue components)
- crane company
- sheet metal and steel fabricators.

Evaluate likely job WHS requirements

Site WHS inductions and adherence to SWMS will apply to almost all Type B installation work. Requirements will always depend on the nature of each installation, but the following notes need particular consideration:
- Asbestos
  - Existing buildings – Asbestos is a Type 1 carcinogenic substance that you must always be on the lookout for. Asbestos-containing materials (ACM) were used in dozens of applications over many decades and to this day many existing sites will still contain ACM. Examples include:
    - roof and wall cladding
    - flue, pipe and ceiling insulation products
    - fire-rated walls and doors.
  - New buildings – Although new buildings and installations are generally free of this hazard, always double check the certification of certain imported products that have been manufactured in countries where controls on asbestos use are lax or non-existent.
- Electrical safety and testing
  - Test for stray current before touching gas appliances and site infrastructure.

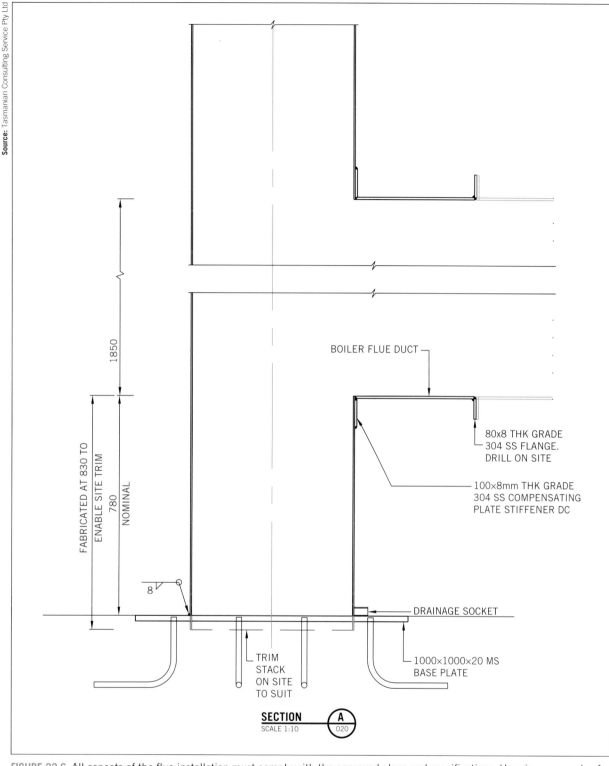

FIGURE 22.6 All aspects of the flue installation must comply with the approved plans and specifications. Here is an example of a flue stack base and appliance connection.

- Look for overhead cables that may be a hazard when manoeuvring flue components.
- Refer to site electrical plan before drilling in slabs and walls.
- Working at heights – Working on Type B flues will almost always involve working at heights and your site evaluation may include:
  - scaffolding – fixed or mobile
  - elevated work platforms
  - roof perimeter scaffolding
  - fall restraint PPE.
- Check that all PPE is serviceable and where relevant check replacement dates on items such as height safety harnesses.

Required tools and test equipment

As with other gas fitting tasks, your choice of tools and equipment will be dependent upon the particular installation requirements and the type of flue system you are working with. As a general guide this may include the following key items:

- crane
- ladders and/or elevated work platform
- assorted power tools relevant to the task
- tools for chemical and mechanical anchoring
- welder
- sheet metal tools for roof flashing installation
- spirit level and plumb bob
- laser levels
- measuring equipment.

Wherever 240 V power tools might be required for a certain task, remember to check that the tool has been tested and take particular note of the retest date on the tag. If it is out of date, let your supervisor know and mark in some manner that it is not to be used.

Identify flue requirements

In this section you will learn about:

- the components that make up a Type B gas appliance flue system including:
    - the basic configuration of natural draught flues found on atmospheric appliances
    - the basic configuration of flues used for forced-draught burner systems
- available and approved materials for flues and the definition of high and low heat loss materials
- flue locations in accordance with the Standard
- how to size natural draught flues for atmospheric Type B appliances.

Flue components – atmospheric burner systems

Before you can move on to learn about more specific flue requirements you need to know the correct terminology for all flue components and their purpose. Refer to **Figure 22.7** to identify each component part.

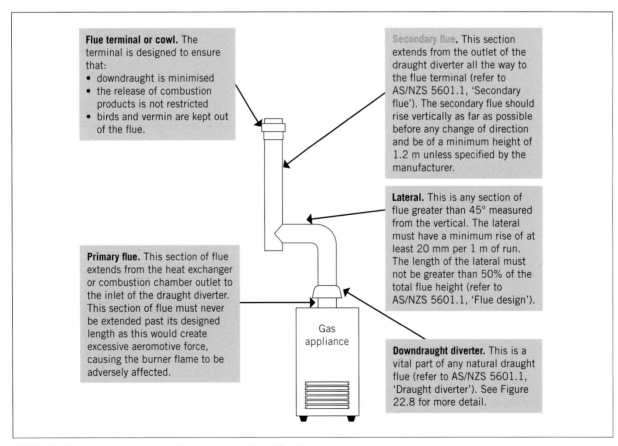

FIGURE 22.7 Natural draught open flue component identification

Downdraught diverter

One of the most important components within a natural draught flue system is the downdraught diverter. The diverter has three key functions:

- to reduce aeromotive force
- to divert any downdraught away from the combustion area
- to dilute the products of combustion.

Review Figure 22.8 to find out more about the downdraught diverter.

What would happen if there were no downdraught diverter?

In almost all cases, natural draught flued appliances would be unable to operate effectively without the use of a downdraught diverter in the flue system. Figure 22.9 shows what would happen if a flue were fitted directly to the gas appliance.

Flue components – forced-draught burner systems

The majority of Type B appliances employ some form of forced-draught burner system (see Figure 22.10). This type of burner incorporates a mechanical combustion fan that provides a forced supply of air for the combustion process.

Combustion products and a calculated percentage of excess air are in turn forced through the flue by this process. Figures 22.11 and 22.12 show how the flue of a forced-draught appliance can differ from an atmospheric appliance.

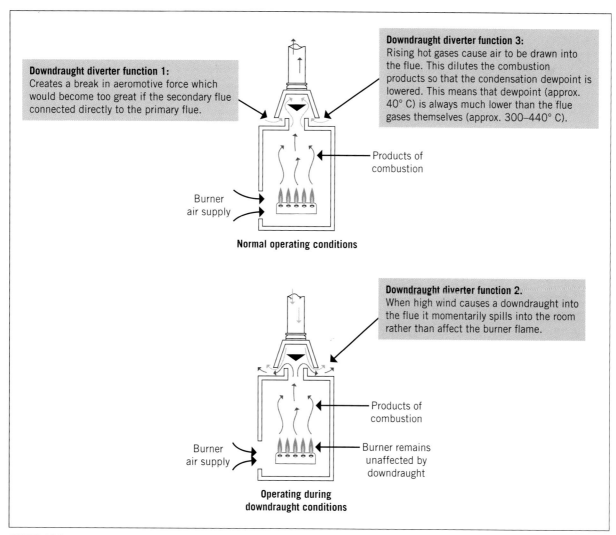

FIGURE 22.8 Functions of a downdraught diverter

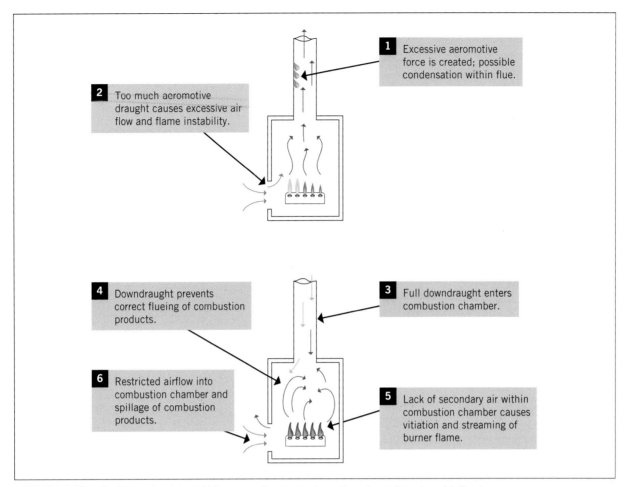

FIGURE 22.9 The effect on natural draught burner performance where there is no downdraught diverter

FIGURE 22.10 This 6-megawatt tube boiler features a package burner with large forced-draught combustion fan

Flue materials

A wide range of flue materials exists for different Type B gas appliance applications. All materials must be approved for use and selected in accordance with AS/NZS 5601.1. and the designer's approved plans.

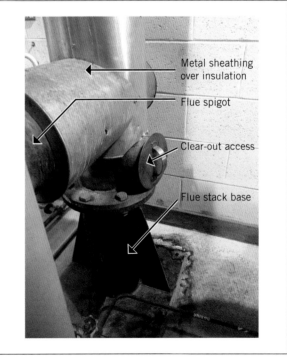

FIGURE 22.11 An older Type B appliance installation showing steel stack base and flue clear-out access

FIGURE 22.12 A typical Type B flue stack serving a forced-draught appliance

Flue material selection will depend on a number of factors:
- *Designer's specifications.* All installations must comply with the designer's very specific instructions. Type B appliances have been granted installation approval by the technical regulator on condition that all components including the flue system are installed as per specification. Where flue materials are prescribed in such a way, you are not permitted to use alternatives without reference to the designer and technical regulator.
- *Durability.* Where sections of a flue installation may be concealed, you will need to ensure that the material you select is corrosion resistant. Some Type B appliances such as pottery kilns and incinerators can produce flue products that will require materials resistant to degradation from certain chemicals and high temperatures. Flues such as these are often specified as mild steel with no protective coating.
- *Flue gas temperature.* Many Type B appliances operate at quite high temperatures. Standard materials suitable for atmospheric burner use are normally only rated up to 300°C. Above this limit, another type of material would be specified.
- *Flue diameter.* When using materials such as mild steel for an atmospheric appliance, you will find that the maximum permissible diameter of the flue depends upon the wall thickness of the material. Similar restrictions apply to stainless steel.
- *Heat loss.* Flue materials should be selected to ensure that the temperature of the flue gases is maintained and not lost prematurely. Some materials cannot be used where the flue exceeds a prescribed height.

Review your Standard and compare the differences between various products and, in particular, note the requirements for flues used in high temperature applications.

REFER TO AS/NZS 5601.1 SECTION 4 'FLUE MATERIAL'

Heat loss in natural draught flue systems

Natural draught flue installations are classed as either 'high heat loss' or 'low heat loss'. This describes the amount of heat lost through the walls of the flue to the surrounding environment. The sizing of flues is to a large degree dependent upon this classification.

High heat loss installations and materials

A high heat loss installation or material is one in which the flue loses a lot of heat through the walls of the flue to the surrounding environment. Depending upon the efficiency and appliance type, the flue gas temperature of many natural draught appliances will range from approximately 100°C to 300°C.

Single wall metal flues are commonly made of materials such as stainless steel, galvanised steel and aluminium. In the right circumstances, a single wall flue offers a cost-effective and adequately performing flue solution. However, you must be aware that high heat loss materials have a number of important disadvantages that need to be taken into account:
- If flue gases lose too much heat, the water vapour, which is normally carried away as one of the products of combustion, will start to condense on the inside of the flue. The temperature at which this happens is called the dew point of the flue gas. For most natural draught flues, the dew point is approximately 40°C. Flue gas condensate is quite acidic, containing both sulfuric and nitric acids. Condensate has a highly corrosive effect upon any flue material such as galvanised or mild steel. This could lead to costly early replacement or a hazardous situation in which a perforated flue allows products of combustion to accumulate inside a building.
- Flue gases rise up and out of the flue because they are hotter and less dense than the surrounding air. If the flue gases lose too much heat through the walls of the flue, this aeromotive force will slow and could stall, almost causing a 'plug' of cold air to form in the flue. When this happens, flue gases may spill from around the draught diverter into the room. As the condition continues, a build-up of flue gases in the combustion chamber will restrict the availability

of secondary air to the burner, causing vitiation and producing carbon monoxide.
- Single wall metal flues (see Figure 22.13) conduct a considerable amount of heat and must be installed in strict compliance with AS/NZS 5601.1 so that clearances from combustible surfaces are maintained. This may add to the installation cost.

FIGURE 22.13 A high heat loss single wall flue (left) and a low heat loss twin wall flue (right)

Note that:
- single wall metal flues installed outdoors are always classed as high heat loss installations
- single wall metal flues installed indoors and not subject to draughts can be classed as low heat loss installations, as long as not more than 1 m of single skin flue is exposed to the outside air. This 1 m rule, although not part of the Standard, is based upon industry and manufacturer experience of many years. If in doubt always consult with the manufacturer and technical regulator.

Low heat loss installations and materials

Low heat loss installations are those in which the flue is insulated against excessive heat loss. This may be due to the nature of the flue material itself or the location of the appliance and flue. Approved low heat loss materials include:
- fire bricks
- autoclaved fibre cement (asbestos free)
- proprietary twin wall flue (see Figure 22.14).

A proprietary twin wall flue is commonly used by gas fitters for smaller natural draught Type B installations. The twin wall flue is made from an outer galvanised steel shell surrounding an internal aluminium core separated with an approximate 5 mm air gap. It is an approved proprietary flue system and must be installed according to the manufacturer's instructions. It offers the gas fitter the following advantages:
- simple twist lock jointing
- corrosion resistance
- able to be installed as close as 10 mm to a combustible surface

FIGURE 22.14 A range of pipe lengths and transition fittings are available in twin wall flue

- can be used in concealed installations
- does not require heat-resistant silicone roof flashings
- available in a wide range of lengths and components
- when installed in accordance with the manufacturer's specifications, it has a long warranty period.

Proprietary-brand twin wall flue systems are available in a range of diameters and lengths (see Figure 22.14).

Note: Twin wall or insulated flues are classed as low heat loss irrespective of whether they are installed outdoors or indoors.

LEARNING TASK 22.4

1. Is a single wall flue installed indoors and not subject to draughts classed as a high heat loss or low heat loss situation?
2. What flue component of a natural draught flue is located between the primary and secondary flues?
3. List at least three materials suitable for a Type B forced-draught appliance where the specifications indicate a working flue temperature of 400°C?

Flue installation and terminal locations

During the planning stages and at the time of installation you must always consider how the appliance is going to be flued. Getting it wrong can be catastrophic, as many fires have been started because of poorly installed flue systems. Adherence to AS/NZS 5601.1 and the relevant manufacturer's instructions will prevent this from happening. When planning for a flue installation you need to consider the following points:
- Is the flue clear of structural members?
- Is any combustible material present and what clearances are required?

- Is any part of the flue going to be concealed and/or inaccessible?
- How will the flue be supported?
- Can the appliance be removed without disturbing the flue itself?
- Where is the flue required to terminate?

Having evaluated your installation, you will find that AS/NZS 5601.1 has a solution to these variables.

Local and statutory approvals for Type B flue terminals

Your Standard details basic terminal locations, but the installation of Type B flues will often need to adhere to additional requirements set by local authorities and statutory entities such as the EPA. The size and combustion product output of some Type B appliances may require flue stacks to terminate at a higher point than would normally be the case and you will need to take this into account when reviewing the installation plans.

Flue termination above a roof line

Flue terminals must be located to ensure that regardless of wind direction the flue will not be subject to downdraughts. AS/NZS 5601.1 provides basic guidelines, but you will also need to account for more extreme installations where building design may cause additional problems. High-pitched roofs and adjacent walls are examples where significant pressure imbalances can lead to downdraught problems.

REFER TO AS/NZS 5601.1 'FLUE TERMINALS'

Required clearances are measured from the building structure to the end of the flue pipe *before* the cowl is fitted. Figures 22.15 and 22.16 illustrate some of the basic clearance requirements where flues are terminated above a roof line.

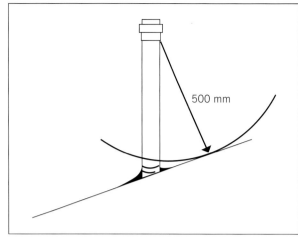

FIGURE 22.15 Clearance for a flue terminal at least 500 mm from the nearest part of a roof

LEARNING TASK 22.5

1. The minimum distance between the end of a flue above a roof and the roof covering shall be no less than:
   a 1 m
   b 500 mm
   c 600 mm
   d 2 m
   e 1.5 m
   f 300 mm
2. List at least four items you need to consider when planning a flue installation?

COMPLETE WORKSHEET 1

Atmospheric gas appliance flue sizing

In this section, you will be introduced to the skills and knowledge required to carry out sizing calculations for

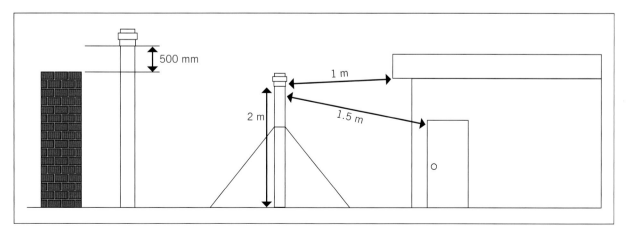

FIGURE 22.16 Clearance for a flue terminal 2 m above a trafficable roof and at least 500 mm above any parapet

simple atmospheric appliance flue installations. In most localities around Australia, plans and specifications for a Type B appliance must be fully approved prior to installation. Therefore, this section on flue sizing is only provided to enable you to confirm the design requirements. Where any discrepancy is identified you need to take your concerns to the designer and/or technical regulator.

Remember, these flue sizing processes are only for atmospheric appliances and do not apply to forced-draught, induced draught or power flue appliances. The following sizing procedures will be covered:
- Individual appliance flues
- Horizontal common flue manifolds – used where a number of appliances are installed at the same level all joined to the same common flue system.

It would be highly unusual to find a multi-appliance Type B installation that would employ a natural draught vertical common flue manifold as found in multistorey buildings and so this form of flue configuration is not covered in this section.

The flue sizing process is described in detail in AS/NZS 5601.1. You should first familiarise yourself with the basic factors affecting flue sizing and design.

REFER TO AS/NZS 5601.1 APPENDIX H1 'FLUE DESIGN FOR APPLIANCES WITH ATMOSPHERIC BURNERS' AND APPENDIX H1.2 'FACTORS INFLUENCING FLUE DESIGN'

Information required before you start flue sizing

Preparation for flue sizing includes not only collating the information required for the calculations but also recording the sizing process in a logical and clear manner. As with any sizing calculation, you need to draw a simple diagram and include the basic information shown in Figure 22.17 before starting.

FROM EXPERIENCE

It is important to remember that a gas fitter is held fully responsible for all work. Where you are asked to install a natural draught flue for a Type B atmospheric appliance, always match the requirements of the appliance with the Standard and the approval documentation. Where anything is not clear always contact the designer and/or technical regulator!

Sizing a simple individual appliance flue

When sizing a natural draught flue for a Type B atmospheric appliance, there is one standard process and two main variations that can be applied. These additional variations are:
- Rule 1 – Additional bends rule
- Rule 2 – Spigot rule

The step-by-step process for flue sizing is detailed in AS/NZS 5601.1. Use this and the following details to size the flue installation in the diagram.

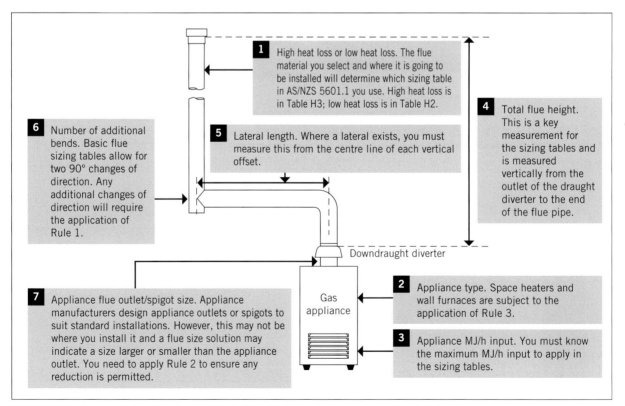

FIGURE 22.17 Confirming the flue size of Type B atmospheric appliances requires all relevant information

HOW TO

SIZE AN INDIVIDUAL APPLIANCE FLUE

Size this simple individual appliance flue by following each box in order from 1 to 5.

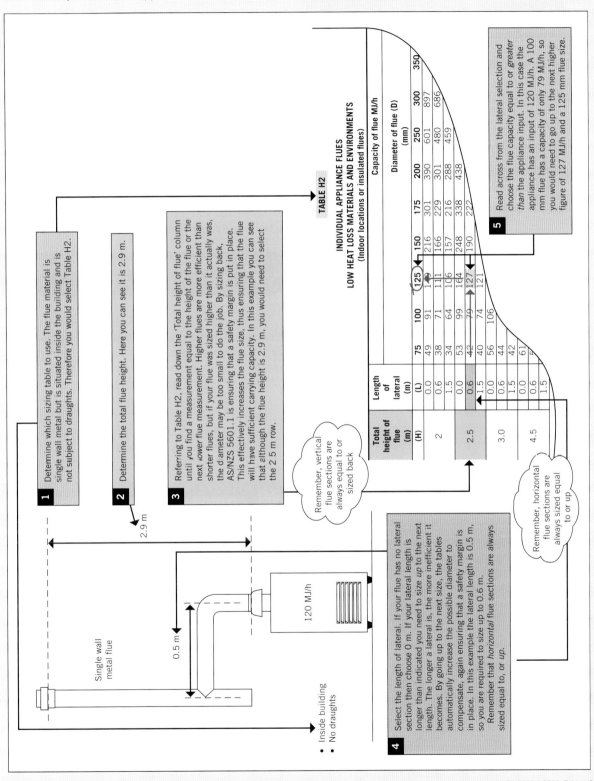

Source: Adapted from Standards Australia AS/NZS 5601.1:2010, Table H2

REFER TO AS/NZS 5601.1 APPENDIX H1.3 'DESIGNING INDIVIDUAL APPLIANCE FLUES'

 COMPLETE WORKSHEET 2

Rule 1 – the additional bends rule
The standard flue sizing procedure has a built-in allowance for up to two 90° changes of direction. Where it is necessary to include *additional* changes of direction, a reduction of the carrying capacity of the flue must be calculated for each bend.

REFER TO AS/NZS 5601.1 APPENDIX H1.2.2 'RESISTANCE TO FLOW OF FLUE GASES'

AS/NZS 5601.1 states that a 10 per cent reduction in flue capacity should be applied for each additional 90° change of direction. However, to remove any ambiguity, it is generally accepted industry practice to apply such a reduction for any change of direction regardless of angle.

To calculate for additional bends follow this procedure:
1. Determine the flue-carrying capacity as per a normal calculation.
2. For a single 10 per cent reduction, multiply the maximum carrying capacity of the flue by 0.9. For two changes of direction or a 20 per cent reduction, multiply the maximum carrying capacity of the flue by 0.8.
3. Apply the reduction calculation until such time as the maximum carrying capacity equals or is greater than the MJ/h input of the appliance.

Because the extra bends will reduce flue efficiency, the maximum carrying capacity is also reduced to compensate, forcing you to size up to a larger capacity and diameter. Remember to apply the reduction calculation to the maximum carrying capacity of the flue and not the MJ/h input of the appliance! Follow the 'How To' example on the following page to see how this all works.

Note: Where additional bends introduce an additional lateral section, add both sections together for a total lateral figure.

 COMPLETE WORKSHEET 3

Rule 2 – the spigot rule
Never assume that the appliance flue outlet size is always going to indicate the flue size itself. Appliances are designed with standard flue outlet or draught diverter sizes to allow for standard installations. Many installations are not standard, and sometimes your calculation will indicate the use of a flue smaller than the appliance outlet size. This is allowed only under strict conditions and you will need to confirm any proposed variation with the technical regulator.

REFER TO AS/NZS 5601.1 APPENDIX H1.3.1 STEP 5

Where your flue size calculation indicates the use of a flue size smaller than the diameter of the appliance flue connection, the smaller size can only be used where the following conditions apply (see Figure 22.18):
- The flue height must be greater than 3 m.
- Flues larger than 300 mm must not be reduced by more than two sizes.
- Flues 300 mm in diameter or less must not be reduced by more than one size.
- A 100 mm flue cannot be reduced at all.

 **COMPLETE WORKSHEET 4**

Sizing horizontal common flue manifolds
Common flue manifolds are used whenever it is necessary or desirable to join multiple appliances together in one combined flue system. You could

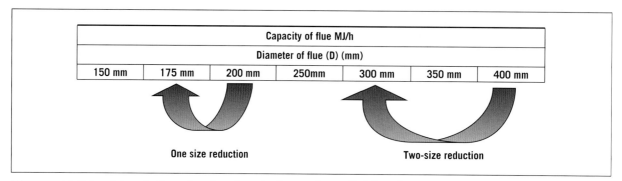

FIGURE 22.18 Determining whether a flue diameter reduction can be applied, based on the conditions listed in AS/NZS 5601.1

HOW TO

CALCULATE FOR ADDITIONAL BENDS

Calculate for additional bends by following the diagram below.

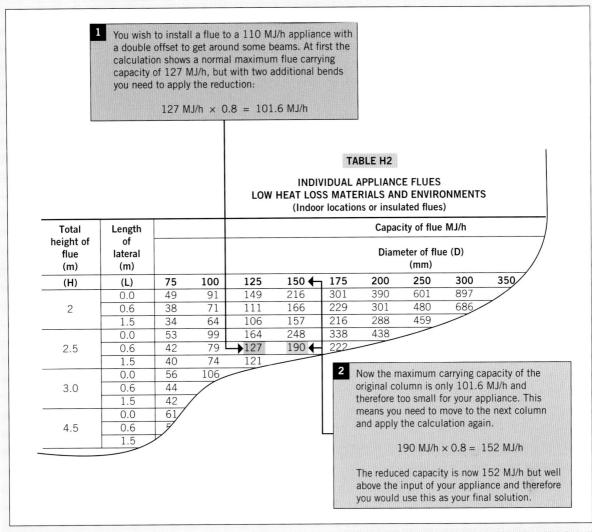

Source: Adapted from Standards Australia AS/NZS 5601.1:2010, Table H2

encounter a need to use common flue manifold systems where multiple appliances are installed in a plant room and there would be some efficiency gained in joining them into the one common flue.

Before learning about common flue manifold systems you need to have a good understanding of individual appliance flueing.

What is a horizontal common flue manifold?

A horizontal common flue manifold is used where a number of appliances are installed closely together at the same level. The manifold itself is a horizontal extension of the lower part of the main common flue. Familiarise yourself with the basic components of a horizontal common flue manifold by reviewing Figure 22.19.

Basic principles and requirements

Before you begin a common flue manifold installation, you must ensure that the basic appliance requirements prescribed in AS/NZS 5601.1 are satisfied.

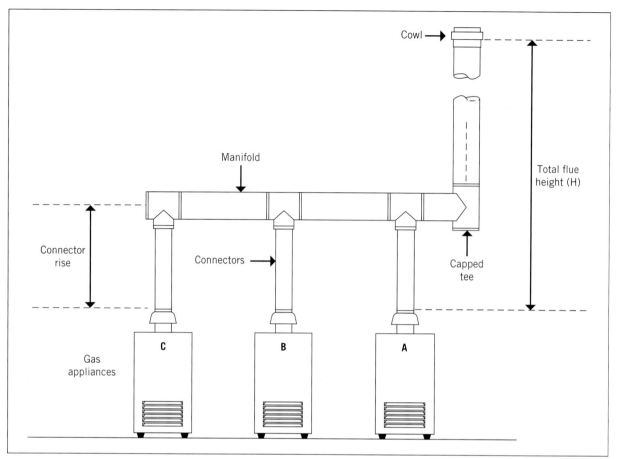

FIGURE 22.19 A horizontal common flue manifold system

Each appliance connected to any common flue manifold must:
- have a flame safeguard system
- have the same burner type. It can be:
  - atmospheric
  - forced draught
  - induced draught
- have sufficient combustion product dilution to ensure that the maximum temperature limits listed in AS/NZS 5601.1 are not exceeded.

REFER TO AS/NZS 5601.1 'COMMON OR COMBINED FLUES'

In addition to these basic appliance requirements, you need to be familiar with some key design parameters that are found in Appendix H of AS/NZS 5601.1 and in the flue product manufacturer's specifications. These common flue considerations are summarised below:
- Appliances designed to operate simultaneously should have the common flue sized as if it were an individual appliance flue using Table H2 or H3 of AS/NZS 5601.1, with the manifold section sized as a lateral.
- Manifold/lateral length must be kept as short as possible and must not exceed 50 per cent of the total flue height.
- Connectors must rise vertically as high as possible before joining any horizontal section.
- Flue connector sizes must be equal to or larger than the draught diverter size.
- The spigot rule does not apply to common flue manifolds.

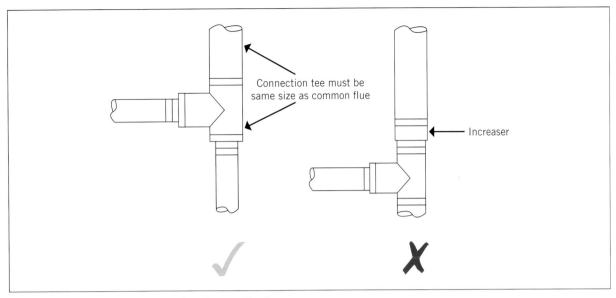

FIGURE 22.20 Selecting the correct size of connection tee

- The range of draught diverter sizes in the one common flue manifold must not exceed the requirements in Table H1 of AS/NZS 5601.1.
- Avoid excessive grade in any horizontal manifold, as it does not improve flue gas movement in manifolds and may cause spillage from the furthest appliances.
- In horizontal manifold systems, appliances with the smallest draught diverter size should be located closest to the vertical section of the flue wherever possible.
- Because connectors are largely vertical sections, they must be sized as equal to or *reduced* to the next lowest figure.
- Connection tees within the manifold or common flue sections must always be the same size as the section of flue above them (see Figure 22.20).

REFER TO AS/NZS 5601.1 APPENDIX H, 'COMMON FLUES (COMBINED AND MULTIPLE FLUES)'

Sizing horizontal common flues

Sizing a horizontal common flue can be broken into two parts:
1. Size the connectors.
2. Size the manifold and common flue.

Within each of these parts are some simple steps you need to take to arrive at a solution.

Learn how to size a horizontal common flue by following each of the steps below in sequence, referring to AS/NZS 5601.1 as you go.

Size the connectors

You would now repeat these steps for each connector and record the results on your worksheet as you go.

Size the manifold and common flue

Now that the connectors are sized, it is time to size the manifold and common flue sections.

COMPLETE WORKSHEET 5

HOW TO

SIZE A HORIZONTAL COMMON FLUE – CONNECTORS

Follow the boxes in the diagram sequence to size a flue.

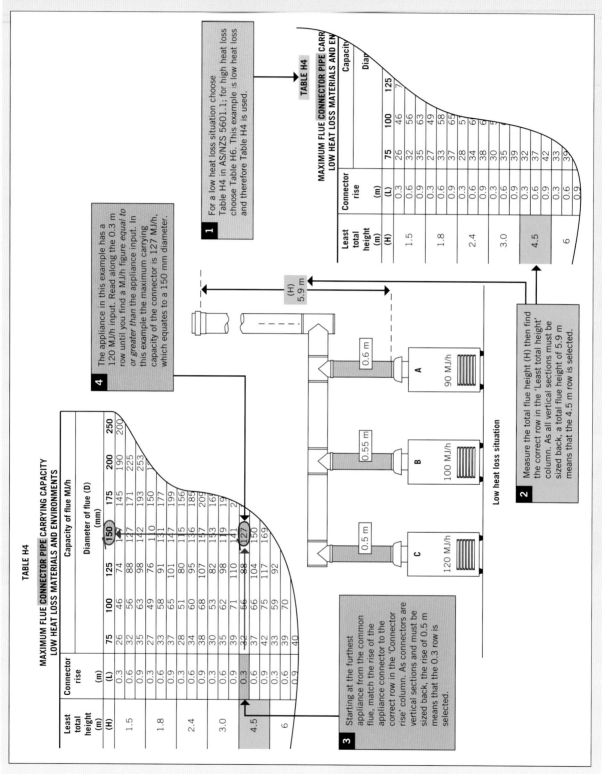

Source: Adapted from Standards Australia AS/NZS 5601.1:2010, Table H4

HOW TO

SIZE A HORIZONTAL COMMON FLUE – MANIFOLD AND COMMON FLUE

Follow the boxes in the diagram in sequence.

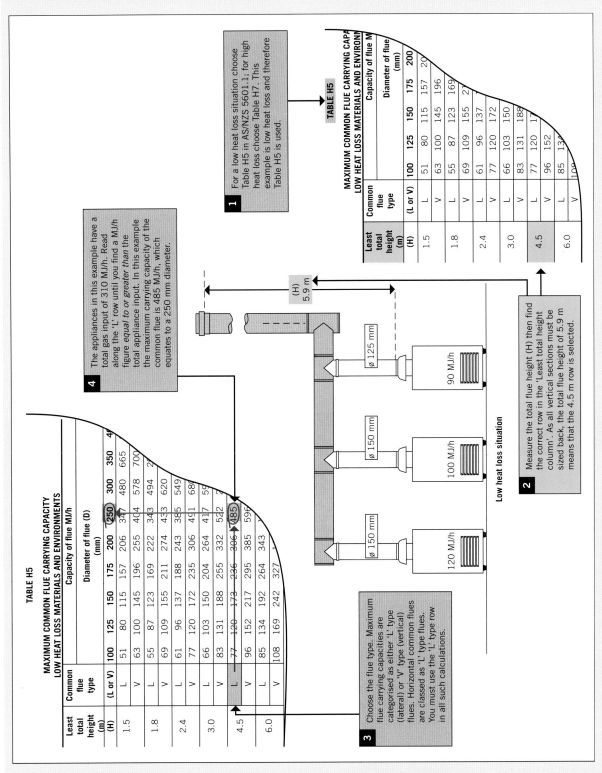

Source: Adapted from Standards Australia AS/NZS 5601.1:2010, Table H5

Record the size of the horizontal common flue in your worksheet.

Install the flue

So far you have learnt where flue terminals may terminate, what materials can be selected and how to size the flue, all in accordance with your Standard. In this section, you will now look at some of the actual installation requirements associated with Type B flue systems.

Confirm flue base and roof penetration locations

Check the plans and carry out a site inspection to ensure that the proposed flue location is suitable. Check that the foundation under the flue stack is capable of supporting the weight of the flue, and use a plumb bob or vertical laser to ensure that the proposed roof penetration point is free from obstructions, such as ceiling joists, purlins or rafters.

Clearances to combustible surfaces

A vital part of any flue installation is to ensure that minimum clearances to combustible surfaces and materials are satisfied in accordance with AS/NZS 5601.1. Unfortunately, many buildings have been destroyed and lives lost because of flues that have been installed incorrectly.

'Combustible surface' is a defined term in AS/NZS 5601.1 and is regularly referred to throughout the standard.

> **REFER TO AS/NZS 5601.1 'COMBUSTIBLE SURFACE'**

Of particular concern is the fact that even relatively low heat applied over a long period of time can cause fires to start many months, or even years, after installation of the appliance and flue. The condition leading up to such fires is known as pyrolysis.

What is pyrolysis?

Pyrolysis is the chemical decomposition of a substance caused by the application of heat. Where sufficient clearance or protection is not provided between the heat source and the surface, ignition and combustion of the substance can occur without the direct application of flames.

To examine this in more detail, consider a timber roof rafter or ceiling joist being subjected to heat from an incorrectly installed single skin flue. Depending upon the species, timber will normally ignite at approximately 260°C. However, exposure to a heat source will cause significant changes to the molecular structure of the timber, with the effect of lowering the ignition temperature as follows:

- At temperatures between 65°C and 100°C, the timber begins to lose structural strength.
- At temperatures above 100°C, the timber begins to dehydrate as chemical bonds begin to break apart, releasing flammable volatiles and tars into the surrounding area.
- As pyrolysis of the timber continues, the ignition temperature of the timber can fall to as low as 100°C! With sufficient air supply and application of the right ignition temperature, the timber is then likely to combust.

Where can pyrolysis occur?

Substances can suffer from pyrolysis wherever an appliance or its flue system is not installed with the mandatory clearances and/or correct physical protection of the surface.

You must always be on the lookout for materials in the vicinity of your flue installation that may ignite and burn. Examples of combustible surfaces that you may encounter near flues and components include, but are not limited to, the following:

- timber in any form including studs, noggins, rafters, joists and battens
- the paper lining of sheet plasterboard
- some thermal insulation products, particularly older sarking and vapour membranes
- some non-metallic exterior cladding and roofing products
- most synthetic surfaces.

> **REFER TO AS/NZS 5601.1 'FLUE INSTALLATION'**

Twin wall flues can be installed as close as 10 mm to any combustible surface, but single wall flues require very specific clearances to both protected and unprotected combustible surfaces. Common problem areas include:

- *flues adjacent to rafters and roof members.* Do not directly clip flues to combustible structures! If you need to support a flue pipe in a roof space, use proprietary hanging clips and where required fabricate lengths of sheet metal folded in an 'L' shape. You can cut these to length and support a flue between roof members without making contact. Larger Type B flues will require the fabrication of custom-made brackets to carry the weight and maintain clearances.
- *ceiling and wall cavity thermal barriers and insulation.* Ensure that where you penetrate any wall or ceiling, the flue pipe is kept well clear of any thermal or vapour wrap as well as any paper-based insulation products.

When protecting any combustible surface use only the methods described in AS/NZS 5601.1. In particular, when you intend to use some form of fire resistant material to provide protection, ensure that it is an approved product. Note that standard cement sheet (commonly known as 'fibro') used by itself is not an approved product for the protection of combustible material, as it conducts too much heat.

Flue socket alignment

Sectional flues must be assembled in the following manner:

- Where the flue section is protected from the weather, then the socket shall face upwards. This is to ensure that any condensation that might form on the inside of the flue can run back down the flue without flowing out from the socket.
- Where the flue section is exposed to the weather, then the socket shall face downwards. This ensures that any rain that falls upon the outside of the flue cannot flow into the socket.

REFER TO AS/NZS 5601.1 SECTION 6 'FLUE JOINTS'

Where necessary, flue sockets should be sealed with a substance suitable for the temperature of the flue gases.

Handling large flues

Most atmospheric burner appliances can be flued with materials and components that come in sections and can often easily be handled into position by one or two workers. Larger appliances are often designed for the installation of long, single section flues. A common solution is a flue made of spiral wound, lock-seamed or welded mild or stainless steel (see Figure 22.21). Although lightweight in respect to their length and diameter, these flues will require a crane and other lifting equipment to lower them into position.

FIGURE 22.21 Spiral wound tube often specified for larger Type B appliance flue installations

Lifting long flue lengths such as these is classed as high-risk work and will require someone with a Rigger's ticket to work with the crane operator. Most crane companies can provide the relevant personnel for the job.

Flue pipe roof penetration flashings

Polymer penetration flashings are a very effective means of flashing flue pipes through both steel and tiled roofs. However, you need to select a flashing rated to resist the expected temperature that the flue pipe is likely to conduct. Standard EPDM polymer flashings are suitable when using a twin wall flue (see Figure 22.22), as very little heat is conducted to the outer sheath of the flue pipe (EPDM is an abbreviation for a synthetic rubber compound called ethylene propylene diene monomer).

FIGURE 22.22 A twin wall gas flue penetration, soaker flashing and standard EPDM synthetic collar

Whenever you need to flash around a single skin metal flue, you must be aware that the flue may become very hot. Standard EPDM polymer flashings are rated to withstand approximately 115°C. Where you suspect that the flue may become hotter than this, it is safer to use an approved high-temperature flashing; these normally have an upper temperature rating of approximately 200°C. Where flues are to carry combustion products at higher temperatures, some form of metallic over-flashing will be required to overlap the roof soaker upstand which has a fixed annular separation from the flue itself.

Good plumbing practice should apply to all your penetration flashings:
- Ensure that your penetration flashing does not impede water flow and if necessary install a soaker flashing that enables water to flow freely on either side of the penetration. Subject to the number of square metres of roof area above larger penetrations, 2 or 3 ribs may need to be removed on either side of the soaker flashing in order to allow sufficient diverted water flow around the upstand in accordance with the SA HB39:2015 Installation code for metal roof and wall cladding.
- Be aware that the grooved seam on single wall sheet metal flues can draw in water through capillary action. Run a bead of silicone up the length of the grooved seam to prevent water ingress.
- In some jurisdictions only a licensed plumber can install roof penetrations and flashings. Check your local requirements first.

Importantly, if your flue penetrates the lap of the roof sheet, a drain notch must be cut into the overlapping sheet immediately upstream of the flashing so that any water within the anti-capillary profile of the sheet is able to drain into the pan without flowing under the flashing and into the building (see Figure 22.23). If this is not clear, ensure you consult your teacher or supervisor to clarify how this is done.

Clean up the job

Your flue installation task is not over until your work area is cleaned up and tools and equipment accounted for and returned to their required location in a serviceable condition.

Clean up your area

The following points are important considerations when finishing the job:
- If not already covered in your site induction, consult with supervisors in relation to the site refuse requirements.
- Cleaning up as you go not only keeps your work area safe it also lessens the amount of work required at the end of the job.
- Ensure any swarf left around your flue roof penetration is swept up to prevent rusting of the roof surface.
- Dispose of rubbish, flue offcuts and old materials in accordance with site-disposal requirements or your own company policy.
- Ensure all materials that can be recycled are separated from general refuse and placed in the appropriate location for collection. On some sites specific bins may be allocated for metals, timber etc. or if not, you may need to take materials back to your business premises.

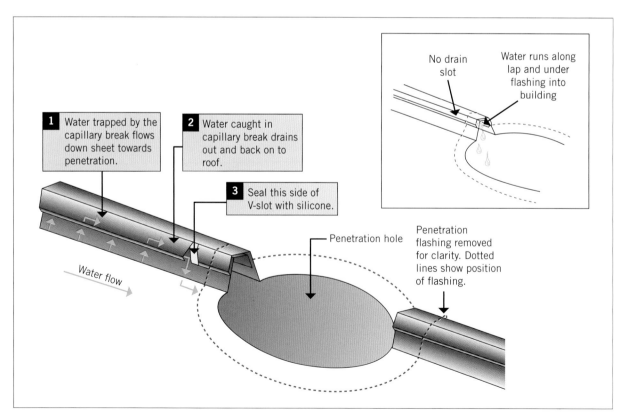

FIGURE 22.23 Cut a small V-slot just upstream of the flue flashing to drain the capillary break into the sheet pan

Check your tools and equipment

Tools and equipment need to be checked for serviceability to ensure that they are ready for use on the job. On a daily basis and at the end of the job check that:
- all items are working as they should with no obvious damage
- high-wear and consumable items such as cutting discs, drill bits and sealants are checked for current condition and replaced as required
- items such as laser levels, measuring devices etc. are working within calibration requirements
- where any item requires repair or replacement, your supervisor is informed and if necessary mark any items before returning to the company store or vehicles.

Complete all necessary job documentation

Most jobs require some level of documentation to be completed. Avoid putting it off till later as you may forget important points or times.

- Ensure all SWMS and/or JSAs are accounted for and filed as per company policy.
- Complete your timesheet accurately.
- Reconcile any paper-based material receipts and ensure electronic receipts have been accurately recorded against the job site and/or number.
- Submit any Completion Notice that may be required by the local authority.

LEARNING TASK 22.6

When cleaning up after a flue installation, you have a number of steel flue offcuts left over. How should these be disposed of?

COMPLETE WORKSHEET 6

SUMMARY

In this unit you have examined the requirements related to the planning, sizing and installation of Type B gas appliance flues including:
- basic combustion principles
- a knowledge of Type B flue types and applications
- job preparation
- specific installation requirements related to Type B flues including terminal location
- sizing of Type B atmospheric flues
- flue installation practice and protection of combustible materials
- clean-up of job and return of equipment.

GET IT RIGHT

The photo below shows an incorrect practice when installing a Type B appliance flue.

Identify the incorrect method and provide reasoning for your answer.

_____
_____
_____
_____

WORKSHEET 1

To be completed by teachers
Student competent ☐
Student not yet competent ☐

Student name: _____

Enrolment year: _____

Class code: _____

Competency name/Number: _____

Task

Review the sections *Overview* through to *Flue installation and terminal locations* and answer the following questions.

1 What three functions does a draught diverter perform?

   i _____

   ii _____

   iii _____

2 If a flue has a vertical height of 6 m measured from the draught diverter to the terminal, what is the maximum permitted length of any lateral section? Provide a reference from AS/NZS 5601 to support your answer.

   _____

   _____

   _____

3 A boiler you are installing inside a plant room requires a natural draught flue. You intend to use single wall galvanised steel of 0.8 mm thickness.

   i What is the maximum flue diameter you can fabricate using this material? Provide a reference from AS/NZS 5601 to support your answer.

   _____

   ii What is the maximum height this flue must not exceed? Provide a reference from AS/NZS 5601 to support your answer.

   _____

   iii Where the flue terminates above the roofline, what is the required clearance between any part of the roof and the end of the flue? Provide a reference from AS/NZS 5601 to support your answer.

   _____

4 What would be the minimum rise required in a lateral measuring 1800 mm long? Provide a reference from AS/NZS 5601 to support your answer.

   _____

5. A pottery kiln flue must be at least 500 mm from an unprotected combustible surface: true or false? Provide a reference from AS/NZS 5601 to support your answer.

   _____

6. Part of your flue installation is to pass through a concealed section of a roof space. Therefore, you are considering having the flue fabricated out of stainless steel. Determine the following requirements:

   i   What grade of stainless steel should you specify to your sheet metal workshop?

   _____

   ii What is the maximum diameter permitted for 0.5 mm thick material?

   _____

WORKSHEET 2

To be completed by teachers
Student competent
Student not yet competent

Student name: _____

Enrolment year: _____

Class code: _____

Competency name/Number: _____

Task

Review the section *Sizing individual appliance flues* and answer the following questions.

1. You now have the opportunity to check and extend your knowledge of basic flue sizing. Complete the following exercise.

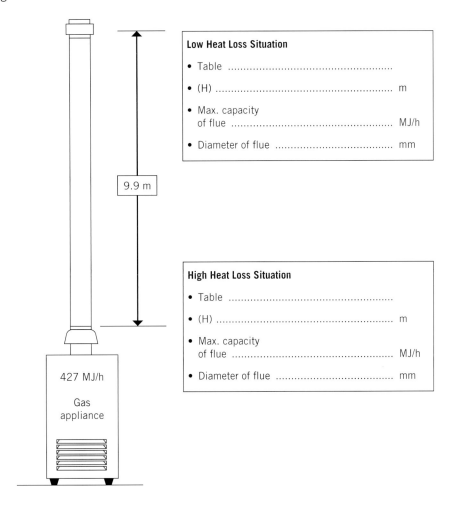

Low Heat Loss Situation

- Table ....
- (H) .... m
- Max. capacity of flue .... MJ/h
- Diameter of flue .... mm

High Heat Loss Situation

- Table ....
- (H) .... m
- Max. capacity of flue .... MJ/h
- Diameter of flue .... mm

2 Complete the following exercise.

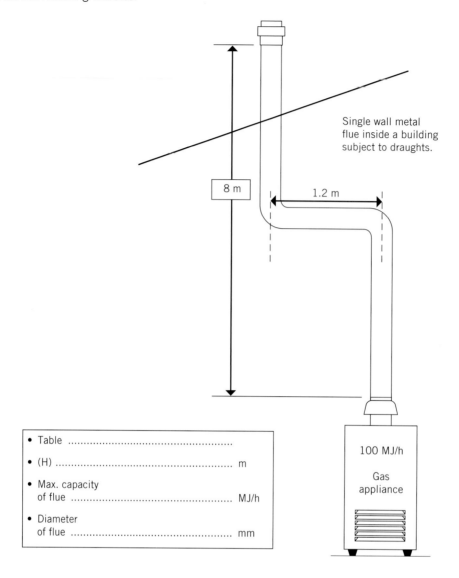

- Table ....
- (H) .... m
- Max. capacity
  of flue .... MJ/h
- Diameter
  of flue .... mm

WORKSHEET 3

To be completed by teachers
Student competent ☐
Student not yet competent ☐

Student name: _____

Enrolment year: _____

Class code: _____

Competency name/Number: _____

Task
Complete the following exercise.

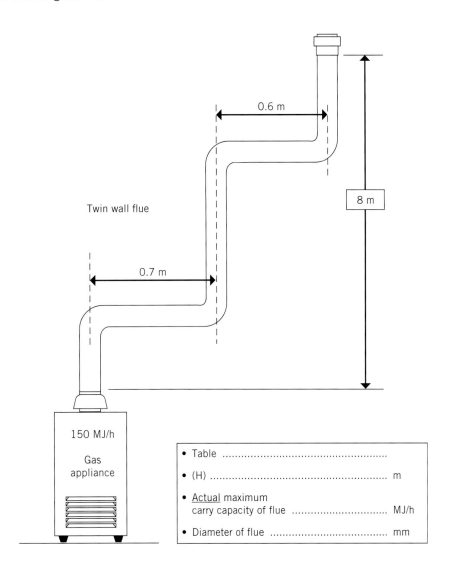

- Table ....
- (H) .... m
- <u>Actual</u> maximum carry capacity of flue .... MJ/h
- Diameter of flue .... mm

INSTALL TYPE B GAS APPLIANCE FLUES 533 GS

WORKSHEET 4

Student name: _____

Enrolment year: _____

Class code: _____

Competency name/Number: _____

To be completed by teachers
Student competent ☐
Student not yet competent ☐

Task
Complete the following exercises.

1

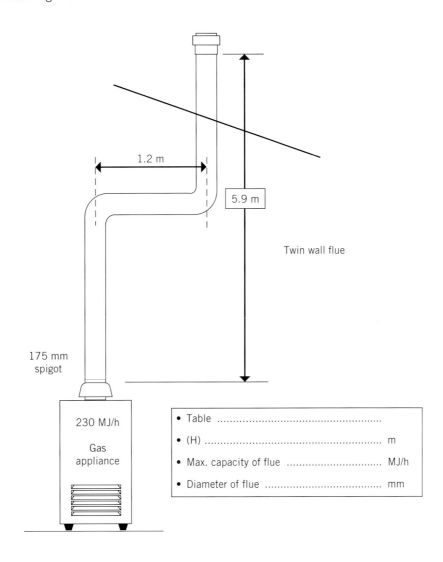

- Table ....
- (H) .... m
- Max. capacity of flue .... MJ/h
- Diameter of flue .... mm

INSTALL TYPE B GAS APPLIANCE FLUES   535 GS

WORKSHEETS 22

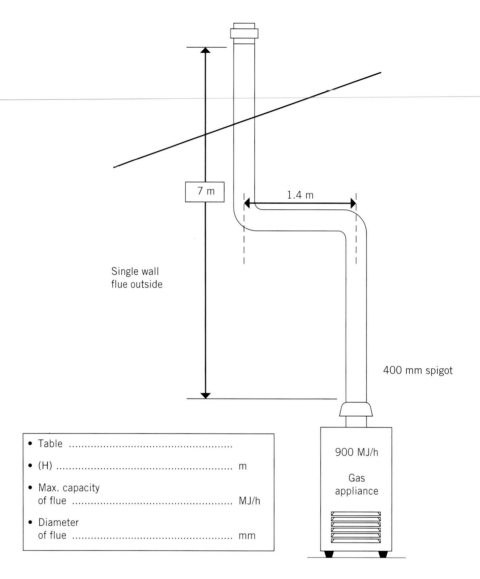

- Table ....
- (H) .... m
- Max. capacity
  of flue .... MJ/h
- Diameter
  of flue .... mm

3

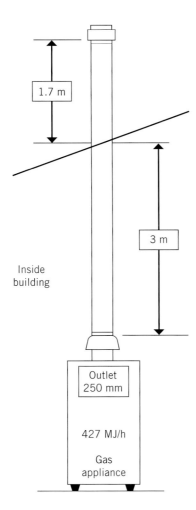

Inside building

- Should this single wall metal installation be classed as a low heat loss or high heat loss situation?
  ....
- Provide reason ....
  ....
  ....
  ....
  ....
  ....
  ....
- Table ....
- (H) .... m
- Max. capacity of flue .... MJ/h
- Diameter of flue .... mm

WORKSHEET 5

Student name: _____

Enrolment year: _____

Class code: _____

Competency name/Number: _____

To be completed by teachers

Student competent ☐

Student not yet competent ☐

Task

Review the section *Sizing horizontal common flue manifolds* and answer the following questions.

1. Now you have an opportunity to apply these procedures to the sizing of a horizontal common flue. Complete the tasks below.

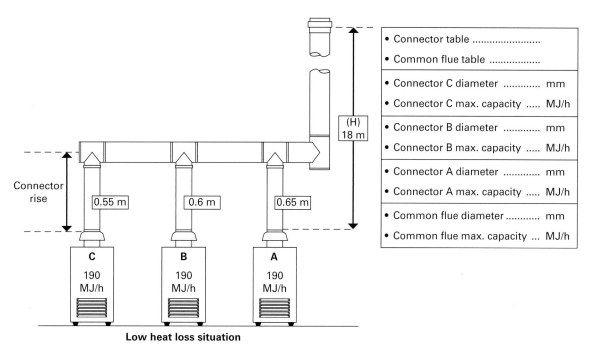

- Connector table ....
- Common flue table ....
- Connector C diameter .... mm
- Connector C max. capacity ..... MJ/h
- Connector B diameter .... mm
- Connector B max. capacity ..... MJ/h
- Connector A diameter .... mm
- Connector A max. capacity ..... MJ/h
- Common flue diameter .... mm
- Common flue max. capacity ... MJ/h

WORKSHEETS 22

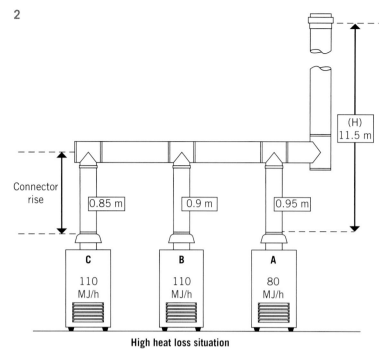

High heat loss situation

- Connector table ....
- Common flue table ....
- Connector C diameter .... mm
- Connector C max. capacity .... MJ/h
- Connector B diameter .... mm
- Connector B max. capacity .... MJ/h
- Connector A diameter .... mm
- Connector A max. capacity .... MJ/h
- Common flue diameter .... mm
- Common flue max. capacity .... MJ/h

WORKSHEET 6

To be completed by teachers
Student competent ☐
Student not yet competent ☐

Student name: _____

Enrolment year: _____

Class code: _____

Competency name/Number: _____

Task

Review the section *Install the flue* and answer the following questions.

1 Where joints in a flue are subjected to the weather, their sockets should face downwards to prevent rainwater ingress. Why is the internal part of the flue required to have upwards facing sockets?

_____

_____

_____

2 List at least four types of combustible material or products you might find around flue installations.

_____

_____

_____

3 Detail one approved method of protecting a combustible surface from the heat of a flue. Provide a reference from AS/NZS 5601 to support your answer.

_____

_____

_____

4 You have cut a penetration through a roof to fit a new gas appliance flue. The penetration intersects the lap of the high profile roof sheeting. Do you cut in an anti-capillary drain V-slot immediately above, or below the penetration flashing?

_____

_____

_____

5 What will be the outcome of failure to remove swarf from around the penetration of a sheet metal roof?

_____

_____

6 Is standard 6 mm cement sheeting a suitable product to separate a single skin metal flue from a timber rafter?

_____

_____

7 What are two tools you could use to plot the exact location of the roof penetration above the centre of the flue stack base?

_____

_____

APPENDIX: ADVANCED SKILLS – FLUE SIZING

Learning objectives

Areas addressed in this Appendix include:
- sizing individual appliance flues
- sizing horizontal common flue manifolds
- sizing vertical common flue manifolds.

This section deals with actual flue sizing and is aimed at those students undertaking higher level Certificate IV gas training as required for full licensing.

Today many manufacturers are supplying pre-sized and approved natural draught flue kits for gas appliances, allowing gas fitters to fit such kits with confidence that the installation will perform as designed. You must remember, however, that this is only applicable to standard installation conditions in relation to flue height and appliance location.

Inevitably you will encounter many installations where this will not be the case. A roof that is higher than normal or has a pitch greater than standard is the most common problem. Unless the manufacturer has a supplementary solution, you have the responsibility to re-size the flue to meet these needs. These skills are what separates a highly professional gas fitter from someone who is just an 'installer'.

In this section, you will be introduced to the skills and knowledge required to carry out sizing calculations for natural draught appliance flue installations, including:

- individual appliance flues. All gas fitters should be able to size an individual flue installation as and when required.
- horizontal common flue manifolds. These are used where a number of appliances are installed at the same level and all joined to the same flue system. Manifolded boilers or commercial hot water cylinders are examples that may suit a horizontal common flue system.
- vertical common flue manifolds. While less common today, such a system could be used in multistorey buildings where an appliance on each floor level is joined to a vertical common flue system.

The flue sizing process is described in detail in AS/NZS 5601.1. You should first familiarise yourself with the basic factors affecting flue sizing and design.

REFER TO AS/NZS 5601.1 APPENDIX H1 'FLUE DESIGN FOR APPLIANCES WITH ATMOSPHERIC BURNERS' AND APPENDIX H1.2 'FACTORS INFLUENCING FLUE DESIGN'

Information required before you start flue sizing

Preparation for flue sizing includes not only collating the information required for the calculations but also recording the sizing process in a logical and clear manner. With any sizing calculation, you need to draw a simple diagram and include the basic information shown in Figure A.1 before starting.

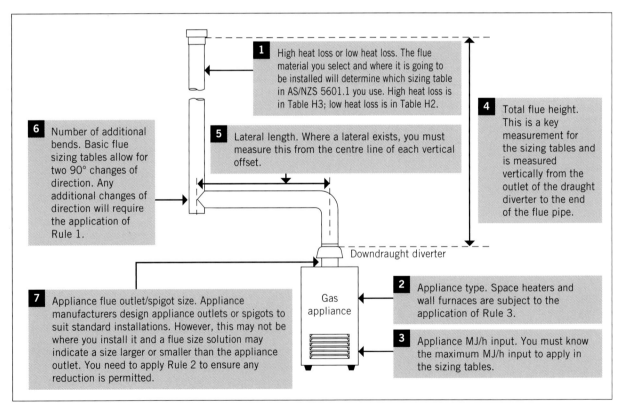

FIGURE A.1 The accurate documentation of correct details for efficient flue sizing

FROM EXPERIENCE

It is important to remember that a gas fitter is held fully responsible for all work. It is vital that you have the ability to size flues so that you can confirm that plans and material schedules will match the requirements of the appliance and the Standard. Do not assume plans are always correct!

Sizing a simple individual appliance flue

Individual appliance flue sizing has one process and three variations:

- Rule 1 – Additional bends
- Rule 2 – Spigot rule
- Rule 3 – Space heater rule

The step-by-step process for flue sizing is detailed in AS/NZS 5601.1 and in the instructions below.

REFER TO AS/NZS 5601.1 APPENDIX H1.3 'DESIGNING INDIVIDUAL APPLIANCE FLUES'

COMPLETE WORKSHEET 1

Rule 1 – the additional bends rule

The standard flue sizing procedure has a built-in allowance for up to two 90° changes of direction. Where it is necessary to include *additional* changes of direction, a reduction of the carrying capacity of the flue must be calculated for each bend.

REFER TO AS/NZS 5601.1 APPENDIX H1.2.2 'RESISTANCE TO FLOW OF FLUE GASES'

AS/NZS 5601.1 states that a 10 per cent reduction in flue capacity should be applied for each additional 90° change of direction. However, to remove any ambiguity, it is generally accepted industry practice to apply such a reduction for any change of direction regardless of angle.

To calculate for additional bends follow this procedure:

1. Determine the carrying capacity of the flue as per a normal calculation.
2. For a single 10 per cent reduction, multiply the maximum carrying capacity of the flue by 0.9. For two changes of direction or a 20 per cent reduction, multiply the maximum carrying capacity of the flue by 0.8.
3. Apply the reduction calculation until such time as the maximum carrying capacity equals or is greater than the MJ/h input of the appliance.

Because the extra bends will reduce flue efficiency, the maximum carrying capacity is also reduced to compensate, forcing you to size up to a larger capacity and diameter. Remember to apply the reduction calculation to the maximum carrying capacity of the flue and not the MJ/h input of the appliance! Follow the example below to see how this all works.

COMPLETE WORKSHEET 2

Rule 2 – the spigot rule

Never assume that the appliance flue outlet size is always going to indicate the flue size itself. Appliances are designed with standard flue outlet or draught diverter sizes to allow for standard installations. Many installations are not standard, and sometimes your calculation will indicate the use of a flue smaller than the appliance outlet size. This is allowed only under strict conditions.

REFER TO AS/NZS 5601.1 APPENDIX H1.3.1 STEP 5

Where your flue size calculation indicates the use of a flue size smaller than the diameter of the appliance flue connection, the smaller size can only be used where the following conditions apply (see Figure A.2):

- The flue height must be greater than 3 m.
- Flues larger than 300 mm must not be reduced by more than two sizes.
- Flues 300 mm in diameter or less must not be reduced by more than one size.
- A 100 mm flue cannot be reduced at all.

COMPLETE WORKSHEET 3

HOW TO

SIZE AN INDIVIDUAL APPLIANCE FLUE

Size a simple individual appliance flue by following each box in order from 1 to 5.

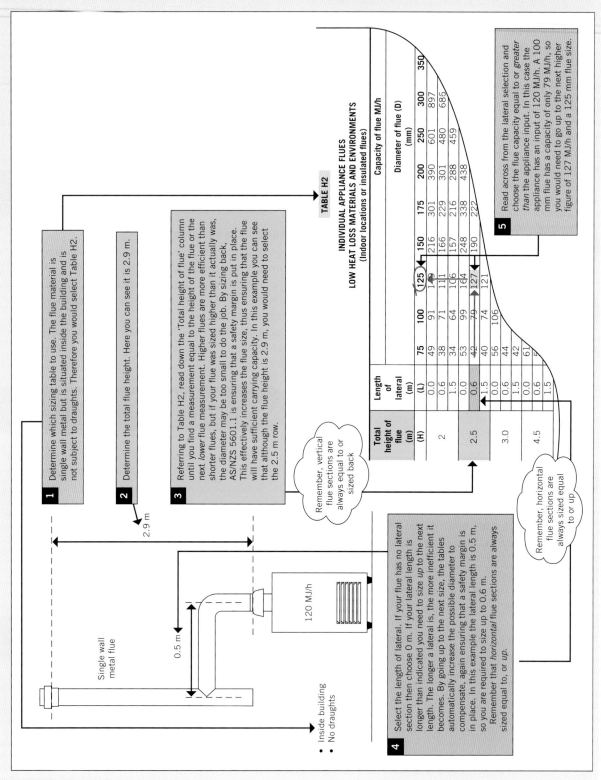

Source: Adapted from Standards Australia AS/NZS 5601.1:2010, Table H2

HOW TO

CALCULATE FOR ADDITIONAL BENDS

Note: Where additional bends introduce an additional lateral section, add both sections together for a total lateral figure.

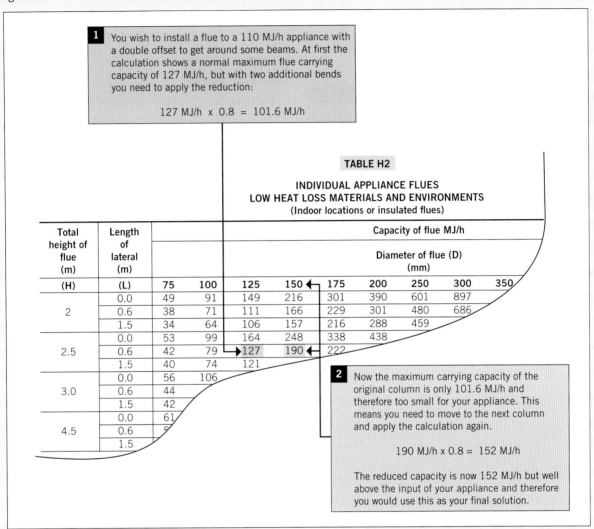

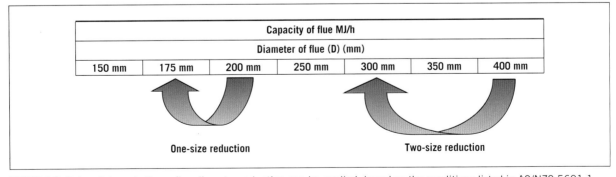

FIGURE A.2 Determining whether a flue diameter reduction can be applied, based on the conditions listed in AS/NZS 5601.1

Rule 3 – the space heater rule

Whenever you are planning to install any space heater or wall furnace, AS/NZS 5601.1 requires that you always consider the appliance MJ/h input to be 40 per cent greater than stated on its data plate. You must add this 40 per cent allowance to the appliance MJ/h input before commencing the calculation. This is done to apply a greater safety margin to such heater installations. Simply multiply the original MJ/h by 1.4 (140 per cent).

COMPLETE WORKSHEET 4

Sizing horizontal and vertical common flue manifolds

Common flue manifolds are used whenever it is necessary or desirable to join multiple appliances together in one combined flue system. You will mainly encounter a need to use common flue manifold systems in multi-residential and commercial gas applications.

Before learning about common flue manifold systems you need to have a good understanding of individual appliance flueing.

EXAMPLE A.1

You are planning to install a 35 MJ/h space heater. Before carrying out the flue calculation you apply the space heater rule:

35 MJ/h × 1.4 = 49 MJ/h

Therefore, you will use 49 MJ/h in your flue calculation.

EXAMPLE A.2

You are planning to install a 42 MJ/h wall furnace. Before carrying out the flue calculation you apply the space heater rule:

42 MJ/h × 1.4 = 58.8 MJ/h

Therefore, you will use 58.8 MJ/h in your flue calculation.

What is a horizontal common flue manifold?

A horizontal common flue manifold is used where a number of appliances are installed closely together at the same level. The manifold itself is a horizontal extension of the lower part of the main common flue. Familiarise yourself with the basic components of a horizontal common flue manifold by reviewing Figure A.3.

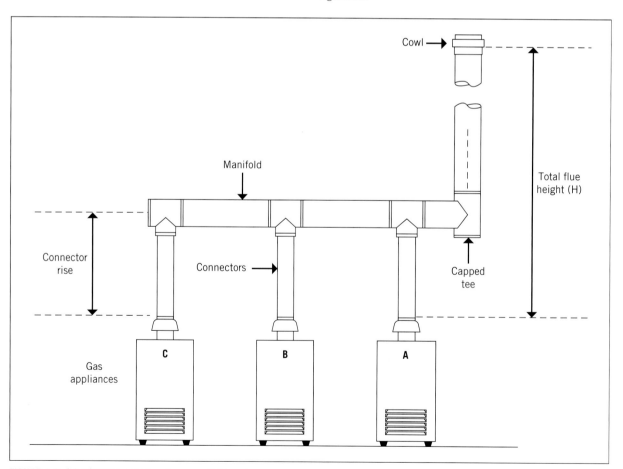

FIGURE A.3 A horizontal common flue manifold system

What is a vertical common flue manifold?

A vertical common flue manifold is used where appliances are installed in a multistorey application and are situated in the same relative position on each individual level. Each appliance is connected to a vertical common flue normally situated in a services duct. Familiarise yourself with the basic components of a vertical common flue manifold by reviewing Figure A.4.

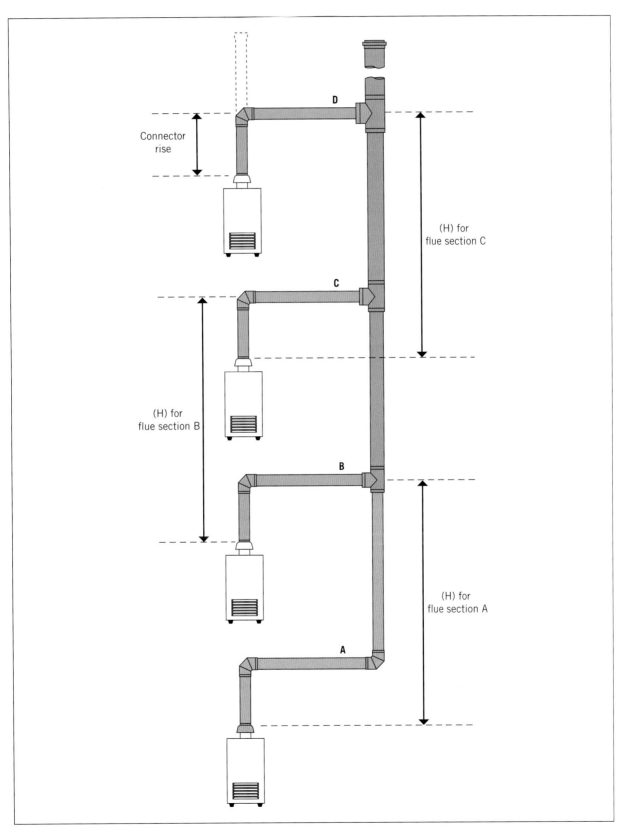

FIGURE A.4 A vertical common flue manifold system

Basic principles and requirements

Before commencing with a common flue manifold installation, you must ensure that the basic appliance requirements prescribed in AS/NZS 5601.1 are satisfied.

Each appliance connected to any common flue manifold must have:
- a flame safeguard system
- the same burner type:
  - atmospheric
  - forced draught
  - induced draught
- sufficient combustion product dilution to ensure that the maximum temperature limits listed in AS/NZS 5601.1 are not exceeded.

REFER TO AS/NZS 5601.1 'COMMON OR COMBINED FLUES'

In addition to these basic appliance requirements, you need to be familiar with some key design parameters that are found in Appendix H of AS/NZS 5601.1 and in the flue product manufacturer's specifications. Both horizontal and vertical common flue considerations are summarised below:
- Appliances designed to operate simultaneously should have the common flue sized as if it were an individual appliance flue using Table H2 or H3 of AS/NZS 5601.1, with the manifold section sized as a lateral.
- Manifold/lateral length must be kept as short as possible and must not exceed 50 per cent of the total flue height.
- Connectors must rise vertically as high as possible before joining any horizontal section.
- Flue connector sizes must be equal to or larger than the draught diverter size.
- The spigot rule does not apply to common flue manifolds.
- The range of draught diverter sizes in the one common flue manifold must not exceed the requirements in Table H1 of AS/NZS 5601.1.
- Avoid excessive grade in any horizontal manifold as it does not improve flue gas movement in manifolds and may cause spillage from the furthest appliances.
- In horizontal manifold systems, appliances with the smallest draught diverter size should be located closest to the vertical section of the flue wherever possible.
- Because connectors are largely vertical sections, they must be sized as equal to or *reduced* to the next lowest figure.
- Connection tees within the manifold or common flue sections must always be the same size as the section of flue above them (see Figure A.5).

REFER TO AS/NZS 5601.1 APPENDIX H 'COMMON FLUES (COMBINED AND MULTIPLE FLUES)'

Sizing horizontal common flues

Sizing a horizontal common flue can be broken up into two parts:
1. Size the connectors.
2. Size the manifold and common flue.

Within each of these parts are some simple steps you need to take to arrive at a solution.

Learn how to size a horizontal common flue by following each of the steps below in sequence, referring to AS/NZS 5601.1 as you go.

Size the connectors

COMPLETE WORKSHEET 5

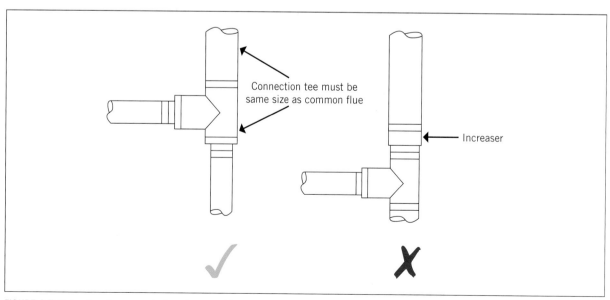

FIGURE A.5 Selecting the correct size of connection tee

HOW TO

SIZE A HORIZONTAL COMMON FLUE – CONNECTORS

Follow the boxes in the diagram sequence.

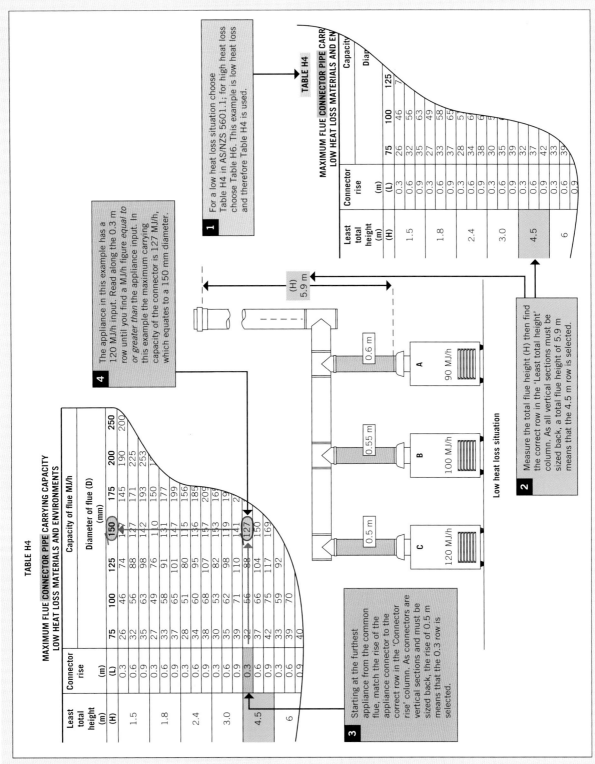

Source: Adapted from Standards Australia AS/NZS 5601.1:2010, Table H4

You would repeat these steps for each connector and record the results on your worksheet as you go.

Size the manifold and common flue

Now that the connectors are sized, it is time to size the manifold and common flue sections.

Sizing vertical common flues

Sizing a vertical common flue can be broken up into three parts:
1. Size the lowest flue section as an individual appliance.
2. Size the connectors.

HOW TO

SIZE A HORIZONTAL COMMON FLUE – MANIFOLD AND COMMON FLUE

Follow the boxes in the diagram sequence.

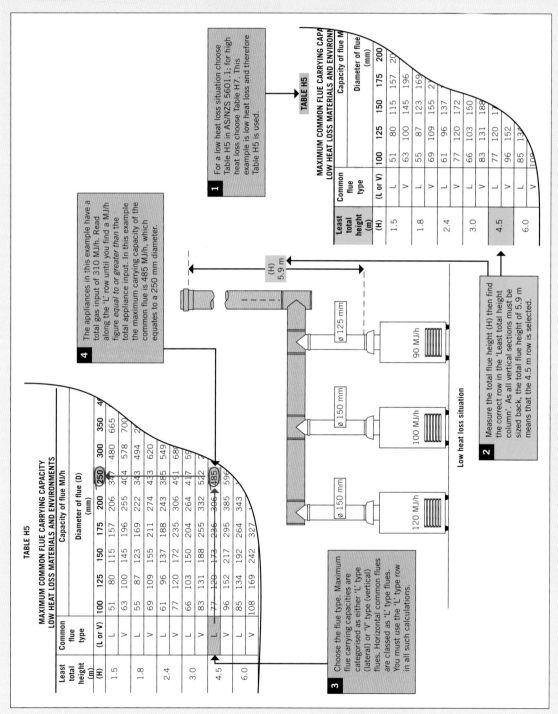

Record the size of the horizontal common flue in your worksheet.

Source: Adapted from Standards Australia AS/NZS 5601.1:2010, Table H5

3 Size the common flue sections.

Within each of these parts are some simple steps you need to take to arrive at a solution. Using the following diagram as an example, follow each step in sequence, referring to AS/NZS 5601.1 as you go.

Size the lowest flue section as an individual appliance

The first part of this process is to size the lowest flue section as an individual appliance. This is shown in orange in the following diagram.

HOW TO

SIZE A VERTICAL COMMON FLUE – THE LOWEST FLUE SECTION

Size the lowest flue section as an individual appliance.

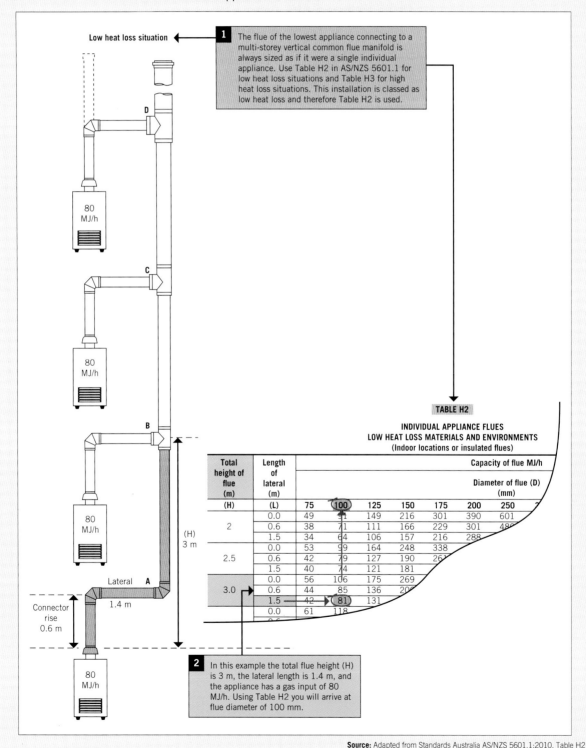

Source: Adapted from Standards Australia AS/NZS 5601.1:2010, Table H2

Size the connectors

Once the lower flue connection has been sized, you can now move on to sizing the connectors at each level. These are shown in blue in the following diagram.

HOW TO

SIZE A VERTICAL COMMON FLUE – SIZE THE CONNECTORS

Size the connectors at each level of a vertical common flue manifold (follow each box in sequence).

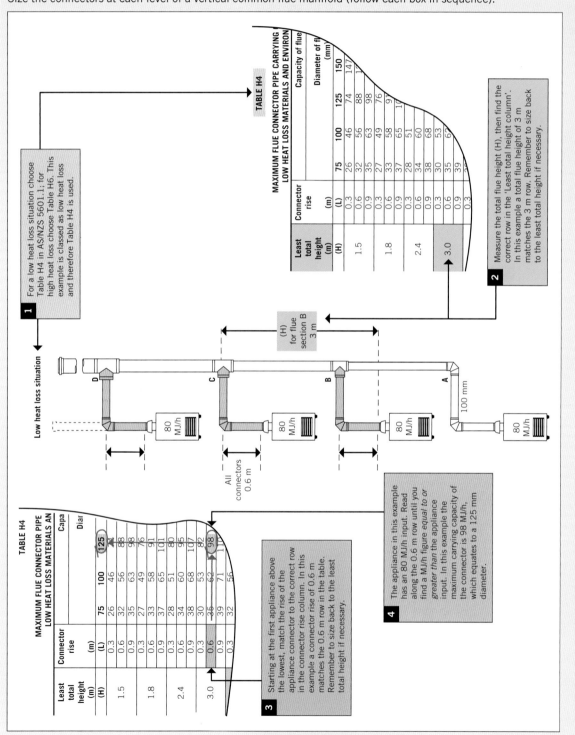

Source: Adapted from Standards Australia AS/NZS 5601.1:2010, Table H4

Repeat these steps for each connector and record in your worksheet as you go.

Size the common flue sections

Now that the connectors are sized, it is time to size the common flue sections.

COMPLETE WORKSHEET 6

HOW TO

SIZE A VERTICAL COMMON FLUE – SIZE COMMON FLUE SECTIONS

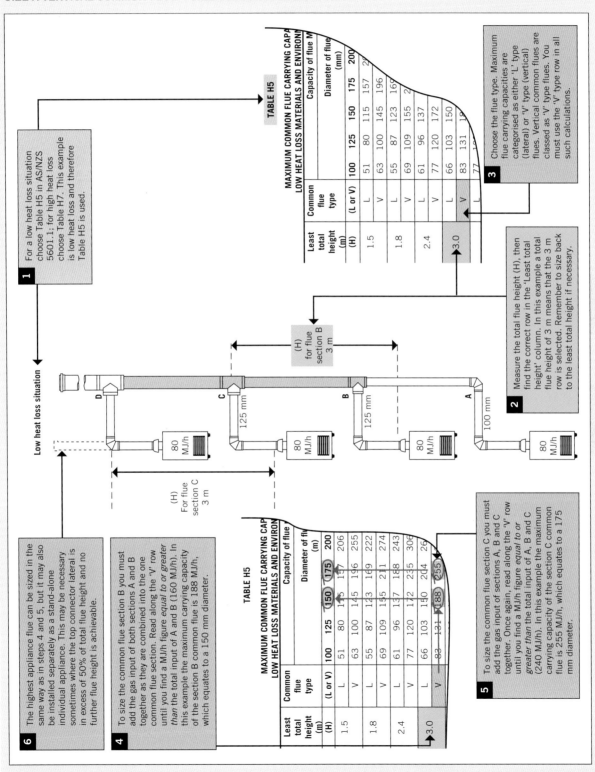

Source: Adapted from Standards Australia AS/NZS 5601.1:2010, Table H5

Record the size of the vertical common flue sections in your worksheet.

APPENDIX: ADVANCED SKILLS – FLUE SIZING **555 GS**

WORKSHEET 1

To be completed by teachers
Student competent ☐
Student not yet competent ☐

Student name: _____

Enrolment year: _____

Class code: _____

Competency name/Number: _____

Task
Review the section *Sizing a simple individual appliance flue* and answer the following questions.

1 You now have the opportunity to check and extend your knowledge of basic flue sizing. Complete the following exercise.

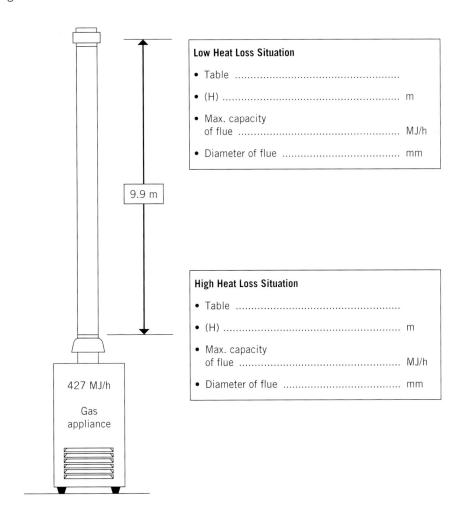

Low Heat Loss Situation
- Table ....
- (H) .... m
- Max. capacity of flue .... MJ/h
- Diameter of flue .... mm

High Heat Loss Situation
- Table ....
- (H) .... m
- Max. capacity of flue .... MJ/h
- Diameter of flue .... mm

APPENDIX: ADVANCED SKILLS – FLUE SIZING

2 Complete the following exercise.

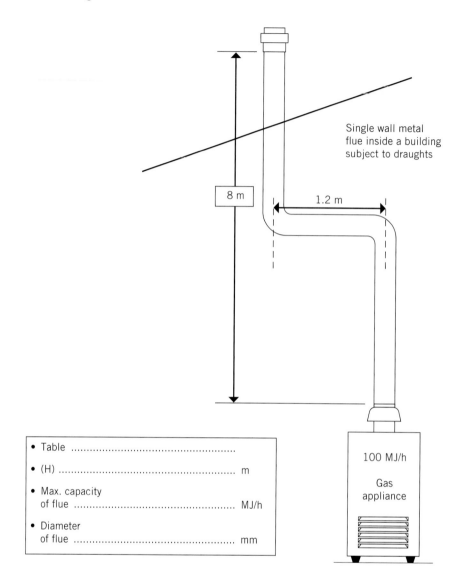

Single wall metal flue inside a building subject to draughts

8 m

1.2 m

100 MJ/h
Gas appliance

- Table ....
- (H) .... m
- Max. capacity of flue .... MJ/h
- Diameter of flue .... mm

WORKSHEET 2

To be completed by teachers
Student competent ☐
Student not yet competent ☐

Student name: _____

Enrolment year: _____

Class code: _____

Competency name/Number: _____

Task
Complete the following exercise.

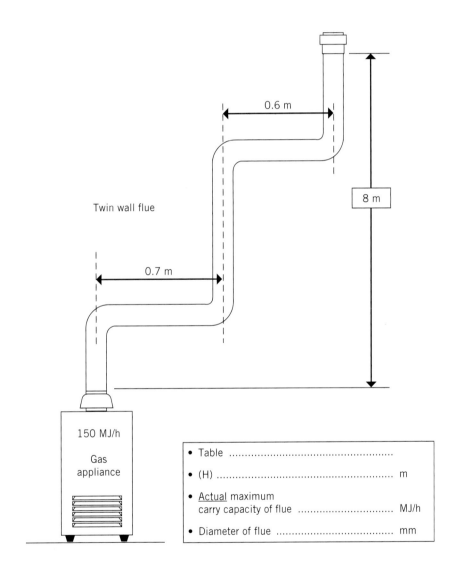

- Table ....
- (H) .... m
- Actual maximum
  carry capacity of flue .... MJ/h
- Diameter of flue .... mm

APPENDIX: ADVANCED SKILLS – FLUE SIZING

WORKSHEET 3

To be completed by teachers
Student competent ☐
Student not yet competent ☐

Student name: _____

Enrolment year: _____

Class code: _____

Competency name/Number: _____

Task
Complete the following exercises.

1

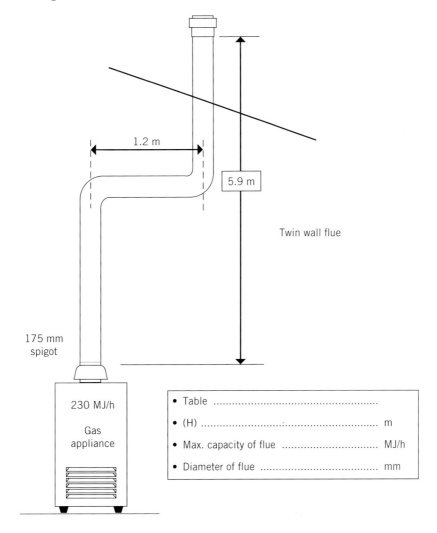

- Table ....
- (H) .... m
- Max. capacity of flue .... MJ/h
- Diameter of flue .... mm

APPENDIX: ADVANCED SKILLS – FLUE SIZING

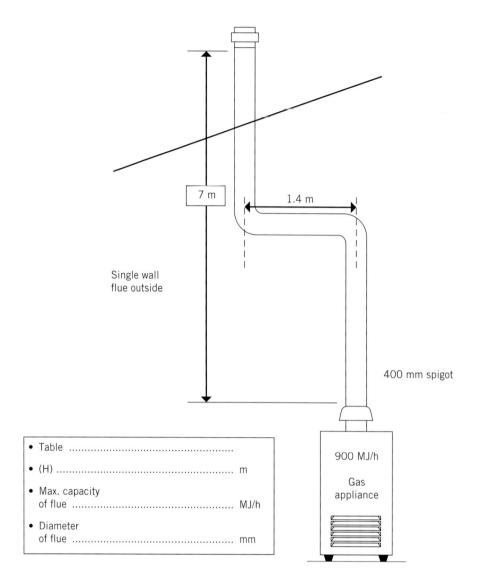

- Table ....
- (H) .... m
- Max. capacity
  of flue .... MJ/h
- Diameter
  of flue .... mm

3

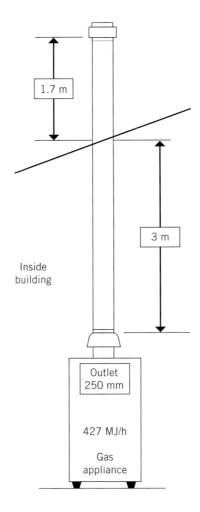

- Should this single wall metal installation be classed as a low heat loss or high heat loss situation?
  ....
- Provide reason ....
  ....
  ....
  ....
  ....
  ....
  ....
- Table ....
- (H) .... m
- Max. capacity of flue .... MJ/h
- Diameter of flue .... mm

WORKSHEET 4

To be completed by teachers
Student competent ☐
Student not yet competent ☐

Student name: _____

Enrolment year: _____

Class code: _____

Competency name/Number: _____

Task
You now have the opportunity to check and extend your knowledge of flue sizing and applying the space heater rule.
Complete the following exercise.

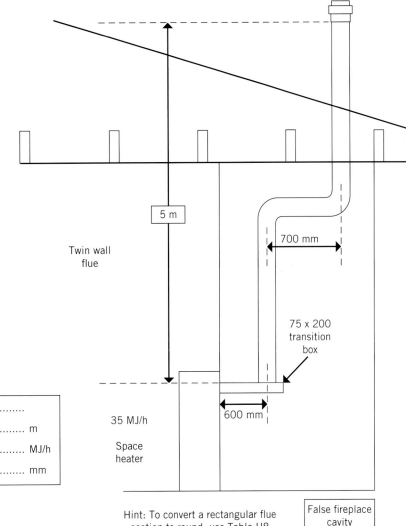

- Table ....
- (H) .... m
- Max. capacity of flue .... MJ/h
- Flue diameter .... mm

WORKSHEET 5

To be completed by teachers
Student competent ☐
Student not yet competent ☐

Student name: _____

Enrolment year: _____

Class code: _____

Competency name/Number: _____

Task
Review the section *Sizing horizontal and vertical common flue manifolds* and answer the following questions.

1 Now you have an opportunity to apply these procedures to the sizing of a horizontal common flue. Complete the tasks below.

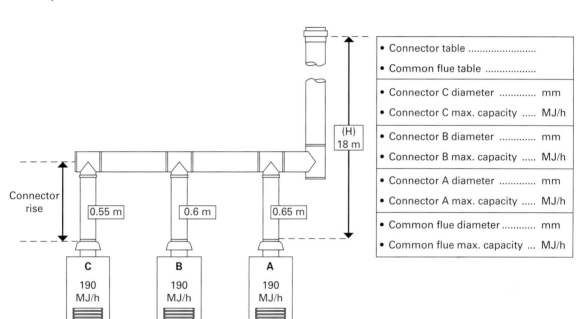

- Connector table ....
- Common flue table ....
- Connector C diameter .... mm
- Connector C max. capacity ..... MJ/h
- Connector B diameter .... mm
- Connector B max. capacity ..... MJ/h
- Connector A diameter .... mm
- Connector A max. capacity ..... MJ/h
- Common flue diameter .... mm
- Common flue max. capacity ... MJ/h

Low heat loss situation

WORKSHEETS

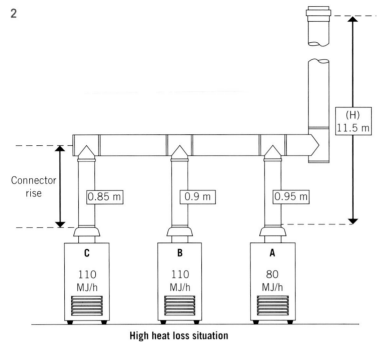

High heat loss situation

- Connector table ....
- Common flue table ....
- Connector C diameter .... mm
- Connector C max. capacity .... MJ/h
- Connector B diameter .... mm
- Connector B max. capacity .... MJ/h
- Connector A diameter .... mm
- Connector A max. capacity .... MJ/h
- Common flue diameter .... mm
- Common flue max. capacity .... MJ/h

WORKSHEET 6

Student name: _____

Enrolment year: _____

Class code: _____

Competency name/Number: _____

To be completed by teachers
Student competent ☐
Student not yet competent ☐

Task

Now you have an opportunity to apply these procedures to the sizing of a vertical common flue. Complete the tasks below.

1.
- Flue A table ....
- Flue A .... mm
- Flue A max. capacity .... MJ/h

- Connector table ....
- Connector B .... mm
- Connector B max. capacity .... MJ/h

- Connector C .... mm
- Connector C max. capacity .... MJ/h

- Connector D .... mm
- Connector D max. capacity .... MJ/h

- Common flue table ....
- Common flue B .... mm
- Common flue B max. capacity .... MJ/h

- Common flue C .... mm
- Common flue C max. capacity .... MJ/h

- Common flue D .... mm
- Common flue D max. capacity .... MJ/h

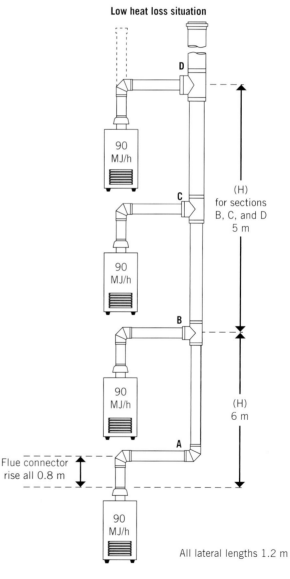

Low heat loss situation

(H) for sections B, C, and D 5 m

(H) 6 m

Flue connector rise all 0.8 m

All lateral lengths 1.2 m

2

- Flue A table ....
- Flue A .... mm
- Flue A max. capacity .... MJ/h
- Connector table ....
- Connector B .... mm
- Connector B max. capacity .... MJ/h
- Connector C .... mm
- Connector C max. capacity .... MJ/h
- Common flue table ....
- Common flue B .... mm
- Common flue B max. capacity .... MJ/h
- Common flue C .... mm
- Common flue C max. capacity .... MJ/h
- Flue D table ....
- Flue D .... mm
- Flue D max. capacity .... MJ/h

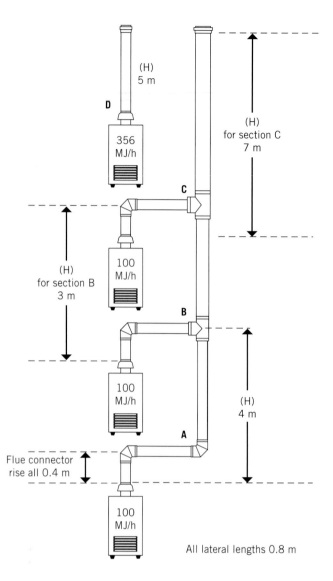

GLOSSARY

A

absolute pressure gauge pressure plus atmospheric pressure.

appliance a device or assembly that uses fuel gas to produce flame, heat, light, power or special atmospheres.

appliance regulator a regulator fitted at the inlet to an individual appliance.

atmospheric burner a burner designed to operate and provide all combustion under standard atmospheric pressure.

Australian Gas Association (AGA) Australian certifying body for gas fitters.

B

balanced flue a flue system where the air for combustion is drawn in and flue gases exhausted from the same terminal.

bayonet fitting an alternative name for a quick-connect device that allows a hose assembly to be connected and disconnected, without the release of gas, to a spring-loaded valve incorporating a bayonet-style twist and lock securing mechanism.

bi-metallic strip an element made from two metals with different expansion rates fused together, and which converts a temperature change into mechanical displacement.

BLEVE (boiling liquid expanding vapour explosion) occurs when a vessel containing LPG ruptures because of fire, corrosion or mechanical impact.

boiling point the point at which the vapour pressure of a liquid equals the environmental pressure being applied to the surface of the liquid.

British Standard Pipe (BSP) threads British standard of pipe thread with both tapered and parallel threads.

British Thermal Unit (BTU) An imperial unit of measurement now replaced by the use of megajoules and kilowatts in Australia.

burner pressure pressure of gas as measured between the outlet of the regulator and burner itself under operating conditions.

C

caravan structure that is or was designed or adapted to be moved from one place to another, whether towed or transported, and is intended for human habitation or use as a workplace; includes a self-propelled recreational vehicle or mobile home. Included is any associated annex and the like, whether permanently or temporarily attached to, or adjoining, the main portion of the structure (AS/NZS 5601.1:2010 © Standards Australia Limited/Standards New Zealand).

city gate plant infrastructure where the trunk main ends and the reticulation main begins.

combination control a gas control component that incorporates two or more control functions within the same assembly.

combination-cooking range up to six commercial cooking appliances using one gas supply, normally combined behind one fascia.

combustible surface any material or object made of, or surfaced with, materials that are capable of being ignited and burned.

combustion the exothermic reaction of a fuel and an oxidiser accompanied by the production of heat.

combustion products produced as a result of the exothermic reaction of a fuel and an oxidiser.

commission to bring an appliance into full working order.

common flue a flue conveying the products of combustion from more than one appliance.

compressed natural gas (CNG) natural gas that is compressed to less than 1 per cent of the volume it normally occupies at standard atmospheric pressure and generally stored in steel cylindrical or spherical containers.

condensate the liquid formed from the condensation of vapour.

condensation formation of liquid when a vapour is cooled to its dew point.

consumer billing meter a meter installed by the network operator at the end of the service line that measures the volume of gas used by the consumer.

consumer piping the complete system of pipes, fittings and components that conveys gas between the meter or first LPG regulator and the gas appliance.

continuous flow water heater a gas appliance with no significant stored volume of water that heats water as it passes through a heat exchanger based on consumer demand; formerly known as an instantaneous water heater.

convection heater a gas appliance that heats air through the action of convection currents passing through or over the appliance heat exchanger.

D

deep fryer a gas appliance designed to cook food in hot oil (generally used in reference to commercial catering equipment).

density mass per unit of volume.

dew point the temperature at which a given volume of air/gas must be cooled for water vapour to condense as liquid.

diaphragm meter a form of gas meter in which passing gas is measured through the cyclical oscillations of internal bellows-type diaphragms.

district regulator stations (DRS) distribution infrastructure designed to reduce high pressure gas supply to a level required within a certain area.

DN diamètre nominal/nominal diameter; refers to the nominal or 'trade' designation of a pipe or tube rather than its actual internal or external diameter.

downdraught diverter an assembly installed at the outlet of the primary flue and designed to protect appliance combustion from the effects of pressure changes brought about by flue downdraught.

duct flexible or rigid passageway designed to convey air to or from a room or gas appliance.

E

electronic flame safeguard a flame supervision system/component that acts to make an appliance safe in the event of flame loss.

enclosure a compartment or space primarily used for the installation of an appliance and associated equipment.

excess air air supplied to a burner in greater quantities than the theoretical minimum requirement.

exchange cylinder an LPG cylinder that when empty is exchanged for a full cylinder by the gas supplier.

F

fan a component of a gas appliance or gas system consisting of rotating vanes or blades that is intended to cause air to flow in a particular direction.

fire triangle a graphical symbol that demonstrates the relationship between the three necessary elements required for combustion: heat, fuel and oxygen.

flame failure device a device that is designed to automatically isolate the supply of gas to a burner in the event of flame loss.

flame lift off a flame that continues to burn while intermittently separated from the burner head.

flame safeguard a device or system that is designed to automatically isolate the supply of gas to a burner in the event of flame loss.

flame speed a measurement of the rate at which a flame propagates through an air–gas mixture.

flammability range the lowest and highest percentages of an air–gas mixture that will support self-sustaining combustion.

flue the system of pipes or passages designed to carry the products of combustion away from the appliance.

flue connection that part of an appliance deemed to be the actual connection point of the secondary flue system (generally the outlet of the draught diverter).

flue cowl a component fitted to the end of the secondary flue designed to prevent entry of water and vermin and retard downdraught, while not restricting the flow of combustion products.

flue gases the products of the combustion process including heat, carbon dioxide and excess air.

flue system the complete primary and secondary flue system including the draught diverter and cowl (terminal).

flue terminal *see* flue cowl.

flued appliance an appliance designed and certified for connection to a flue assembly.

forced-draught burner a burner system designed to provide combustion air/oxygen under pressure.

free ventilated area the cross sectional area in mm^2 of the actual unobstructed free gaps within a vent component.

fuel gas a gas that can be burned to produce thermal energy.

G

gas consumption the amount of gas supplied to an appliance measured as gas input in megajoules per hour.

gas input the amount of gas introduced into the appliance measured in megajoules per hour (MJ/h).

gasometer a specific holding vessel designed to hold a large volume of gas for storage and a buffer supply (typically towns gas).

gas-tight any part of the gas system pressure tested to ensure that pressure drop does not exceed the limit set by the technical regulator (generally now regarded to be no loss of pressure).

gauge pressure absolute pressure minus atmospheric pressure.

griller a gas appliance designed to cook food by radiant heat.

H

heat energy in transit from a high-temperature object/substance to a low-temperature object/substance.

heat exchanger a component within a gas appliance designed to transfer heat to fluid or air.

heat value the amount of heat (MJ) released during the complete combustion of $1\ m^3$ of gas in air.

holding gas a gas used on a temporary basis until the introduction and supply of a new type of gas.

hose assembly a certified assembly of hose and end-fittings designed to provide a flexible connection between the appliance inlet and the supply point.

I

induced draught burner a burner system designed to provide combustion air via mechanical suction in the combustion chamber.

injector a component within a gas appliance designed and sized to measure a specific quantity of gas to the burner and inspirate primary air into the mixing tube.

in-situ cylinder an LPG cylinder that remains on site and is topped up with a fill hose from an LPG tanker truck.

inspirate the effect of drawing air into the burner through the passage of the gas stream past the primary air-port orifice.

instantaneous water heater the traditional term used for a gas appliance with no significant stored volume of water that heats the water as it passes through a heat exchanger. Although still referred to in the AS 5601, this term is gradually being replaced by the technically more correct term 'continuous flow water heater'.

interlock a device or system that only allows the operation of a function or process through the dependent operation of another part of the system.

International System of Units (SI) the modern metric system of measurement based upon the 'metre–kilogram–second' unit convention. It is the most widely used measurement system in the world.

interrupter screw screw situated in the throat of a burner that is used as a form of primary air control by screwing it into the path of the gas stream issuing from the injector.

isolation valve a valve designed to shut-off and fully isolate a section of pipe or appliance.

J

jet *see* injector.

jointing compound a certified substance used to seal the threads of gas fittings and pipes (also thread paste).

L

lateral the nominally horizontal offset section of a secondary flue (generally between 0° and 45° from the horizontal).

liquefied petroleum gas (LPG) the name given to a mixture of hydrocarbon gases: principally propane and butane.

lock out the safety shutdown of a gas supply system that requires manual reset.

lock up where a regulator main valve shuts off the supply of gas at the inlet orifice.

lower explosive limit (LEL) the lowest concentration of a gas at which ignition of an air–gas mixture will take place; normally measured as a percentage of gas in air.

M

manometer a pressure-measuring device that generally refers to liquid-column hydrostatic instruments containing either water or mercury, but also includes digital devices.

maximum over-pressure the maximum pressure at which either the whole or part of an installation and appliance remains safe.

meniscus the characteristic concave or convex curve in the surface of a fluid held within a vessel.

meter device used to measure the quantity of a substance as it passes through.

meter bend a proprietary brand component assembly that allows multiple position connection between the two fixed points of a gas meter outlet and the inlet to the consumer piping.

minimum inlet pressure the prescribed minimum gas pressure that must be available at the inlet of each appliance with all appliances operating at maximum gas consumption.

mixing tube that section of a burner assembly between the burner throat and the burner head that is sized in both length and diameter to bring about adequate mixing of the gas and combustion air.

N

natural convection circulation brought about as a result of differences in density between warmer and cooler fluids/air.

natural draught flue a flue that operates through the natural convective rise of the flue gases.

natural gas (NG) a hydrocarbon consisting principally of methane.

NPT National Pipe Thread Taper; a US standard of tapered threads.

O

operating pressure the gas pressure which any part of the installation is subjected to under normal operating conditions.

over-pressure protection the requirement to prevent the pressure within a gas system from exceeding a pre-determined point.

oxygen depletion device (OD pilot) a pilot burner designed to facilitate the shut-off of gas supply to a burner when the concentration of oxygen in the ambient air supply drops to a pre-determined point.

oxygen depletion pilot (OD pilot) flame supervision system that will act to shut off the gas supply to the burner in the event available oxygen falls below pre-determined limits.

P

permanent joint a fitting or joint that cannot readily be disassembled.

piezoelectric ignition a burner ignition device that produces a high voltage spark when a lead zirconate titanate (or quartz) crystal is compressed between an internal hammer and anvil.

pigtail a flexible connection between an LPG cylinder and the cylinder regulator.

plant room a room designed to accommodate one or more appliances in which the appliances can be fully maintained, and that is not normally occupied or frequented for extended periods.

POL (Prest-O-Lite) fitting a commercial name for a specific left-hand threaded union used as a fuel gas connection to cylinders.

power flue a system in which the products of combustion are removed from the appliance and flue by a fan situated in the flue itself.

pressure relief valve a device designed to relieve pressure in excess of a pre-determined setting.

primary air the air drawn into the burner by the passage of the gas stream past the primary air port.

primary air port the primary air intake port for an atmospheric burner; usually adjacent to the injector.

primary flue that section of flue extending between the combustion chamber and the inlet to the draught diverter.

purge the process of removing air with gas or gas with air/inert gas to ensure that a hazardous air–gas mixture does not exist in pipework or an appliance.

pyrolysis the chemical decomposition of an organic substance by the application of low-level heat.

Q

quick-connect device a spigot and socket assembly that enables the connection of an appliance with a hose assembly without the use of tools.

R

radiant–convection heater a gas appliance that combines the features of a convection and radiant heater in the one appliance.

radiant heater a gas appliance that provides heat through radiation.

range a cooking appliance usually larger than a standard domestic cooker, often incorporating a double oven. A commercial upright cooker is normally known as a range.

rated working pressure the stipulated working pressure measured while the burner is operating.

refinery an industrial plant where crude oil and gas are separated and treated following underground extraction.

register the covering of an air opening or the locating ring around the flue penetration of a ceiling.

regulator a device that reduces a higher incoming gas pressure to a pre-determined outlet pressure.

relative density (RD) the ratio of density of a substance to the density of another reference material.

relief device a device designed to prevent an excess of temperature or pressure.

reticulated gas gas that is supplied to consumers through a network of underground pipes up to the consumer billing meter.

reticulation main the system of underground pipes that carries the gas from the city gate to the consumer.

retort an airtight chamber in which a substance (e.g. coal) is heated in the absence of oxygen to produce gaseous products for subsequent processing.

room any normally habitable room space, for example, this could include a kitchen, lounge area, bedroom, office or other workspace.

room-sealed appliance an appliance for which the air for combustion and the products of combustion remain separate from the internal air supply of a building.

rotary meter a form of gas meter in which passing gas is measured through the counter rotation of two internal lobed impellers.

S

SAE threads a US standard of thread type and pitch; SAE stands for Society of Automotive Engineers.

salamander a commercial catering grill similar to an oven but featuring an overhead burner assembly and no doors.

secondary air the supply of combustion air surrounding the flame/burner head.

secondary flue that section of flue extending from the outlet of the draught diverter to the flue terminal.

service line the gas supply line between the reticulation main and the consumer billing meter.

service regulator a regulator that reduces reticulation main pressure before the start of the consumer piping.

SI *see* International System of Units.

simulated natural gas (SNG) a mixture of LPG and air designed to give the gas similar burning characteristics to natural gas.

solenoid valve valve held open by the magnetic action of an electrically energised coil, which will shut off with the removal of the electrical supply.

specific gravity the density of a substance divided by the density of water at 0°C.

spillage where products of combustion are not adequately exhausted from the flue system and spill from the appliance around the draught diverter.

static pressure the pressure within the consumer piping when the gas is on but the gas appliances are turned off.

storage water heater a gas appliance that is designed to heat a stored volume of water.

sub-meter a meter that is used to measure gas consumption of appliances or installations downstream of the main consumer billing meter.

T

tailpipes short sections of pipe fitted at the lowest point in a consumer piping system to facilitate the removal of condensate; usually found on older towns gas installations.

tempered liquefied petroleum gas (TLP) a mixture of LPG and air designed to give the gas similar burning characteristics to towns gas.

thermal conduction generally referred to as the transfer of thermal energy through solid objects.

thermocouple a device that produces an electric current through the heating of the junction of two dissimilar metals.

thermostat an automatic device designed to alter the flow of gas to an appliance in response to changes in temperature.

towns gas (TG) a fuel gas produced from the heating of coal or hydrocarbons in a retort.

transmission main *see* trunk main.

trivet a metal frame located over cooktop burners to support a cooking vessel.

trunk main the high-pressure pipeline that carries the gas from the refinery to the point of distribution.

U

upper explosive limit (UEL) the highest concentration of a gas at which ignition of an air–gas mixture will take place; normally measured as a percentage of gas in air.

V

valve a device designed to control or shut off the supply of gas or liquid.

valve train an assembly of valves, components, unions and regulators required to correctly operate a gas appliance.

vent line a pipe that conveys the possible release of gas from a regulator relief vent or breather to the atmosphere.

ventilation the supply of adequate air to the appliance for combustion and operation.

venturi the constriction at the throat of a burner that causes an increase in air velocity as it enters the burner.

W

well head components and infrastructure that sits atop an oil or gas well and controls its extraction under pressure.

wet gas a fuel gas that carries a percentage of water vapour and condensable hydrocarbons.

working pressure the pressure read between the appliance regulator and the burner while it is working (effectively same as test point pressure).

INDEX

A
absolute pressure 19
absorption 113
accuracy 125
    alignment with zero mark for 15–16
acetylene 46
adaptor fittings 158
additional bends rule 514, 516–17
adhesive tapes 162
adjustment range 370, 447
adventitious air 244, 324
adventitious ventilation 411
aeration 349
air 3, 6, 45, 82, 242, 406
    air–fuel–heat interaction 82
    constituents 243
    lack of 82, 504
    requirements for combustion 46, 243
    at sea level atmospheric pressure 45
air flow 251–2, 321, 393, 411, 474
air inspiration *see* inspiration
air purge 226, 235
air supply 86–7, 93, 244, 252, 286, 411
    adequate supply, importance 93
    direct to outside; from adjacent room 244
    restricted 82
air–gas mixture 45–6, 84–6, 229, 312, 464
    proportions 503
    rapid 85
    re-ignition 90
alarms 59, 414
aldehydes 83, 243, 312, 425, 504
alternating current (AC) 62, 64, 66, 469
alternative current path 69
ambient air supply 93
ambient temperature 113–14, 166, 242
ammonia 58
amperes/amps 62
animals 4
annealed copper loops 276, 278, 390
    usage note 277–8
annealed copper tubing 390, 408
appliance burners *see* burners
appliance data plates 13, 17, 21, 94, 98, 126–7, 130, 243, 342, 434
appliance evaluation steps 484–5
appliance installation
    disconnection 273–4
    identifying requirements 272–3
    locations 273
appliance load 200
appliance operation
    customer information 350
    settings 348–9
appliance regulators 187–9, 193, 464
appliance servicing 18
appliances 6, 13–14, 341, 385, 394, 405
    bringing to correct working condition *see* commissioning appliances
    connections 347, 408, 411
    cost-effective solutions 82
    diagnoses 68
    gas input as MJ per hour 13
    identifying approved/certified 126–7
    identifying requirements 339–44
    incorrect gas use 97–8
    isolation and connection 273–4
    maintenance equipment 458
    overall condition 485
    piping to furthest *see* main run
    pressures available across each 19, 128, 132
    purging through 168–9, 348
    in roof spaces/under floors 253
    supply on; appliances off for lock-up pressure 20
    wear and tear 82
    on wheels/castors 273
    see also commissioning appliances
arcing 429
asbestos 246, 290
    WHS requirements 506
AS 3688 149
AS/NZS 1596 210, 360
AS/NZS 1869 159–60
AS/NZS 5601 45–6, 93, 128, 131, 149–50, 157–9, 162–4, 168, 187, 201, 203–4, 209, 211, 226, 229–30, 242–5, 247, 254, 272–3, 286–92, 294, 312, 315, 317, 319–23, 345–6, 362–3, 366–7, 370, 384, 386, 388–90, 392–6, 406–11, 413–14, 427, 432, 442–3, 448, 458, 463, 510–14, 518, 522
    allowable pressure drop 130, 132–4
    conversion chart 21
    defined terms/definitions 18–19, 199–200, 205
    higher pressure drop 133–4
    LPG installation application 210
    piping system installation 148–70
    selecting products 125
asphyxiates 48
asphyxiation 57–8, 60, 273, 280–1, 463
aspirator bulbs 157
assessment 57, 148, 155–6, 230
assumption 341, 362, 368, 385, 462
atmosphere 5, 258, 312, 405
atmospheric burners/systems 503–4, 508–9
atmospheric conditions 82, 84
atmospheric gas appliance flue sizing 513–21
atmospheric gas testing 405
atmospheric pressure 15–17, 112–13, 116, 313
    gas flow pressure greater than 19
    sea level 45
audible alarms 59
Australia, integrated gas network 30
Australian Gas Association (AGA) 125, 149, 314, 343, 506
    approval numbers 127–8
Australian Standards *see* AS; AS/NZS ranges
automatic change-over regulators 406
automatic change-over valves 370
automotive fuel 6
'backbone' main 33

B
backfilling 154
balanced flues 313
    terminal locations 319
barrier mesh 229
bars 14
baseline test 258, 324
basic combustion principles 503–4
basic electrical safety 61–6, 70
batteries 480
bayonet fittings *see* quick-connect devices
bedding 152–3
bi-metal strips 474
bi-metallic operation principles 465, 472
biogas/industry 5
'boat' 404
body, the, effects of current levels on 63
body attitude 63
boilers 502
boiling liquid expanding vapour explosion (BLEVE) 115–16
boiling points 112–14, 116–17
bomb calorimeter 48
bonding straps 65, 69–70, 165, 167, 345, 348, 428–9, 460
boots 63–4, 148, 425
bourdon gauges 18, 159
brain, the 60
brain damage 60
branches, branches 231
brazed fittings 151
brazing 46, 391
breather vents 198–9, 201
    domestic installations 203
    versus relief devices 202–3
    requirements 203
breathing apparatus 60
British Thermal Units (BTU) 13
BSP thread 158, 430–1
bubbles 58, 113, 228, 432
buffer supply 280
building plans/specs 148
buildings 150, 257, 366
Bunsen burners *see* partial pre-mix burners
burner design, primary/secondary classifications 84–7
burner flames
    flame speed = air–gas mixture speed 46–7
    retaining *see* flame retention

burner pressure 19–21, 348
burners 19–20, 87, 91, 274, 508–9
    burning back 46–7
    classifications 1 and 2 84–91
    combustion air supply 86–7
    functions and types 84
    heads 89
    performance 318
    purging through open burner 168–9, 292, 347
burning velocity 46–7
butane 5, 35, 44, 82, 112, 116–17
bypass flame 476, 479
by-products 31, 47

C

cables 154–5
calmness 58
Camloc-type fittings 228
cancer 83, 504
capacity 370, 405–6, 447
capillary break 324
capillary fittings 160
capping off 165
caravan LPG installation 361, 368, 384–96
    appliance types and locations 388–9
    caravan chassis 392
    'caravan' defined 384
    domestic–commercial distinction 384
    prep work 390–1
    system layout 385–9
caravan piping 389–90
carbon 83, 94, 243, 312, 425, 504
carbon dioxide 4–5, 44, 59–60, 82, 425, 503
carbon monoxide 4, 44, 46, 48, 60–1, 82–3, 94–5, 97, 125, 243, 312, 316, 324, 405, 425, 503–4, 512
carbon monoxide detectors 462
carbonising (of coal) 3
carcinogenic (cancer-causing) substances 83, 504
cardio-pulmonary resuscitation (CPR) procedures 60
carrying capacity, MJ 134
cast iron burners 88
cast iron upright gas cooker 475
catalytic reforming technologies 4
ceiling penetrations 321–2
ceiling thermal barriers 322, 522
cement sheeting 290, 322
central flues 280–1
certification 405
certification/certified products 18–19, 21, 125–7, 149–50, 160, 272, 276–8, 341, 343, 502, 506
chemical hazards 56
chemical properties 31–2, 35
chemicals/reactions 82–3, 99
chimneys 322–3
city gate 32–3
class A, B, C, D hoses 159–60
clearance 154–5, 273, 275, 286–7, 289–90, 319–21, 368–9, 512–13, 522
    to non-diesel fuel filler cap/vent 389
clips 278, 392, 411
closed circuits 471
clothing (protective) 148

coal 4, 44
    'carbonising' 3
co-axial flues 313–14
codes/coding 57, 125
cold-start choke 474
colour 94–5, 97–9
colour coding 125
combination controls 482–3
combination cooking ranges 279
combined radiant/convection heating 285
combustible surfaces 290, 512
    clearance 286, 321, 522
    defined 289
    protecting 259, 272–3, 288–91, 328
combustion 12–13, 45, 56–7, 82, 243
    from air + fuel + heat 82
    basic principles 425
    effect of inadequate flueing on 503–4
    efficient burners 20
    inadequate flueing effects 312
    inconsistent 96
combustion air supply 86–7, 93, 242
    mechanical means 86
    point of air–gas mix 84–6
combustion chambers 84–6, 503
combustion characteristics, mixing to create 6
combustion conditions 474
combustion fans 61, 313
combustion principles 82–3, 503–4
combustion products 82–3, 255, 286, 312, 513
combustion system 486
combustion zones 93, 97
commercial catering equipment 278–9, 384
commercial consumers 32
commercial installations 273, 278, 361, 502
commercial kitchens 254
commissioning appliances 3, 21, 46–7, 82, 93, 168, 209–13, 272–94, 347–8, 395–6, 413
    instructions 82
    procedures 272
    steps 292
common flue 517, 518
common flue manifold installation, principles/requirements 517–18
common trenches 154–5
communication 229, 340, 506
community 148
comparison 13
    fuel gases constituents 44–6
    of scale and water level 15–16
compatibility 97–8, 159, 162–3
complete combustion 82, 127, 244, 503
    products 82, 312, 425, 503
compliance 57, 125, 153–4, 201, 341–5, 392, 458, 463, 511
    assessment 57, 155–6, 230
compliance plates 396, 415
components 98, 189, 315, 508–10
    certification of 125
    key installation requirements 148–9
compressed natural gas (CNG) 6
compression 480
compression flare fittings 409
compression unions 160–1
concavity (of water level) 15

concealed locations 150–2
    pressure above/below 7 kpa 150–3
condensate 3
condensation 312, 316, 503, 511
condensing flue water heater 282
condensing flues 281
conductors 62
confined spaces 58, 63, 385, 394, 405
connections 347, 370, 408, 411
connectors 520
consistency 12, 31–2, 35, 96, 186
constant pressure appliance regulators 187–92
    operation 190–1
construction and property services sector 56
construction plans 148
consumer billing meters 12, 34–5, 424
consumer demand 32
consumer instruction labels/plates 395–6, 414
consumer piping 19, 128, 408
    piping–hose connectors 160
    pressurising 165–7
    purge 226–36
    regulators 201
    sizing 125–36, 148
    tests 163
consumer-owned cylinders 360–1
consumption *see* gas consumption
consumption patterns 442
contaminated atmospheres 56, 99, 160, 242, 255, 409, 448
continuity test 68
continuous flow water heaters 273, 280, 465
control measures 56–7, 186–214
control panels 61, 414
control systems 464
control valves 372
control/regulating equipment installation 209–13
convection heating 283–4
convective circulation 283
conversions 4, 21, 96–7, 432, 462
    provisos 272
    of units *see* unit conversions
cooking appliances 273, 408
cooktop burners 87
cooktops 274–5
cooperation 340, 506
copper 62, 148, 157, 209
copper alloy 149–50
copper pigtails 449
copper pipe 409
copper risers 34, 154
copper tubing 125–6, 203
corrosion 316, 386–7, 430, 511
corrosion resistance 406
cost-effective appliance solutions 82
costs 4, 6, 512
cover, depth of 152–3
covered areas 320
crimped fittings 151–2
cross-linked polyethylene (PEX) 125–6, 148, 209
cross-ventilation 320, 411
crude oil 5
crust (Earth) 4, 31
cubic metres to litres 12

current 62–3, 506
    alternative paths for 69
    effects on the body 63
    testing for stray 63, 67, 69, 158, 340–1, 428–9, 485, 506
cyclic capacity 231
cyclic volume 432
cylinder capacity 406
cylinder locations 385, 387, 405–7
    installation requirements 366–71
cylinder manifolds 369
cylinder pressure 114
cylinder storage 406
cylinder testing 361
cylinders 35–6, 114, 201, 228, 360–2, 371–2, 407
    '100-pound' cylinder 13
    clearance and exclusion zones 368–9
    coatings 387
    commercial use 13
    compartments 388, 406–7
    general site requirements 366
    LPG 5–6, 13, 114
    maximum aggregate capacity of 370
    mountings and restraints 387–8
    overfilled 114
    standard sizing 360, 363
    steel 5–6, 13, 35, 386
    volume 361
    see also tanks

D

damage 408
danger 44, 48, 60, 463
dangerous atmospheres 405
data 129
    for sizing 134–5
data plates *see* appliance data plates
death 60, 63, 69, 125, 229, 429
decanting 361
decompression 31
decorative flame effect fires 253, 287–8
    classifications 1 and 2 287
    installation requirements 287
    ventilation 287–8
demand 6, 32, 364
    continuous change in 186
    high profile 135
    niche 5
density 44–5, 203, 272
designer specifications 511
dew point 281, 316, 511
dezincification resistant (DR) 150
diaphragm meters 425–6
diaphragms 201–4
diesel 6
digital manometers 17–18, 158–9, 395, 413, 430, 460
digital multimeters 459
digital thermometers 460
direct current (DC) 62, 469
direct-fired vaporisers 444
direction 132
Directory of AGA Certified Products 127–8, 149, 343
dirt 448
dispersal 45
displacement 48, 202, 229

district regulator stations (DRS) 33
DLK formula 362–3
    percentage determination 362–3
DN size 155
DN union 157, 162
documentation 136, 164, 170, 214, 236, 245, 259–60, 294, 328, 350, 372, 396, 415, 435, 450, 487–8, 525
    interpretation 148
    signed off 506
domestic consumers 32
domestic cookers 274–7
domestic oven burners 90–1
domestic ranges 276
downdraught diverters 315–16, 318, 509–10
    functions 317
downdraught problems 513
downstream regulators 206–7
drain openings 368, 388, 407
drainage box ventilation 412
draught diverters 252
drawbar weight 386
drilling 4
dual power supply 384–5, 404–5
dual stage LPG regulators 193–7, 199, 405
    operation 194–6
    outlet pressure increases/decreases 195–6
    over-pressure relief valve 196–7
dual stage manual change-over regulators 391, 410
duckbill burners 90–1
ducted central heating 284–5
Ducted forced convection heating 284
due process 125
durability 315, 511
dust 244, 246, 255, 288, 409
dust extraction equipment 248
dust masks 148
duty of care 57, 156, 463

E

Earth 31
earth stakes 62, 64–5
earth/earthing 61–2, 65
efficiency 82, 93, 127, 148, 154, 280, 347
    of combustion 20
    correct supply and pressure for 14–15
    pipe sizing 125
efficiency comparisons 13
elbows 130
electric shock 61–3, 65, 69
electrical appliances 13
electrical cables 154–5
electrical hazards 62, 148, 246, 338, 463
    causes 64–6
electrical safety 165, 167, 200, 235, 404, 425, 462–3, 506–7
    basic 61–6, 70
    before/during work 344, 428–9
electrical services, working in close proximity to 56
electrical terminology 62
electrical test equipment 66–70, 158
electrical testing 462, 506–7
electrical theory 63
electrical tracing 256–7
electricians (licensed) 62

electricity 480
    basic laws 62
    flow 62
    grid 5
electrocution 61, 63
electrolysis 159, 430
electromagnets 465–8
electronic ignition 480–2
elevated cookers 276–7
elevated work platforms 248
elevation view 130
emergency services 58
emergency situations 6
emissions 4
employees 56
    documentation interpretation 148
employers 56
enclosed spaces 95, 408
enclosures 244, 250–1
energy 5, 113
    from heat 12–13, 48
    SNG alternative 6
energy cut-out systems (ECO)/devices 464, 471–2
energy sources, concentrated 48
energy-efficiency standards 82
engagement 160
engineering controls 57
engines 6
environment 5
Environment Protection Authority (EPA) guidelines 506, 513
EPDM flashings 523
equipment 57, 66–70, 136, 157–8, 186–209, 226–9, 246–8, 278–9, 288, 338–9, 371, 384, 390–1, 409, 428, 448, 450, 508, 525
    disconnect; reconnect 344
    test operation 346
    working in close proximity to 56
erosion 133
error/mistake 128, 200
evacuation 58
evaluation 56–7, 484–5
evaporation 113
excavations, working in and around 56
excess air 82
exchange cylinders 35, 200, 361, 366–7
    versus in-situ fill 369
exclusion zones 210–11, 368–9
exclusions 404
exhaust hoods 255
expansion 465
explosion 45, 57, 114, 389, 463
    see also lower explosive limit; upper explosive limit
exposure (to CO) 60–1
extensions 148, 155, 162–3
external appliance balanced flues 313
extinguishability 87
extraction 3–5, 31, 324
extraction fans 254

F

fan operation 349
fan-assisted, balanced flues 242, 320
fan-assisted space heater flue installation 325–7

fan-driven air inlets 244
fan-forced gas oven 276
fast-burning gas 84
fatigue 465
fault conditions 66–7
faults 68, 148, 156, 409
    actions on finding 69, 463
fibro 290, 322
final connection tests 163, 168, 346
final installation tests 235
fire extinguishers 229
fire triangle 45
fire-resistant materials 273, 276, 290
fires 57, 97, 114, 389, 463
first aid response 60–1
first and second stage LPG regulators 197
fittings 149, 158, 160–3, 228
fixings 159
flame characteristics 92–4
    exceptions to rule 94
    'good' flame defined 93–4
flame chilling 83
flame effect fires 94, 287–8
flame failure operation 349
flame fault diagnosis 94–9
    checklist 99
    excessively large flames 97
    flame lift-off 87, 90, 96
    orange flames 99
    streaming flame 95
    undersized flames 98
    unstable 87
    yellow tips 94, 243
flame loss 464, 466
flame rectification systems 466–8
flame retention
    hardware 90–1, 96
    methods 87–8
    parts and rings 87–9
    recirculation 88–90
flame retention design 87–91
flame safeguard devices 169, 293, 348, 389, 408
flame speed 46–7, 84
flame supervision devices (FSD) 464
flame-failure devices (FFD) 96, 464–6
flame-failure system 286
flames 477–8
flammability range 45–6
flammable substances 4, 48, 226
flare purging 227
flare stacks 226
flared compression fittings 149–51, 160–1
flares 160–2, 167
flashings 323, 523
fleets 6
flotation 44–5
flow capacities 13
flow conditions 19
flow graphs 128
flow meters 228–9
flue base locations 522
flue cowl 323
flue gases 281, 316, 319, 511
    temperature 315, 323, 511–12
flue installation 312–24, 391–6, 522–3
    clean-up 524–5
    example 507

identifying requirements 315, 508
instal and test 321–3
materials 314
prep work 314–15, 506–8
regulatory/statutory requirements 506, 513
flue pipe roof penetration flashings 323, 523–4
flue sizing 513–21
    Appendix 544–55
    pre-requisite information 514
    simple individual appliance flue 514–15
flue socket alignment 523
flue spillage 258, 294, 324
flue terminals 281, 319–21, 512–13
    above roof line 319, 513
flue-assisted flue installations 323–4
flued appliance (spillage) testing 324–8
flued appliances 258, 286
flueing 286
flueless appliances 257
    prohibited locations 273, 464
flueless space heaters 254, 285, 411
flues 312, 319, 502–4
    adjacent to rafters/roof members 321, 522
    components 315, 508–10
    configurations 324
    diameter 315, 511
    handling large 523
    inadequate 312, 503–4
    materials 315–16, 510–11
    operational check 486–7
    testing 257–9
    types 312–14
flying debris 246
footwear 63–4
force 14, 278
forced combustion air 314
forced draught burners/systems 86, 509–10
fossil fuels 4
free ventilated areas 244, 393, 412
freestanding cookers 275
freezing 32
friction 127
fuel 45, 60, 82, 312, 503
    air–fuel–heat interaction 82
fuel gas purge 226, 233
fuel gases 3–6, 87, 226, 243
    constituents; characteristics 44–8
    controlled release 226
    discontinued *see* towns gas
    hydrocarbon-based 12–13
    leaks 405
    purge calculations/planning 231–2
    see also liquefied petroleum gas
fuel vapour 389
full pre-mix burner 85

G

galvanised steel 157, 209, 512
gas 3, 5, 57, 84
    delivery 84
    fields 4
    for fuel *see* fuel gases
    gas-to-energy sites 5
    levels (in vessels) 3–4
    safe use of 125

gas appliances *see* appliances
gas consumption 12–13, 19, 128, 364
    consumption patterns 37
    high pressure buffer 32
    maximum 132
    readings 35
gas cookers, protection 291
gas detectors 59, 61, 229, 258, 413
gas distribution 33
    gas network 31–2
    LPG 35–7
    natural gas 30–5
gas distribution systems 35
gas distributors 32, 226
    authority for service regulator adjustment 34
gas fitters 3–4, 66, 82, 94, 112, 148, 226, 288, 432, 443, 449
    common conversions 21
    scope of work 30
gas flow 162, 186, 228
    pressure greater than atmospheric 19
    section 132
gas industry
    terms 12–21
    workplace safety 56–70
gas inlets 469
gas input 13, 21, 89, 126
    rating 130
Gas Installation Standards *see* AS/NZS 5601
gas isolation 200
gas leaks 57–60
    causes 57
    detection; response 57–8
    odorant addition to alert 32, 48
gas load 130, 132–3, 154–5, 385–9, 405–8
    requirements 150
gas meters, reading 432
gas nozzles 85–6
gas pipe sizing *see* pipe sizing
gas piping system testing 163–5
gas pressure 32–3, 92
    for concealed location piping 150
    correct pressure, importance 93–4, 96–8
gas pressure control equipment 186–214
    identifying requirements 200–1
    see also regulators
gas pressure measurement 15–18, 32
    other forms 14
    terms used 18–20
gas pressure testing 346
gas rate 243
gas refrigerators 394–5
gas reticulation pressures 32–3
gas storage 3–4
    cylinders 5–6, 35, 114–15
    LPG 112–15, 360–72, 442–50
    pressurised 6
    propane 114
gas sub meters installation 424–35
    install and test 428–34
    system requirement determination 424–7
gas supply 481
    buffer 32
    capacity 186
    correct 14, 97
    coverage estimates 363–5, 445–6

cut off on fan failure 244
depressurising 167
high-pressure (source to city) 14
isolation 235, 345
'locking up' 194
match with appliance 341
multistage systems 135
outage 6
pressure 130
on for static measurement 20
gas tightness tests 395, 413
gas type 200, 340–1, 458, 462
gas types 21, 84, 87, 92, 129
compatibility with appliance 97–8
gas use estimates 37
gas vapour 443
gas velocity 133
gas welding 160
gaseous by-products 31, 47
gases, basic gas/combustion principles 425
gaslights 273
gasometers 3
gas-tight flared joints 161–2
gas-tight installations 57, 163, 166, 168, 230, 346, 413, 464
gauge assemblies 361
gauge pressure 16–17, 19
general compliance assessment 57
general-purpose outlet (GPO) 64, 68
gimbals 408
glasses 148, 424, 429
global warming 4
gloves 148
graphs/graphing 128
greenhouse gas 231
greenhouse gas emissions 4–5, 15
grinders 246
gross potential draw-off 364–5, 445

H

haemoglobin 59–60
hand pumps 157, 165, 167
hard hats 148
hazard identification 56, 148
hazardous combustion 82
hazards 45, 47, 56–7, 59–62, 64–6, 82, 97, 99, 128, 132, 246, 324, 338, 428, 463, 503
hearing protection 148, 424
heat 4, 12–13, 45, 112–13, 280, 312, 425, 503
air–fuel–heat interaction 82
consistent output 31–2
source 47
heat exchangers 32, 283, 471–2
heat loss 316–18, 511–12
heat value 48, 92, 243, 425
heavier-than-air gases 45
height access 315
heights 56, 507
HEPA-rated vacuum equipment 248
high heat loss installations 316–18, 511–12
high resistance earth fault 68
high temperature cut-out 471
high-pressure pipe installation requirements 408
high-pressure piping 408
installation requirements 389

hinged protective hoods 367
hobs 275
holding gas 6
homogenous mixtures 84
horizontal common flue manifolds 516–18
sizing 519–21
hose assemblies 159–60, 273, 276–8, 345, 392, 408, 411
maximum working pressure/temperature 345
hosing 157–8, 228, 408
hot surface ignition 482–3
hydraulically pressed fittings 151
hydrocarbons 4–5, 35, 44, 82, 243
see also liquefied petroleum gas
hydrogen 44
hypoxia 60

I

ice 443
ignition 45, 68, 116, 276, 321, 482–3
ignition sources 368
elimination 58
ignition systems 61, 479–82, 487
ignition temperature 47, 522
imperial measurement system 13–14
conversions 21
cubic feet per revolution 12
inconsistencies 12
pounds 13
impurities 3
in situ fill cylinders 35, 201
inbuilt ovens 276
incandescent flame 84
incense sticks 259, 328
inches 17
inches of mercury (inHg or ″Hg) 14
inches water gauge (″WG)/scale 17
incomplete combustion 82–3, 96, 125, 132, 503–4
causes 82–3
products 83, 312, 504
index counters 426
index faces 433–4
indirect-fired vaporisers 444–5
indoor appliances 273
induced combustion air 314
induced draught burners 86
industrial appliances 14
industrial burners 85
industrial consumers 32
industrial installations 502
industrial LPG regulators 197–8
industry requirements 148
inert gas 229
inert gas purge 226, 234
purging quantities 231–2
inert gases 60, 82, 226
inertia 92
information 128–30, 132
on data plates 21, 126–7
infrastructure 30–7
injectors 89
correct size, importance 91–2, 97
'flash back' to 98
gas type, pressure and orifice size 92
injury 69, 125, 229, 429
inlet connections 370

inlet pressure 19, 127, 370, 447
inlets 244, 469
in-situ cylinders/tanks 35–7, 361
sites 367
inspection 57, 245, 247, 385
cameras 247, 409–10
site 148
inspiration 86–7, 243–4
instability 87–8, 90, 243
installation 57, 186–214, 312–24, 424–35, 502, 511–12
accessibility 150–2
care while conducting 4
certified components 125
compliance 341–4
existing 291
gas rate determination 432–3
general/specific requirements 272–7, 463
obtain building plans/specs 148
pre-installation checks 506
prep work 208–9, 288, 371, 448
purging through appliance 168
section gas flow 132
sectional (pipe) measurements 130–2
selecting products 125
volumes 226
also under specific installation
installation characteristics 149
installation tests 163, 235, 346
appliances connected; no gas 166–7
instantaneous water heaters *see* continuous flow water heaters
insulation 322, 522
insulators 63–4
interlock 473
internal appliances 242–4
International System of Units (SI) 154–5
base 10 and multiples 12
interpretation 148, 279
interrupter screws 89, 486
inversion 469
isolation 57–8, 168, 200, 235, 273–4, 345, 430
isolation valves 125, 167, 431
upstream 201

J

jab saws 246
jet drills 461
jets 92
job safety analysis (JSA) 170, 200, 245, 338, 425
joints/jointing 150–2, 160–2
joule 13

K

kiln burners 87–8
kilograms 12–13
kilopascal 14
kilowatts 13
kinetic energy 113

L

labels/labelling 35, 130–1, 231, 395–6
laboratory burners 87–9, 93
labour 4

landfill gas 5
large volume gas installations 226
lath walls 257
leakage tests 57, 163, 167–8, 230, 235, 344, 346, 430–1, 463–4
legislation 57
let-down station *see* city gate
LGP gas systems installation 409–14
    identifying requirements 443
    test and commission 413
LGP storage installation 360–72
    instal and test 371–2
licence 56, 148, 272, 506
    scope of 502
linear expansion 465
lint 255
liquefied petroleum gas (LPG) 3, 44, 46, 48, 84, 96, 226, 272, 404, 462
    air–gas proportion 503
    applications 112
    basics 5–6, 112–17
    connection 37
    delivery 35–7
    distribution 35–7
    heat value 243
    liquid weight in kilograms 12–13
    minimum inlet pressure 19, 127
    NG–LPG differences 205
    nominal burner pressure 93
    odorant addition 48
    physical characteristics 5–6, 112
    pipe identification 162
    relative density 92, 406
    source 35
    storing 112–15, 360–72
    tank installation 36–7
    tempered *see* tempered liquefied petroleum
    transporting 115
liquid expansion thermostats 478
liquid withdrawal 361, 443
liquids 4, 13, 35, 112–13
loam 153
lock out 281
lock-up condition 19
lock-up pressure 20, 168, 346
lock-up tests 196
low flame adjustments 349
low heat loss installations 317–18, 512
low pressure bourdon gauges 18, 159
lower explosive limit (LEL) 45, 48, 58–9, 229
low-pressure compensating regulators 192–3
low-pressure pipe installation requirements 408–9
low-pressure piping 408
    installation requirements 389–90
LPG calculations 360, 442–4
LPG gas storage 112–15, 442–50
    aggregate storage capacity exceeding 500 litres/less than 8kL 442–50
    aggregate storage capacity up to 500 litres 360–72
    identifying requirements 361–2, 443
LPG regulators 197, 370–1, 447
LPG system installation 384–96, 404–15, 442–9
    clean-up 396
    cylinders 360–2, 386–8
    identifying requirements 384–5, 404–5
    test and commission 395–6
LPG tank connections 449
LPG tank storage systems, instal and test 448–9
LPG tanks 442
luminous burners *see* post-mix burners

M

magnetic fields 66
main run 130–3, 231
maintenance 32, 148, 155, 163, 483–7, 512
    identifying requirements 458–64
    procedure 485–6
maintenance procedure 485–7
manometers 15–17, 157–8, 229, 395, 413, 430, 460
    reading 15–16, 20
manual change-over (MCO) valves 370
manual change-over regulators 406
manual shut off-valves 279
manufacturer's instructions/specs 20–1, 82, 91, 98, 167, 245, 272–3, 277–8, 315, 347, 385, 389, 406, 408, 485, 512
marine craft gas detection system requirements 413–14
marine craft LPG installation 386, 404–15
    clean-ups 415
    prep work 409
    selection and location 406–7
    system layout 405–8
marine craft piping 408–9
marine craft ventilation 411–13
    formula 412
marine fuel 6
masonry removal 257
mass 13–14
materials 4, 130–2, 135–6, 154–6, 170, 273, 276, 314–16, 389–90, 408–9, 428
    best quality 156
    'fit for purpose' 149
    high heat loss 316–17, 511–12
    identifying approved/certified 125–6, 150
    low heat loss 317–18, 512
    for piping 148–9
    protecting 156
    quantity details 155
mathematical principles 432
maximum gas consumption 132
maximum gas input 21, 130
maximum over-pressure 19, 205
measurement 12–21, 130–2, 136, 148
    see also units
mechanical air supply 86–7
mechanical float valves 361
mechanical ventilation 244, 252–3, 324
    rules 252
megajoules per hour (MJ/h) 154–5
megapascal 14
MEN (multiple earthed neutral) system 64–7
meniscus 15
mercaptans 48
mercury 14, 31

mesh screens 412
metallic pipes/risers 153–4
metals 465
meter bends 430
meter outlet connections 431
meter sizing fill
meter support 430
meters 12, 424
    accessibility and connection 430
    applications 424
    matching to system specifications 425–6
    selection 426–7
meters locations 427
    recessed 427
methane 4, 31, 44, 82, 231
    other sources 5
millibars 14
minimum inlet pressure 19, 127–8, 134
mixing tubes 86, 89
MJ carrying capacity 134
mobile homes/workplaces LPG installation 384–96
model (appliances) 21
modulating rod and tube thermostats 475
moisture 3, 31
molecules 113
monitor regulators 206–7
motion 14
movement 410, 465, 481
multilayer construction piping system 125–6, 162–3, 209, 344
    materials 149, 157
multi-meters 68
multimeters 459
multistage gas supply systems 135, 197, 200
multistage LPG systems 197–8
'mushroom' style ventilation 412

N

naphtha 4
natural convection 283
natural draught burners 86
natural draught flues/systems 284, 312–13, 320–1, 503–4, 508–12
    components 316
natural gas appliances 187, 202
natural gas (NG) 3–5, 44–8, 84, 96–7, 226, 272, 462
    air–gas proportion 503
    compressed *see* compressed natural gas
    distribution 30–5
    heat value 243
    LEL; UEL 45
    minimum inlet pressure 19, 127
    NG–LPG differences 205
    nominal burner pressure 93
    odorant addition 48
    pipe identification 162
    pipelines 30
    relative density 92
    reserves of 4
    separation from crude oil 31
    simulated *see* simulated natural gas
natural ventilation 243–4
    air sources 244

natural ventilation installation 242–60
    identifying requirements 248–50
neutral bars 64–5
neutral–earth polarity fault 68
nitrogen 44, 82, 229–30, 243, 503
    cautions for use 60
nitrogen cylinders 228
'no earth' problem 68
noise 98, 116, 133, 246
noise hazards 56
nominal capacity 406
non-contact voltage detectors 66, 68, 256, 340
non-renewable resources 4
non-thermostatically controlled space heaters 285
noxious fumes 244
nozzle-mix burners 85–6
nylon 153, 157, 209

O

odorant 32, 48
odour 48, 58, 83
ohms 62
oil 4–5, 44
oil crisis 6
oil production 31
open burners 168–9, 292, 347
open circuits 471
operating conditions, normal 19, 82, 84, 165–6, 347, 414, 509–10
    also under specific condition
operating pressure 14, 19, 167–8, 205, 346
OPSO devices 211–13
organic compounds 83
organic matter 5
outdoor appliances 242, 273
outlet pressure 370, 447
outside atmosphere 202, 204, 235
overalls 148
overhead radiant heaters 286–7
    installation requirements 287
overhead radiant tube heaters 255–6
    installation requirements 286
overheat conditions 464
over-pressure conditions 19
    protection requirements determination 205–8
over-pressure protection/devices 205–8
over-pressure relief valves, operation 196–7
over-pressure shut-off devices 207–8
ownership 32
oxidation 12–13, 82–3
oxidisers 82, 312, 503
oxy-acetylene 166
oxygen 44–5, 59–60, 82, 243, 503
    displacement of 48
oxygen depleted atmosphere 405
oxygen depletion 464
    pilot sequence 474
oxygen depletion pilots/devices (OD pilots) 464, 473–4

P

partial pre-mix burners 84–6
parts ordering 21
pascal 14
peak-load vaporisation requirements 442

penetrations 321, 323, 411, 524
percentage 45, 48, 443–4
permanent joints 151–2
permanent ventilation 250–2
personal protective equipment (PPE) 57, 148, 338, 424–5, 459, 507
perspective view 130
pest protection 211
Pete's Plugs 212
petroleum 4, 6
physical characteristics 3, 5–6, 44–8, 84–7, 112, 129, 442
piezoelectric ignition system 479–80
pigtails 361, 369, 371–2, 391, 408, 443, 447
pilot cleaners 461
pilot lights 170, 349, 405, 465, 480–1
pipe design 130, 150, 391–2, 410–11
pipe engagement 160
pipe fixings 390
pipe identification 162
pipe leakage testing 235
pipe sizing 127–33, 148, 386, 406, 513–21
    each section 133
    job requirements 125
    methodology 128–30
    of multistage gas supply systems 135
    procedure 130–3
    of proprietary brand piping systems 134–5
    required information 128–30, 132
    resources 128
    tables 128–9, 132–3
    worksheet *example* 129
pipe volumes 163–5
pipeline operators 32
pipework installation 391–2
    multiple 392
    pipe design and location 391–2, 410–11
pipework tests 163, 346, 395, 413
    no appliances connected 165–6
piping 389–90
    accessibility 150
    branches 130
    in concealed locations 150–2
    disconnection 69–70
    of gas 4
    key installation requirements 148
    material 130
    plastic 34, 153
    versus tubing 125
    underground requirements 152–4
    see also consumer piping
piping identification 162
piping support/fixings 159
piping system installation 148–70
    design and location 150, 201
    identifying requirements 148–50
    maintenance/extension of existing 148, 155, 162–3
    material quantities 154–6
    preparing for 156–8
    purging procedures 168–70
    testing 159–68
piping systems
    multistage 135
    proprietary brand 134–5
    testing 163–5

plan view 130
plans/planning 338, 340
    approved versions 128, 148
    for flue installation 506, 512–13
    information sources 245
    with others 339–40
    for purging 200, 230–2
    for ventilation 245–6
plant rooms 244, 251–2
plants 4
plaster 257
plaster saws 246
plastic coating 409
plastic piping 34
    underground use 153–4
poisonous substances 4, 46, 60–1, 83
POL adaptors 449
polarity 68
pollutants 15
polyethylene (PE)/service lines 34, 149, 153, 157, 209
poor combustion 82, 125, 425
portable appliances 6
portable gas detectors 61
positive displacement meters 425
post-mix burners 84
pounds 13
pounds per square inch (psi) 14
power flue 284
power flues 313–14
power point testers 68–9, 459
power supply 384–5, 404–5
power tools 315, 428
praxis (theory–practice blend) 148
pre-aerated burners 86–7, 89
    combustion zones 93
pre-maintenance checks 458–64
press-fit crimped fitting systems 390, 409
pressure 4, 14, 19–20, 62, 92–3, 313, 447, 462
    accumulative loss 127
    CNG storage under 6
    correct *see* operating pressure
    equalising 114
    low-, medium- and high-pressures 32–3
    maximum *see* over-pressure conditions; rated working pressure
    minimum *see* minimum inlet pressure
    regulation in multiple stages 135
    at sea level 14
    transmission pressure 32
    variations 186
pressure control fundamentals 186–200
pressure drops 166, 194, 395, 406
    allowable 130, 132–4
    due to small sizing 125
    mandatory requirements 132
    parameters for increased 134
    standard-approved application of higher 133–4
pressure measurement instruments 15
pressure relief valves 116, 371–2, 448–9
pressure switches 472–3
pressure testing 19, 157–9, 163–5, 344, 346, 395, 413, 449, 464
    instruments 209
    points 201–2
pressure-reducing let-down station 32

pre-work leakage test 57
pre-work system pressure test 344
price 4
primary air port (primary air) 85–7, 93–4, 96
primary flues 280
probable daily consumption 364, 445–6
probable simultaneous demand 364, 446
probes 59, 466, 480
products 5–6, 82–3, 125, 312, 503–4
    approved 150
    proprietary branded 134–5
    see also by-products; certified products; combustion products
prohibited appliances 394, 408
propane 5–6, 35, 44, 82, 112–14, 360, 364
    boiling point 113–14, 442
properties 129
proprietary brackets 430
proprietary brand piping systems 125–6, 134–5, 151–2, 162
proprietary twin wall flues/systems 317–18, 512
protection 205–8, 211, 259, 272–3, 288–91, 328, 424
protective clothing 148
protective hoods 35
PTFE (Teflon) tape 160
pumps 157, 165, 167
purge buckets 227–8
purge calculation, worksheets/planning 231–2
purge hose 228
purge regulator assembly 229
purge stack 227
purge zones 231, 235
purging 347
purging operations 60, 235–6, 348, 432–4
    calculations and planning 230–2
    consumer piping 226–36
    identifying requirements 229
    large volume gas installations requirements 226
    prep work 226–7
    purge and test 233–5
    purge types 226
    through appliance 292, 348
    tools and equipment 227–9
PVC-HI (high impact PVC) 149, 153
pyrolysis
    defined 289, 522
    occurrence locations 289, 321–2, 522

Q
quality assurance 154
quick release fittings 228
quick-connect devices 160, 273, 392, 411
quotations 148, 463

R
radiant heaters/heating 255–6, 282–3, 286–7
radiant–convection heaters 285
range 370, 447
    of flammability *see* flammability range
rated working pressure 19, 205
rating 21, 130

ratio (air–gas) 46, 243
raw materials 4, 48
reactions 82–3
records/recording 129–31, 346
rectification 469, 471
refinery 31–2
refining process 3–5
refuelling 6
regulator bodies 410
regulator capacity 385
regulator connections, direct valve to 449
regulator installation 186–214, 391, 410
    not within buildings 201
    prep work 208–9
regulator outlet pressure 191–2
regulator vents 35
regulator–OPSO assembly 207–8
regulators 135, 167, 186–200, 202, 209–13, 385, 464
    'chatter' 203
    choice of 391, 410
    design and location 201
    filters 199
    mounting and location 391, 410
    outlet pressure 191–2, 195–6, 201
    pressure reduction 186
    regulator creep 196
    selection and capacity 405–6
    service life 199
    two in one *see* dual stage LPG regulators
    types 187–200
    valve effect 192–3
    venting requirements determination 201–4
regulatory requirements 502
re-ignition 90
relationships
    between burner classifications 84
    performance–injector orifice size 91
    total consumption–air volume 243
    volume–cylinder pressure drop 230
relative density (RD) 44–5, 92, 203, 272, 406
relief valves/vents 114, 116, 196–7, 202–3, 206
renewable resources 5
residual currents devices (RCDs) 463
resistance 62–3
    path of least 62, 65
    variable 89
resources 128
    captured waste as 5
    also under resources type
respirators 148
respiratory hazards 59–61
retention ports 89
re-test dates 428
reticulated gases 3, 6, 30, 462
    volume calculations for consumption 12
reticulation mains 32–4
reticulation networks 424
reticulation pressure 32
retorts 3–4
reversion fittings 162–3
risers 154
risk assessment 148
risk control 56–7

risk elimination 57
risk mitigation 56, 63–4, 425
roar 116
rock 4
roof penetrations locations 522
rooms 244
room-sealed appliance 286
room-sealed appliances 242, 280–1, 284
rotary hammer drills 247
rotary meters 425–6
rough-in tests 165, 168
rulers (steel) 15–16

S
SAE thread 158
safe work method statements (SWMS) 56–7, 170, 200, 245, 338, 425
safety 64, 82, 96, 125, 148, 200, 235, 245, 327, 458, 462–3, 502
    general evaluations 56–7
    safe to touch 340
    in the workplace 44–8, 56
    see also work, health and safety
safety boots 148, 425
safety devices 464–74, 487
safety equipment 226–7
safety factor 231
safety glasses 148, 424, 429
safety shut-off systems, vent sizing for 204
SAI Global 125, 149, 314, 506
sandy loam 153
saturation temperature 113
scale rule 128
scales 17
    central manometer scale 15–16
    direct versus indirect [manometer] 17
screws 89
sea level 14, 45, 112
sealants 160–1, 235
secondary air supply 85, 93, 95, 312, 503–4, 512
    contaminated 99
securing clips 278
sediment 4
self-sealing test points/fittings 201, 212
self-sustaining combustion 47, 82, 84, 312
sensors 414
separation 31, 154–5
    see also clearance
serial numbers 21
service lines 33–4
service regulators 34, 198–9
serviceability (tools/equipment) 156, 170, 214, 235–6, 259, 294, 328, 350, 372, 396, 409, 415, 434–5, 448, 450, 487, 525
services 154–5
servicing 18, 21
shape 94–5, 97–9
sheath 466
sheet lap anti-capillary drain 323
shock 61–3, 65, 69
short circuit 65–6
shutdown 340, 344, 465
shut-off valve 279
SI (International System of Units)
    see International System of Units
signage 229, 388, 396

silicone hose 157–8
simulated natural gas (SNG) 6, 462
    minimum inlet pressure 19, 127
single residential premises 273
single stage LPG regulators 199–200, 385, 405
single wall metal flues 512
site conditions 458
site inspection 148, 156, 245, 385
site requirements 6, 148, 366
sizing 386, 406, 519–21
    variation rules 514, 516
    see also pipe sizing
slow-burning gas 84, 87
smell 32, 48, 58, 83, 243
smoke candles 252, 258, 328, 461
smoke generation options 258–9, 327–8
smoke match 258, 328
smoke test 258, 324
snap-action rod and tube thermostats 475–6, 479
soap and water solution 58, 158, 168, 227–8, 346, 432
solenoid valves 61, 414, 469, 472
solids 4, 31
soot 83, 94–5, 97, 243, 312, 425, 504
Sourdillon pilot burner 474
space 114, 442
space heater installation 322
    installation requirements 286
space heaters 273, 282–6, 313
specific gravity 44–5
specifications 148, 506, 511
    manuals and 338
spigot rule 514, 516
spillage 243, 258, 294, 323–4, 511–12
spillage test 258, 294, 324
spreaders 89
spring-loaded valve 160
stability 47, 84, 87, 93, 166–7, 504
stainless steel 406
stairwells 366
standard atmospheric pressure 112–13
standard components 228–9
standard reference tables 230–1
standard test procedures, less than 30 L pipe volume 165–8
standard thread 163
standards 57, 82
    standard sizing tables 128–9
    also under specific standard; *see also* AS/NZS series
standing pilot ignition 480
static (lock-up) pressure 19–20
static pressures 19
steam 113
steel 157, 209, 386, 406, 512
steel risers 34
stickers 162, 201, 396
storage hot water burners 89
storage hot water systems 83, 504
storage of gas *see* gas storage
storage water heaters 20, 242, 280–2, 394, 476
    installation requirements 281
stray current 63, 67, 69, 158, 340–1, 428–9, 485, 506
stress points 277–8

strip-metal fatigue 465
stud finders 247
stud finding 256–7
sub-meter installation 427–8
    prep work 427–8
sub meter purge 432–4
    purge requirements 432
sub-meters 226, 424–35
    applications 424
    enclosure ventilation 427
    isolation requirements 430
    purge 230
suffocation 48
supervised gas appliances 464
supervision control 469, 471
supply
    multistage 135
    outage 6
supply main 32–3
supply pressure 92, 200, 458, 462
    continuous change in 186
    pressure drop match 130
support devices 159
surface area 114, 116, 193, 362
    for cooking 275
sustainability 5
switchboard 64
switches 471–3
symbols 62, 205
system leakage tests 463–4
Système International (SI) units
    see International System of Units

T
tables 128–9, 132, 230–1, 245, 360
    shaded areas 133
tailpipes 3
tank connections 449
tank installation
    clean-ups 449–50
    LPG regulator choice 447
    requirements 446–7
tank pressure relief 448–9
tank storage, calculating requirements 444–7
tank testing 442–3
tanker-truck access 442
tanks 114
    basics 442–3
    in groups 446–7
    large yet small-range CNG 6
    locations 446–7
    versus multiple cylinders 36
    service life 442–3
    in-situ 35–6
    small tanks 442
    steel 35
    see also cylinders
tapes 160, 162
tapping saddles 34
technical regulators 149, 163–4, 226, 228, 272–3, 280, 342, 404, 413, 463, 506
technology 4
temperature 3, 47, 112–14, 166–7, 242, 315, 323, 465, 511, 522
    differentials 465, 478
    limitations 273
temperature control 474–9

temperature multipliers 362, 443
temperature-control systems 487
tempered liquefied petroleum (TLP) 6
terminal locations 319–21, 512–13
    above roof line 319, 513
    local/statutory approvals for 513
terminology 19–20, 62
    defined and industry 18–20, 187
    nautical terms 404
test equipment 165–6, 315, 339–40, 428, 508
test instruments 66–9, 157, 348
test procedures 163
test tees 157, 167
test times 166–8
    greater than 30 L 165
    multi-branch installations 164–5
    single run installation 164
tests/testing 57, 62, 65–70, 125, 157–68, 196, 230, 233–5, 256–9, 315, 321–8, 339–40, 344, 346, 371–2, 395–6, 413, 430, 459, 485, 506–7
    names indicating test nature 19
    test points 201
    also under specific test
thermal conduction 290
thermal loss 281
thermal mass 323
thermocouples 465–8
thermoelectric flame-failure devices 465–6
thermoelectric safety device 170
thermopiles 466
thermostatically controlled space heaters 285
thermostats 349, 464, 474–9
thread sealant/restrictions 160–1, 163
tight-wire method 368
tilt switches 472
tonne 37
tools 156–7, 208–9, 246–8, 288, 338–9, 371, 390–1, 409, 428, 448, 450, 508, 525
    for purging 227–9
torches 460
torsional twisting forces 278
total water capacity 361
towns gas (TG) 3–4, 44–8, 84, 96–7
    applications 3
    odorant addition 48
toxic substances 44, 83, 504
toxicity 48
trace gases 243
training 60, 134–5, 428
transmission main *see* trunk main
transport (LPG) 115
trenches 152–4
trivets 274
trunk main 32
trust 148
turbulence 90, 127
turn-down ratio 426–7
twin cylinder exchange installations 366–7
twin wall flues 321–2, 522
twin wall gas flue penetration 323
two-way radios 229
Type A gas appliance flue installation 312–28
Type A gas appliance installation 272–94
    install and commission 288
    prep work 288

Type A gas appliance ventilation 259–60
  basics 242–4
  installation 242–60
  prep work 244–8
  specific calculations 253–6
  testing 256–9
Type A gas appliances 18–19, 46–7, 84, 86, 91, 93–4, 272, 347, 471, 473–4, 487–8
  design 82
  disconnect; reconnect 338–50
  maintenance 458–88
  prep work 338–40
  regulator application 187–9
Type A gas fitters 30
Type B flue terminals, local/statutory approvals for 513
Type B gas appliance flue installation 502–25
Type B gas appliances 502–3, 511
  regulatory State/Territory requirements 502
  regulatory/statutory requirements 514

U

unburnt gas 57, 96
under bench ovens 276–7
underground electrical cables 154–5
underground piping requirements 152–4
  depth of cover and bedding 152–3
underground trace wire 154
unflued water heaters 280
Uniform Shipping Laws Code of Australia 404
unit conversions 14, 21
units 12–21
  also under specific unit; *see also* imperial measurement system; International System of Units
unsafe conditions 464–5
unsupervised gas appliances 464
upper explosive limit (UEL) 45
upright cookers 275–6, 278
upstream regulators 206
U-tube manometers 158
U-tubes *see* water manometers

V

vacuum 3
value 48, 92, 243, 425
valves 160, 167, 206, 279, 361, 370–2
  poor selection 274
  regulator valve effect 192–3
vaporisation 5–6, 35, 37, 114, 444
  formulas and temperature multipliers 362, 443

vaporisation calculation
  DLK calculation percentage 443–4
  DLK formula 362–3, 443–4
vaporisation capability 442
vaporisation capacity 364, 446
vaporisers 444
vapour 5–6, 35, 37, 82, 99, 112–13, 244, 255, 312, 316, 361–2, 389, 425, 443, 503–4
vapour space 114
vapour withdrawal 443
vapour-proof compartments 406
variable resistance screws 89
variable volume gas container *see* gasometers
velocity 86–7, 90, 133
vent cut-outs 256
vent interconnection 209–10
vent lines 201, 203, 212
vent location 411–13
vent sizing 204
vent terminal exclusion zone 210
vent terminal location 210
ventilation 46, 58, 60, 82, 150–2, 242–52, 287–8, 320, 391–6, 411–13, 427, 503
  additional needs 253–4
  air quality 244
  inadequate 243
  method 1 prior to AS/NZS 5601 248–50
  method 2 after AS/NZS 5601 251–2
  of room/enclosure via adjacent room 250–1
  specific calculations 253–6
  terms 244–5
  tools and equipment 246–8
venting 235
venturi 86, 89, 94
vibration 408, 410
video inspection borescopes 461
vitiation 95, 312, 316, 504, 512
volt sticks 66–8, 256, 340, 429, 459
voltage 62–3
voltage detectors 66, 68
volume 12, 97–9, 163, 226, 230–1, 361
volume calculations 12

W

wall cavity thermal barriers 322, 522
wall ovens 276
wall penetrations 321–2
warning signs 388
waste products
  LPG as 6
  waste-to-energy 5
water 3, 44
  boiling point 113
  heating 476–7

water capacity (WC) 360–1, 442
water gauge manometers 229, 413
water gauges *see* water manometers
water heaters 242, 274, 280, 313, 394, 476
water manometers 15, 430, 460
water services 154, 162
water vapour *see* vapour 281, 312, 316, 425, 503–4
watts 13
weight 12–13, 386
welded fittings 151
welding 46
well head 31
wet gas 3
wetted surface area 114, 116, 362
wiring layouts 384–5, 405
wiring systems 64, 68
witness hole 151–2
work 429–30
  approach 338
  avoid working alone 64
  clean-ups 135–6, 170, 213–14, 235–6, 259–60, 294, 328, 350, 372, 415, 434–5, 485, 487–8
  communication with all parties 340, 506
  deliberate manner of 63–4
  electrical safety before/during 428–9
  in older buildings 257
  pre-installation checks 506
  preparing for under specific job/task
  pre-work leakage test 57
  safe habits 425
  scale of 148
work, health and safety (WH&S)
  considerations 148, 200, 338, 384–5, 404–5, 424–5
  requirements 506
  see also safety; workplace safety
work instructions 506
work practices 226
  good practice 201, 323, 524
  non-compliance procedures 463
work practices/protocols 63–4, 245
  document storage protocols 136
  good practice 132
  poor practices 69
working at heights 56, 246, 507
working pressure 17, 486
workplace safety 56–70
  see also safe work method statements
workplaces, mobile 384–96

Z

zero mark 15–16